ALGAE AND HUMAN AFFAIRS

Also sponsored by the Phycological Society of America

Handbook of phycological methods
Culture methods and growth measurements
Edited by Janet R. Stein

Handbook of phycological methods
Physiological and biochemical methods
Edited by Johan A. Hellebust and J. S. Craigie

Handbook of phycological methods
Developmental and cytological methods
Edited by Elisabeth Gantt

Handbook of phycological methods
Ecological field methods: macroalgae
Edited by Mark M. Littler and Diane S. Littler

Experimental phycology
A laboratory manual
Edited by
Christopher S. Lobban, David J. Chapman, and Bruno P. Kremer

ALGAE
AND HUMAN AFFAIRS

EDITED BY
CAROLE A. LEMBI
Purdue University

J. ROBERT WAALAND
University of Washington, Seattle

SPONSORED BY THE PHYCOLOGICAL SOCIETY OF AMERICA, INC.

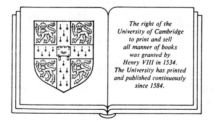

The right of the
University of Cambridge
to print and sell
all manner of books
was granted by
Henry VIII in 1534.
The University has printed
and published continuously
since 1584.

CAMBRIDGE UNIVERSITY PRESS

CAMBRIDGE

NEW YORK ● NEW ROCHELLE ● MELBOURNE ● SYDNEY

Published by the Press Syndicate of the University of Cambridge
The Pitt Building, Trumpington Street, Cambridge CB2 1RP
32 East 57th Street, New York, NY 10022, USA
10 Stamford Road, Oakleigh, Melbourne 3166, Australia

First published 1988

Printed in the United States of America

Library of Congress Cataloging-in-Publication Data

Algae and human affairs / edited by Carole A. Lembi, J. Robert Waaland
: sponsored by the Phycological Society of America, Inc.

p. cm.

Includes indexes.

ISBN 0-521-32115-8

1. Algae – Utilization. I. Lembi, Carole A. II. Waaland, J.
Robert. III. Phycological Society of America.
SH390.7.A44 1988
589.3′6 – dc19   88-11679
CIP

British Library Cataloguing in Publication Data

Algae and human affairs

1. Algae. Production & use
I. Lembi, Carole A. II. Waaland, J.
Robert III. Phycological Society of
America
635.9′393

ISBN 0 521 32115 8

Contents

[v]

ADVERSE IMPACTS OF ALGAE

THE FUTURE OF ALGAE IN HUMAN AFFAIRS

INDEXES

Editors' preface

The initial impetus to organize a book on the role algae play in human affairs came during a symposium sponsored by the Phycological Society of America on "Environmental roles of algae and their impact on society" (Indiana University, Bloomington, August 1981). At that time it was becoming apparent that the productivity associated with algae in natural communities could be harnessed commercially into productivity tailored to human specifications. A driving force in this development was the petroleum shortage experienced in the 1970s and the emergence of biotechnology. The rapid increase in research and development of algal biomass cultivation technology for specialty chemicals, food, and energy is a testament to the concept that algae can play new and different roles in relation to humanity. As of this writing (early 1988), a (likely) temporary excess of petroleum supplies has pushed crude oil prices low enough to dampen domestic enthusiasm for cultivating algae as biomass for economical fuels. However, the investigation of algae for specialty uses and products such as hydrocolloids, carotene, gourmet foods, and diatomaceous earth continues at an exciting pace.

The impact of algae on human society is not limited to those that we harvest or grow for their economically useful products. From early in biotic evolution, algae played essential roles with significant and lasting impacts on the biosphere. Through photosynthesis they poured oxygen into the atmosphere billions of years ago. Their fossil remains have produced black pools of petroleum deep within the earth and the famous white chalk cliffs at Dover. Directly, algae are the major source of food for many aquatic organisms from copepods to abalone and, indirectly, they are essential links in food chains of creatures such as lobster and salmon. In their most infamous roles they sometimes are nuisance organisms, their blooming billions polluting water supplies with foul tastes, odors, and even toxins, and when they find their way into the flesh of shellfish they leave a paralyzing, deadly toxin for imprudent consumers of oysters, clams, and mussels. In such nefarious roles they cause millions of dollars of economic disruption.

We believe algae have a bright future and that ways will be found to minimize their adverse effects and maximize their positive impacts on human society. However, such developments will require fresh, imagi-

[vii]

native, and innovative thinking. Fortunately, in many areas this process is well under way. For example, the highly efficient light-harvesting pigments of certain seaweeds are now being coupled to antibodies where they become brilliant fluorescent labels for antigens in cell biology and medical diagnosis. In certain localities, the high productivity of microscopic freshwater algae has been used to harness solar energy and treat sewage. We hope that reference to these examples will stimulate interest in reading many or all the chapters that recount the interreactions of algae with humans and will lead to the recognition that algae have a past, present, and future partnership with people.

The editors would like to thank the contributors for their patience, cooperation, and gracious compliance with our editorial suggestions. Special thanks are due the members of the board of advisors, each of whom contributed much appreciated advice and editorial help: William R. Barclay (Solar Energy Research Institute, Golden, CO), James S. Craigie (Atlantic Research Laboratory, Halifax, Canada), W. Marshall Darley (University of Georgia), Janet R. Stein, and Michael J. Sullivan (Mississippi State University). We also thank the Phycological Society of America for its support, both moral and financial. The secretarial assistance of Diana VanHorn and Sandy Geisler is greatly appreciated.

Carole A. Lembi
J. Robert Waaland

ALGAE AND THEIR ROLE IN NATURAL ECOSYSTEMS

1. The algae: an overview

MARILYN M. HARLIN

Department of Botany, University of Rhode Island, Kingston, RI 02881

W. MARSHALL DARLEY

Department of Botany, University of Georgia, Athens, GA 30602

CONTENTS

Although the term *algae* has no formal taxonomic standing, it is nevertheless used to refer to a vast array of primitive, generally aquatic and photosynthetic organisms – fascinating in their diversity, beauty, and utility. Algae are polyphyletic and range in size from picoplankton only 0.2–2.0 μm in diameter to giant kelps with fronds up to 60 m in length. Various species have colonized virtually every sunlit habitat in the biosphere. The exquisite forms of algae have been viewed with wonder since the invention of the light microscope (and, more recently, the scanning electron microscope) and may also be appreciated with the unaided eye as evidenced by the collections of artistically arranged pressed seaweeds popular since the Victorian era. The usefulness of algae has been recognized even longer and is a major theme of this book.

Alga (plural = algae) is the Latin word for seaweed. There is little evidence that the Romans appreciated these plants for their aesthetic value. Chapman (1970) reported that Roman ladies in the time of the Caesars used a rouge extracted from *Fucus*, and other uses were recorded as early as the fourth century when Palladius wrote of using seaweed as a partial substitute for manure. In 1665 King Charles II granted a charter to "his loving subjects" for the use of lands along shores "below which the tides ebb and flow" (mean high tide). One of the uses of this right was "the gathering of seaweed" for agricultural use (Wheeler and Hartman 1893). In Rhode Island, farmers collected their valuable drift algae in oxcarts, which led to the present-day right-of-way of "one ox-cart width," and shore access easements of such width are still included in land records. In the nineteenth century Rhode Island farms relied heavily on this seaweed manure, which was used fresh, composted, or scattered as the ash from burned seaweed. To this day seaweeds and soil algae are used in agriculture to build soil, add nutrients (Metting, Rayburn, and Reynaud, Chapter 14), and reduce weeds and insects.

It is unlikely that the Romans considered freshwater algae more than "slime" on a pond or slippery patches on a rock. Most certainly, the importance of algal primary production in the biosphere would have been unrecognized. Even today, the general public has a limited aware-

[4]

ness of algae, although it is unlikely that any reader has passed a single day without using one to several products from algae, from toothpaste to ice cream.

This book deals not only with the commercial industries based on useful applications of algae but with other aspects that have an impact on human well-being, such as the roles of algae in the environment and situations in which algae are considered to be a nuisance. Most discoveries involving algae require an understanding of basic biology. The objective of this introductory chapter is to provide an appreciation for the natural history of these organisms: how they evolved, what they are, where they live, and what they do.

I. The origins of algae

The first organisms on earth arose more than 3.5 billion years ago in an oceanic, organic soup that had been synthesized by abiotic, high-energy processes (lightning, ultraviolet radiation, and pressure shock). The atmospheric environment was anaerobic and contained methane (CH_4), hydrogen (H_2), and ammonia (NH_3). These earliest cells were prokaryotic in structure and heterotrophic in nutrition, obtaining their energy through anaerobic metabolism involving fermentation of the abundant organic molecules in their environment. Biosynthetic pathways evolved as the environment became increasingly deficient in usable abiotically produced organic molecules. Important innovations included nonphotosynthetic carbon dioxide fixation, nitrogen fixation, and the ability to synthesize porphyrins and carotenoids and use light energy to produce adenosine triphosphate (ATP) (Margulis 1981).

Photosynthesis evolved first in anaerobic bacteria; the selective pressure was probably the continuing decline in abiotically produced organic matter. The primary photosynthetic pigment in extant anaerobic photoautotrophic bacteria is either bacteriochlorophyll or bacteriorhodopsin. The electron donor is H_2S, H_2, or an organic molecule (not H_2O). Oxygen is not released, and only a photosystem-I-like reaction center is involved. The cyanobacteria (traditionally called blue-green algae) evolved 1.5 billion years later, probably from organisms resembling modern, nonsulfur, purple photosynthetic bacteria. Cyanobacteria contain a different chlorophyll (chlorophyll *a*), use H_2O as an electron donor, release O_2 and use both photosystems I and II. The same photosynthetic process is found in all photosynthetic eukaryotes today. That cyanobacteria evolved from anaerobic photosynthetic bacteria is suggested by the fact that some cyanobacteria facultatively use sulfide (H_2S or Na_2S) as an electron donor under anaerobic conditions (Padan 1979). In 1976 R. A. Lewin described a new photosynthetic, O_2-evolving prokaryote, which he called *Prochloron* (Lewin 1976, 1984). In structure this green unicell closely resembles cyanobacteria, but its photo-synthetic apparatus is exceptional among prokaryotes in that it contains chlorophylls *a* and *b* and lacks the phycobiliproteins characteristic of cyanobacteria. It is more like a green algal chloroplast than any other free-living prokaryote.

A. Photoautotrophs and the evolving biosphere

The release of O_2 by these new photoautotrophs and its eventual accu-mulation in the atmosphere (beginning approximately 2 billion years ago) resulted in the most profound and enduring impact algae have ever exerted on life on this planet (Margulis and Sagan 1986). (Note that *algae* is used in its traditional sense to include the cyanobacteria or blue-green algae.) Oxygen was a poison to all organisms then living except the O_2-evolving autotrophs. The anaerobes had to adapt if they were to survive. Some merely stayed in anaerobic environments and are found living there to this day as obligate anaerobes. Others developed meta-bolic systems to rid themselves of free O_2. Perhaps bioluminescence evolved in this way. Aerobic metabolism (the Krebs cycle and the elec-tron transport chain) evolved as another way to deal with O_2 and even profit from it with the much more efficient production of ATP. Fur-thermore, increasing levels of oxygen in the atmosphere led to the pro-duction of ozone in the stratosphere, which shielded living organisms from the mutagenic effects of ultraviolet radiation, thus permitting the continued development of surface-dwelling life forms.

The accumulation of atmospheric oxygen also coincided with the appearance of compartmentalized eukaryotic cells, which is thought to have occurred approximately 1.5 billion years ago (Margulis and Sagan 1986). An old theory beginning in the late 1880s with Schemper (see Wilson 1925) has received a great deal of support recently from the fields of cell and molecular biology. This theory suggests that the eukaryotic cell evolved in a very complex, interacting community of microbes by a serial sequence of symbiotic associations followed by joint evolution of the symbiotic association.

The host organism for this symbiosis was likely a wall-less, phagocytic, anaerobic or microaerophilic heterotroph with fermentative respiration and was the putative source of the cytoplasm and nucleoplasm in mod-ern eukaryotic cells. This host cell might have been similar to modern-day thermoacidophiles, such as *Thermoplasma,* which have recently been recognized as a distinct group of prokaryotes (Lake et al. 1984). The endomembrane system (nuclear envelope, endoplasmic reticulum, Golgi apparatus) of the eukaryotic cell is likely to have come from the plasma membrane of the host cell (Cavalier-Smith 1975). The selective pressure for a separate nuclear entity might have been a more reliable

Table 1-1. *Types of mitochondrial cristae in eukaryotes*

Tubular cristae		Flattened cristae	
Bacillariophyceae	Xanthophyceae	Chlorophyta	Higher animals
Chloromonadophyceae		Cryptophyceae	Ascomycetes
Chrysophyceae	Bicosoecida	*Cyanophora*	Basidiomycetes
Eustigmatophyceae	Ciliophora	Euglenophyceae	Choanoflagellida
Phaeophyceae	Myxomycetes	Rhodophyceae	Chytridiomycetes
Prymnesiophyceae	Oomycetes	Higher plants	Kinetoplastida
Pyrrhophyta	Opalinata		

Source: After Stewart and Mattox 1980.

separation mechanism for the chromosomes that were probably attached to the plasma membrane as they are in modern prokaryotes.

B. Symbiotic origin of organelles

One or more of the other organelles characteristic of eukaryotic cells (mitochondria, plastids, and flagella) are thought to have arisen as a prokaryotic symbiont that evolved into the present-day organelle. Evidence from molecular biology is "persuasive" and "overwhelming" for the endosymbiotic origin of the mitochondrion and plastid, respectively (Gray 1983). Evolutionary reduction of the symbiont has involved the loss of a cell wall, the loss of redundant genes, and the transfer of many, but not all, of the genes coding for organellar proteins to the eukaryotic nucleus. These proteins, once synthesized in the cytoplasm, are transported into the organelle (see Weeden 1981). The outer membrane of these organelles is presumably derived from the host phagocytic vacuole, and the inner membrane from the plasma membrane of the endosymbiont. Cristae and thylakoids are derived from the invagination of the endosymbiont's plasma membrane. A symbiotic origin for the flagellum is much more doubtful. Because of the diversity of prokaryotes involved, the acquisition of organelles probably occurred several times with several distinct species as symbiotic partners (Margulis 1981).

Mitochondria probably arose from an endosymbiont related to purple, nonsulfur, photosynthetic bacteria (Rhodospirillaceae) with aerobic respiration. The origin of mitochondria almost certainly predates the origin of chloroplasts and led to a great diversity of fully eukaryotic, heterotrophic unicellular organisms. In modern eukaryotes a given group has mitochondria with either tubular cristae or flattened (lamellar) cristae, but not both (Table 1-1). To explain this, Stewart and Mattox (1984) suggested that mitochondria originated from two separate, but related, lineages within the Rhodospirillaceae. Note that chloroplasts in eukaryotes with flattened cristae contain either chlorophyll *b* or phycobilins, whereas chloroplasts in eukaryotes with tubular cristae have chlorophyll *c* and are usually yellowish or brownish, always lack chlorophyll *b* and phycobilins, and, with the exception of the Pyrrhophyta, are surrounded by endoplasmic reticulum.

The symbiotic origin of plastids from symbiotic photosynthetic prokaryotes is likely a more recent event, has occurred several times, and has involved two (or possibly three) different prokaryotes and several different heterotrophic eukaryotes. Green algal plastids are likely derived from green prokaryotes *(Prochloron)*, whereas red algal plastids are likely derived from blue-green prokaryotes (cyanobacteria). Plastids in the chlorophyll-*c*-containing algae could have evolved from a yellow-green prokaryote, although no such organism is known at present. The occurrence of a functional, chlorophyll-*c*-like pigment in primitive green algae (Brown 1985; Wilhelm et al. 1986) suggests that yellow-green plastids could have evolved from a green prokaryote.

The correlation between chloroplast pigments and type of mitochondrial cristae (Table 1-1) suggested to Stewart and Mattox (1980) that there may have been selective pressure for a genetic "fit" between the host, as evidenced by the type of mitochondria present, and the "prospective" chloroplast. If true, this proposal suggests that the "yellow" chloroplasts found in cells with tubular cristae have an origin independent of *Prochloron* and the cyanobacteria. On the other hand, could it be that the chloroplast endoplasmic reticulum (CER) surrounding these chloroplasts is a mechanism for dealing with the genetic mismatch between an originally green (and now yellow-green) plastid and a host cell with tubular cristae?

In several groups, plastids are thought to have evolved in a eukaryotic alga that was later engulfed as a symbiont by another eukaryote (i.e., a host with an endosymbiont, which in turn has an endosymbiont). Progressive reduction of the eukaryotic endosymbiont resulted in the loss of everything except the plasma membrane and the chloroplast. Again, such symbioses follow the cristae–pigment correlation outlined in Table 1-1. Thus, a heterotrophic euglenoid ancestor engulfed a green alga, and all that remains today is the double-membrane-bound chloroplast of the green alga now surrounded by a third membrane, possibly the green alga's plasma membrane (Gibbs 1978). Symbiotic associations with other photosynthetic eukaryotes is likewise thought to be responsible for the acquisitions of plastids in dinoflagellates (Tomas and Cox 1973; Wilcox and Wedemayer 1985) and cryptomonads (McKerracher and Gibbs 1982).

According to Margulis (1981), eukaryotic flagella might have evolved from a symbiotic association between a spirochete (or a wall-less spiro-plasma) and the evolving eukaryotic cell. The symbiotic origin for the flagellum is not generally accepted, but it would be very useful to have more information on the molecular biology of microtubule organizing centers and on the presence of tubulin, microtubules, and their nuclea-tion centers in spirochetes or other prokaryotes. In any event, the nearly universal 9 + 2 organization of the flagellum suggests it had a very early origin, probably before the mitochondrion. Perhaps we should not expect an organelle of such antiquity to still retain an obvious genome and protein synthesizing system as the chloroplast and mitochondrion have done. Stewart and Mattox (1980) also support the idea that the flagellum evolved before mitosis but state that an endosym-biotic origin is "a concept entirely unsupported by evidence or devel-opmental studies."

C. Evolution of higher plant forms

From the symbiotic associations that led to the internal structure of eukaryotic cells to the aggregation of eukaryotic algal cells into more complex morphological forms, algae are considered to have paved the way for the development of more advanced forms of plant life. Most authorities agree that the ancestors of land plants would be classified today as green algae (Graham 1985) and that the adaptation to the land environment occurred approximately 400 million to 450 million years ago during the Silurian period (reviewed by Gensel and Andrews 1987). Evidence for the linkage of land plants with the charophycean green algae includes the presence in both groups of a phragmoplast (micro-tubules oriented at right angles to the direction of new cross-wall for-mation), similar flagellated reproductive cells containing a multilayered structure associated with a microtubular root (Stewart and Mattox 1975), and the photorespiratory enzyme, glycolate oxidase (Frederick, Gruber, and Tolbert 1973). Fossil evidence of the order Charales dates back to 400 million years (Schopf 1970) and of *Parka* and *Pachytheca,* which strongly resemble the extant charophyte *Coleochaete,* to the Upper Silurian–Lower Devonian period (Niklas 1976). The retention and cov-erage of the zygote on the parental haploid body in *Coleochaete* is thought to be indicative of one of the first steps in the evolution of the multicellular diploid generation and alternation of generations charac-teristic of higher plant forms (Graham 1984). Certainly further investi-gations into the molecular and biochemical as well as the structural and developmental features of the charophycean algae will result in contin-ued progress toward our understanding of plant evolution.

Clearly, there are numerous, fascinating questions to be answered by developmental and molecular biologists once they wean themselves from laboratory weeds (e.g., *Chlamydomonas reinhardtii* and *Euglena gra-*

cilis) and study the incredible diversity presented by other algae. Not only will such information help us understand the origins of the algae, but it will also help in another uniquely human endeavor, that of classifying these organisms, in a manner, we hope, that will reflect their evolutionary relationships.

II. Classification of the algae

The major groups (divisions and/or classes) of algae have been distinguished traditionally on the basis of pigmentation, cell wall composition, flagellar characteristics, and storage products. More recently a variety of ultrastructural characteristics have been recognized as useful and have contributed to the continuing state of flux that has characterized the classification of algae for a number of years. The following discussion is presented more as an introduction to the biology of the major groups of algae than as a defense of this particular classification scheme. The reader is referred to Bold and Wynne (1985) for more information concerning the structure, reproductive characters, and classification of algae.

The cyanobacteria are prokaryotic in cell structure and therefore are not included among the algae by some scientists (e.g., Sieburth 1979; Margulis 1981). Phycologists, however, have traditionally called them blue-green algae, and most continue to consider them algae, albeit prokaryotic algae (Cyanophyta: Bold and Wynne 1985), citing the presence of chlorophyll-*a*-dependent, oxygen-evolving photosynthesis, their algal-like forms and ecological relationships. In structural diversity, blue-greens range from unicells to branched and unbranched filaments to unspecialized colonial aggregations and are possibly the most widely distributed of any group of algae. They are: (1) planktonic, occasionally forming blooms in eutrophic lakes and as important components of the recently discovered picoplankton in both marine and freshwater systems; (2) benthic, as dense mats on soil or in mud flats and hot springs, as the "black zone" high on the seashore, and as relatively inconspicuous components in most soils; and (3) symbiotic, in lichens, cycads, sponges, and other systems. The ability of some blue-greens to fix nitrogen is unique among the algae and makes them a very important part of the earth's nitrogen cycle. Blue-greens interact with humans as food (e.g., *Spirulina;* Jassby, Chapter 7), when blooms produce toxins (Gorham and Carmichael, Chapter 16), and in symbiotic nitrogen fixation in rice cultivation (Metting et al., Chapter 14).

Among eukaryotic algae, the evolutionary relationships are especially confused because classification is based on characteristics of not only the host cell but also one or more symbionts – and all these partners do not necessarily share a common evolutionary history. For example, using the flagellum as one of the most ancient of the eukaryotic organelles, Moestrup (1982) recognized six major lines of evolution (called divisions here), distinguished on the basis of ultrastructure: Chlorophyta, Euglenophyta, Cryptophyta, Pyrrhophyta, Prymnesiophyta, and Heterokontophyta. The nonflagellate Rhodophyta constitutes a seventh division. Using a combination of characteristics – flagella, pigments, storage compounds, and life histories (Table 1-2) – Bold and Wynne (1985) hold a more conservative view of relationships.

The green algae (Chlorophyta) are probably the most structurally diverse group of algae with many types of unicells, colonies, filaments, siphons, and thalloid forms. Recent ultrastructural studies have established four lines of evolution from the primitive unicellular condition (Mattox and Stewart 1984). The Chlorophyceae are primarily freshwater forms; this class contains most of the species of green algae including *Chlamydomonas* and the volvocine line, *Chlorella* and most other chlorococcalean forms, a number of branched and unbranched filaments, and the familiar Oedogoniales. Most of the marine green algae belong to the Ulvophyceae, a generally tropical and subtropical group containing a number of relatively large forms (e.g., *Ulva, Codium, Valonia, Halimeda*), even though plasmodesmata (a prerequisite for tissue differentiation and specialization) never evolved in this class. The Charophyceae contains common freshwater forms such as *Spirogyra,* ornate desmids, and charophytes and is probably ancestral to green land plants (Graham 1984). The Pleurastrophyceae is a small, recently recognized class containing the flagellate *Tetraselmis* (= *Platymonas*) as well as the lichen symbionts *Trebouxia* and *Pseudotrebouxia.*

The euglenoids (Euglenophyta) resemble the green algae in pigmentation but are dissimilar in every other respect, a phenomenon explained in section I. The group contains mostly flagellate unicells, which are especially abundant in habitats rich in organic matter; a number of species are colorless heterotrophs. In spite of the popularity of *Euglena gracilis* as an experimental organism, most members of this group have not yet been cultured.

The cryptomonads (Cryptophyta) are a small group of flagellated unicells often abundant near the bottom of the euphotic zone in lakes. Their evolutionary relationships with other algae are particularly obscure because of the unique combination of features found in these algae, namely, chlorophyll *c,* flattened mitochondrial cristae, and phycobilin pigments.

The dinoflagellates (Pyrrhophyta) are common planktonic (especially in the sea) flagellated unicells with a number of bizarre morphologies. The group is best known for the often toxic blooms or "red tides" (Steidinger and Vargo, Chapter 15) they produce in coastal waters. Some species are bioluminescent, and a number of species are nonpigmented

Table 1-2. *Features of major groups of algae discussed in this book*

Division (group)	Pigments	Thylakoid arrangement	Photosynthate storage	Cell wall chemistry	Flagella and insertion
Chlorophyta (green algae)	Chlorophyll *a,b* carotenoids	Many/stack, variably associated	Starch (± oil)	Cellulose and other polymers	2, 4, or more (usually smooth, usually apical)
Phaeophyta (brown algae)	Chlorophyll *a,c* β-carotene fucoxanthin	3/stack	Laminarin, mannitol	Cellulose, alginic acid	Usually 2 (1 tinsel, 1 whiplash lateral)
Chrysophyta (golden and yellow-green algae)	Chlorophyll *a,c* several carotenoids and xanthophylls including fucoxanthin	3/stack	Laminarin, oil	Cellulose, silica, $CaCO_3$ chitin, or absent	1–3, unequal or equal apical (tinsel, whiplash, or both)
Pyrrhophyta (dinoflagellates)	Chlorophyll *a,c* β-carotene and several xanthophylls	3/stack	Starch (± oil)	± Cellulose in plates	2 unequal (1 trailing, 1 girdling)
Rhodophyta (red algae)	Chlorophyll *a* (± d), phycobilins	Single, not associated	Floridean starch	Cellulose, xylans, sulfated galactans (± calcification)	Absent
Cyanophyta (blue-green algae)	Chlorophyll *a* phycobilins carotenoids	No chloroplast, single thylakoids in cytoplasm	Granules resembling glycogen	Assorted amino acids	Absent

Source: After Bold and Wynne 1985; note that Moestrup (1982) would join the Phaeophyta and Chrysophyta in one division, the Heterokontophyta.

saprotrophs or parasites. The group is thought to be primitive and very ancient; the nucleus with its permanently condensed chromosomes and unique type of mitosis is especially interesting, and so is the cell covering of the armored species.

The haptophytes (Prymnesiophyta) are mostly golden-brown, motile and nonmotile unicells with a special appendage called a haptonema. They have only recently been recognized as a separate division, perhaps best known for those species (coccolithophorids) that produce calcareous scales known as coccoliths. The toxin-producing *Prymnesium parvum* is a problem in fish culture ponds in Israel

The Heterokontophyta (*sensu* Moestrup) brings together groups that share a common flagellar structure and a number of other features but that previously had been classified separately. This division as defined by Moestrup (1982) includes the yellow-green algae (Tribophyceae = Xanthophyceae), the chrysophytes (Chrysophyceae), the diatoms (Bacillariophyceae), the brown algae (Phaeophyceae), and two lesser-known small classes. Another way of grouping these classes is to include Tribophyceae, Chrysophyceae, and Bacillariophyceae in the Chrysophyta and Phaeophyceae in the Phaeophyta based on pigments (golden and brown, respectively) and flagella (apical and lateral, respectively).

The unicellular diatoms are very common, are often abundant, and are easily recognized by their highly ornamented siliceous cell walls. The pennate (bilaterally symmetrical) forms are usually benthic on soil, mud, and other submerged surfaces and are especially common epiphytes on aquatic vascular plants and other algae. Centric (radically symmetrical) forms are typically planktonic and are often united into filamentous colonies. Geologic accumulations of the siliceous frustules are mined as diatomaceous earth, which has many economic uses (George, Chapter 13).

The brown algae are almost entirely marine and range in size from microscopic filaments (*Ectocarpus*) to the rockweeds (Fucales) that may dominate north temperate rocky shores to the largest and most structurally complex of the algae, the kelps (Laminariales). *Macrocystis pyrifera*, the giant kelp, is harvested commercially for the phycocolloid alginic acid, which has a long history in our food industry (Lewis, Stanley, and Guist, Chapter 9). The *Sargassum* of the Sargasso Sea, originates as drift from coastal plants and is the only oceanic macroalga.

The red algae (Rhodophyta) are the third group (along with the brown and green algae) that contributes to the "seaweed" flora. Some red algae are microscopic and even unicellular. A few grow in swift currents in freshwater streams (Sheath and Hymes 1980) but are often overlooked due to their small size. Most, however, are conspicuous, multicellular seaweeds. Understanding the complex life histories of red algae is one of the more challenging topics in undergraduate phycology courses, and control of the life history of *Porphyra* has proved to be of major economic importance to the nori industry of Japan, China, Korea, and we hope the United States (Mumford and Miura, Chapter 4). Carrageenan and agar are commercially valuable polysaccharides obtained from certain red algae (Lewis et al., Chapter 9). The earlier theory that the pigmentation in red algae was a special adaptation to the low irradiances in deep water has now been disputed (Ramus 1983).

III. Algal productivity

Primary productivity is the rate at which plants synthesize organic matter from inorganic carbon and the radiant energy captured from sunlight. Of the total (gross) algal primary productivity, a portion is respired to support algal metabolism, a portion is excreted as organic carbon, and the remainder contributes to an increase in algal biomass. The latter two portions constitute net primary production available to the rest of the community or for human utilization. The respiratory loss in algae is not often measured but, because of the small amount of heterotrophic tissue to support, is generally much less than the approximately 50% loss found in vascular plants. Recent studies (Fogg 1983) suggest that the amount of organic carbon excreted by phytoplankton is usually less than 20% of net productivity although earlier studies reported much higher losses. In seaweeds, reports of organic carbon release range from 1–4% (Harlin and Craigie 1975; Brylinsky 1977) to 40% (Sieburth 1969).

Primary productivity may be reported in terms of carbon fixed on an area or volume basis (e.g., $g\ C\cdot m^{-2}\cdot day^{-1}$, or $g\ C\cdot m^{-3}\cdot day^{-1}$) or as a specific growth rate (e.g., $g\ C\cdot g\ C^{-1}\cdot day^{-1}$, which is often abbreviated day^{-1}). Productivity is also reported in energy units (e.g., $kcal\cdot m^{-2}\cdot year^{-1}$). Crisp (1975) published conversion factors that may be used to convert estimates from one system of units to another.

Productivity should not be confused with standing crop biomass because the latter does not take into consideration the time factor and the often high growth rates of algae. In an algal bloom that has exhausted the local nutrient supply, for example, standing crop may be high but productivity would be low. Conversely, in a planktonic system with abundant grazing zooplankton, the standing crop may be low but could have a high specific growth rate and productivity due to a rapid turnover of nutrients. In the mass culture of algae, productivity is often called yield ($g\cdot m^{-2}\cdot day^{-1}$). Maximum yields are usually obtained at intermediate densities and growth rates because maximum growth rates usually occur at lower densities, and high densities usually result in slower growth rates as competition for light and nutrients limit the growth rates.

Estimates of primary productivity are obtained by measuring the increase in biomass over a period of time or by measuring the exchange of a molecule involved in photosynthesis. Direct measurements of biomass by weight are possible for seaweeds but are seldom used for phytoplankton due to the presence of relatively large amounts of dead organic matter. Chlorophyll, which degrades rapidly in the environment, is often measured as an estimate of algal biomass, although the carbon:chlorophyll *a* conversion factor varies considerably. The release of oxygen in the "light- and dark-bottle method" has been used extensively since 1927 to measure both gross and net primary productivity and now may be used with an oxygen electrode for rapid oxygen measurement. The uptake of radiocarbon (^{14}C) is much more sensitive than the oxygen method and has tended to replace it in recent phytoplankton research. A great many *ifs, ands,* and *buts* are associated with all the methods of measuring primary productivity, and the reader is encouraged to investigate the recent literature before commencing such studies (e.g., Parsons, Maita, and Lalli 1984; Littler and Littler 1985; Harris 1986).

A. Global distribution of primary productivity

The distribution of primary production in the world's oceans is shown in Figure 1-1 (Rodin, Bazilevich, and Rozov 1975). In estuaries and coastal regions productivity is high because of abundant nutrients from river runoff and upwellings. Productivity in coral reefs is also high, but here the explanation lies in rapid recycling of nutrients within the ecosystem (Littler and Littler, Chapter 2). The important contribution of seaweeds in bays and coastal waters was long underestimated (Mann 1973, 1978) because only standing crop and not annual growth rate (turnover time) of these algae had been measured. Although primary productivity over much of the ocean is relatively low because of nutrient limitation, the contribution of the productive areas plus the fact that the oceans cover three-fourths of the earth's surface (lakes and rivers cover less than 1%) add up to a total primary production in aquatic systems estimated to be between one-third (Rodin et al. 1975) and one-half (Leith and Whittaker 1975; Smayda, personal communication) of total global productivity.

The data in Table 1-3 compare terrestrial and aquatic systems in terms of plant biomass and productivity. Terrestrial plant biomass is largely wood (82%), even though forest cover is only 39% (Walter 1985), and has a relatively slow turnover time with annual production amounting to only 7% of the biomass. In contrast, much of the oceanic plant biomass is in the form of single-celled phytoplankton with a rapid turnover and an annual production amounting to over 300 times their biomass. From another viewpoint, on a global scale algal systems are more

Fig. 1-1. Pattern of distribution for primary production in the world's oceans. (From Rodin et al. 1975. Reproduced with permission of the National Academy of Sciences.)

Table 1-3. *Distribution of global primary productivity (mt = metric tons dry wt)*

	Area (km² × 10⁶)	Plant biomass		Primary production	
		Total (mt × 10⁹)	Average (mt·ha⁻¹)	Total (mt·ha⁻¹)	Average (mt·ha⁻¹·y⁻¹)
Land excluding glaciers	133	2,400	180	172	12.8
Lakes and rivers	2	0.04	0.2	1	5
Oceans	361	0.17	0.005	60	1.7

Source: From Walter 1985, after Bazilevich, Rodin, and Rozov 1970.

Table 1-4. *Net primary productivity (NPP) of major habitats in the biosphere*

Terrestrial ecosystem	NPP (gC·m⁻²·y⁻¹)	Aquatic ecosystem	NPP (gC·m⁻²·y⁻¹)
Tropical rain forest	988	Swamp and marsh	1,350
Tropical seasonal forest	720	Algal bed and reef	1,167
Temperate evergreen forest	580	Estuaries	714
Temperate deciduous forest	543	Upwelling zones	250
Savanna	407	Lake and stream	200
Boreal forest	358	Continental shelf	162
Woodland and shrubland	318		
Cultivated land	293		
Temperate grassland	267		

Source: After Woodwell et al. 1978, in Bott 1983.

productive than highly productive terrestrial systems (e.g., tropical rain forest, average cultivated land) (Table 1-4). With aquaculture techniques, these productivities can be even higher.

B. Factors affecting algal productivity

Of the large number of environmental parameters that affect algal productivity, irradiance and nutrient supply are probably the most important in nature. Changes in these parameters, in conjunction with temperature changes, lead to seasonality in both primary and secondary productivity. The typical response of photosynthesis as a function of potentially limiting environmental factors is presented in Figure 1-2 (Cobb and Harlin 1976). Under conditions in which a specific parameter is limiting photosynthesis (A), modification of that parameter will

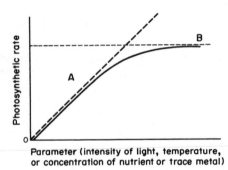

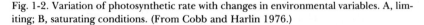

Fig. 1-2. Variation of photosynthetic rate with changes in environmental variables. A, limiting; B, saturating conditions. (From Cobb and Harlin 1976.)

affect algal photosynthesis whereas modification of other parameters will have little or no effect. Modifications in parameters that are saturating for photosynthesis (B) will also have little effect. It is important, therefore, to understand the factors limiting productivity in the system under study so that the most efficient use of resources may be applied to enhance, redirect, or curtail productivity as the situation warrants. Such response curves, obtained in laboratory studies, also have been used to predict productivity in nature (Dugdale 1967; Eppley, Rogers, and McCarthy 1969; Steele 1974). It is becoming increasingly clear, however, that factor interactions (see Goldman and Mann 1980; Healey 1985) may have significant effects on the response and must be taken into account.

Algal growth is necessarily limited to surface waters due to the rapid attenuation of light energy in the water column. Although the 1% of surface irradiance level has been used for years to define the lower limit of the euphotic zone, recent work suggests that the lower limit might be more realistically placed at the 0.1% level or even lower (Richardson, Beardall, and Raven 1983). Attached seaweeds, which are not subject to periodic removal from the euphotic zone (as are phytoplankton), have recently been found growing (slowly, no doubt) at depths in excess of 200 m at less than 0.01% of ambient surface irradiation (Littler et al. 1985; Littler and Littler, Chapter 2). Because the longer wavelengths of light are absorbed more readily than the shorter wavelengths, the quality of light also changes with depth. Light quality is known to be important for gametogenesis in some kelps (Lüning 1980) and in phototropic responses in algae (Buggeln 1974; Waaland, Nehlsen, and Waaland 1977). Photoperiodic responses controlling reproduction and frond development are known in many algae (Dring 1984).

Nutrient limitation accounts for the low primary productivity of the subtropical oceanic gyres (Fig. 1-1). The persistent thermocline and stable water column in these regions prevents the nutrient-rich deeper water from mixing with the nutrient-deficient surface waters of the euphotic zone. Productivity might not be as low as depicted in Figure 1-1, however. Recent studies (e.g., Laws et al. 1984) suggest that the nutrient pool in such systems is turned over very rapidly by a microplankton food web – so rapidly that it has probably been missed in many of the standard measurements of primary productivity. One theory holds that the algal cells in these systems have an enhanced nutrient uptake capacity that quickly allows them to accumulate relatively large amounts of nutrients whenever they encounter temporary "micropatches" of nutrients (Goldman and Glibert 1982; Scavia et al. 1984). This strategy may allow algae to grow close to their maximum rates even in waters where nutrient concentrations in bulk samples are near or below the limit of detection by chemical means. The more productive areas of the ocean (Fig. 1-1) are usually associated with regions of upwelling and mixing where "new" nutrients are brought to the surface. In general, nitrogen is more often limiting in marine systems, and phosphorus is more often limiting in freshwater systems. Silicon deficiency may be important in limiting spring diatom blooms in both systems.

C. Producer/consumer relationships

Primary production by algae and aquatic vascular plants, along with externally derived mineral nutrients and organic matter, supports the consumers in aquatic systems and also provides habitat. The dissolved organic carbon released by phytoplankton and seaweeds presumably is consumed by bacteria, but the percentage contribution of this material to bacterial production is still in doubt. Of the living algal organic matter consumed directly by zooplankton (e.g., ciliates and copepods) and benthic consumers (e.g., worms, sea urchins, gastropods), a significant portion is left over as particulate organic matter. This material arises from "sloppy" feeding in which the consumer, large or small, fails to ingest all of the material collected and from fecal material that may contain living algal cells and/or spores. Much of the algal particulate production, especially in seaweeds, is liberated due to the natural death and/or erosion of seaweed tissues such as the receptacles of *Ascophyllum* (Josselyn and Mathieson 1978) and the blade tips of *Laminaria* (Mann 1973). All of this nonliving particulate organic matter becomes detritus, which is colonized by bacteria and fungi and supports another food chain containing detritivores. The food value of detritus is often improved by the presence of the decomposing microorganisms.

Many of the macroconsumers of seaweeds (e.g., herbivorous fishes, gastropods, and sea urchins) exhibit a definite selectivity in the species of seaweeds they consume. Littler and Littler (1980) and Steneck and Watling (1982) have grouped seaweeds into functional-form groups that reflect, among other characteristics, differences in their suscepti- bility to grazing (see also Littler and Littler, Chapter 2). The rapidly growing filamentous and foliose forms tend to be rapidly consumed by most herbivores, whereas many of the calcareous, crustose, and thicker forms tend to be tougher, less nutritious on a per weight basis, and more resistant to grazing. Chemical deterrents (Sun, Paul, and Fenical 1983; Littler, Taylor and Littler 1986; Harlin 1987), life history refuges (Lubchenco and Cubit 1980), and spatial refuges are also known to pro- tect algae from grazers. Grazing by voracious sea urchins decimates kelp beds (Leighton, Jones, and North 1966; Breen and Mann 1976; Fore- man 1977; Dean, Schroeter and Dixon 1984). In outdoor aquaculture, these herbivores can constitute a major threat.

IV. Integrating scientific disciplines

The subject matter of this book exists at the interface between "basic" or "pure" science and "applied" or "practical" science. All too often these disciplines are perceived and funded as distinct endeavors when, in fact, a continuum exists and a natural synergism comes into play when these two approaches are allowed to interact. Although the applied aspects of algae are emphasized in this book, the reader should be aware that an understanding of the basic biology of algae has been crucial to many of the developments in algal technology (or phycotech- nology: Nonomura, Chapter 21) and that many of these applied uses of algae have stimulated and will continue to stimulate additional research into fundamental questions relating to metabolism, growth, reproduc- tion, development, and ecology of algae. It is also important that sci- entists engaged in different areas of basic biological research commu- nicate more effectively with one another. For example, molecular biologists would do well to consider the ecological significance of their findings, and ecologists would benefit from understanding the molecu- lar mechanisms underlying phenomena they observe. Some examples of integrated approaches follow.

As our technological society produces new and increasing amounts of pollutants, there is a greater demand for organisms that can be used to test our water supplies for the presence of toxins. Living material is especially useful for at least two reasons: (1) Organisms respond to poi- sons in complex effluents in which it would be difficult to identify all possible pollutants, and (2) organisms integrate the effects of the toxin over time whereas chemical assays need to be repeated at intervals. Sci- entists at the Environmental Protection Agency (EPA) in Rhode Island recently developed a toxicity test for pollutants in coastal waters. Dr. Richard Steele, a dedicated phycologist who has cultured algae for over two decades, had a "feeling for the organisms" and was familiar with their life histories. Drs. Dennis Hanisak and Glen Thursby are experimental botanists who contributed their expertise in quantitative experimentation and provided focus for the research. After a number of seaweeds had been tested at different stages in their life histories, it was discovered that sexual reproduction was the most sensitive and practical end point. Work has been completed for using the red alga, *Champia parvula,* to test the toxicity of single compounds or complex effluents (Thursby and Steele 1986). *Champia* is found throughout the world, grows well at room temperature, is manageable, and has a short life cycle (Thursby, Steele, and Kane 1985). Another test species, the kelp *Laminaria saccharina,* has the additional advantage of being ecologically and economically important. Work on this species began with field-collected material (Steele and Hanisak 1978) and has progressed to relying on laboratory cultures of male and female gametophytes. *Laminaria* will soon join *Champia* as one of the standard test species at EPA's Environmental Research Laboratory in Narragansett, Rhode Island. The development of toxicity test procedures on both of these species was built upon a century of basic research on algal life histories (Bigelow 1887).

The concept of *r*- and *K*-selected species in terrestrial organisms was derived from the logistic equation for growth, where *r* represents the biotic potential of the individual and *K* represents the carrying capacity of the system, a term that includes the detrimental effects of the other individuals in the population on that potential (environmental resistance) (Ricklefs 1973). This concept has proved useful in identifying a series of related characteristics in a spectrum of life history strategies (Pianka 1978). The *r*-selected species tend to be short-lived, fast-growing opportunists and are typically found in unstable environments with little competition. *K*-Selected species are found in more stable environments where competition is more intense; they grow more slowly, live longer, and reach reproductive maturity later. These concepts are currently being modified and applied to phytoplankton (Kilham and Kilham 1980; Harris 1986) and seaweeds (Littler and Littler 1980; Steneck and Watling 1982; Littler, Littler, and Taylor 1983). The applied payoff of this basic research relates to the fact that most agronomically important plants are *r*-selected species, namely, domesticated weeds. We should understand the life history strategies of the algal species we intend to exploit commercially in order to make the most efficient use of the resources involved in these projects. For example, in many cases *r*-selected features such as rapid growth and early reproduction might be favorable attributes, whereas in other cases *K*-selected features such as grazer resistance and phycocolloid production might be more desirable.

In natural ecosystems, stability may be provided by interactions among the many different species present and the diverse environment. Ecosystems managed by humans are simplified to a greater or lesser extent to make more of that system's production available for human use. Much of the cost of agriculture is associated with delivering nutrients and controlling pests. Although we associate the term *cost* with economics, we must also remember the ecological costs of agriculture – pollution of other ecosystems, loss of habitat, and extinction of species. Once we understand (through basic research) how natural systems operate, it becomes clear that they provide many "services," such as predator control and recycling of nutrients. The challenge we face as we increasingly develop (through applied research) aquatic systems for human use is to learn from the mistakes of terrestrial management and develop systems that utilize natural processes as much as possible. The results will be well worth the effort – economically, ecologically, and aesthetically.

This book begins by describing the role of algae in two natural environments: the virtually pristine biotic reefs of the tropical oceans and the heavily impacted Laurentian Great Lakes system of North America. The direct and indirect effects algae have on the human experience in just these two systems aptly illustrate the overall importance of algae to our global environment. The book then introduces the reader to direct uses of algae and their products – from culture to market. The nuisance aspect of algae is also presented. Note that one person's livelihood can be another's nuisance (for example, a nori culture in front of a vacation home), or, vice versa, nuisance algae (e.g., a windrow of seaweeds rotting on a bathing beach) can become someone else's livelihood (a source of garden fertilizer). Finally, this book looks into the future of algae in industry, biotechnology, and outer space. Although the large number of authors testifies to the amount of commercial activity involving algae, each contributor knows algae primarily as organisms – beautiful, fascinating, and responsive to natural laws.

V. References

Bazilevich, N. I., L. E. Rodin, and N. N. Rozov. 1970. Untersuchungen der biologischen Producktivität in geographischer Sicht. V. *Tag. Geogr. Ges.* USSR, Leningrad 1970 (in Russian). In Walter 1985.

Bigelow, R. P. 1887. On the structure of the frond in *Champia parvula* Harv. *Proc. Amer. Acad. Arts Sci.* 23, 111–21.

Bold, H. C., and M. J. Wynne. 1985. *Introduction to the Algae: Structure and Reproduction.* 2nd ed. Prentice-Hall, Englewood Cliffs, NJ. 706 pp.

Bott, T. L. 1983. Primary productivity in streams. In *Stream Ecology: Application and Testing of General Ecological Theory,* ed. J. R. Barnes and G. W. Minshall, pp. 29–53. Plenum Press, New York.

Breen, P. A., and K. H. Mann. 1976. Destructive grazing of kelp by sea urchins in eastern Canada. *J. Fish. Res. Bd. Can.* 33, 1278–83.

Brown, J. S. 1985. Three photosynthetic antenna porphyrins in a primitive green alga. *Biochim. Biophys. Acta* 807, 143–6.

Brylinsky, M. 1977. Release of dissolved organic matter by some marine macrophytes. *Mar. Biol.* 39, 213–20.

Buggeln, R. G. 1974. Negative phototropism of the haptera of *Alaria esculenta* (Laminariales). *J. Phycol.* 10, 80–2.

Cavalier-Smith, T. 1975. The origin of nuclei and of eukaryotic cells. *Nature* 256, 463–8.

Chapman, V. J. 1970. *Seaweeds and Their Uses.* 2nd ed. Methuen and Co., London. 304 pp.

Cobb, J. S., and M. M. Harlin. 1976. *Marine Ecology: Selected Readings.* University Park Press, Baltimore. 546 pp.

Crisp, D. J. 1975. Secondary productivity in the sea. In *Productivity of World Ecosystems.* Proc. Symp. Aug. 31–Sept. 1, 1972, Seattle, pp. 71–89. National Academy of Sciences, Washington, DC.

Dean, T. A., S. C. Schroeter, and J. D. Dixon. 1984. Effects of grazing by two species of sea urchins (*Strongylocentrotus franciscanus* and *Lytechinus unamesus*) on recruitment and survival of two species of kelp (*Macrocystis pyrifera* and *Pterygophora californica*). *Mar. Biol.* 78, 301–13.

Dring, M. J. 1984. Photoperiodism and phycology. In *Progress in Phycological Research,* vol. 3, ed. F. E. Round and D. J. Chapman, pp. 159–92. Biopress, Bristol.

Dugdale, R. C. 1967. Nutrient limitation in the sea: dynamics, identification, and significance. *Limnol. Oceanogr.* 12, 685–95.

Eppley, R. W., J. N. Rogers, and J. J. McCarthy. 1969. Half-saturation constants for uptake of nitrate and ammonium by marine phytoplankton. *Limnol. Oceangr.* 14, 912–20.

Fogg, G. E. 1983. The ecological significance of extracellular products of phytoplankton photosynthesis. *Bot. Mar.* 26, 3–14.

Foreman, R. E. 1977. Benthic community modification and recovery following intensive grazing by *Strongylocentrotus drobachinsis. Helgoländ. Wiss. Meeresunters.* 30, 468–84.

Frederick, S. E., P. J. Gruber, and N. E. Tolbert. 1973. The occurrence of glycolate dehydrogenase and glycolate oxidase in green plants: an evolutionary survey. *Plant Physiol.* 52, 318–23.

Gensel, P. G., and H. N. Andrews. 1987. The evolution of early land plants. *Amer. Sci.* 75, 478–89.

Gibbs, S. P. 1978. The chloroplasts of *Euglena* may have evolved from symbiotic green algae. *Can. J. Bot.* 56, 2883–9.

Goldman, J. C., and P. M. Glibert. 1982. Comparative rapid ammonium uptake by four species of marine phytoplankton. *Limnol. Oceanogr.* 27, 814–27.

Goldman, J. C., and R. Mann. 1980. Temperature-influenced variations in speciation and chemical composition of marine phytoplankton in outdoor mass cultures. *J. Exp. Mar. Biol. Ecol.* 46, 29–39.

Graham, L. E. 1984. *Coleochaete* and the origin of land plants. *Amer. J. Bot.* 71, 603–8.

Graham, L. E. 1985. The origin of the life cycle of land plants. *Amer. Sci.* 73, 178–86.

Gray, M. W. 1983. The bacterial ancestry of plastids and mitochondria. *BioScience* 33, 693–9.

Harlin, M. M. 1987. Allelochemistry in marine macroalgae. *CRC Crit. Rev. in Plant Sci.* 5, 237–49.

Harlin, M. M., and J. S. Craigie. 1975. The distribution of photosynthate in *Ascophyllum nodosum* as it relates to epiphytic *Polysiphonia lanosa. J. Phycol.* 11, 109–13.

Harris, G. P. 1986. *Phytoplankton Ecology: Structure, Function, and Fluctuation.* Chapman and Hall, London. 384 pp.

Healey, F. P. 1985. Interacting effects of light and nutrient limitation on the growth rate of *Synechococcus linearis* (Cyanophyceae). *J. Phycol.* 21, 134–46.

Josselyn, M. N., and A. C. Mathieson. 1978. Contribution of receptacles from the fucoid *Ascophyllum nodosum* to the detrital pool of a north temperate estuary. *Estuaries* 1, 258–61.

Kilham, P., and S. S. Kilham. 1980. The evolutionary ecology of phytoplankton. In *The Physiological Ecology of Phytoplankton,* ed. I. Morris, pp. 571–97. Blackwell Scientific Publications, Oxford.

Lake, J. A., E. Henderson, M. Oakes, and M. W. Clark. 1984. Eocytes: a new ribosome structure indicates a kingdom with a close relationship to eukaryotes. *Proc. Natl. Acad. Sci. USA* 81, 3786–90.

Laws, E. A., D. G. Redalje, L. W. Haas, P. K. Bienfang, R. W. Eppley, W. G. Harrison, D. M. Karl, and J. Marra. 1984. High phytoplankton growth and production rates in oligotrophic Hawaiian coastal waters. *Limnol. Oceanogr.* 29, 1161–9.

Leighton, D. L., L. G. Jones, and W. J. North. 1966. Ecological relationships between giant kelp and sea urchins in southern California. *Proc. Int. Seaweed Symp.* 5, 141–53.

Leith, H., and R. H. Whittaker. 1975. *Primary Productivity of the Biosphere.* Springer-Verlag, New York.

Lewin, R. A. 1976. Prochlorophyta as a proposed new division of algae. *Nature* 261, 697–8.

Lewin, R. A. 1984. *Prochloron:* a status report. *Phycologia* 23, 203–8.

Littler, M. M., and D. S. Littler. 1980. The evolution of thallus form and survival strategies in benthic marine macroalgae: field and laboratory tests of a functional form model. *Amer. Nat.* 116, 25–44.

Littler, M. M., and D. S. Littler (eds.). 1985. *Handbook of Phycological Methods. Ecological Field Methods: Macroalgae.* Cambridge University Press, Cambridge. 617 pp.

Littler, M. M., D. S. Littler, S. M. Blair, and J. N. Norris. 1985. Deepest known plant life discovered on an uncharted seamount. *Science* 227, 57–9.

Littler, M. M., D. S. Littler, and P. R. Taylor. 1983. Evolutionary strategies in a tropical barrier reef system: functional-form groups of marine macroalgae. *J. Phycol.* 19, 229–37.

Littler, M. M., P. R. Taylor, and D. S. Littler. 1986. Plant defense associations in the marine environment. *Coral Reefs* 5, 63–71.

Lubchenco, J., and J. Cubit. 1980. Heteromorphic life histories of certain marine algae as adaptations to variations in herbivory. *Ecology* 61, 676–87.

Lüning, K. 1980. Critical levels of light and temperature regulating the gametogenesis of three *Laminaria* species. *J. Phycol.* 16, 1015.

McKerracher, L., and S. P. Gibbs. 1982. Cell and nucleomorph division in the alga *Cryptomonas*. *Can. J. Bot.* 60, 2440–52.

Mann, K. H. 1973. Seaweeds: their productivity and strategy for growth. *Science* 182, 975–81.

Mann, K. H. 1978. *Ecology of Coastal Waters: A Systems Approach.* University of California Press, Berkeley. 322 pp.

Margulis, L. 1981. *Symbiosis in Cell Evolution.* W. H. Freeman and Co., San Francisco. 419 pp.

Margulis, L., and D. Sagan. 1986. *Microcosmos: Four Billion Years of Evolution from our Microbial Ancestors.* Summit Books, New York. 301 pp.

Mattox, K. R., and K. D. Stewart. 1984. Classification of the green algae: a concept based on comparative cytology. In *Systematics of the Green Algae,* ed. D. E. G. Irvine and D. M. John, pp. 29–72. Systematics Association special vol. no. 27. Academic Press, London.

Moestrup, Ø. 1982. Flagellar structure in algae: a review with new observations particularly on the Chrysophyceae, Phaeophyceae (Fucophyceae), Euglenophyceae, and *Reckertia. Phycologia* 21, 427–528.

Niklas, K. J. 1976. Morphological and ontogenetic reconstruction of *Parka decipiens* Fleming and *Pachytheca* Hooker from the Lower Old Red Sandstone, Scotland. *Trans. Roy. Soc. Edinburgh* 69, 483–99.

Padan, E. 1979. Facultative anoxygenic photosynthesis in cyanobacteria. *Ann. Rev. Plant Physiol.* 30, 27.

Parsons, T. R., Y. Maita, and C. M. Lalli. 1984. *A Manual of Chemical and Biological Methods for Seawater Analyses.* Pergamon Press, New York. 173 pp.

Pianka, E. R. 1978. *Evolutionary Ecology.* 2nd ed. Harper and Row, New York. 397 pp.

Ramus, J. 1983. A physiological test of the theory of complementary chromatic adaptation. II. Brown, green, and red seaweeds. *J. Phycol.* 19, 173–8.

Richardson, K., J. Beardall, and J. A. Raven. 1983. Adaptation of unicellular algae to irradiance: an analysis of strategies. *New Phytol.* 93, 157–91.

Ricklefs, R. E. 1973. *Ecology.* 2nd ed. Chiron Press, New York. 966 pp.

Rodin, L. E., N. I. Bazilevich, and N. N. Rozov. 1975. Productivity of the world's main ecosystems. In *Productivity of World Ecosystems.* Proc. Symp. Aug. 31–Sept. 1, 1972, Seattle, pp. 13–26. National Academy of Sciences, Washington, DC.

Scavia, D., G. L. Fahnenstiel, J. A. Davis, and R. G. Kreis, Jr. 1984. Small-scale nutrient patchiness: some consequences and a new encounter mechanism. *Limnol. Oceanogr.* 29, 785–93.

Schopf, J. W. 1970. Pre-cambrian micro-organisms and evolutionary events prior to the origin of vascular plants. *Biol. Rev.* 45, 319–52.

Sheath, R. G., and B. J. Hymes. 1980. A preliminary investigation of the freshwater red algae in streams of southern Ontario, Canada. *Can. J. Bot.* 58, 1295–318.

Sieburth, J. McN. 1969. Studies on algal substances in the sea. III. The production of extracellular organic matter by littoral marine algae. *J. Exp. Mar. Biol. Ecol.* 3, 290–309.

Sieburth, J. McN. 1979. *Sea Microbes.* Oxford University Press, New York. 491 pp.

Steele, J. H. 1974. *The Structure of Marine Ecosystems.* Harvard University Press, Cambride, MA. 128 pp.

Steele, R. L., and M. D. Hanisak. 1978. Sensitivity of some brown algal reproductive stages to oil pollution. *Proc. Int. Seaweed Symp.* 9, 181–90.

Steneck, R. S., and L. Watling. 1982. Feeding capabilities and limitation of herbivorous molluscs: a functional group approach. *Mar. Biol.* 68, 299–319.

Stewart, K. D., and K. R. Mattox. 1975. Comparative cytology, evolution, and classification of the green algae with some consideration of the origin of other organisms with chlorophylls *a* and *b*. *Bot. Rev.* 41, 104–35.

Stewart, K. D., and K. R. Mattox. 1980. Phylogeny in phytoflagellates. In *Phytoflagellates*, ed. E. R. Cox, pp. 433–62. Elsevier/North-Holland, Amsterdam.

Stewart, K. D., and K. R. Mattox. 1984. The case for a polyphyletic origin of mitochondria: morphological and molecular comparisons. *J. Mol. Evol.* 21, 54–7.

Sun, H. H., V. J. Paul, and W. Fenical. 1983. Avrainvilleol, a brominated diphenylmethane derivative with feeding deterrent properties from the tropical green alga *Avrainvillea longicaulis*. *Phytochemistry* 22, 743–5.

Thursby, G. B., and R. L. Steele. 1986. Comparison of short- and long-term sexual reproduction tests with the marine red alga *Champia parvula*. *Envir. Toxicol. Chem.* 5, 1001–18.

Thursby, G. B., R. L. Steele, and M. E. Kane. 1985. The effect of organic chemicals on growth and reproduction in the marine red algae *Champia parvula*. *Envir. Toxicol. Chem.* 4, 797–805.

Tomas, R. N., and E. R. Cox. 1973. Observations on the symbiosis of *Peridinium balticum* and its intracellular alga. I. Ultrastructure. *J. Phycol.* 9, 304–23.

Waaland, S. D., W. Nehlsen, and J. R. Waaland. 1977. Phototropism in a red alga, *Griffithsia pacifica*. *Plant Cell Physiol.* 18, 603–12.

Walter, H. 1985. *Vegetation of the Earth and Ecological Systems of the Geo-biosphere.* 3rd ed. Springer-Verlag, New York. 318 pp.

Weeden, N. F. 1981. Genetic and biochemical implications of the endosymbiotic origin of the chloroplast. *J. Mol. Evol.* 17, 133–9.

Wheeler, H. J., and B. L. Hartman. 1893. *Sea-weeds, Their Agricultural Value: The Chemical Composition of Certain Species.* Agricultural Experiment Station of the Rhode Island College of Agriculture and Mechanic Arts, Kingston, RI.

Wilcox, L. W., and G. J. Wedemayer. 1985. Dinoflagellate with blue-green chloroplasts derived from an endosymbiotic eukaryote. *Science* 227, 192–4.

Wilhelm, C., I. Levarty-Weiler, I. Weidemann, and A. Wild. 1986. The light harvesting system of a *Micromonas* species (Prasinophyceae): the combination of three different chlorophyll species in one single chlorophyll-protein complex. *Phycologia* 25, 304.

Wilson, E. B. 1925. *The Cell in Development and Heredity.* Macmillan Press, New York.

Woodwell, G. M., R. H. Whittaker, W. A. Reiners, G. E. Likens, C. C. Delwiche, and D. B. Botkin. 1978. The biota and the world carbon budget. *Science* 199, 141–6.

2. Structure and role of algae in tropical reef communities

MARK M. LITTLER AND DIANE S. LITTLER

Department of Botany, National Museum of Natural History, Smithsonian Institution, Washington, DC 20560

CONTENTS

Contribution No. 192 of the Smithsonian Marine Station at Link Port, FL.

I. Introduction

Tropical reef ecosystems consist of spectacular biological communities on limestone bases, derived mainly from the fossilized remains of coelenterate corals and calcareous algae, and are found within the 22°C (north–south) isotherms throughout the world's oceans (Fig. 2-1). Because of the long geological history of environmental constancy within tropical zones, reef systems have developed an extremely high level of biological diversity, including many specialized organisms. The calcareous skeletal material of coral animals (aragonite) supplies much of the structural bulk, and the calcite cement produced by coralline algae (Rhodophyta) consolidates this material and other debris to contribute to reef formation. In addition, nonarticulated coralline algae may form an outer intertidal algal ridge that buffers wave forces and prevents erosion and destruction of the more delicate corals and softer organisms typical of back-reef habitats. A diverse group of calcified green algae (Chlorophyta) belonging to the orders Caulerpales and Dasycladales deposit the aragonite form of calcium carbonate, which is responsible for much of the sand and lagoonal sediment within the back reef and deeper areas. Tropical reefs are remarkable for their development of massive structure and high primary productivity. As we will show, algae are responsible for much of the former and all of the latter.

On a global scale, reef ecosystems comprise a quantitatively significant geographical, geochemical, and biological resource. Their total area, fisheries yield, calcium mass balance, and primary productivity have been calculated (Lewis 1977; Smith 1978) and found to be substantial. For example, total reef area is approximately $6 \times 10^5 \ \mathrm{km}^2$, which is about 0.17% of the world's ocean area and about 15% of the shallow sea floor within the 0–30 m depth range (Smith 1978). Although taken from this relatively small region of the world's oceans, the reef fisheries potential is estimated (Smith 1978) to be about 9% (approximately $6 \times 10^9 \ \mathrm{kg \cdot y^{-1}}$) of the total oceanic fisheries yield. Further, shallow reefs are conservatively estimated (Smith 1978) to precipitate about 6×10^{12} mol $CaCO_3$ annually, which is equal to about 50% of the yearly calcium input to the world's oceans. Reefs retain a signif-

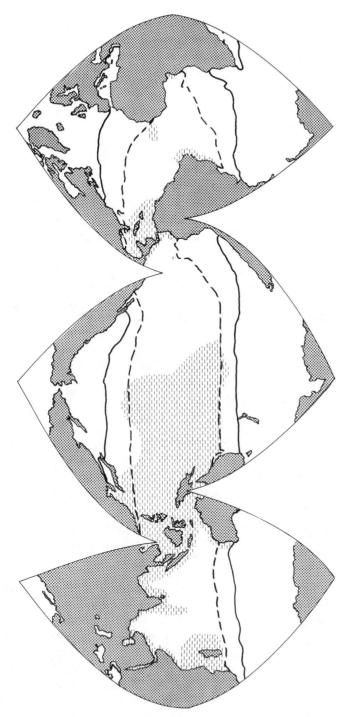

Fig. 2–1. Distribution of reefs and relation to controlling sea surface temperature. Solid bold line is extended 21°C isotherm, broken bold line is restricted 21°C isotherm, and lightly shaded areas are generalized areas containing the majority of coral reefs. (Modified from Wells 1957.)

Table 2-1. *Productivity figures for open ocean phytoplankton vs. reef populations*

Type of algae	Productivity (g $C \cdot m^{-2} \cdot d^{-1}$)
Open-sea phytoplankton	0.1–0.35
Reef community	5.0–10.0
Encrusting coralline algae	2.7
Fleshy and filamentous algae	3.3
Coral zooxanthellae	11.0

Source: Marsh 1976; Wanders 1976.

icant fraction of precipitated sediments by means of their wave-resistant structures and lagoons and constitute a major global sink for calcium sediments.

Biotic reefs represent some of the most productive natural ecosystems known (Westlake 1963; Lewis 1977), although they mainly occur in nutrient-poor waters. Shallow reef flats characteristically have gross production rates 50 to 100 times higher than the productivity of the adjacent open ocean (Table 2-1). The anomalously high organic productivity of reefs is attributable to: (*a*) extensive spatial rugosity (dramatically increasing metabolic surface area), (*b*) continual nutrient input from water currents flowing across these surfaces, (*c*) specialized benthic nitrogen-fixing blue-green algae and bacteria, and (*d*) biological systems that efficiently retain nutrients by recycling. Four major groups of sessile photosynthetic organisms are responsible for the bulk of reef productivity (Fig. 2-2A–F): (1) microfilamentous turf algae, (2) symbiotic unicellular algae within hermatypic corals, (3) frondose macroalgae, and (4) coralline algae. In some soft bottom habitats, seagrasses, algal epiphytes, and siphonaceous Chlorophyta (Fig. 2-2F) contribute substantially to overall productivity.

The average living coral colony on Eniwetok atoll contains three times as much plant as animal tissue (Odum and Odum 1955), of which only about 6% is contributed by zooxanthellae, the remainder being composed of microfilamentous green algae within the coral skeletal structure. On reefs not dominated by corals, coralline algae and various small filamentous algae usually compose the majority of cover (e.g., standing stocks of < 0.27 kg $\cdot m^{-2}$: Brawley and Adey 1977). Recolonization of new substrata by microfilamentous algae is extremely rapid on reefs. For example, after corals were killed by the predatory starfish *Acanthaster* on Guam reefs, blue-green algae became established within 24 hours, and a diverse filamentous community of algae was dominant after 26 days (Belk and Belk 1975). Larger frondose algae occur abun-

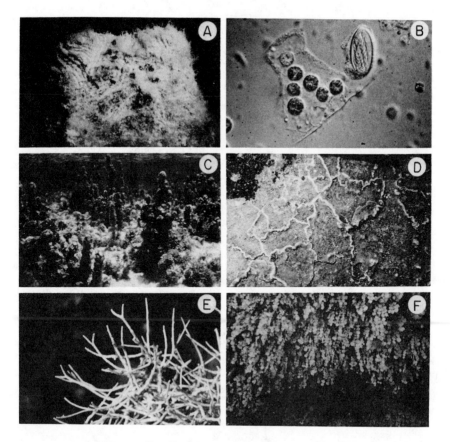

Fig. 2–2. A. Microfilamentous algal turf scraped by herbivorous parrotfish. B. Symbiotic zooxanthellae within a hermatypic coral (photo courtesy of William Fitt and Robert Trench). C. Frondose algal populations of *Turbinaria* and *Sargassum*. D. Nonarticulated (crustose) coralline alga. E. Articulated coralline alga. F. The siphonaceous calcifying chlorophyte *Halimeda*.

dantly on reef flats (Doty 1971; Wanders 1976), unstructured sand plains (Earle 1972; Hay 1981a), or deepwater sites (Littler et al. 1985), with typical mean algal standing stocks of 3.0–3.5 kg·m^{-2} (Doty 1969) and unusually rich locations (e.g., Martinique: Connor and Adey 1977) ranging up to 10 kg·m^{-2}. Thus, although temperate algal standing stocks tend to be higher (Taylor 1960), some tropical reef habitats can support algal biomass levels of a comparable high magnitude.

All of the coral and algal groups compete for space, nutrients, and light (e.g., Littler and Doty 1975); any one assemblage can predominate under specific environmental conditions. For this reason, and despite its wide usage, the popular term *coral reef* can be misleading unless it

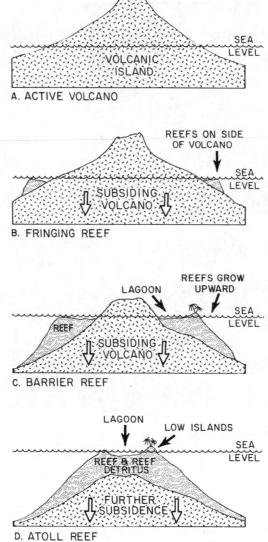

Fig. 2–3. Subsidence sequence proposed by Darwin leading from B (fringing reef) to C (barrier reef) to D (atoll reef formation).

refers specifically to those systems dominated by coelenterate corals. Consequently, in the general sense, *biotic reef* (Womersley and Bailey 1970) or simply *reef system* is preferable.

There are three reef types, based on their location – fringing, barrier, and atoll (Fig. 2-3) – and all have the same basic ecological zones.

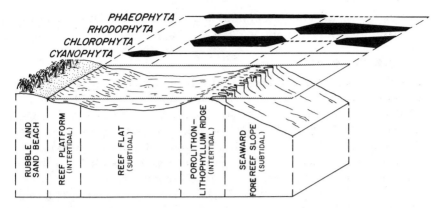

Fig. 2–4. The generalized primary habitats of the tropical marine divisions of calcareous macroalgae. The relative extents of zonal configurations vary markedly from reef to reef.

Related to these basic reef configurations are patch reefs, barrier bank reefs, and table reefs. The most seaward portion of a typical reef (Fig. 2-4) is the outer fore reef slope, which grades upward to the shallow fore reef and the reef crest. Where wave action is consistently high, the reef crest develops into an intertidal algal ridge (often including a spur-and-groove configuration) generally dominated by the coralline algal genera *Porolithon* and *Lithophyllum*. The most massive algal ridges are found on Pacific atolls although they are present intertidally on any reef system consistently subjected to high wave action. Behind the algal ridge is the shallow back reef, which is the product of limestone-boring organisms that cause the disintegration of calcareous material. In this habitat are the slower-growing corals, various coralline algae, and frondose algae. The shallow back reef immediately behind the algal crest usually grades upward toward the shoreline to form a reef flat and/or platform where storms may cast calcareous sediments, rubble, and boulders. This material accumulates, particularly on windward reefs, to form consolidated terrestrial surfaces on which islands can develop.

II. An overview of shallow-water algae

Although it is difficult to generalize, there is a tendency for the various calcareous and noncalcareous groups of algae to predominate within different reef habitats. The distribution of frondose algae, calcareous algae, and corals appears to be related directly to biological factors such as competition and grazing as well as physical factors, including nutrient levels, wave action, irradiance, and temperature. As will be discussed, much of the pattern seems to be generated by competition mediated by the interaction of variations in nutrient availability and disturbances by herbivorous fishes and wave action.

A. Noncalcareous forms

Frondose macroalgal abundances (Fig. 2-2C) are normally reduced on reefs by herbivorous fishes and sea urchins (Littler and Doty 1975; Hay 1981b; Hatcher and Larkum 1983). The inconspicuousness of microfilamentous algae (Fig. 2-2A) on shallow reef-front systems also is thought (Littler and Doty 1975; Borowitzka 1981) to result primarily from intensive grazing by the numerous herbivores and omnivores inhabiting these spatially heterogeneous habitats. Where there is much turbulence or little topographic relief on tropical reefs, herbivore activity is reduced (Brock 1979; Hay, Colburn, and Downing 1983), thereby enabling reasonably large standing stocks of macrophytes *(Sargassum, Turbinaria, Acanthophora)* to develop (Doty 1971; Wanders 1976). Some chemically well-defended macroalgae (e.g., *Halimeda, Stypopodium, Dictyota;* see Norris and Fenical 1982) can be abundant where grazing is high. Such macroalgal populations may contribute a major portion of the total primary productivity of tropical reefs (Rogers and Salesky 1981). However, most evidence (Odum and Odum 1955; Marsh 1976; Carpenter 1985) indicates that the sparse mats of fast-growing, opportunistic filamentous algae result in the very high primary productivity per unit area of biotic reefs. For example, on St. Croix reefs dominated by the coral *Acropora palmata* (where larger macrophytes are rare), filamentous algae, although present only in very low biomass ($180–270 \text{ g} \cdot \text{m}^{-2}$), are responsible (Brawley and Adey 1977) for 70–80% of the productivity of the shallow fore and back reefs ($5–7 \text{ g net C fixed} \cdot \text{m}^{-2} \cdot \text{d}^{-1}$). Proportionately, sparse filamentous mats are considerably more productive per unit of algal biomass than are dense stands of the larger macroalgae. This can be related to functional morphological differences (Littler 1980; Littler, Littler, and Taylor 1983a) where opportunistic species with high surface-to-volume ratios typically outproduce those having a coarser morphology. Variations of this sort are due (Littler and Littler 1980) to differential allocation of materials to photosynthetic versus structural tissues. Herbivorous fish, by their scraping mode of feeding, continuously provide new substrata and thereby select for opportunistic microalgal forms (Fig. 2-2A), as well as long-lived scrape-resistant coralline algae (Littler and Doty 1975; Wanders 1976).

Fixation of atmospheric nitrogen by blue-green algae such as *Calothrix crustacea* (Wiebe, Johannes, and Webb 1975; Mague and Holm-Hansen 1975) within filamentous microalgal assemblages also is an important feature that enhances reef productivity and nutrition. The greater productivity of benthic reef communities versus planktonic oceanic systems (Table 2-1; Lewis 1977), is in large part due to such nitrogen fixation within mixed filamentous algal communities (Wiebe et al. 1975; Mague and Holm-Hansen 1975), as well as unusually efficient nitrogen and phosphorus recycling within the symbiotic populations (Johannes et al. 1972). Other reef organisms, also closely associated with blue-green algae and showing high nitrogen fixation rates, include various macroalgal populations (Capone, Taylor and Taylor, 1977) and corals (Crossland and Barnes 1976). Such blue-green algal associations fix nitrogen at levels (985 kg·ha^{-1}·y^{-1}: Wiebe et al. 1975) equal to those recorded for the richest nitrogen-fixing terrestrial systems known (e.g., alfalfa fields).

Another important component of reef biology includes algae that cause breakdown of reef structure. Duerden (1902) recognized that penetrating or boring algae play a critical role in bioerosional processes. The commonest of diagenic algae are blue-green algae (see Nadson 1900; Purdy and Kornicker 1958) that attack skeletal materials differentially; the aragonitic coral skeletons are most susceptible, and the denser calcitic deposits of coralline algae are most resistant. Several distributional surveys of rock boring by microalgae (Green 1975; May, Macintyre, and Perkins 1982) documented the importance of this group on Caribbean reef systems. Weber-van Bosse (1932), during an extensive systematic study of penetrating algae in the Indo-Pacific, recorded 20 species distributed among the Cyanophyta, Chlorophyta, Phaeophyta, and Rhodophyta. Thirty-three species of carbonate-boring tropical algae have been reported from China (Chu and Wu 1983), two of which are active at depths exceeding 100 m. Much research remains to be done with this very interesting group of endolithic marine plants.

B. Calcareous algae

Calcareous algae are universally recognized (James and Macintyre 1985) as important contributors to both the bulk and frame structures of the majority of reef limestone deposits. Such deposits often have been associated with petroleum reserves, and this relationship has drawn the attention of geologists, paleobiologists, and others to calcifying seaweeds. Important historical roles of algal taxa as sediment producers and reef builders have been chronicled thoroughly by Wray (1971, 1977).

The order of prominence for reef-forming organisms in providing bulk during the development of the reef at Funafuti atoll, Ellice Islands (8°30'S, 179°10'E), was estimated as follows (Finckh 1904): (1) nonarticulated corraline algae (Rhodophyta), (2) the green alga *Halimeda* (Chlorophyta), (3) foraminifera, and (4) corals. Subsequent ecological work (e.g., Womersley and Bailey 1970; Littler 1971) and paleontological studies (e.g., Easton and Olson 1976) have substantiated the predominant role of coralline algae in cementing coarse and fine-grained sediments produced by calcareous green algae, molluscs, and foraminifera, along with the bulkier deposits provided by hermatypic corals (see also James and Macintyre 1985).

Some of the advantages of algal calcification postulated by Littler (1976) include mechanical support, resistance to sand scour, wave shock, and grazing, as well as protection against fouling epiphytes (by means of carbonate sloughing) and photoinhibition due to intense irradiance. Also, by providing their own substrata, calcareous algae may increase the stability and quality of their attachment sites.

The calcifying Rhodophyta grow on solid substrata intertidally and subtidally down to at least 268 m (Fig. 2-4; Littler et al. 1985, 1986) but attain maximum abundances in shallow turbulent areas. Coralline algae (Fig. 2-2D, E), in contrast to most fleshy algae, are relatively low primary producers, as documented by studies of their functional morphology (Littler and Littler 1980, 1984a; Littler, Littler, and Taylor 1983). Marsh (1970) was the first to note the low productivity of nonarticulated coralline populations (0.66 g net $C \cdot m^{-2} \cdot d^{-1}$ at Eniwetok atoll) and surmised that these algae are more important in reef building than in primary production. In his review, Lewis (1977) concluded that coralline algae have rates (Table 2-1) of primary production that are much lower than mean values for reefs as a whole. It is interesting to note that calcification rates appear to differ little among reef-flat communities consisting of diverse kinds of calcifiers (Smith 1973; Wanders 1976), whether they are corals, nonarticulated coralline algae, or turfs of articulated coralline algae. For the frame-building Corallinaceae, the available information (Littler et al. 1985, 1986) indicates an algal group able to grow at greater depths in weaker light than other primary producers. *Porolithon* and certain *Lithophyllum* species that dominate algal ridges (Fig. 2-4; e.g., *P. onkodes, P. pachydermum, P. craspedium, P. gardineri, L. kotschyanum, L. congestum, L. moluccense*) are somewhat exceptional in that they can withstand considerable desiccation and exposure to the highest sunlight irradiances.

The calcareous Chlorophyta predominate mainly in protected shallow areas on soft bottoms, often in association with seagrasses, as well as on the deeper reef slopes in the case of *Halimeda* (Hillis-Colinvaux 1977; Littler et al. 1985). In summarizing the ecological research on reef-building algae, the published data for the sediment-producing shallow Chlorophyta indicate an assemblage adapted to calm, soft-bottom habitats (which are unsuited for most other macroalgae) throughout subtropical to tropical zones. Few quantitative studies exist on any aspect of the ecology of the calcareous Chlorophyta with the exception of the widely studied genus *Halimeda* (see review by Hillis-Colinvaux 1980). Numerous *Halimeda* species are abundant on protected back-reef and fore-reef habitats, occurring over a broad depth range on both hard and soft substrata. Other psammophytic (sand-dwelling) forms are associated with shallow seagrasses and mangroves. Natural *Halimeda* populations of 100 plants·m^{-2} are common (Hillis-Colinvaux 1980), and densities up to 500 m^{-2} are found on some Jamaican reefs. Recently, spectacular, large banklike mounds composed of living *Halimeda* and its sediments (dating to 5,000 years B.P.) have been discovered (Davies and Marshall 1985) in the back-reef portion of the Great Barrier Reef, Australia. Skeletal sand-sized components from some tropical Atlantic peripheral reef and lagoonal sediments are composed of up to 44% and 77% *Halimeda* fragments, respectively (Milliman 1974; Orme 1977). An excellent, but largely overlooked, review of the importance of *Halimeda* as a reef-forming organism throughout the tropical Pacific was published (Chapman and Mawson 1906) over 80 years ago.

C. Coral-algal symbiosis

The major reef-building invertebrates belong to the phylum Cnidaria, classes Anthozoa and Hydrozoa, and have dinoflagellate microalgae (zooxanthellae) living within their tissues as primary producers (Fig. 2-2B). This symbiosis usually results in a net daily excess in primary productivity for shallow colonies (e.g., about 11 net g C·m^{-2}·d^{-1}: Wanders 1976). Zooxanthellae are derived from at least two genera of dinoflagellates, including *Symbiodinium* and *Amphidinium* (see review by Trench 1981), and by their symbiosis gain protection from zooplankton predators and obtain a wide variety of nutrients from the coral host. The nutrients obtained directly include phosphate and a variety of nitrogenous compounds such as ammonium, uric acid, guanine, alanine, and adenine. The coelenterate acquires algal photosynthates in the form of glycerol (currently estimated at up to 98% of the total carbon produced; see Muscatine, Falkowski, and Dubinsky 1983) and glucose (Battey and Patton 1984) as well as oxygen. Transfers of nutrients and energy sources between algal symbionts and their hosts (e.g., many corals, foraminifera, sponges, and molluscs) are important pathways in reef community recycling, representing a major mechanism enabling survival in oligotrophic waters (Muscatine and Porter 1977). The contribution of organic carbon from zooxanthellae to their hosts is significant, often estimated (Muscatine et al. 1983) at over 100% of the level of carbon required to meet the host's metabolic requirements. Under adverse conditions (i.e., low light), the amount of photosynthetically fixed carbon available to the host decreases, in spite of photoadaptation by the zooxanthellae (Muscatine et al. 1983; Porter et al. 1984). Algal symbiosis is energetically of great benefit when food is scarce but of less benefit when food resources are plentiful (Hallock 1985). Filamentous, boring green algae (Chlorophyta), mostly of the genus *Ostreobium*, also occur within the calcified coral skeletons and together with the zooxanthellae provide a very significant portion of the living biomass of a coral. For example (Muscatine et al. 1983), zooxanthellae contribute as much as 15% of total protein of coelenterate symbioses, and boring algal biomass can exceed that of the animal host (Odum and Odum 1955). However, the biology of the boring forms is poorly known. Endozoic algae embedded in the skeletons of living corals have not proven (Kanwisher and Wainwright 1967) to be significant primary producers on reefs.

III. An overview of deepwater algae

Submersible vessels have greatly expanded our knowledge of distributional limits for marine organisms; however, macroalgae have received only incidental attention until recently. For example, crustose red and filamentous green algae were seen by Lang (1974) as deep as 175 m at Discovery Bay, Jamaica. Adey and Macintyre (1973) cite the unpublished observations of J. C. Lang and J. W. Porter for the maximum depths for crustose coralline algae (200 m) on steep Belizean and Bahamian reef walls.

Recently, a record depth (268 m) for an attached living marine macrophyte was reported (Littler et al. 1985) during an extensive floristic-ecological survey of a seamount off San Salvador Island, Bahamas (Fig. 2-5). Studies from a submersible in conjunction with shore-based productivity measurements (Jensen et al. 1985; Littler et al. 1985, 1986) revealed unsuspected abundances and potential importances of other deepwater tropical macroalgae. Four zonal assemblages occurred over the depth range from 81 m to 268 m and consisted of a *Lobophora*-dominated group (Phaeophyta, 81–90 m), a *Halimeda* assemblage (Chlorophyta, 90–130 m), a *Peyssonnelia* group (Rhodophyta, 130–189 m), and a crustose coralline zone (Rhodophyta, 189–268 m).

The overall deepwater flora of the Bahamas is composed of unique deepwater taxa combined with shallow-water species characteristic of shaded low-light conditions. Species of Chlorophyta (e.g., *Johnson-sealinkia profunda* at 200 m) are the deepest-growing frondose algae, followed by Phaeophyta (e.g., *Lobophora variegata* at 100 m). Encrusting calcareous Rhodophyta extend below all other forms of macrophytes. The zonational pattern observed (Fig. 2-5) – namely, reds > greens > browns with increasing depth – is quite similar to that recorded (Larkum, Drew, and Cosset 1967) on the south coast of Malta. Off Malta, the shallower zone to 15 m is dominated by brown algae; green algae (mainly *Halimeda*) are then most abundant down to 75 m, and red algae (particularly *Peyssonnelia* and *Lithophyllum*) dominate below the 80–90 m transitional depth.

Dominant members of the diverse multilayered macrophyte commu-

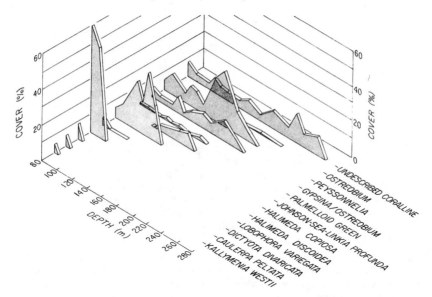

Fig. 2–5. Distribution and abundance patterns of the major deepwater plant cover on San Salvador seamount, Bahamas.

nity on top of San Salvador seamount (81 m depth) showed net productivity levels comparable to shallow-water seaweeds although receiving only 1–2% of the light energy available at the surface. Productivity rates reported for shallow carbonate reef systems range from 0.06 to 0.72 g $C \cdot m^{-2}$ of substratum $\cdot h^{-1}$ (approximated from daily rates) as follows: intertidal, blue-green algal-dominated substrate at Eniwetok atoll = 0.06–0.22 g $C \cdot m^{-2} \cdot h^{-1}$ (Bakus 1967); macroalgal-dominated habitats in the Canary Islands = 0.15–0.30 (Johnston 1969); photosynthetic coral and algal turfs at Eniwetok = 0.16–0.72 (Smith 1973); a coralline algal-dominated fringing reef in French Polynesia = 0.38 (Sournia 1976); and fore- and back-reef algal turf systems = 0.5–0.7 (Brawley and Adey 1977). The mean rate for the seamount plateau, 0.066, falls within the lower end of this range. Littler et al. (1986) conclude that deepwater macroalgal communities produce at rates comparable to some shallow reef systems but lower than most seagrass beds (see Littler et al. 1985) or typical carbonate reef-flat habitats. It is significant that other physiological processes (i.e., calcification) for four deepwater *Halimeda* species (Jensen et al. 1985) also showed similar rates to those reported (Borowitzka and Larkum 1976) for shallow forms of the genus.

The great abundance and considerable productivity of macroalgae in the previously unknown deep-sea realm underscores their potential

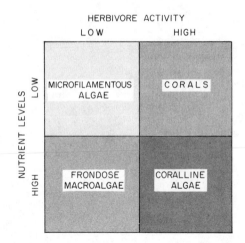

Fig. 2–6. Diagrammatic representation of the relative dominance paradigm. Potentially predominant space-occupying groups of photosynthetic reef organisms are emphasized as a function of long-term nutrient levels and disturbance. Hermatypic corals are hypothesized to be the competitive dominants, and coralline algae are projected as the poorest competitors but the most physically resistant group. Grazing is considered the more important direct controller of algal standing stocks on undisturbed reefs, whereas nutrients set the potential upper limits to biomass. (From Littler and Littler 1985.)

widespread importance to the marine food web and reef biogenesis in other clear tropical waters. As more plant specialists gain access to deep-sea communities, the role of benthic algae will become clearer and new botanical theories and concepts based on *in situ* research will replace older ones derived from shipboard studies and remote sensing.

IV. Factors regulating algal distribution

Variations in the levels of grazing and wave shock (physical disturbance) and limiting or toxic nutrient levels (physiological stress) hypothetically are major factors that lead to spatial segregation of coral-, coralline-, or fleshy algal-dominated communities between or within habitats, or they may lead to temporal separations important during succession and reef biogenesis (Fig. 2-6). Corals, though preyed upon by a few omnivorous fishes and sea urchins, generally gain primacy under intensive herbivory (Brock 1979), moderate levels of wave shear, and very low nutrient concentrations (Bakus 1969). The decrease in coelenterate coral cover, relative to macroalgae (Doty 1971) and coralline algae (Littler 1971), on the reef flat at Waikiki, Oahu, over a 45-year span may be attributable to increases in nutrients from eutrophication due to municipal sewage effluents. With an increase in the amount of nutrients, the growth of short-lived filamentous and leafy algae is favored over the slower-growing corals, and the latter become endangered by competition for space unless they can attain a refuge in size (Birkeland 1977). Townsley (cited in Doty 1969) found that some Hawaiian corals (e.g., *Porites compressa*) are extremely sensitive to increased phosphate concentrations. Orthophosphates are known (Simkiss 1964) to inhibit $CaCO_3$ crystal formation at concentrations as low as 0.01 μM and can inhibit production of external skeletal material in marine animals. The 50% reduction in calcification rate observed (Kinsey and Domm 1974; Kinsey and Davies 1979), following experimental fertilization of a patch reef on the Great Barrier Reef, was partly attributed to phosphate poisoning.

Coralline algae, on the other hand, predominate in areas of moderate to heavy grazing (or heavy wave shear) and are not inhibited by moderate to high levels of nutrient enrichment (Fig. 2-6). However, high phosphate concentrations also are known (Brown, Ducker, and Rowan 1977) to reduce calcification and growth of some temperate articulated coralline algae.

Eutrophic waters, where herbivory and wave-shearing forces are low, tend to favor large populations of frondose macroalgae that can ultimately overgrow and kill both coralline algae (Littler and Doty 1975) and corals (Banner 1974). When nutrient levels are low and grazing activity is low to moderate (Fig. 2-6), microfilamentous forms with greater surface area to volume ratios tend to predominate (e.g., as in damselfish territories [Vine 1974; Brawley and Adey 1977; Hixon and Brostoff 1982]). Birkeland (1977) also noted that filamentous and fleshy algae outcompete corals (which are inhibited) under elevated nutrient levels. He concluded that small, fast-growing algae are not just opportunists that depend on disturbances to release space resources from established longer-lived populations but, rather, become the superior competitors when provided with abundant nutrients. Shifts from coral dominance to fleshy algal dominance have been related to excess nutrient increases for reefs off Venezuela (Weiss and Goddard 1977), on the Abaco reef system, Bahamas (Lighty, Macintyre, and Neumann 1980), and in Kaneohe Bay, Hawaii (Banner 1974; Smith et al. 1981).

Nitrogen is generally thought (Hatcher and Larkum 1983) to be the nutrient most often limiting tropical marine algal growth rates. Phosphorus, due to its effective recycling, traditionally has been considered less likely to be in short supply (Pilson and Betzer 1973). However, recent evidence (Lapointe 1983) is beginning to reveal that phosphorus is limiting under more widespread conditions than previously thought. The reason for phosphorus limitation (Simkiss 1964) is phosphate binding by calcium carbonate particles that are abundant in nearly all reef environments and the presence of nitrogen-fixing blue-green algae, making nitrogen relatively plentiful in many benthic reef environments.

Table 2-2. *Herbivore resistance mechanisms shown by benthic macrophytes*

Resistance mechanisms	Examples	References
	Noncoexistence escapes	
1. *Temporal escapes* Short or alternating life histories	Annuals (e.g., *Liagora, Trichogloea*) and seasonally alternating heteromorphic forms (*Porphyra, Gigartina*)	Littler and Littler 1980; Lubchenco and Cubit 1980
Opportunistic colonization of unpredictable new substrata	Thin, sheetlike, and filamentous forms (e.g., *Polysiphonia, Centroceras Enteromorpha*)	Littler and Littler 1980
2. *Spatial escapes* Refuge habitats	Intertidal habitats (*Ahnfeltia, Chnoospora*), high-energy environments (*Sargassum*), sand plains (*Gracilaria*), and crevices (*Amansia*)	Earle 1972; Ogden et al. 1973; Brock 1979; Hay 1981a; Hay et al. 1983
Association with unpalatable organisms	Next to toxic or stinging organisms such as *Stypopodium, Gorgonia* and *Millepora* (*Liagora, Laurencia*)	Littler, Taylor, and Littler 1986; Littler et al., 1987b
Association with carnivorous predators	Next to grouper or snapper lairs (*Acanthophora*)	Randall 1965; Ogden et al. 1973
Association with territorial animals	Within damselfish territories (e.g., *Polysiphonia*)	Brawley and Adey 1977; Hixon and Brostoff 1982
	Coexistence defenses	
1. *Crypsis*	Inconspicuous forms (some *Gracilaria*) and small forms (*Plectonema*)	Gaines and Lubchenco 1982
2. *Mimicry*	Coral mimics (*Eucheuma arnoldii*)	Kraft 1972; Littler, Taylor and Littler 1983

Table 2-2. *(cont.)*

Resistance mechanisms	Examples	References
3. *Structural defenses*		
Morphologies that minimize accessibility	Upright large algae (e.g., *Turbinaria*), branched crustose corallines *(Neogoniolithon),* and turf-forming colonies *(Halimeda)*	Steneck and Watling 1982; Hay 1981a
Textures that inhibit manipulation and feeding	Encrusting species *(Lobophora)* and spiny forms *(Turbinaria)*	Nicotri 1980
Materials that lower food quality	$CaCO_3$ in corallines *(Porolithon)* and calcareous green algae *(Penicillus, Neomeris)*	Littler 1976
4. *Chemical defenses*		
Toxins, digestion inhibitors, unpalatable substances	*Halimeda, Caulerpa, Stypopodium, Dictyota*	Norris and Fenical 1982; Paul and Fenical 1983
Lowered energetic content	Coralline algae *(Amphiroa)*	Paine and Vadas 1969; Littler, Littler, and Taylor 1983
5. *Satiation of herbivores*	Rapid replacement of lost tissues *(Ulva)*	Littler and Littler 1980; Borowitzka 1981

Enrichment studies (i.e., fertilization) of reef communities have shown (Kinsey and Domm 1974; Kinsey and Davies 1979) substantial enhancement of photosynthesis, but because of methodological limitations no significant changes could be detected in macrophyte standing stocks.

Of the two factors, nutrients and grazing, the latter is probably most important in directly determining algal distributions. Several herbivore-resistance mechanisms are often expressed by a given algal species and are summarized in Table 2-2. The synthesis of toxic secondary organics by algae increases dramatically in herbivore-rich subtropical and tropical reef systems (Fenical 1980) compared to temperate coastal environments. This correlation has fostered the rather speculative literature on algal chemical defenses, as pointed out by Norris and Fenical (1982). With the exceptions of several studies on temperate algae (e.g., Steinberg 1985), these contentions have rarely been evaluated by ecologically relevant experiments. The stage has now been set for more precisely controlled field studies involving hypothetical chemical-defense compounds used on natural reef populations at normal levels, with artifactual problems eliminated, to establish the adaptive significance of algal secondary metabolites.

V. Human influences

Reefs have developed over long geological time periods under relatively constant conditions. This stability has resulted in the evolution of an astounding abundance and diversity of specialized plant and animal life, which makes these areas fascinating subjects for research. Human interest is readily stimulated by the beauty, mystique, and complexity of reef organisms, and in the past human activities have caused little damage as they have been, for the most part, in harmony with reef ecosystems. However, this condition in recent years has undergone rapid change.

Reef formations have a long history of use by humans as a source of food, materials, and recreation. Ciguatera poisoning, frequently associated with reefs throughout the tropical world, often results in death to humans who have eaten carnivorous fishes that contain ciguatera toxins. It is now known that the ciguatera-type toxins are derived from the benthic dinoflagellate *Gambierdiscus toxicus* (Bagnis et al. 1979). The flattened cells of this organism are most often attached to highly branched filamentous algal thalli but also occur on primary substrata. When such cells are inadvertently eaten by grazing fish, the result is that two toxic substances, maitotoxin and ciguatoxin, tend to be accumulated and retained in the food chain. The toxins, with time, become greatly concentrated in the higher trophic levels, particularly by carnivorous fishes.

Because of its complex nature, the reef environment is particularly vulnerable to deleterious external influences. There are two basic environmental concerns: resource utilization and pollution. The most obvious activities of humans that affect the marine algae of tropical reefs are indirect and include the harvesting of fish, the gathering of molluscs, the collecting of sea urchins, and the hunting of herbivorous marine turtles. These activities, where excessive, may reduce the populations of key species of herbivores, thereby eliminating the constraints on certain algal populations and leading to unusual predominances by one or several forms. However, ecological release from competition or predation and compensatory expansion of a particular herbivore group following the reduction of another can occur. For example (Hay and Taylor 1985), on overfished reefs, disturbance is ameliorated when sea urchin populations increase and expand, keeping localized ratios of corallines, frondose macroalgae, microfilamentous forms, and corals near their previous levels. When human overexploitation of gastropods and urchins as well as fishes occurs, frondose algal stocks can become quite extensive (e.g., the 1–2 m deep Waikiki reef flat: Doty 1971). Perturbations are usually restricted to those sections of the reef and adjoining coastal habitats most conveniently accessible to man.

The construction of seawalls and the building of groins, platforms, and extensions of the land over reef flats for human habitation all affect the normal current patterns and sedimentological processes. The cutting of terrestrial forest communities greatly increases sediments that wash onto reef systems. Such sedimentation effects include scouring, smothering, and lowering of light levels, all of which are deleterious to both plant and animal stocks. Ship groundings occur frequently (Hatcher 1984; Littler et al. 1987a) with devastating physical results. Algal community structure resulting from such stochastic physical disturbances (including hurricanes: Mah and Stearn 1985) may be permanently changed to some alternate steady state (Hatcher 1984).

Tropical algae have been utilized by man as food, medicines, poultices on wounds, ceremonial objects, and for ornaments. In the Philippines, Malyasia, Indonesia, China, and Japan, farming of seaweeds on shallow reef flats is now undertaken on a relatively broad scale (see Lewis, Stanley and Quist, Chapter 9). However, it is not known what effects such increases in algal standing stocks and shading have on the natural biota. Also, numerous wild algal populations are commercially harvested as local food sources and for colloidal extracts (see Abbott, Chapter 6). The gathering of calcareous coralline algal nodules, dredging, or dynamite blasting of sizable reef blocks as land fill, to be transformed and used as mortar, or for liming acidic tropical soils has obvious (but unmeasured) detrimental effects.

Protection of reefs is becoming increasing difficult because the areas in which they are found, particularly near the coast, are used as sites for tourism, industry, and resource exploitation. There appear to be several ways in which excess nutrient enrichment (i.e., eutrophication due to fertilizer runoff, sewage, etc.) can decrease coral fitness causing reef biogenesis to decrease and bioerosion to increase. For example, declines in water transparency due to phytoplankton blooms may limit photosynthetic production and growth (Tomascik and Sander 1985), phosphate may inhibit calcification (Simkiss 1964; Kinsey and Domm 1974), and suspension-feeding reef borers may become disproportionately abundant (Smith et al. 1981). In association with nutrient enrichment due to sewage effluents, *Dictyosphaeria* and other algae overgrew and killed many of the more luxuriant coral communities throughout Kaneohe Bay, Hawaii (Banner 1974; Smith et al. 1981). Six years after termination of the sewage discharge (Maragos, Evans, and Holthus 1985), *D. cavernosa* declined whereas corals showed a remarkable recovery, corroborating the prediction that sewage eutrophication is a major stress to lagoon corals but a stimulant to algal growth.

Already many of the world's reefs have been exposed to different degrees of sewage pollution and oil spills; spear fishing and fish trapping have been so intensive that the larger reef fishes have almost disappeared near many of the heavily populated regions. Despite extensive research, human influence on reef environments continues to outpace our understanding of the ecological changes that are taking place.

VI. Future outlook

We still know very little about the physiological ecology, population biology, and community dynamics of algae that affect the ecology and biogenesis of biotic reefs. Especially lacking have been biological studies of reef-boring macroalgae and deepwater macrophytes. As pointed out by Littler (1972), and still true almost two decades later, areas of investigation concerning the role of bioenergetics, heterotrophy, allelopathy, interspecific and intraspecific competition, recruitment, natality, and mortality phenomena remain virtually untouched for most tropical algae. Until we begin to understand processes at these levels, it will be difficult to make definitive statements concerning the role of algae, or other organisms, in the dynamics of tropical reef communities.

Submersibles have greatly expanded our knowledge of the distributional patterns and abundances of deep-sea organisms; however, macroalgae have, until recently, received only anecdotal attention. As more phycologists gain access to deep-sea communities, the diversity, abundances, and production levels of benthic algae will be revealed and new oceanographic theories based on *in situ* research will replace older dogma derived from less efficient remote sampling from shipboard.

Until recently, few workers had directed their efforts toward determining the functional and ecologial roles of algae on living reefs. Philosophical approaches that yield mechanistic and predictive insights are needed to supplement the more traditional descriptive approaches and methodologies. Taxonomically sound, reproducible, conceptually based, experimental analyses of ecological processes in the field have been particularly few. Historical reasons for this scarcity would seem to be taxonomic difficulties, an intimidating array of life-history complexities, and a lack of quantitative ecological techniques with sufficient resolution. Fortunately, taxonomic difficulties are now surmountable (Adey and Macintyre 1973), and various descriptive (Stoddart and Johannes 1978; Rützler and Macintyre 1982) as well as experimental (e.g., Wanders 1976; Hay 1981a; Littler, Littler, and Taylor 1983; Littler, Taylor, and Littler 1983; Hatcher and Larkum 1983; Littler and Littler 1984b) ecological methods are available. With these obstacles posing less difficulty, advancement beyond the observational descriptive level toward more theoretical issues is rapidly taking place, although our knowledge of algal biology still lags behind that available for most other reef organisms.

Because seaweeds are recognized by their external and internal morphology, it has proven instructive to relate morphological groups (Littler 1980) with measurements of functional properties, relative to the environmental parameters normally experienced. Recently, research focusing on the functional significance of morphology has provided the focal point for the postulation of testable new theory concerning the ecological significance of anatomical variability and has enabled reef ecologists to consider a broad array of evolutionary phenomena for tropical seaweeds. Structure at all levels of organization ultimately governs all biological activities, and since morphological/anatomical features can be measured relatively accurately and easily, functional/morphological groups have proven to be extremely useful (Littler and Littler 1980, 1984a; Steneck and Watling 1982; Littler, Littler, and Taylor 1983; Littler, Taylor, and Littler 1983) for the interpretation of complex biotic reef ecosystems.

We are at a stage where descriptive (correlative) and mechanistic (i.e., experimental/causative) approaches must be combined to produce more conceptual theoretical themes, which will accelerate our predictive understanding of algal roles in reef biology. Experimental approaches are increasingly being utilized with the result that general ecological theories now are being modified by studies of tropical algal biology (see reviews by Lubchenco and Gaines 1981; Gaines and Lubchenco 1982; Littler and Littler 1984b). A burgeoning awareness of the attractiveness of tropical marine plants as experimental organisms for the elucidation of ecological and biogenic processes offers exciting prospects for the future of reef research.

VII. References

Adey, W. H., and I. G. Macintyre. 1973. Crustose coralline algae: a reevaluation in the geological sciences. *Geol. Soc. Amer. Bull.* 84, 883–904.

Bagnis, R., J.-M. Hurtel, S. Chanteau, E. Chungue, A. Inoue, and T. Yasumoto. 1979. Le dinoflagelle *Gambierdiscus toxicus* Adachi et Fuduyo; agent causal probable de la ciguatera. *C. R. Acad. Sci. Paris* 289, 671–4.

Bakus, G. J. 1967. The feeding habits of fishes and primary production at Eniwetok, Marshall Islands. *Micronesica* 3, 135–49.

Bakus, G. J. 1969. Energetics and feeding in shallow marine waters. *Int. Rev. Gen. Exp. Zool.* 4, 275–369.

Banner, A. H. 1974. Kaneohe Bay, Hawaii: urban pollution and a coral reef ecosystem. In *Proceedings of the Second International Coral Reef Symposium*, vol. 2, ed. A. M. Cameron, B. M. Campbell, A. B. Cribb, R. Endean, J. S. Jell, O.

A. Jones, P. Mather, and F. H. Talbot, pp. 685–702. Great Barrier Reef Committee, Brisbane.

Battey, J. F., and J. S. Patton. 1984. A reevaluation of the role of glycerol in carbon translocation in zooxanthellae-coelenterate symbiosis. *Mar. Biol.* 79, 27–38.

Belk, M. S., and D. Belk. 1975. An observation of algal colonization on *Acropora aspera* killed by *Acanthaster planci*. *Hydrobiologia* 46, 29–32.

Birkeland, C. 1977. The importance of rate of biomass accumulation in early successional stages of benthic communities to the survival of coral recruits. In *Proceedings: Third International Coral Reef Symposium,* ed. D. L. Taylor, pp. 15–21. Rosenstiel School of Marine and Atmospheric Science, University of Miami, Miami, FL.

Borowitzka, M. A. 1981. Algae and grazing in coral reef ecosystems. *Endeavour* 5, 99–106.

Borowitzka, M. A., and A. W. D. Larkum. 1976. Calcification in the green alga *Halimeda*. IV. The action of metabolic inhibitors on photosynthesis and calcification. *J. Exp. Bot.* 27, 894–907.

Brawley, S. H., and W. H. Adey. 1977. Territorial behavior of threespot damselfish *(Eupomacentrus planifrons)* increases reef algal biomass and productivity. *Envir. Biol. Fish.* 2, 45–51.

Brock, R. E. 1979. An experimental study on the effects of grazing by parrotfishes and role of refuges in benthic community structure. *Mar. Biol.* 51, 381–8.

Brown, V., S. C. Ducker, and K. S. Rowan. 1977. The effect of orthophosphate concentration on the growth of articulated coralline algae (Rhodophyta). *Phycologia* 16, 125–31.

Capone, D. G., D. L. Taylor, and B. F. Taylor. 1977. Nitrogen fixation (acetylene reduction) associated with macroalgae in a coral-reef community in the Bahamas. *Mar. Biol.* 40, 29–32.

Carpenter, R. C. 1985. Relationships between primary production and irradiance in coral reef algal communities. *Limnol. Oceanogr.* 30, 784–93.

Chapman, F., and D. Mawson. 1906. On the importance of *Halimeda* as a reef-forming organism: with a description of the *Halimeda*-limestones of the New Hebrides. *J. Geol. Soc. Lond.* 62, 702–11.

Chu, H., and B. T. Wu. 1983. Studies on the lime-boring algae from China. I. The lime-boring algae collected from the Xisha Islands. In *XIth International Seaweed Symposium,* p. 46. Institute Oceanology Academia Sinica, Qingdao, China (abstract only).

Connor, J. L., and W. H. Adey. 1977. The benthic algal composition, standing crop, and productivity of a Caribbean algal ridge. *Atoll Res. Bull.* 211, 1–15.

Crossland, C. J., and D. J. Barnes. 1976. Acetylene reduction by corals. *Limnol. Oceanogr.* 21, 153–5.

Davies, P. J., and J. F. Marshall. 1985. *Halimeda* bioherms – low energy reefs, Northern Great Barrier Reef. In *Proceedings of the Fifth International Coral Reef Congress,* vol. 5, ed. V. M. Harmelin and B. Salvat, pp. 1–7. Antenne Museum–EPHE, Moorea, French Polynesia.

Doty, M. S. 1969. *The Ecology of Honaunau Bay, Hawaii.* Botanical Science Paper no. 14, p. 221. University of Hawaii, Honolulu.

Doty, M. S. 1971. Physical factors in the production of tropical benthic marine algae. In *Fertility of the Sea,* vol. 1, ed. J. D. Costlow, Jr., pp. 99–121. Gordon and Breach, New York.

Duerden, J. E. 1902. Boring algae as agents in the disintegration of corals. *Bull. Amer. Mus. Nat. Hist.* 16, 323–32.

Earle, S. A. 1972. The influence of herbivores on the marine plants of Great Lameshur Bay, with an annotated list of plants. In *Results of the Tektite Program: Ecology of Coral Reef Fishes,* Science Bulletin 14, ed. B. B. Collette and S. A. Earle, pp. 17–44. Los Angeles County Natural History Museum, Los Angeles.

Easton, W. H., and E. A. Olson. 1976. Radiocarbon profile of Hanauma Reef, Oahu, Hawaii: reply. *Geol. Soc. Amer. Bull.* 87, 711–9.

Fenical, W. H. 1980. Distributional and taxonomic features of toxin-producing marine algae. In *Pacific Seaweed Aquaculture,* ed. I. A. Abbott, M. S. Foster, and L. F. Eklund, pp. 144–51. California Sea Grant College Program, Institute of Marine Resources, University of California, La Jolla.

Finckh, A. E. 1904. Biology of the reef-forming organisms at Funafuti Atoll, Section VI. In *The Atoll of Funafuti,* pp. 125–50. Report of the Coral Reef Committee, Royal Society, London.

Gaines, S. D., and J. Lubchenco. 1982. A unified approach to marine plant–herbivore interactions. II. Biogeography. *Ann. Rev. Ecol. Syst.* 13, 111–38.

Green, M. A. 1975. *Survey of Endolithic Organisms from the Northeast Bering Sea, Jamaica, and Florida Bay.* M.S. thesis, Department of Geology, Duke University, Durham, N.C. 112 pp.

Hallock, P. 1985. Why are larger foraminifera large? *Paleobiology* 11, 195–208.

Hatcher, B. G. 1984. A maritime accident provided evidence for alternate stable states in benthic communities on coral reefs. *Coral Reefs* 3, 199–204.

Hatcher, B. G., and A. W. D. Larkum. 1983. An experimental analysis of factors controlling the standing crop of the epilithic algal community on a coral reef. *J. Exp. Mar. Biol. Ecol.* 69, 61–84.

Hay, M. E. 1981a. Herbivory, algal distribution, and the maintenance of between-habitat diversity on a tropical fringing reef. *Amer. Nat.* 118, 520–40.

Hay, M. E. 1981b. Spatial patterns of grazing intensity on a Caribbean barrier reef: herbivory and algal distribution. *Aquat. Bot.* 11, 97–109.

Hay, M. E., T. Colburn, and D. Downing. 1983. Spatial and temporal patterns in herbivory on a Caribbean fringing reef: the effects on plant distribution. *Oecologia* 58, 299–308.

Hay, M. E., and P. R. Taylor. 1985. Competition between herbivorous fishes and urchins on Caribbean reefs. *Oecologia* 65, 591–8.

Hillis-Colinvaux, L. 1977. *Halimeda* and *Tydemania:* distribution, diversity, and productivity at Enewetak. In *Proceedings: Third International Coral Reef Symposium,* ed. D. L. Taylor, pp. 365–70. Rosenstiel School of Marine and Atmospheric Science, University of Miami, Miami, FL.

Hillis-Colinvaux, L. 1980. Ecology and taxonomy of *Halimeda:* primary producer of coral reefs. *Adv. Mar. Biol.* 17, 1–327.

Hixon, M. A., and W. N. Brostoff. 1982. Differential fish grazing and benthic community structure on Hawaiian reefs. In *Fish Food Habits Studies: Proceed-*

ings of the Third Pacific Workshop, ed. G. M. Cailliet and C. A. Simenstad, pp. 249–57. Washington Sea Grant Program, University of Washington, Seattle.

James, N. P., and I. A. Macintyre. 1985. Carbonate depositional environments: modern and ancient. 1. Reefs: Zonation, depositional facies, diagenesis. *Colorado School of Mines Quarterly* 80, 1–70.

Jensen, P. R., R. A. Gibson, M. M. Littler, and D. S. Littler. 1985. Photosynthesis and calcification in four deep-water *Halimeda* species (Chlorophyceae, Caulerpales). *Deep-Sea Res.* 32, 421–64.

Johannes, R. E., et al. 1972. The metabolism of some coral reef communities. *BioScience* 22, 541–3.

Johnston, C. S. 1969. The ecological distribution and primary production of macrophytic marine algae in the eastern Canaries. *Int. Rev. ges. Hydrobiol. Hydrograph.* 54, 473–90.

Kanwisher, J. W., and S. A. Wainwright. 1967. Oxygen balance in some reef corals. *Biol. Bull.* 133, 378–90.

Kinsey, D. W., and Davies, P. J. 1979. Effects of elevated nitrogen and phosphorus on coral reef growth. *Limnol. Oceanogr.* 24, 935–40.

Kinsey, D. W., and A. Domm. 1974. Effects of fertilization on a coral reef environment: primary production studies. In *Proceedings of the Second International Coral Reef Symposium,* vol. 1, ed. A. M. Cameron, B. M. Campbell, A. B. Cribb, R. Endean, J. S. Jell, O. A. Jones., P. Mather, and F. H. Talbot, pp. 49–66. Great Barrier Reef Committee, Brisbane.

Kraft, G. T. 1972. Preliminary studies of Philippine *Eucheuma* species (Rhodophyta). 1. Taxonomy and ecology of *Eucheuma arnoldii* Weber-van Bosse. *Pac. Sci.* 26, 318–34.

Lang, J. C. 1974. Biological zonation at the base of a reef. *Amer. Sci.* 62, 272–81.

Lapointe, B. E. 1983. Nutrient limited seaweed growth in nearshore waters of the Florida Keys: nitrogen and phosphorus. *EOS, Trans. Amer. Geophys. Union* 64, 105 (abstract only).

Larkum, A. W. D., E. A. Drew, and R. N. Crosset. 1967. The vertical distribution of attached marine algae in Malta. *J. Ecol.* 55, 361–71.

Lewis, J. B. 1977. Processes of organic production on coral reefs. *Biol. Rev.* 52, 305–47.

Lighty, R. G., I. G. Macintyre, and A. C. Neumann. 1980. Demise of a Holocene barrier-reef complex in the Northern Bahamas. *Geol. Soc. Amer. Bull.* 12, 471.

Littler, M. M. 1971. Standing stock measurements of crustose coralline algae (Rhodophyta) and other saxicolous organisms. *J. Exp. Mar. Biol. Ecol.* 6, 91–9.

Littler, M. M. 1972. The crustose Corallinaceae. *Oceanogr. Mar. Biol. Ann. Rev.* 10, 311–47.

Littler, M. M. 1976. Calcification and its role among the macroalgae. *Micronesica* 12, 27–41.

Littler, M. M. 1980. Morphological form and photosynthetic performances of marine macroalgae: tests of a functional/form hypothesis. *Bot. Mar.* 22, 161–5.

Littler, M. M., and M. S. Doty. 1975. Ecological components structuring the seaward edges of tropical Pacific reefs: the distribution, communities, and productivity of *Porolithon. J. Ecol.* 63, 117–29.

Littler, M. M., and D. S. Littler. 1980. The evolution of thallus form and survival strategies in benthic marine macroalgae: field and laboratory tests of a functional form model. *Amer. Nat.* 116, 25–44.

Littler, M. M., and D. S. Littler. 1984a. Relationships between macroalgal functional form groups and substrata stability in a subtropical rocky-intertidal system. *J. Exp. Mar. Biol. Ecol.* 74, 13–34.

Littler, M. M., and D. S. Littler. 1984b. Models of tropical reef biogenesis: the contribution of algae. In *Progress in Phycological Research,* vol. 3, ed. F. E. Round, pp. 323–64. Biopress, Bristol.

Littler, M. M., and D. S. Littler, 1985. Factors controlling relative dominance of primary producers on biotic reefs. In *Proceedings of the Fifth International Coral Reef Congress,* vol. 4, ed. C. Gabrie and B. Salvat, pp. 35–9. Antenne Museum–EPHE, Moorea, French Polynesia.

Littler, M. M., D. S. Littler, S. M. Blair, and J. N. Norris. 1985. Deepest known plant life discovered on an uncharted seamount. *Science* 227, 57–9.

Littler, M. M., D. S. Littler, S. M. Blair, and J. N. Norris. 1986. Deepwater plant communities from an uncharted seamount off San Salvador Island, Bahamas: distribution, abundance, and primary productivity. *Deep-Sea Res.* 33, 882–92.

Littler, M. M., D. S. Littler, J. N. Norris, and K. E. Bucher. 1987a. *Recolonization of Algal Communities Following the Grounding of the Freighter* Wellwood *on Molasses Reef, Key Largo National Marine Sanctuary. NOAA Tech. Memor. Ser.* Nos. MEMD 15, National Oceanic and Atmospheric Administration, Washington, DC. 42 pp.

Littler, M. M., D. S. Littler, and P. R. Taylor. 1983. Evolutionary strategies in a tropical barrier reef system: functional-form groups of marine macroalgae. *J. Phycol.* 19, 229–37.

Littler, M. M., D. S. Littler, and P. R. Taylor. 1987b. Animal-plant defense associations: effects on the distribution and abundance of tropical reef macrophytes. *J. Exp. Mar. Biol. Ecol.* 105, 107–21.

Littler, M. M., P. R. Taylor, and D. S. Littler. 1983. Algal resistance to herbivory on a Caribbean barrier reef. *Coral Reefs* 2, 111–18.

Littler, M. M., P. R. Taylor, and D. S. Littler. 1986. Plant defense associations in the marine environment. *Coral Reefs* 5, 63–71.

Lubchenco, J., and J. Cubit. 1980. Heteromorphic life histories of certain marine algae as adaptations to variations in herbivory. *Ecology* 61, 676–87.

Lubchenco, J., and S. D. Gaines. 1981. A unified approach to marine plant–herbivore interactions. I. Populations and communities. *Ann. Rev. Ecol. Syst.* 12, 405–37.

Mauge, T. H., and O. Holm-Hansen. 1975. Nitrogen fixation on a coral reef. *Phycologia* 14, 87–92.

Mah, A. J., and C. W. Stearn. 1985. Changes produced by a hurricane on a fringing reef (Barbados). In *Proceedings of the Fifth International Coral Reef Congress,* vol. 4, ed. C. Gabrie and B. Salvat, pp. 189–94. Antenne Museum–EPHE, Moorea, French Polynesia.

Maragos, J., C. Evans, and P. Holthus. 1985. Reef corals in Kaneohe Bay six years before and after termination of sewage discharges (Oahu, Hawaiian Archipelago). In *Proceedings of the Fifth International Coral Reef Congress*, vol. 2, ed. C. Gabrie, J. L. Toffart, and B. Salvat, p. 236. Antenne Museum–EPHE, Moorea, French Polynesia.

Marsh, J. A., Jr. 1970. Primary productivity of reef-building calcareous red algae. *Ecology* 51, 255–63.

Marsh, J. A., Jr. 1976. Energetic role of algae in reef ecosystems. *Micronesica* 12, 13–21.

May, J. A., I. G. Macintyre, and R. D. Perkins. 1982. Distribution of microborers within planted substrates along a barrier reef transect, Carrie Bow Cay, Belize. In *The Atlantic Barrier Reef Ecosystem at Carrie Bow Cay, Belize*, vol. 1, *Structure and Communities*, Smithsonian Contributions to the Marine Sciences, no. 12, ed. K. Rützler and I. G. Macintyre, pp. 93–107. Smithsonian Institution Press, Washington, DC.

Milliman, J. D. 1974. *Recent Sedimentary Carbonates: Marine Carbonates, Part I.* Springer-Verlag, New York.

Muscatine, L., P. G. Falkowski, and Z. Dubinsky. 1983. Carbon budgets in symbiotic associations. In *Endocytobiology*, vol. II, ed. H. E. A. Schenk and W. Schwemmler, pp. 646–58. Walter de Gruyter, Berlin.

Muscatine, L., and J. W. Porter. 1977. Reef corals: mutualistic symbioses adapted to nutrient-poor environments. *BioScience* 27, 454–60.

Nadson, G. 1900. Die perforierenden (kalkbohrenden) Algen und ihre Bedeutung in der Natur. *Botanica Scripta* 18, 1–40.

Nicotri, M. E. 1980. Factors involved in herbivore food preference. *J. Exp. Mar. Biol. Ecol.* 42, 13–26.

Norris, J. N., and W. Fenical. 1982. Chemical defense in tropical marine algae. In *The Atlantic Barrier Reef Ecosystem at Carrie Bow Cay, Belize*, vol. 1, *Structure and Communities*, Smithsonian Contributions to the Marine Sciences, no. 12, ed. K. Rützler and I. G. Macintyre, pp. 417–31. Smithsonian Institution Press, Washington, DC.

Odum, H. T., and E. P. Odum. 1955. Trophic structure and productivity of a windward coral reef community on Eniwetok Atoll. *Ecol. Monogr.* 25, 291–320.

Ogden, J. C., R. A. Brown, and N. Salesky. 1973. Grazing by the echinoid *Diadema antillarum* Philippi: formation of halos around West Indian patch reefs. *Science* 182, 715–17.

Orme, G. R. 1977. Aspects of sedimentation in the coral reef environment. In *Biology and Geology of Coral Reefs*, vol. 4, *Geology 2*, ed. O. A. Jones and R. Endean, pp. 129–82. Academic Press, New York.

Paine, R. T., and R. L. Vadas. 1969. Calorific values of benthic marine algae and their postulated relation to invertebrate food preference. *Mar. Biol.* 4, 79–86.

Paul, V. J., and W. Fenical. 1983. Isolation of halimedatrial: chemical defense adaptation in the calcareous reef-building alga *Halimeda*. *Science* 221, 747–9.

Pilson, M. E. Q., and S. B. Betzer. 1973. Phosphorus flux across a coral reef. *Ecology* 54, 581–8.

Porter, J. W., L. Muscatine, Z. Dubinsky, and P. G. Falkowski. 1984. Primary production and photoadaptation in light and shade-adapted colonies of the symbiotic coral *Stylophora pistillata*. *Proc. Roy. Soc. Lond. B* 222, 161–80.

Purdy, E. G., and L. S. Kornicker. 1958. Algal disintegration of Bahamian limestone coasts. *J. Geol.* 66, 97–9.

Randall, J. E. 1965. Grazing effect on seagrasses by herbivorous reef fishes in the West Indies. *Ecology* 46, 255–60.

Rogers, C. S., and N. H. Salesky. 1981. Productivity of *Acropora palmata* (Lamarck), macroscopic algae, and algal turf from Tague Bay Reef, St. Croix, U.S. Virgin Islands. *J. Exp. Mar. Biol. Ecol.* 49, 179–87.

Rützler, R., and I. G. Macintyre (eds.). 1982. *The Atlantic Barrier Reef Ecosystem at Carrie Bow Cay, Belize. I. Structure and Communities.* Smithsonian Contributions to the Marine Sciences, no. 12. Smithsonian Institution Press, Washington, DC.

Simkiss, K. 1964. Phosphates as crystal poisons of calcification. *Biol. Rev.* 39, 487–505.

Smith, S. V. 1973. Carbon dioxide dynamics: a record of organic carbon production, respiration, and calcification in the Eniwetok reef flat community. *Limnol. Oceanogr.* 18, 106–20.

Smith, S. V. 1978. Coral-reef area and the contributions of reefs to processes and resources of the world's oceans. *Nature* 273, 225–6.

Smith, S. V., W. J. Kimmerer, E. A. Laws, R. E. Brock, and T. W. Walsh. 1981. Kaneohe Bay sewage diversion experiment: perspectives on ecosystem responses to nutritional perturbation. *Pac. Sci.* 35, 279–397.

Sournia, A. 1976. Oxygen metabolism of a fringing reef in French Polynesia. *Helgoländer Wiss. Meeresunters.* 28, 401–10.

Steinberg, P. D. 1985. Feeding preferences of *Tegula funebralis* and chemical defenses of marine brown algae. *Ecol. Monogr.* 55, 333–49.

Steneck, R. S., and L. Watling. 1982. Feeding capabilities and limitation of herbivorous molluscs: a functional group approach. *Mar. Biol.* 68, 299–319.

Stoddart, D. R., and R. E. Johannes (eds.). 1978. *Coral Reefs: Research Methods.* UNESCO, Paris. 581 pp.

Taylor, W. R. 1960. *Marine Algae of the Eastern Tropical and Subtropical Coasts of the Americas.* University of Michigan Press, Ann Arbor.

Tomascik, T., and F. Sander. 1985. Effects of eutrophication on reef-building corals. I. Growth rate of the reef-building coral *Montastrea annularis*. *Mar. Biol.* 87, 143–55.

Trench, R. K. 1981. Cellular and molecular interactions in symbioses between dinoflagellates and marine invertebrates. *Pure Appl. Chem.* 53, 819–35.

Vine, P. J. 1974. Effects of algal grazing and aggressive behaviour of the fishes *Pomacentrus lividus* and *Acanthurus sohal* on coral-reef ecology. *Mar. Biol.* 24, 131–6.

Wanders, J. B. W. 1976. The role of benthic algae in the shallow reef of Curaçao (Netherlands Antilles). I. Primary production in the coral reef. *Aquat. Bot.* 2, 235–70.

Weber-van Bosse, A. 1932. Algues. *Mem. Mus. R. Hist. Natur. Belg.* 6, 1–27.

Weiss, M. P., and D. A. Goddard. 1977. Man's impact on coastal reefs: an example from Venezuela. *Amer. Assoc. Pet. Geol. Stud. Geol.* 4, 111–24.

Wells, J. W. 1957. Coral Reefs. In *Treatise on Marine Ecology and Paleoecology*, vol. 1, *Ecology*, ed. J. W. Hedgpeth, pp. 609–31. Geological Society of America, Memoir 67, National Academy of Sciences, Washington, DC.

Westlake, D. F. 1963. Comparisons of plant productivity. *Biol. Rev.* 38, 385–425.

Wiebe, W. J., R. E. Johannes, and K. L. Webb. 1975. Nitrogen fixation in a coral reef community. *Science* 188, 257–9.

Womersley, H. B. S., and A. Bailey. 1970. Marine algae of the Solomon Islands. *Phil. Trans. Roy. Soc.*, ser. B, 259, 257–352.

Wray, J. L. 1971. Algae in reefs through time. *Proc. N. Amer. Paleont. Soc. Convention* (Chicago, 1969), 1, 1358–73.

Wray, J. L. 1977. *Calcareous algae.* Elsevier, Amsterdam. 185 pp.

3. Algae and the environment: the Great Lakes case

EUGENE F. STOERMER

Great Lakes Research Division, University of Michigan, Ann Arbor, MI 48109

CONTENTS

I. Introduction

The Laurentian Great Lakes are major physiographic features of North America. To most North Americans they are *the* Great Lakes. Although many other bodies of fresh water also deserve the appellation "great lakes," the Laurentian Great Lakes of North America are a unique ecological system and a unique resource. They are the largest single, continuously connected body of fresh water on earth. Taken together, they constitute nearly 20% of the standing fresh water on earth (Chandler 1964). A significant portion of the populations of the United States and Canada lives in states and provinces bordering the lakes, and the well-being of these people is affected by the lakes in various ways.

The Great Lakes furnished the primary mode of entry into the heartland of North America for early European explorers and later settlers. Because their drainage basins are contiguous to the Mississippi River system to the south and west they formed the easiest mode of transport for explorers and traders. This route was followed by successive waves of European settlers, and, after construction of the Erie Canal, a continuous waterway was available from the East Coast of the United States to the Midwest. Today, they continue to provide one of the major arteries for transport of bulk cargo and, since construction of the St. Lawrence Seaway, are directly connected to the shipping routes of the world.

A consequence of this transportation artery was the establishment of population centers on the lakes' shores. The earliest of these were way-points and trading centers. Besides the stimulus to commerce furnished by transportation, the immediate environs of the lakes provided numerous other stimuli for economic and population growth. The relatively rich soils in areas covered by previous high stands of the lakes (Hough 1958) were a source of valuable timber, and after the timber was cleared agriculture flourished. Rivers entering or connecting the lakes provided a ready source of mechanical and, later, electrical hydropower. The lakes themselves were an almost perfectly exploitable resource, providing a rich harvest of fish and a seemingly inexhaustible supply of high-quality water. These favorable conditions eventually led to development of major industrial and population centers. These include the Buffalo–

[58]

Hamilton–Toronto metroplex on Lake Ontario, the Cleveland–Toledo–Detroit industrial strip on Lake Erie, and the Gary–Chicago–Milwaukee region of southern Lake Michigan. The growth of these centers and their satellites holds the potential for the development of a "megalopolis" – a virtually continuous urban-industrial complex stretching from Toronto to Milwaukee.

Of perhaps equal importance, the shores of the lakes provide a favored environment for habitation and enjoyment. The attraction of the lakeshore is easily measured by the current premium on lakeshore property. One of the region's major industries, tourism and recreation, involves the Great Lakes. At present, two areas of Great Lakes shoreline have been designated as national seashores, and the area has numerous national, state and provincial parks.

The Great Lakes region thus provides a very striking example of the potential for conflict between economic growth and preservation of the factors that make such growth possible. This conflict is currently being played out under greatly changed priorities. The regional growth that occurred between 1850 and 1950 was largely fueled by the convergence of exploitable resources of the region facilitated by cheap transportation. Sometime in the 1960s this formula for success began to come apart. The economic and social distress of the great industrial cities of the region during the 1970s has received sufficient coverage in the popular press that details need not be retold here. The important point is that economic advantage has shifted from exploitable resources to preservable and renewable resources. Even during the recent period of severe economic distress and consequent population loss, certain areas within the region have continued moderate population growth and increased economic activity. Without exception these areas provide the environmental and educational resources attractive to high technology. There is little reason to doubt that this trend will continue into the foreseeable future.

Unfortunately, during the grand period of regional growth the potential for population growth and economic development to cause destruction of the natural resource base of the Great Lakes was little appreciated. The lakes were widely regarded as an infinite and convenient receptacle for the wastes generated by a growing population and its industrial base. There were signs that all was not well with the system: visibly gross pollution, which would be considered intolerable today (MacKay 1930), and the decline of certain valued fish stocks, notably salmon, which were entirely eliminated from Lake Ontario before 1900 (Parsons 1973). Fish communities provide about the only example of reliable continuous records of changes in biotic components of the Great Lakes ecosystem. This record indicates continued deterioration of many native stocks, culminating in their demise and replacement by exotic populations during the 1940s through the 1960s (Beeton 1969). In retrospect it is rather curious that these well-documented problems were almost universally ascribed to proximate causes, and the remedies attempted were directed to these factors. Declines in valued fish stocks were blamed on factors such as overfishing, physical habitat destruction, competition with exotic populations, or some combination of these problems. Apparently the concept that perturbations of populations at the top of the food chain might reflect fundamental changes in the ecosystem did not come through, even though remedy of the apparent proximal causes seldom provided relief.

Aside from some exploratory records (e.g., Bailey 1842; Ehrenberg 1846; Briggs 1872), many early reports on the algae of the lakes came from studies of situations regarded as local pollution problems (e.g., Vorce 1882; Thomas and Chase 1887). Although the problem continued to grow, there was apparently little appreciation that the increased wastes generated by population growth and industrialization could seriously modify the desirable qualities of the Great Lakes. In Lake Michigan at least one problem, usually involving either the fishery or water supplies, was severe enough to generate a study of some sort in every decade since 1860. However, most problems were perceived as being local and temporary. Not until population growth, increased sewering, and the introduction of phosphate detergents in the 1940s and 1950s caused very widespread and grossly visible problems was the potential for highly undesirable modification of the Great Lakes ecosystem realized. Public concern resulted in studies, largely carried out during the past 30 years, which have a least partially illuminated the causes of, and possible solutions to, ecological problems in the Great Lakes system. Those problems that achieved public awareness during the past few decades resulted from fundamental changes in the lakes' ecosystem that were set in motion very early in the history of European settlement of the region.

Algae have played an important role in some of the highly visible and publicly perceived problems in the Great Lakes. They also played a role in other problems that are less obvious but more complex and more critically important to maintaining the Great Lakes ecosystem in its most desirable and useful condition. Before we explore specific cases, it is necessary to outline some basic physical and chemical characteristics of the Great Lakes that control, in both the immediate and long-term sense, the life processes of organisms that inhabit them.

II. Physical and chemical characteristics of the Great Lakes

Unlike most very large lakes, the Laurentian Great Lakes are geologically very recent (Hutchinson 1957; Herdendorf 1982). Formed during the most recent Pleistocene glaciation, they reached their present con-

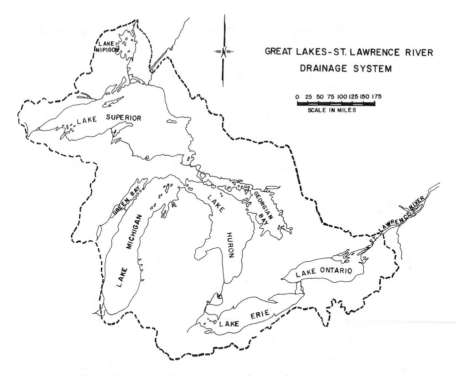

Fig. 3–1. Generalized map of the Laurentian Great Lakes and their drainage basin.

figuration only during the past few thousand years. Due to their distance from other, more ancient great lakes, there apparently was little interchange of organisms adapted to the particular conditions characteristic of very large bodies of fresh water. Due to their recent formation, the populations that do inhabit the Laurentian Great Lakes have not had sufficient time to achieve optimal adaptive evolution. As a result, the biota of the lakes is, to some degree, maladapted. Consequently, lake communities lack the finely tuned and resilient structure characteristic of ocean communities or communities in small lakes and are particularly susceptible to perturbation. One of the scientific tragedies of human intervention in the ecology of the lakes is that some very interesting experiments in adaptive radiation were prematurely terminated.

The surface area of the Great Lakes is very large (Fig. 3-1; Table 3-1) and relatively large compared to their total drainage area. Approximately 30% of the water entering the lakes arrives as rainfall directly on the lake surface. In addition, a significant portion of the land drainage comes from landscapes containing a large proportion of bedrock types, which are very resistant to leaching so that dissolved constituents in the

Table 3-1. *Morphometric and hydrologic data on the Laurentian Great Lakes*

	Superior	Michigan	Huron	Erie	Ontario	Total
Area, water surface $(10^3 km^2)$	82.4	58.0	59.6	25.7	19.7	245
Area, drainage basin $(10^3 km^2)$	207	176	188	87.4	90.1	745
Ratio of drainage basin to water surface	2.51	3.02	3.16	3.27	4.63	3.12
Mean depth (m)	148	84	53	17	86	
Volume $(10^3 km^3)$	12.2	4.87	3.53	0.458	1.64	22.7
Outflow $(km^3 \cdot y^{-1})$	65	49	159	175	208	
Renewal time (years)	190	100	22	3	8	
Outflow $(10^3\ m^3 \cdot sec^{-1})$	2.07	1.56	5.04	5.55	6.63	

Source: Adapted from Schelske 1975.

water of the Great Lakes bear the signature of continental rainfall. In terms of major nutrients controlling algal growth, they are deficient in phosphorus and silicon but relatively rich in nitrogen. These basic nutrient-loading parameters likely controlled both the original composition of the algal flora and the floristic succession as the nutrient balance was modified by human activities.

The above characteristics make the Great Lakes particularly susceptible to atmospheric inputs. Although the waters of the lakes are relatively base-rich and thus buffered from the immediate effects of acidic deposition, they are easily affected by other atmospheric contaminants such as toxic metals (Edgington and Robbins 1976) and persistent toxic organic compounds (Andren and Strand 1981).

In these very large systems, any change in dissolved solids takes place relatively slowly, particularly in the upper lakes. Because of their large volume relative to hydraulic inputs and outputs (Table 3-1), a response, measurable as an increase or decrease in concentration, occurs slowly, even in the case of conserved substances. In the case of nutrients, which are actively sequestered by the biota, considerable increases in loadings may not be reflected by measurable increases in concentration of the nutrient dissolved in the water.

It is also apparent from the data in Table 3-1 that the lower lakes eventually receive materials added to the upstream system, provided the materials are not removed by biological activity or precipitated and deposited in the sediments.

Climate also is important in controlling algal occurrence and abundance, and in the Great Lakes the interplay between regional climate and physical limnology of the system accounts for some of the successional changes that have taken place. The northernmost part of the Great Lakes system borders on the boreal forest, but the southernmost part extends well into the temperate deciduous forest biome. Their large volume gives the lakes tremendous heat-storage capacity, and their heating and cooling significantly lag behind those of the surrounding land mass. Because of this, the lakes significantly modify local climatic conditions. Many fruit crops are grown on the leeward shores of the lakes because the cold water mass moderates early spring temperature maximums and retards flowering beyond the latest spring killing frosts. During the fall, heat stored in the lakes moderates cold extremes and permits maturation of the crops. As any resident of Buffalo, New York, knows, this is not without price. During the winter, cold winds traversing the lakes are charged with moisture from the still relatively warm lake surface and deposit significantly enhanced snowfall on their leeward shores.

Like most deep, temperate lakes, the Laurentian Great Lakes are dimictic with periods of winter circulation and summer stratification (Hutchinson 1957). Unlike most lakes, however, the waters of the Great Lakes are so transparent that significant algal production can take place below the summer thermocline (Brooks and Torke 1977). This leads to interesting differences between the indigenous floras of the Great Lakes and those of most smaller systems. In the Great Lakes, a subsurface habitat characterized by low to moderate nutrient supply, low light, and cold temperatures is available throughout the year. This refugium allows many species with primarily boreal distribution and species that usually occur only during the winter in smaller lakes to exist all year in the Great Lakes. Prior to human modification of the system, such species were apparently dominant elements of the flora because nutrients controlling algal production were reduced to limiting levels in the epilimnion soon after stratification in most parts of the system. Certain regions within the system, mostly large shallow bays such as lower Green Bay of Lake Michigan, Saginaw Bay of Lake Huron, and the western basin of Lake Erie can be periodically wind-mixed during the summer. These regions have and apparently always have had (Taft and Kishler 1973; Burns 1985) considerably different algal floras than the open waters of the large lakes. These differences were greatly enhanced in the modern era because these regions, which have the largest natural nutrient supply, now receive the greatest anthropogenic loadings.

Another consequence of the very large volume of these lakes and their resultant thermal inertia, which modifies the distribution and seasonal succession of the algal flora, is the so-called thermal bar (Rodgers 1965; Huang 1972). During the spring warming period, inflowing rivers and the shallow nearshore waters heat much more rapidly than offshore waters, which remain below 4°C. As a result, a region of convergence and sinking called the thermal bar develops around the 4°C surface isotherm. Its effect on the algal flora is to separate the lake into two distinct regions. The nearshore waters are relatively warm and well supplied with nutrients from river inflows during the spring runoff maximum. The phytoplankton flora circulates through a relatively shallow, well-lighted water column. Phytoplankton populations in offshore waters during this time are limited by low nutrients, temperature, and light due to circulation through a deep water column. As a result, dramatic differences in the composition and abundance of the phytoplankton flora are found landward and lakeward of the spring thermal bar (Stoermer 1968; Lorifice and Munawar 1974). A similar situation also develops during the fall cooling period, but its effects are less pronounced.

Thus, the Great Lakes have a more complex regional and seasonal (Figs. 3-2 and 3-3) structure to their algal associations than exists or is resolved in smaller freshwater systems. This complexity is further complicated by other physical conditions characteristic of large systems, such as the presence of large-scale upwelling events (Strong et al. 1974; Haffner et al. 1984) and persistent current patterns (Mortimer 1974). In the modern era floristic differences resulting from these underlying physical factors, particularly the different physiography of certain large bays and the spring thermal bar, have been enhanced by anthropogenic inputs to the nearshore regions of the lakes. Large differences in the composition, abundance, and productivity of phytoplankton communities in the nearshore and offshore regions of the lakes now occur (Holland 1969; Ladewski and Stoermer 1973). There is good reason to suspect, however, that significant differences existed in the indigenous floras of different regions of the lakes (Taft and Kishler 1973).

III. Algae and ecosystem problems

It is unlikely that we will ever be able to reconstruct the original abundance of some algal groups, notably microflagellates (Munawar and Munawar 1975) that do not form cysts, and picoplankton (Sicko-Goad and Stoermer 1984). The earliest preserved samples of algae from the Great Lakes that I am aware of are collections taken by Bailey in 1839 and sent to the German micropaleontologist C. G. Ehrenberg (Stoermer and Ladewski 1982). Although Ehrenberg did not consider the material he investigated to be plants (in fact, he quit working on them when it was proven that they were plants [Boling 1976]), he described several

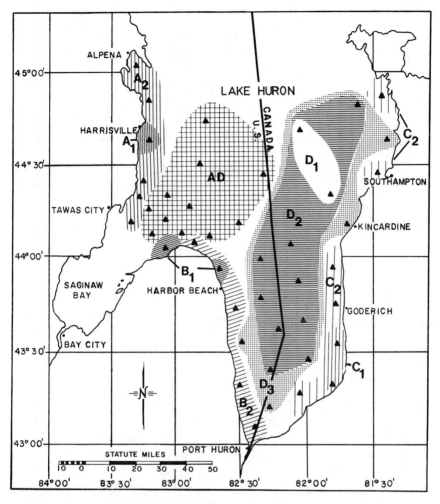

Fig. 3–2. Regions of Lake Huron containing similar phytoplankton communities during the summer of 1980. Zonation is based on average conditions during the entire sampling period. Black triangles indicate location of stations sampled. (After Stoermer et al. 1983.)

species of diatoms (Ehrenberg 1846, 1854). From this beginning until today the diatom flora has remained the best known and described part of the lakes' flora (Stoermer and Kreis 1978). Part of this is due to the rather unusual course of development of phycological work on the lakes. Although the region surrounding the Great Lakes contains a number of major universities, academic phycologists, with a few notable exceptions, tended to shun work on the lakes until the recent era. Lack of suitable facilities to undertake meaningful studies of such large, complex, and physically difficult systems tended to turn the interests of most

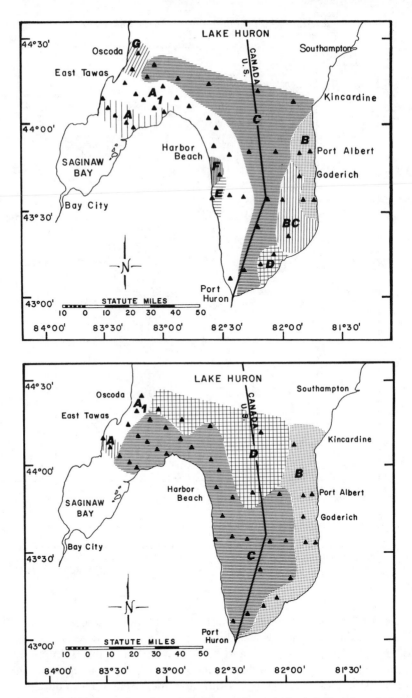

Fig. 3–3. *Top:* Regions of southern Lake Huron that contained similar phytoplankton communities in late June 1976. At this time phytoplankton distribution was strongly controlled by the spring thermal bar. The 4° surface isotherm closely approximates the boundary of zone *C* except in the far northeastern segment of the study area, where shore-

academic phycologists to more tractable problems. The early work that was carried out was largely undertaken by a remarkable group of microscope enthusiasts and natural historians, most of whom were interested in diatoms. Most of these studies were undertaken in response to what were conceived as important practical water pollution problems (Vorce 1882; Thomas and Chase 1887). Perhaps the most important contributions of these workers were the permanent slides and preserved samples, which can be recovered and verified today.

Their work was also the beginning of a curious phenomenon that continues, to some extent, to the present day. Most of the impetus (and funding) for research on the Great Lakes comes from practical concerns with water quality. As a result, research has been focused on immediate solutions of large-scale ecological problems rather than on fundamental observations. A surprising amount of the phycological research done on the lakes was accomplished in response to practical problems by people who were not primarily trained as phycologists. As a result, our knowledge of the Great Lakes algal flora has a curious "inverted" quality. Certain studies that are usually considered fundamental, such as basic floristic studies, are largely lacking. On the other hand, extended ecological systems models have been constructed. Despite, or perhaps because of, the unconventional approach some unified understanding of ecological processes as they affect, and are affected by, algal populations in the lakes is beginning to emerge. I would like to review our present understanding of the system and how we arrived at it, because I feel that it provides some interesting perspectives on science, including one of its more esoteric branches, phycology, and its applications to human needs. As is common with scientific problems, our current understanding of algal ecology in the Great Lakes has consisted of three fairly distinct stages, which I will term *discovery, elucidation,* and *synthesis.* At the present time it is doubtful if any of these stages is complete, and our current synthesis must be regarded as a rapidly moving target. I will, howver, attempt to project possible future developments from the evidence presently at hand.

A. *Discovery*

In the Great Lakes, discovery of severe problems long preceded most planned remedies. As discussed in Section I, local pollution problems were evident in the 1870s in Lake Erie and the 1880s in Lake Michigan.

line inputs are minimal, and the far southern segment. (After Stoermer and Kreis 1980.) *Bottom:* Regions of southern Lake Huron that contained similar phytoplankton communities in August 1976. At this time phytoplankton distribution was strongly controlled by upwelling on the Canadian coast and reversal of normal current patterns. (After Stoermer and Kreis 1980.)

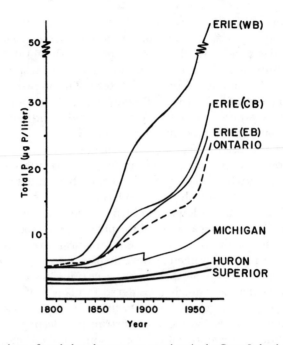

Fig. 3–4. Hindcast of total phosphorus concentrations in the Great Lakes based on model simulations. The sharp inflection in the curve for Lake Michigan resulted from the Chicago diversion, which removed sewage loadings from this major source. (After Chapra 1977.)

Indeed, one of the most effective control actions, the Chicago Sewage Diversion (Fig. 3-4), which removed sewage loading from the region's largest city from the Great Lakes, was undertaken in response to this problem and was completed very early in the twentieth century. Although the full implications of this and other actions that diverted nutrients from the system were not fully appreciated, there was continuing concern about deteriorating conditions in the lakes and sporadic attempts at remedial actions.

It was not until the late 1950s that certain problems became so severe and visible that public opinion was aroused and demands for remedial action became politically feasible. Three particular nuisances caused by algae became more or less celebrated causes of the environmental movement of the late 1960s and early 1970s. All three were symptoms of the same underlying cause, and in all three cases expression of the problem was modified by peculiarities of the Great Lakes system.

The first problem was the spring blooms of certain colonial species of *Stephanodiscus* (Fig. 3-5). These blooms greatly reduced water clarity and

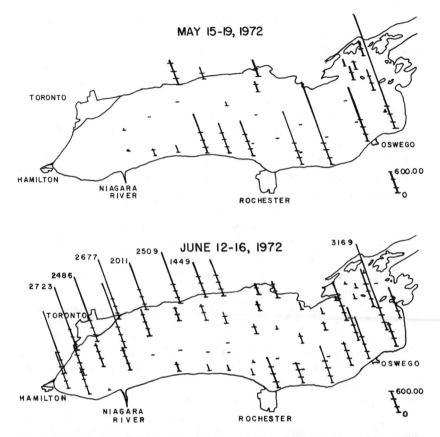

Fig. 3–5. Distribution of *Stephanodiscus binderanus* in Lake Ontario in May and June 1972. Vertical scale indicates abundance (cells·ml^{-1}). The introduction of this diatom to the Great Lakes system was widely noted because it causes problems at municipal water filtration plants during the spring. (After Stoermer et al. 1974.)

caused problems with undesirable odors and short filter runs at municipal water filtration plants around the lakes (Brunel 1956; Vaughn 1961; Schenk and Thompson 1965).

The second most noticeable problem was an obvious increase in both the phytoplankton quantity in some areas of the Great Lakes system and the change in species composition, which caused objectionable surface accumulations, particularly of certain blue-green algae. In the most seriously affected parts of the system these summer blooms also caused objectionable "earthy" or "musty" odors in water supplies (Bierman and Dolan 1981).

The third notable algal nuisance was the so-called *Cladophora* problem. It may be difficult for most phycologists to conceive of *Cladophora* as a menace, but it was widely regarded as such in the Great Lakes (Bellis and McLarty 1967; Auer 1982; see also Lembi, O'Neal, and Spencer, Chapter 18). Extremely dense growths occurred locally in Lakes Michigan and Huron and generally in Lake Erie and, particularly, Lake Ontario. Late summer or fall storms often stripped these luxuriant growths from their rock substrate and deposited them on shore where they decayed and caused gross aesthetic nuisances (see also Lembi et al., Chapter 18).

Although these problems were painfully and publicly evident, there was genuine debate about when they had begun and how significant a departure from previous conditions they represented. It must be remembered that at this time very little in the way of phycological or basic limnological information was available for comparison. As a result, considerable effort was devoted to retrospective studies designed to detect trends. Because few useful published records existed, data on phytoplankton abundance and water chemistry routinely taken at water filtration plants around the lakes were studied. Although these records were intended primarily as a practical aid to water treatment technique and not as a scientific tool, they did provide useful information. Records of phytoplankton abundance in the raw water supplies of the cities of Chicago (Damann 1960), Cleveland (Davis 1964), and Toronto (Schenk and Thompson 1965) provided convincing evidence of a clear trend toward both increased algal abundance and altered seasonal succession. Chemical data from the same sources (Beeton 1965; Powers and Ayers 1967) showed a clear pattern of increase in nonnutrient ions in the waters of Lakes Erie, Michigan, and Ontario. The available data proved too crude to demonstrate clearly trends in nutrients, although Powers and Ayers (1967) noted an apparent downward trend in dissolved silica in Lake Michigan, which provided one clue to eventual elucidation of nutrient trends in the lakes. From a phycological standpoint, water filtration plant records proved to be inadequate, since changes in species composition of the flora were poorly resolved. Reanalysis of the few available historic samples from Lake Erie (Hohn 1969) and Lake Michigan (Stoermer and Yang 1969) provided some insight into this problem. It showed that a major change in composition of the flora had occurred, at least in its diatom component as diatoms were the only organisms preserved well enough to be analyzed. In particular these studies demonstrated that the lakes had been invaded successfully by a series of species usually found in highly eutrophic environments. Work on quantitative and dynamic aspects of the problem proceeded more slowly. Although some pioneering limnological studies contained quantitative estimates of phytoplankton abundance (Burkholder 1929; Chandler 1940, 1942, 1944; Verduin 1952, 1954) and production (Saunders, Trama and Bachmann 1962; Saunders 1964; Parkos, Olson, and Odlaug 1969), large-scale studies of the Great Lakes using modern methods were not reported until the 1970s (e.g., Fee 1973).

B. Elucidation

From the beginning increased phosphorus was strongly suspected as the cause of the grossly visible algal overabundance in the Great Lakes. Of the major plant nutrients present, phosphorus was considered most likely to control algal growth. This conclusion was reinforced by comparison of phosphorus concentrations in areas with severe algal nuisances versus those areas of the upper lakes where such problems did not occur (Dobson, Gilbertson, and Sly 1974). These primitive arguments were sufficiently compelling that phosphorus control measures were advocated before the full extent of the problem was known. At this point the issue moved from the scientific arena into the ill-defined ground between research findings and practical implementation. It was apparent that limiting phosphorus inputs to the Great Lakes would be an extensive, expensive, and politically and economically sensitive undertaking. Control measures, specifically removal or limitation of phosphorus content in washing detergents, would adversely affect the economic interests of certain industries. At the national level this conflict of interests stimulated the so-called limiting nutrient controversy (Likens 1972), which resulted in extensive and intensive research on factors controlling primary production in lakes.

Since the condition of the Great Lakes was a prime ecological issue, greatly increased resources became available to initiate large-scale field and laboratory investigations in Canada and the United States. The field studies provided a much clearer picture of the basic limnology of the Great Lakes (Dobson et al. 1974) and phytoplankton abundance–productivity relationships (Vollenweider, Munawar, and Stadelmann 1974). From the standpoint of classical phycological studies, detailed information became available on the spatial and temporal distribution of phytoplankton populations in Lake Erie (Munawar and Munawar 1976), Lake Ontario (Stoermer et al. 1974), southern Lake Huron (Stoermer and Kreis 1980; Stoermer, Sicko-Goad, and Frey 1983), northern Lake Michigan (Stoermer 1984), and Lake Superior (Munawar and Munawar 1978). Checklists of certain elements of the flora were published (Stoermer and Kreis 1978), and compilations of the ecological tendencies of important species in plankton (Stoermer 1978) and periphyton (Stoermer 1980) communities became available.

The effects of nutrient enrichment on natural phytoplankton communities were investigated in field (Schelske, Rothman, and Simmons

1978) and laboratory (Schelske et al. 1974) studies. Nutrient competition between individual species or within natural assemblages was also extensively studied (Kilham 1986). Because of practical problems associated with *Cladophora* overgrowths in certain regions of the Great Lakes, considerable research was undertaken on attached algal communities characteristic of eutrophied nearshore regions (summarized in Auer 1982). Although the less productive but more diverse benthic communities characteristic of the upper lakes (Stoermer 1975) have received less attention, several studies of variations in community structure with depth were carried out in Lake Michigan (Stevenson and Stoermer 1981; Stoermer 1981; Kingston et al. 1983). Several attempts were made to incorporate this information into models of ecosystem function. These range from first-order models of the effect of phosphorus loading on algal standing crop in lakes with different morphometric characteristics (Vollenweider 1968) to more complex models of phytoplankton (e.g., Bierman et al. 1980; Bierman and Dolan 1986a, b) and periphyton (Auer and Canale 1980) response.

The immediate result of these extensive field and laboratory investigations was to reinforce the conclusion that phosphorus is indeed the nutrient primarily controlling production in the Great Lakes, as it is in most smaller freshwater lakes (Schindler 1974, 1977). Perhaps more importantly, this research uncovered an aspect of the problem peculiar to the Great Lakes that had not been widely appreciated previously. Field observations (Dobson 1981) and experimental studies (Schelske and Stoermer 1971, 1972; Schelske et al. 1974) showed that, in long-residence systems such as the Great Lakes, phosphorus loadings control not only the growth rate and standing crop of phytoplankton but also the availability of other nutrients. Because of their unusual hydraulic budget, increased anthropogenic loadings cause gross changes in N:P:Si ratios in Great Lakes waters. Schelske (1975) pointed out that annual depletion of nitrogen and silicon from the water column provides an index of eutrophication in these large lakes. Unlike most temperate lakes, increased phosphorus loading first results in secondary limitation by silicon. Nitrogen becomes secondarily limiting only in areas that receive very heavy phosphorus loadings. These differences in nutrient availability may cause considerable changes in the composition of the algal flora. Secondary silicon limitation causes a shift from dominance by diatoms and chrysophytes to dominance by species that do not require silicon. Where nitrogen becomes limiting, nitrogen-fixing cyanophytes become dominant. In the Great Lakes this has occurred mainly in large bays and shallow regions such as western Lake Erie (Ogawa and Carr 1969), Green Bay of Lake Michigan (Vanderhoef et al. 1972, 1974), and Saginaw Bay of Lake Huron (Bierman et al. 1980)

and results from a combination of increased external phosphorus loading and minimal internal nitrogen reserves. In the offshore waters of
the Great Lakes, succession in the phytoplankton flora is further
affected by physical characteristics. As the flora of the surface waters
changes and increases in abundance due to increased phosphorus supplies, light limitation may eventually eliminate the subthermocline niche
and result in extirpation of a large element of the indigenous flora (Moll
and Stoermer 1982). Populations particularly affected are diatoms,
which have their maximum growth period during the summer in oligotrophic lakes, particularly members of the genus *Cyclotella* (Hutchinson 1967).

During the past few years we have gained a much improved perspective on the timing of these changes. The availability of radiometrically
dated cores has allowed reconstruction of changes in phytoplankton
species that leave siliceous microfossils in Lake Erie (Frederick 1981;
Harris and Vollenweider 1982), Lake Ontario (Stoermer, Wolin, et al.
1985a), and Lake Superior (Stoermer, Kociolek, et al. 1985). These
studies indicate that the original phytoplankton flora of the Great Lakes
was somewhat different than had been imagined from the earliest available historic records. Prior to European settlement of the region the
siliceous microfossil assemblages deposited in offshore sedimentation
basins were composed of a surprisingly large number of taxa that grew
in benthic habitats, planktonic species that grew during the winter or
below the summer thermocline, and relatively few species that usually
grew in the epilimnion. The epilimnetic species were largely chrysophytes, with a small representation of the "oligotrophic *Cyclotella* flora"
(Hutchinson 1967). This probably indicates that nutrient limitation
severely limited growth in the summer epilimnion and that all of the
lakes were ultraoligotrophic. The settlement horizon in all cores investigated to date was marked by drastic change in both the composition
and the abundance of siliceous microfossils. In the Bay of Quinte, an
arm of Lake Ontario, perturbation of the algal flora apparently dates
to the very earliest European settlement (Stoermer, Wolin, et al.
1985b), and species that usually thrive under eutrophic conditions
appear very early in the postsettlement record. Apparently many populations that we think of as being indicative of oligotrophic conditions
in the modern context became abundant only after the lakes had been
modified by human activities. These species subsequently were extir-
pated by further changes in the ecology of Lake Erie (Harris and Vollenweider 1982) and Lake Ontario (Stoermer, Wolin, et al. 1985a),
although they still persist in Lake Superior. Another interesting aspect
of diatom succession in the Great Lakes is the observation that the most
persistent species are those which, in addition to optimizing growth

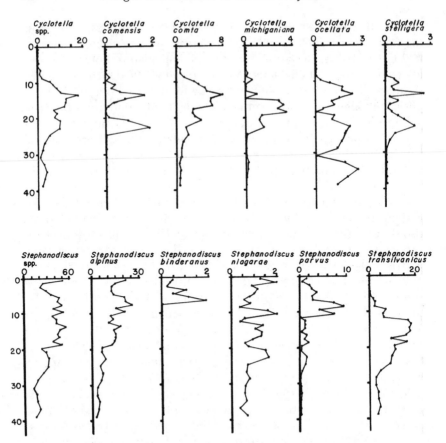

Fig. 3–6. Abundance (10^6 valves·g^{-1} dry wt) of selected *Cyclotella* and *Stephanodiscus* species versus depth (cm) in Lake Ontario sediments that contain material deposited between 1704 and 1979. (After Stoermer, Wolin, et al. 1985a.)

under winter conditions, apparently adjust their morphology to accommodate reduced silicon availability (Theriot and Stoermer 1984; Stoermer, Wolin, et al. 1985c).

C. Current understanding

These findings have several important implications. They indicate that modification of primary producer communities in the Great Lakes was even more extensive than was generally realized (Fig. 3-6). Further, these changes began much earlier than had been previously appreciated (Fig. 3-7). This is important for management because modification of higher trophic levels, including the once important commercial fishery of the Great Lakes, has generally been attributed to factors resulting

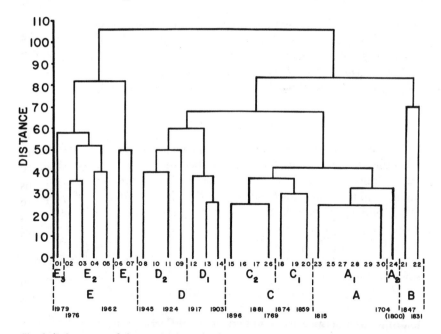

Fig. 3–7. Summary of changes in the Lake Ontario diatom flora based on cluster analysis of data derived from specimens preserved in radiometrically dated sediments. Large changes in composition of the flora are associated with settlement of the region between ca. 1815 and 1850 and severe eutrophication after 1945. (After Stoermer, Wolin, et al. 1985a).

directly from human activities such as overfishing, introduction of exotic species, and pollution by toxic chemicals. The current information suggests that these factors, though important, likely worked their havoc on stocks that were already stressed by fundamental changes in ecosystem structure and perhaps would have eventually collapsed from this cause alone. It is significant that success in restoration of indigenous fish stocks has been poor, even when the supposed cause of their demise has been removed. Although an extended historical debate continues between advocates of the "environmental" versus the "overfishing" explanation for decline of fish stocks in Lake Erie (Arnold 1969), surprisingly little direct evidence is available. Aside from an intriguing report by Hohn (1966), little is known about first feeding of freshwater larval fish on phytoplankton, which has proven to be crucial for some marine stocks (Lasker 1975). Information on the importance of phytoplankton food quality to zooplankton growth and reproduction is also lacking for the Great Lakes situation, although this is known to be important in some cases (Arnold 1971).

Most importantly, these results indicate that it will be necessary to retreat further from current nutrient loadings than is presently envisioned if the Great Lakes ecosystem is to be restored to the most desirable condition. Truly effective nutrient management will likely involve control of so-called diffuse sources from land runoff and the atmosphere as well as loadings from sewage and industrial sources.

IV. What lies ahead?

It is quite clear that the long-standing trend toward ever-increasing phosphorus loadings to the Great Lakes has been reversed (Dobson 1981) and that the effects of this reduction are visible in changes in the phytoplankton flora of Lake Erie (Nicholls, Standen, and Hopkins 1980) and Lake Ontario (Stoermer, Wolin, et al. 1985a). The future direction and eventual extent of these changes are a matter of conjecture. It is unlikely that we will see the algal flora of the Great Lakes return to any of its previous floristic associations. There are several reasons for this.

The nutrient balance that existed in the Great Lakes prior to human intervention will not be restored rapidly. Even if excess phosphorus loadings were entirely eliminated, it would take considerable time to replenish silicon concentrations to their previous levels because of the very long hydraulic residence time of the upper lakes. It is also likely that some increases in external loadings, as a practical matter, will be virtually impossible to reverse. As pointed out previously, the Great Lakes are particularly susceptible to atmospheric loadings, and there is ample evidence that such loadings have been grossly modified by human activities, even in regions most remote from major sources.

As previously noted, considerable changes in conservative ion concentrations have occurred in the Great Lakes. Although the concentrations of chloride, sulfate, and other conserved substances are still relatively low, the invasion of biological species usually found in high-salinity environments is remarkable. The more obvious examples include unplanned introduction of marine fishes, such as the sea lamprey and the alewife, which have caused severe problems, and deliberate introduction of species such as Pacific salmon. Such modifications at the apex of the trophic pyramid can have effects that resonate down through the system and cause significant changes in species composition and processes at the primary-producer level (Scavia et al. 1986). Although the number of salinity-tolerant invertebrates and algae that have become established in the Great Lakes is much larger, these introductions have received less attention. Perhaps the most conspicuous algal invader is *Bangia* (Lin and Blum 1977; Sheath and Cole 1984). It has become adapted to growth in the Great Lakes and now forms a conspicuous element of attached communities in certain areas of the lakes.

Several halophilic species of diatoms (Hasle and Evensen 1976; Wujek and Graebner 1980) are common in the plankton flora, and occasional occurrences of exotic periphyton species (Wujek and Welling 1981) have been noted. How these populations became established in the Great Lakes is still a matter of conjecture. Brine discharges in some local areas provide salinity levels approaching seawater. Exotic populations may have been transported to these regions in the ballast water of ocean ships that visit Great Lakes ports, became established in local high-salinity areas, and subsequently adapted to lower salinity levels and became more generally distributed. This radiation was likely possible because native species were stressed by changing conditions, thus reducing competition.

Of great concern presently are loadings of persistent synthetic compounds, which enter the Great Lakes from a variety of sources and accumulate in the biota. Although the effects of most of these compounds on algal populations are poorly known, it is highly unlikely that they pass through this component of the ecosystem without effect.

The important point is that significant changes in both the basic chemistry of the Great Lakes and their biota have occurred, in addition to process modifications forced by altered nutrient inputs. It is clear that reversal of changes in conservative ion concentration will be very slow, if possible at all under modern conditions. The response of the exotic biota to any future alterations of conditions is a matter of conjecture, but present indications are that at least some species have become adapted to the Great Lakes habitat and will not be immediately replaced. Thus, the future flora and fauna of the Great Lakes will almost certainly be different from any that has existed previously and will exist in an environment different from that experienced by any of their progenitors. It thus appears that the ecology of algae in the Great Lakes is irretrievably linked to human affairs.

V. References

Andren, A. W., and J. W. Strand. 1981. Atmospheric deposition of particulate organic carbon and polyaromatic hydrocarbons to Lake Michigan. In *Atmospheric Pollutants in Natural Waters*, ed. S. J. Eisenreich, pp. 459–79. Ann Arbor Science, Ann Arbor, MI.

Arnold, D. E. 1969. The ecological decline of Lake Erie. *N.Y. Fish and Game J.* 16, 27–45.

Arnold, D. E. 1971. Ingestion, assimilation, survival, and reproduction by *Daphnia pulex* fed on seven species of blue-green algae. *Limnol. Oceanogr.* 16, 906–20.

Auer, M. T. (ed.). 1982. Ecology of filamentous algae. *J. Great Lakes Res.* 8, 1–237.

Auer, M. T., and R. P. Canale. 1980. Phosphorus uptake dynamics as related to mathematical modeling of *Cladophora* at a site in Lake Huron. *J. Great Lakes Res.* 6, 1–7.

Bailey, J. W. 1842. A sketch of the infusoria of the Family Bacillaria, with some account of the most interesting species which have been found in a recent or fossil state in the United States. *Amer. J. Sci. Arts* 42, 88–105.

Beeton, A. M. 1965. Eutrophication of the St. Lawrence Great Lakes. *Limnol. Oceanogr.* 10, 240–54.

Beeton, A. M. 1969. Changes in the environment and biota of the Great Lakes. In *Eutrophication: Causes, Consequences, Correctives,* pp. 150–87. National Academy of Science, Washington, DC.

Bellis, V. J., and D. A. McLarty. 1967. Ecology of *Cladophora glomerata* (L) Kütz. in southern Ontario. *J. Phycol.* 3, 57–63.

Bierman, V. J., Jr., and D. M. Dolan. 1981. Modeling of phytoplankton-nutrient dynamics in Saginaw Bay, Lake Huron. *J. Great Lakes Res.* 7, 409–39.

Bierman, V. J., Jr., and D. M. Dolan. 1986a. Modeling of phytoplankton in Saginaw Bay. I. Calibration phase. *J. Envir. Eng.* 112, 400–14.

Bierman, V. J., Jr., and D. M. Dolan. 1986b. Modeling of phytoplankton in Saginaw Bay. II. Post-audit phase. *J. Envir. Eng.* 112, 415–29.

Bierman, V. J., Jr., D. M. Dolan, E. F. Stoermer, J. E. Gannon, and V. E. Smith. 1980. *The Development and Calibration of a Spatially Simplified Multi-class Phytoplankton Model for Saginaw Bay, Lake Huron.* Great Lakes Planning Study Contribution no. 33. Great Lakes Basin Commission, Ann Arbor MI. 126 pp.

Boling, R. 1976. *Das Leben und Werk Christian Gottfried Ehrenberg.* Veroffentl. Delitzscher Gesch. Herausg. Kreismuseum, Delitzsch. 61 pp.

Briggs, S. A. 1872. The Diatomaceae of Lake Michigan. *Lens* 1, 41–4.

Brooks, A. J., and B. G. Torke. 1977. Vertical and seasonal distribution of chlorophyll *a* in Lake Michigan. *J. Fish. Res. Bd. Can.* 34, 2280–7.

Brunel, J. 1956. Addition du *Stephanodiscus binderanus* a la flore diatomique de l'Amerique du Nord. *Naturaliste Canadien* 83, 89–95.

Burkholder, P. R. 1929. Microplankton studies of Lake Erie. *Bull. Buffalo Soc. Nat. Sci.* 14, 73–93.

Burns, N. M. 1985. *Erie: The Lake that Survived.* Rowman and Allanheld, Totawa, NJ. 320 pp.

Chandler, D. C. 1940. Limnological studies of western Lake Erie. I. Plankton and certain physical-chemical data on the Bass Islands region, from September, 1938, to November, 1939. *Ohio J. Sci.* 40, 291–336.

Chandler, D. C. 1942. Limnological studies of western Lake Erie. III. Phytoplankton and physical-chemical data from November, 1939, to November, 1940. *Ohio J. Sci.* 42, 24–44.

Chandler, D. C. 1944. Limnological studies of western Lake Erie. IV. Relation of limnological and climatic factors to the phytoplankton of 1941. *Trans. Amer. Microsc. Soc.* 63, 203–36.

Chandler, D. C. 1964. The St. Lawrence Great Lakes. *Verh. Internat. Verein. Limnol.* 15, 59–75.

Chapra, S. C. 1977. Total phosphorus model for the Great Lakes. *J. Envir. Eng. Div.* 103, 147–61.

Damann, K. E. 1960. Plankton studies of Lake Michigan. II. Thirty-three years of continuous plankton and coliform bacteria data collected from Lake Michigan at Chicago, Illinois. *Trans. Amer. Microsc. Soc.* 79, 397–404.

Davis, C. C. 1964. Evidence of the eutrophication of Lake Erie from phytoplankton records. *Limnol. Oceanogr.* 9, 275–83.

Dobson, H. H. 1981. Trophic conditions and trends in the Laurentian Great Lakes. *W.H.O. Water Qual. Bull.* 6, 146–51, 158–60.

Dobson, H. H., M. Gilbertson, and P. G. Sly. 1974. A summary and comparison of nutrients and related water quality in Lakes Erie, Huron, and Superior. *J. Fish. Res. Bd. Can.* 31, 731–8.

Edgington, D. N., and J. A. Robbins. 1976. Records of lead deposition in Lake Michigan sediments since 1800. *Envir. Sci. Technol.* 10, 266–74.

Ehrenberg, C. G. 1846. Neue Untersuchungen über das kleinste Leben als geologisches Moment. *Ber. Akad. Wiss.* [Berlin] 1845, 53–88.

Ehrenberg, C. G. 1854. *Mikrogeologie: Das Erde und Felsen schaffende Wirken des unsichtbar kleinen selbständigen Lebens auf der Erde.* Voss, Leipzig. 374 pp.

Fee, E. J. 1973. A numerical model for determining integral primary production and its application to Lake Michigan. *J. Fish. Res. Bd. Can.* 30, 1447–68.

Frederick, V. R. 1981. Preliminary investigation of the algal flora in the sediments of Lake Erie. *J. Great Lakes Res.* 7, 404–8.

Haffner, G. D., M. L. Yallop, P. D. N. Hebert, and M. Griffiths. 1984. Ecological significance of upwelling events in Lake Ontario. *J. Great Lakes Res.* 10, 28–37.

Harris, G. P., and R. A. Vollenweider. 1982. Paleolimnological evidence of early eutrophication in Lake Erie. *Can. J. Fish. Aquat. Sci.* 39, 618–26.

Hasle, G. R., and D. L. Evensen. 1976. Brackish water and freshwater species of the diatom genus *Skeletonema*. II. *Skeletonema potamos* comb. nov. *J. Phycol.* 12, 73–82.

Herdendorf, C. E. 1982. Large lakes of the world. *J. Great Lakes Res.* 8, 379–412.

Hohn, M. H. 1966. Analysis of plankton ingested by *Stizostedium vitreum vitreum* (Mitchill) fry and concurrent vertical plankton tows from southwestern Lake Erie, May 1961 and May 1962. *Ohio J. Sci.* 66, 193–7.

Hohn, M. H. 1969. Qualitative and quantitative analysis of plankton diatoms, Bass Island area, Lake Erie. *Bull. Ohio Biol. Surv., N.S.* 3, 1–211.

Holland, R. E. 1969. Seasonal fluctuations of Lake Michigan diatoms. *Limnol. Oceanogr.* 14, 423–36.

Hough, J. L. 1958. *Geology of the Great Lakes.* University of Illinois Press, Urbana. 313 pp.

Huang, J. C. K. 1972. The thermal bar. *Geophys. Fluid Dyn.* 3, 1–25.

Hutchinson, G. E. 1957. *A Treatise on Limnology.* Vol. 1, *Geography, Physics, and Chemistry.* John Wiley, New York. 1,015 pp.

Hutchinson, G. E. 1967. *A Treatise on Limnology.* Vol. 2. *Introduction to Lake Biology and the Limnoplankton.* John Wiley, New York. 1,115 pp.

Kilham, S. S. 1986. Dynamics of Lake Michigan natural phytoplankton communities in continuous cultures along a Si:P gradient. *Can J. Fish. Aquat. Sci.* 43, 351–60.

Kingston, J. C., R. L. Lowe, E. F. Stoermer, and T. B. Ladewski. 1983. Spatial and temporal distribution of benthic diatoms in northern Lake Michigan. *Ecology* 64, 1566–80.

Ladewski, T. B., and E. F. Stoermer. 1973. Water transparency in southern Lake Michigan in 1971 and 1972. In *Proceedings of the Sixteenth Conference on Great Lakes Research*, pp. 791–807. Internat. Assoc. Great Lakes Res, Ann Arbor, MI.

Lasker, R. 1975. Field criteria for survival of anchovy larvae: the relation between inshore chlorophyll maximum layers and successful first feeding. *Fish. Bull.* 73, 453–61.

Likens, G. E. (ed.). 1972. *Nutrients and Eutrophication: The Limiting Nutrient Controversy.* Special Symposium vol. 1. American Society of Limnology and Oceanography, Allen Press, Lawrence, KS.

Lin, C. K., and J. L. Blum. 1977. Recent invasion of a red alga *(Bangia atropurpurea)* in Lake Michigan. *J. Fish. Res. Bd. Can.* 34, 2413–16.

Lorifice, G. J., and M. Munawar. 1974. The abundance of diatoms in the southwestern nearshore region of Lake Ontario during the spring thermal bar period. In *Proceedings of the Seventeenth Conference on Great Lakes Research*, pp. 619–28. International Association, Great Lakes Research, Ann Arbor, MI.

MacKay, H. H. 1930. Pollution problems in Lake Ontario. *Trans. Amer. Fish. Soc.* 60, 297–305.

Moll, R. A., and E. F. Stoermer. 1982. A hypothesis relating trophic status and subsurface chlorophyll maxima of lakes. *Archiv. Hydrobiol.* 94, 425–40.

Mortimer, C. H. 1974. Lake hydrodynamics. *Mitt. Internat. Verein. Limnol.* 20, 124–97.

Munawar, M., and I. F. Munawar. 1975. Abundance and significance of phytoflagellates and nannoplankton in the St. Lawrence Great Lakes. *Verh. Internat. Verein. Limnol.* 19, 705–23.

Munawar, M., and I. F. Munawar. 1976. A lakewide study of phytoplankton biomass and its species composition in Lake Erie, April–December, 1970. *J. Fish. Res. Bd. Can.* 33, 581–600.

Munawar, M., and I. F. Munawar. 1978. Phytoplankton of Lake Superior 1973. *J. Great Lakes Res.* 4, 415–42.

Nicholls, K. H., D. W. Standen, and G. J. Hopkins. 1980. Recent changes in the nearshore phytoplankton of Lake Erie's western basin at Kingsville, Ontario. *J. Great Lakes Res.* 6, 146–53.

Ogawa, R. E., and J. F. Carr. 1969. The influence of nitrogen on heterocyst production in blue-green algae. *Limnol. Oceanogr.* 14, 342–51.

Parkos, W. G., T. A. Olson, and T. O. Odlaug. 1969. *Water Quality Studies on the Great Lakes Based on Carbon Fourteen Measurements on Primary Productivity.* University of Minnesota Water Resources Center, Bull. no. 17. 121 pp.

Parsons, J. W. 1973. *History of Salmon in the Great Lakes.* U.S. Bureau of Sport Fisheries and Wildlife, Tech. Paper no. 68.

Powers, C. F., and J. C. Ayers. 1967. Water quality and eutrophication trends in southern Lake Michigan. In *Studies on the Environment and Eutrophication of Lake Michigan*, ed. J. C. Ayers and D. C. Chandler, pp. 142–78. University of Michigan, Great Lakes Res. Div. Spec. Rep. no. 30.

Rodgers, G. K. 1965. The thermal bar in the Laurentian Great Lakes. In *Proceedings of the Eight Conference on Great Lakes Research*, pp. 358–63. University of Michigan, Great Lakes Res. Div. Publ. no. 13.

Saunders, G. W. 1964. Studies of primary productivity in the Great Lakes. In *Proceedings of the Sixth Conference on Great Lakes Research*, pp. 122–9. University of Michigan, Great Lakes Res. Div. Publ. no. 11.

Saunders, G. W., B. Trama, and R. W. Bachmann. 1962. *Evaluation of a Modified C-14 Technique for Shipboard Estimation of Photosynthesis in Large Lakes*. University of Michigan, Great Lakes Res. Div. Publ. no. 8. 61 pp.

Scavia, D. L., G. L. Fahnenstiel, M. S. Evans, D. J. Jude, and J. T. Lehman. 1986. Influence of salmonid predation and weather on long-term water quality trends in Lake Michigan. *Can. J. Fish. Aquat. Sci.* 43, 435–43.

Schelske, C. L. 1975. Silica and nitrate depletion as related to rate of eutrophication in Lakes Michigan, Huron, and Superior. In *Coupling of Land and Water Systems*, ed. A. D. Hasler, pp. 277–98. Springer-Verlag, New York.

Schelske, C. L., E. D. Rothman, and M. S. Simmons. 1978. Comparison of bioassay procedures for growth-limiting nutrients in the Laurentian Great Lakes. *Mitt. Internat. Verein. Limnol.* 21, 65–80.

Schelske, C. L., E. D. Rothman, E. F. Stoermer, and M. A. Santiago. 1974. Responses of phosphorus limited Lake Michigan phytoplankton to factorial enrichments with nitrogen and phosphorus. *Limnol. Oceanogr.* 19, 409–19.

Schelske, C. L., and E. F. Stoermer. 1971. Eutrophication, silica, and predicted changes in algal quality in Lake Michigan. *Science* 173, 423–4.

Schelske, C. L., and E. F. Stoermer. 1972. Phosphorus, silica, and eutrophication of Lake Michigan. In *Nutrients and Eutrophication: The Limiting Nutrient Controversy*, Special Symposium vol. 1, ed. G. E. Likens, pp. 157–71. American Society of Limnology and Oceanography, Allen Press, Lawrence, KS.

Schenk, C. F., and R. E. Thompson. 1965. Long-term changes in water chemistry and abundance of plankton at a single sampling location in Lake Ontario. In *Proceedings of the Eighth Conference on Great Lakes Research*, pp. 197–208. University of Michigan, Great Lakes Res. Div. Publ. no. 13.

Schindler, D. W. 1974. Eutrophication and recovery in experimental lakes: implications for lake management. *Science* 184, 897–9.

Schindler, D. W. 1977. Evolution of phosphorus limitation in lakes. *Science* 195, 260–2.

Sheath, R. G., and K. M. Cole. 1984. Systematics of *Bangia* (Rhodophyta) in North America. I. Biogeographic trends in morphology. *Phycologia* 23, 383–96.

Sicko-Goad, L., and E. F. Stoermer. 1984. The need for uniform terminology concerning phytoplankton cell size fractions and examples of picoplankton from the Laurentian Great Lakes. *J. Great Lakes Res.* 10, 90–3.

Stevenson, R. J., and E. F. Stoermer. 1981. Quantitative differences between benthic algal communities along a depth gradient in Lake Michigan. *J. Phycol.* 17, 29–36.

Stoermer, E. F. 1968. Nearshore phytoplankton populations in the Grand Haven, Michigan, vicinity during thermal bar conditions. In *Proceedings of the Eleventh Conference on Great Lakes Research*, pp. 137–50. Internat. Assoc. Great Lakes Res. Ann Arbor, MI.

Stoermer, E. F. 1975. Comparison of benthic diatom communities in Lake Michigan and Lake Superior. *Verh. Internat. Verein. Limnol.* 19, 932–8.

Stoermer, E. F. 1978. Phytoplankton as indicators of water quality in the Laurentian Great Lakes. *Trans. Amer. Microsc. Soc.* 97, 2–16.

Stoermer, E. F. 1980. *Characteristics of Benthic Algal Communities in the Upper Great Lakes.* U.S. Environmental Protection Agency, Ecological Research Series EPA-600/3-80-073. USEPA Environmental Research laboratory, Duluth, MN. 72 pp.

Stoermer, E. F. 1981. Diatoms associated with bryophyte communities growing at great depth in Lake Michigan. *Proc. Iowa Acad. Sci.* 88, 91–5.

Stoermer, E. F. 1984. *Limnological Characteristics of Northern Lake Michigan, 1976.* Pt. 2, *Phytoplankton Population Studies.* University of Michigan, Great Lakes Res. Div. Spec. Rep. no. 95., pp. 127–245.

Stoermer, E. F., M. M. Bowman, J. C. Kingston, and A. L. Schaedel. 1974. *Phytoplankton Composition and Abundance in Lake Ontario during IFYGL.* University of Michigan, Great Lakes Res. Div. Spec. Rep. no. 53. 373 pp.

Stoermer, E. F., J. P. Kociolek, C. L. Schelske, and D. J. Conley. 1985. Siliceous microfossil succession in the recent history of Lake Superior. *Proc. Acad. Nat. Sci. Phila.* 137, 106–18.

Stoermer, E. F., and R. G. Kreis, Jr. 1978. A preliminary checklist of diatoms (Bacillariophyceae) from the Laurentian Great Lakes. *J. Great Lakes Res.* 4, 149–69.

Stoermer, E. F., and R. G. Kreis, Jr. 1980. *Phytoplankton Composition and Abundance in Southern Lake Huron.* University of Michigan, Great Lakes Res. Div. Spec. Rep. no. 65. 382 pp.

Stoermer, E. F., and T. B. Ladewski. 1982. Quantitative analysis of shape variation in type and modern populations of *Gomphoneis herculeana. Nova Hedw., Beih.* 73, 347–86.

Stoermer, E. F., L. Sicko-Goad, and L. C. Frey. 1983. *Effects of Phosphorus Loading on Phytoplankton Distribution and Certain Aspects of Cytology in Saginaw Bay, Lake Huron.* Final Report to Office of Research and Development, U.S. Environmental Protection Agency, Duluth, MN. 163 pp.

Stoermer, E. F., J. A. Wolin, C. L. Schelske, and D. J. Conley. 1985a. An assessment of changes during the recent history of Lake Ontario based on siliceous microfossils preserved in the sediments. *J. Phycol.* 21, 57–76.

Stoermer, E. F., J. A. Wolin, C. L. Schelske, and D. J. Conley. 1985b. Post settlement diatom succession in the Bay of Quinte, Lake Ontario. *Can. J. Fish. Aquat. Sci.* 42, 754–67.

Stoermer, E. F., J. A. Wolin, C. L. Schelske, and D. J. Conley. 1985c. Variations in *Melosira islandica* valve morphology in Lake Ontario sediments related to eutrophication and silica depletion. *Limnol. Oceanogr.* 30, 416–20.

Stoermer, E. F., and J. J. Yang. 1969. *Plankton Diatom Assemblages in Lake Michigan.* University of Michigan, Great Lakes Res. Div. Spec. Rep. no. 47. 168 pp.

Strong, A. E., H. G. Stumpf, J. L. Hart, and J. A. Pritchard. 1974. Extensive summer upwelling on Lake Michigan during 1973 observed by NOAA-2 and ERTS-1 Satellites. In *Proceedings of the Ninth International Symposium on Remote*

Sensing of the Environment, pp. 923–32. Environmental Research Institute of Michigan, Ann Arbor.

Taft, C. E., and W. J. Kishler. 1973. *Cladophora as Related to Pollution and Eutrophication in Western Lake Erie.* Water Resources Center, Ohio State University, Columbus. Published in cooperation with the Center for Lake Erie Area Research. 103 pp.

Theriot, E. C., and E. F. Stoermer. 1984. Principal components analysis of character variation in *Stephanodiscus niagarae* Ehrenb.: morphological variation related to lake trophic status. In *Proceedings of the Seventh International Diatom Symposium,* pp. 97–111. Otto Koeltz, Koenigstein.

Thomas, B. W., and H. H. Chase. 1887. Diatomaceae of Lake Michigan as collected during the last 16 years from the water supply of the city of Chicago. *Notarisia* 2, 328–30.

Vanderhoef, L. N., B. Dana, D. Enerich, and R. H. Burris. 1972. Acetylene reduction in relation to levels of phosphate and fixed nitrogen in Green Bay. *New Phytol.* 71, 1097–1105.

Vanderhoef, L. N., C.-Y. Huang, R. Musil, and J. Williams. 1974. Nitrogen fixation (acetylene reduction) by phytoplankton in Green Bay, Lake Michigan, in relation to nutrient concentrations. *Limnol. Oceanogr.* 19, 119–25.

Vaughn, J. C. 1961. Coagulation difficulties of the south district filtration plant. *Pure Water* 13, 45–9.

Verduin, J. 1952. Photosynthesis and growth rates of two diatom communities in western Lake Erie. *Ecology* 33, 163–8.

Verduin, J. 1954. Phytoplankton and turbidity in western Lake Erie. *Ecology* 35, 550–61.

Vollenweider, R. A. 1968. *Scientific Fundamentals of the Eutrophication of Lakes and Flowing Waters, with Particular Reference to Nitrogen and Phosphorus as Factors in Eutrophication.* OECD Rep. DAS/CSC/68.27, Paris. 182 pp.

Vollenweider, R. A., M. Munawar, and P. Stadelmann. 1974. A comparative review of phytoplankton and primary production in the Laurentian Great Lakes. *J. Fish. Res. Bd. Can.* 31, 739–62.

Vorce, C. M. 1882. Microscopic forms observed in the waters of the Lake Erie. *Trans. Amer. Microsc. Soc.* 5, 187–96.

Wujek, D. E., and M. Graebner. 1980. A new freshwater species of *Chaetoceros* from the Great Lakes region. *J. Great Lakes Res.* 6, 260–2.

Wujek, D. E., and M. L. Welling. 1981. The occurrence of two centric diatoms new to the Great Lakes. *J. Great Lakes Res.* 7, 55–6.

ALGAE FOR FOOD AND FOOD SUPPLEMENTS

4. *Porphyra* as food: cultivation and economics

THOMAS F. MUMFORD, JR.

THOMAS F. MUMFORD, JR.

Division of Aquatic Lands, Washington Department of Natural Resources, Olympia, WA 98504

AKIO MIURA

Phycology Laboratory, Tokyo University of Fisheries, 4-5-7 Konan, Minato-ku, Tokyo 108, Japan

CONTENTS

[87]

I. Introduction

The human relationship with *Porphyra* (Rhodophyta) is perhaps closer than with any other alga. Most cultures with access to *Porphyra* use it as a food or dietary supplement. However, the use and cultivation of *Porphyra* have reached a pinnacle in the Orient. The cultivation of *Porphyra* in Japan during the 1986 production year yielded ¥ 10.35 trillion (about US $450 million) of value to the farmer and employed over 20,000 people at the farming level (*Nori Times* 1986). The retail value is about $1 billion. This makes the *Porphyra* industry in Japan, which produces only about 60% of the world's total, the world's highest-valued nearshore fishery.

Woessner (1981) described a model for domestication of algae consisting of three preconditions and four stages. The preconditions are (1) availability, (2) utilizability, and (3) discovery. The four stages of domestication are (1) collection from natural populations, (2) resource management, (3) husbandry, and (4) full cultivation. *Porphyra* is probably the most domesticated of the marine algae as it is available in most temperate marine intertidal zones, is easily utilized, and is well into the fourth stage of domestication. Driven by the economic impetus of domestication, the biology of the genus *Porphyra* is becoming the best understood of any seaweed.

We will focus on the uses of *Porphyra* by humans and the economic and political arena in which the plant is grown. Present trends and important new discoveries will be emphasized in our discussion.

II. Food uses

The use of *Porphyra* as a food is universal in areas where the genus is found. Through centuries of trial and error, food gatherers discovered that *Porphyra* not only tastes good and is nutritious but enhances the flavor of other foods. In some cultures it was used incidentally or only in times of need; in others its use reached a high degree of sophistication. For example, criteria for quality and use in China and Japan parallel those applied to wines or cheeses in Western cultures.

A. China, Japan, Korea, and the Philippines

In China the use of *Porphyra* ("zicai") dates to at least A.D. 533–44 (Tseng 1981), and it continues to be highly valued in soups and as flavoring in many dishes.

In Japan, the traditional diet consisted largely of rice, fish, and *nori* – the Japanese word for seaweeds in general. *Amanori* is specifically those foods made from certain *Porphyra* species. The selection of amanori in the diet was not accidental. It contains essential amino acids, vitamins, and minerals (see Section III). Although this diet has been partially abandoned in recent years for a more diverse Western-style diet, it provided healthy foods free of many of the sugars and fats that are implicated in health problems with other diets. Arasaki and Arasaki (1983) give a delightful and detailed account of the historical use of seaweeds in Japan.

In Korea processed *Porphyra* is known as *kim* or *laver*. It was first cultivated in 1623–49 at Tae-in Island, Chun-ra-nam-do, where fishermen, having found floating bamboo twigs on which *Porphyra* grew, initiated their own culture by planting bamboo twigs in the sea (Kang and Ko 1977). Production has increased greatly during the last 10 years (Fig. 4-1B) with most of the production being consumed domestically.

A similar technique using vertical poles to culture *Porphyra* has been employed in northern Luzon in the Philippines. The alga is used as a foodstuff; no production figures are available (Michanek 1975).

B. West Coast of North America, Hawaii, and New Zealand

Porphyra perforata was collected and eaten by most of the coastal Northwest Amerindians from Washington to southeastern Alaska (Turner and Bell 1971, 1973; Turner 1973; Williams 1979; Roland and Coon 1984). It was dried and stored for later use or, sometimes, chewed and then fermented in boxes and partially dried for later use. The Haida Indians of the Queen Charlotte Islands still collect *Porphyra perforata,* dry it, and trade it with tribes in the interior of British Columbia. Today *Porphyra* is collected by people of Oriental background along most of the coast from California to British Columbia. They have found *P. perforata, P. abbottae, P. pseudolanceolata,* and *P. torta* to be good substitutes for the *Porphyra* they ate in their or their parents' native lands.

For Hawaiians of Polynesian ancestry, an undescribed *Porphyra* is one of the traditional delicacies that in the past was often reserved for royalty (Abbott and Williamson 1974). With the immigration of many oriental people to Hawaii, *Porphyra* use in sushi and soups became widespread.

The Maoris of New Zealand also traditionally used *P. columbina* as a food, calling it *karengo* (Chapman and Chapman 1980). Today it is

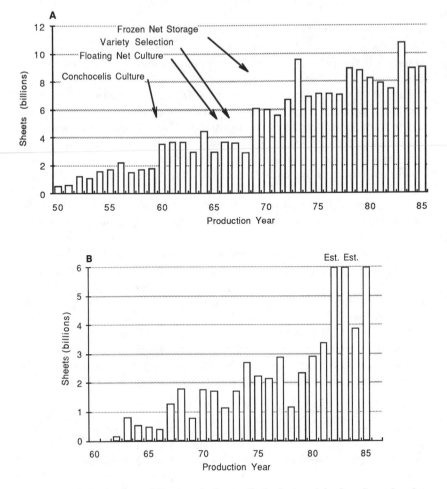

Fig. 4-1. (A) Production of *Porphyra* in Japan. Each sheet weighs 3 g. Several major advances in technology have resulted in dramatic increases in production. (B) Production of *Porphyra* in Korea. Each sheet weighs approximately 2–2.5 g.

served after boiling to near disintegration, with the resulting mush fried in butter.

C. British Isles

Porphyra laciniata traditionally has been collected and used in a variety of ways (e.g., "laver bread") in the British Isles, especially those areas with strong Celtic roots. It is usually fried with butter. In Ireland, it is collected and made into a jelly by stewing or boiling, then used later (Chapman and Chapman 1980).

III. Food value

The food value of *Porphyra* comes not so much from the bulk of nutrients it provides but rather from the essential vitamins and minerals it contains. Its high protein content notwithstanding, people seldom consume enough in a day to provide a meaningful amount of protein. In Japan, the annual per capita consumption of nori was 60 sheets (180 g) in 1967 (Miura 1975), which is equivalent to 0.15 g protein per day or 0.2% of a human's daily requirement.

The quality and value of *Porphyra*, especially in Japan, depends on a variety of visual and organoleptic factors. Interestingly, *Porphyra* considered to be of the highest quality also is of the highest nutritional value. In Japan the following are considered to be desirable characteristics of nori (Noda and Iwada 1978):

1. A black color tinged with green accompanied by good luster. The black color comes from the total absorption of light by the chlorophyll, carotenoid, and phycobilin pigments. Luster is a surface phenomenon and depends upon rapid growth of the *Porphyra*.

2. Superior flavor. The best-tasting nori contains high levels of the 5′ -nucleic acids guanylic acid (0.10 mg·g^{-1}) and inosinic acid (0.8 mg·g^{-1}) and the flavor-inducing amino acids taurine (4.8 mg·g^{-1}), asparagine (2.5 mg·g^{-1}), glutamic acid (10 mg·g^{-1}), and alanine (18 mg·g^{-1}). There are also low levels of the free sugars floridoside, iso-floridoside, glucose, galactose, and the polysaccharides mannose, arabinose, xylose, and ribose. A variety of aromatic organic compounds give the characteristic odor of nori, which also contributes to its taste. In superior nori, the balance among all these compounds is influenced by culture region, nutrients, weather conditions, culture methods, growing season, and other rearing conditions.

3. Softness when placed in the mouth. This quality depends on the area in which *Porphyra* is grown. Moderate amounts of freshwater influence, siltation, and calm conditions make a more tender nori. The season and sequence of harvesting also are critical. Early season and first harvests from nets are most desirable.

4. Sweetness. This characteristic comes from a balance of the flavor components mentioned above, especially the free sugar derivative iso-floridoside.

5. Persistence of the green color which develops during toasting. Chlorophyll pigments are more stable than the phycobilins and carotenoids when heated, resulting in a green-colored product called *yaki-nori* in Japanese. Plants from open ocean or pole culture seem to make a superior product in this regard. Low moisture or oxygen-free storage is essential in maintaining good quality (Oohusa 1984).

Table 4-1. *Composition of Porphyra tenera (dry wt basis)*

	Water (%)	Crude protein (%)	Fat (%)	Carbohydrates (%)	Fiber (%)	Ash (%)
Low quality	13.4	29.0	0.6	39.1	7.0	10.9
Medium quality	11.1	34.2	0.7	40.5	4.8	8.7
High quality	11.4	35.6	0.7	39.6	4.7	8.0
Amino acids	2.4–4.0 mg·g^{-1}			Free sugars	0.5–2.5	mg·g^{-1}
Carotenoids	0.16–0.18 mg·g^{-1}			Phycocyanin	38–45	mg·g^{-1}
Chlorophyll	0.55–0.69 mg·g^{-1}			Phycoerythrin	30–40	mg·g^{-1}

Source: Arasaki and Arasaki 1983.

Table 4-2. *Amino acid nitrogen distribution in proteins of Porphyra tenera*

Amino acid	Amino acid-N/ protein-N (%)	Amino acid	Amino acid-N/ protein-N (%)
Alanine	7.4	Methionine	1.7
Arginine	16.4	NH$_3$	5.1
Aspartic acid	7.0	Phenylalanine	3.9
Cystine	0.3	Proline	6.4
Glutamic acid	7.2	Serine	2.9
Glycine	7.2	Threonine	4.0
Histidine	1.4	Tryptophan	1.3
Isoleucine	4.0	Tyrosine	2.4
Leucine	8.7	Valine	6.4
Lysine	4.5		

Source: Fujiwara-Arasaki et al. 1984.

The major constituents of *Porphyra* are shown in Table 4-1. Fresh weight:commercial dry weight ratio is usually 7:1 (83–86% water). Commercially "dry" *Porphyra* contains 11.1–13.4% moisture but is best kept below 8% (Oohusa 1984).

The total protein content of commercially grown *Porphyra* species ranges from 30% to 50%. Generally, the higher grades contain greater amounts of protein. The amino acids found in the proteins are given in Table 4-2. The measure of value of essential amino acids is E:T (sum of essential amino acids, in mg·g^{-1} protein-N). Arasaki and Arasaki (1983) found an E:T value of 3,206 for *Porphyra*. This value is about the same as that for whole eggs and is considerably more than the base value of 2,250 recommended by the Food and Agriculture Organization/World Health Organization of the United Nations (FAO/WHO). The degree

Table 4-3. *Vitamin content of Porphyra tenera*

								Folic
	A (IU)	B_1 (mg)	B_2 (mg)	Niacin (mg)	C (mg)	B_6 (mg)	B_{12} (g)	acid (g)
High quality	44,500	0.25	1.24	10.0	20	—	—	—
Medium quality	38,400	0.21	1.00	3.0	20	1.04	13.29	8.8
Low quality	20,400	0.12	0.89	2.6	20	—	—	—

Source: Arasaki and Arasaki 1983.
Note: Dash means data not available.

of digestibility of *Porphyra* protein is between 72% and 78% (Matsuki 1960; Fujiwara-Arasaki, Mino, and Kuroda 1984).

The vitamin content of *Porphyra* (Table 4-3) is extraordinarily high and is of major nutritional significance for human consumption. O-ohusa (1984) showed that, unless great care is taken to keep moisture content low and exclude oxygen from the dried sheets of *Porphyra,* vitamin C content falls drastically in three months.

During processing into sheets, most salt is washed from the *Porphyra,* resulting in a low sodium content of usually 1%, but up to 3–4% wet weight. Other major minerals include (in $mg \cdot g^{-1}$ dry wt) calcium, 4.7; iron, 0.23; and iodine, 0.005 (Arasaki and Arasaki 1983).

IV. Other uses or potential uses

A. Phycobiliprotein source

Porphyra contains proteinaceous water-soluble pigments called phycobilins. At least three types are found in the genus (Ragan 1981): R-phycoerythrin, C-phycocyanin, and allophycocyanin. R-Phycoerythrin is valuable as a fluorophore used on immunofluorescent probes for and analogues of biological macromolecules (Glazer, Oi, and Stryer 1985; Haughland 1985). *Porphyra yezoensis* phycoerythrin is particularly valuable because of its stability and high fluorescence (see also Nonomura, Chapter 21). Current prices for purified phycobiliproteins range from US \$15 to $50 \cdot mg^{-1}$ (G. Cyzewski, personal communication). Phycoerythrin may constitute up to 8% of the dry weight (1.14% wet) of the plant; however, purification costs and a small market limit the production of this valuable pigment.

Table 4-4. *Yield of Porphyra and Macrocystis*

Yield (sheets·harvest^{-1}·saku^{-1})	Production time (mo·y^{-1})	Yield (dafmt^{d}·ha^{-1}·y^{-1})
Porphyra[a]		
300 (Japanese av.)	5.5 (Japanese av.)	4.3
1,500 (Washington max.)	5.5	21.4
7,000 (Korean max.)	5.5	99.8
300	11.0 (Washington)	8.5
1,500	11.0	42.7
7,000	11.0	199.6
(Commercial yields)[b]		32.0
(Commercial yields)[b]		18.2
Macrocystis[c]		
Short term (potential yields)		100–50
Long term (realistic)		25

[a]Assumptions: saku = area of standard Japanese net (18 m × 1.5 m) = 27.0 m^2; 3.0 gdw/sheet; 8% ash; 2.76 dafg (dry ash-free grams)/sheet. Growth cycle is 14-day growth, first harvest, 10-day regrowth, second harvest, and replace net; average 12 days per harvest; possible 30.42 harvests per year.
[b]From Electric Power Res. Inst. 1979.   [c]From Bird, in press.
[d]Dry ash free metric tons.

B. Biomass

The yields from commercial *Porphyra* cultivation are in the middle to high end of the ranges of those found for most algal biomass production systems (Table 4-4). This is proven commercial production, however, and with the exception of *Laminaria* cultivation, no other plant system is more productive. It should be noted that production in these systems is maximized not for biomass but for a high-quality, edible product. On the other hand, the value of *Porphyra* as a biomass source would be only a fraction of its value as food or for other products.

C. Phycocolloids

The cell walls of the blade (gametophyte) phase of *Porphyra* contain the nongelling polysaccharide porphyran, which consists of several chains of alternating 1,3-linked beta-D-galactose and 1,4-linked alpha-L-galactose units (Rees and Conway 1962; Morrice et al. 1984). Alkali pretreatment changes the sulfated portion into anhydrogalactose, producing a low-quality agar suitable for food use. Commercial edible species of *Porphyra* were selected for thin cell walls and hence contain little porphyran. By using low-grade surplus nori or by selecting for strains with thick cell walls, *Porphyra* could become a major source of agar (Yaphe 1984).

D. Medical

A diet containing *P. tenera* was reported to lower the incidence of intestinal cancer in rats and mammary cancer in mice (Yamamoto and Maruyama 1984, 1985). These effects seemed to be correlated to the sulfate ester content of the wall polysaccharides. Sakagami et al. (1982) showed *Porphyra* to be effective against stomach ulcers.

With the rise in popularity of diets that reduce the risk of cancer and the recent discovery that beta-carotene is effective in lowering risk of a wide variety of cancers, the use of *Porphyra* could become more widespread. One sheet of high-grade nori contains 27% of the U.S. recommended daily allowance of vitamin A as beta-carotene.

Abe and Kaneda (1972) found that a diet of *Porphyra* significantly lowered blood cholesterol levels in rats. The active substance was found to be in the betain group, especially beta-homobetain.

V. Development of cultivation methods

A. History

As with most crop plants, *Porphyra* cultivation began with the casual collection of wild specimens. Methods for controlling the amount, predictability, and quality of plants gradually evolved through time (Woessner 1981; Arasaki and Arasaki 1983). Chinese records dating from A.D. 533 to 544 (Tseng 1981) show that this highly valued wild plant was collected from rocks. About 200 years ago, the first steps of cultivation began by "rock clearing" – improving the substrate for the settlement of spores by removing competing organisms such as seaweeds and barnacles. This procedure was used until the twentieth century when methods using nets became common.

In Japan, the earliest use of *Porphyra* probably dates from the period that its use became common in China. *Porphyra* was first cultured commercially in Tokyo Bay in the 1650s. The method used was to place twigs or branches of bamboo at the correct tidal height at the critical time in the fall when the spores were present, thus mimicking what probably had been noticed to be the timing of the natural occurrence of the plants. The plants grew on the branches in the fall and early winter and were picked by hand and chopped with knives into a slurry. The slurry was poured on bamboo mats, which were dried in the winter sun. The technique was labor-intensive, and the season was short (Okasaki 1971). The format of the product in Japan, an 18 × 21 cm sheet weighing 3 g, was set during these early years and has not changed since. A crew of eight people could make 1,500 sheets per day using this method in the 1960s (Inayoshi, personal communication); today, machines can make that many sheets or more in one hour with one or two operators. The practice of cultivation spread only slowly from Tokyo (Arasaki and Arasaki 1983).

During the first two decades of the twentieth century, methods were devised to replace the branches with more effective and productive mats or nets. The mats were initially made from thin bamboo shoots, which were replaced by netting made from a variety of natural fibers. During the 1960s synthetic fibers were first used, and today netting is made from a mixture of polypropylene fibers for strength and Vinylon or Cremona fibers, a hydrophilic fiber made from polymerized polyvinyl alcohol to which spores attach readily. In Japan, the netting is a standard size and mesh (18 × 1.5 m net with 30 cm stretched mesh).

The sites originally valued for *Porphyra* culture were close to river mouths where some freshwater influence and high nutrient levels were found. The method of placing brush at the right height intertidally required large areas, usually shallow muddy or sandy bays. When mats and nets began to be used, they were suspended from bamboo poles driven into the muddy bottom, thus allowing somewhat deeper areas to be utilized. *P. tenera* was the species favored for this type of culture. The technique mimics the conditions under which the plants are found naturally in the high intertidal zone. Periods of both drying and freezing occur naturally, and the heights of the nets are adjusted carefully for the correct amount of exposure to kill competing organisms (see Miura 1975, p. 286, for a diagram). The pole method of cultivation is still widely used in addition to the floating method described below. The pole method makes the best-tasting and softest nori although production is inconsistent, depending on the rainfall/salinity regime, and the color may not be sufficiently dark.

The technique of culturing the plants on netting placed horizontally on floating rafts was introduced in the late 1960s in Japan. It was discovered that blades, upon reaching a size of 2–3 cm, no longer required emersion, and if fouling organisms could be controlled, this method produced a dark but tougher product. Its main advantage was that any deepwater area could be utilized for production. Production greatly expanded whereas before it had been limited because many of the shallow areas previously preferred for pole production had become polluted, filled for upland uses, or used for other purposes.

B. Discovery and utilization of the life cycle

As in all aquaculture methods, the impetus is to gain more and more control over the reproductive biology of the organism. It is remarkable that so much *Porphyra* was produced before its reproductive biology was elucidated. Until the early 1950s, although cultivation details were well worked out, there was no idea of where the spores came from. Hence, the seeding process could not be controlled. Nets were seeded in the few areas known to produce blades and then transferred to other growing sites.

Drew (1949) discovered that spores from a blade of *P. umbilicalis* germinated and grew into a piece of shell that she had fortuitously placed in the culture dish. She recognized the resulting filamentous plant as *Conchocelis rosea.* This discovery was the catalyst needed in the already heated research climate of Japan and China where a solution to this cultivation problem was desperately sought. Although others had noted the "abnormal" germination of carpospores, none had recognized that a vast difference exists between the alternate generations. Indeed, the difference is so great that the algae were thought to belong to different subclasses, the Bangiophycidae and Florideophycidae (Bold and Wynne 1985).

Within a few years the life cycle was completed for *P. tenera* (Kurogi 1953a, b). Kurogi (1953a), however, noted that Migita (1951) perhaps was the first to complete the life cycle. It was not until 1960 that the commercial use of the conchocelis phase was sufficiently widespread to influence overall production. This advance in techniques permitted a farmer to seed nets at his own site when he wanted and, most importantly, to seed with a known genetic stock.

In the early 1970s the practice of strain selection to improve production and quality became common in Japan. The traditional agronomic method of selecting plants for desired traits was practiced, and, as shown by Miura (1979), such traits as blade length could be selected for quickly and effectively. At least 32 varieties of *P. tenera* and *P. yezoensis* have been named, and hundreds more have been selected by individuals and cooperatives for use in particular areas of Japan (Inayoshi, personal communication). Ten varieties were declared in the public domain in 1980. Since then, new varieties are protected by law, and users must pay royalties to the originator. The selection of cultivars was one of the main reasons for improvement of production. In new areas being developed for nori culture such as in Korea and Washington State, it is essential that suitable strains for each site be developed (Mumford et al. 1985).

The Japanese government recently expressed concern (Kito 1977 and personal communication) that the use of highly selected strains may lead to genetic uniformity and hence to great potential for crop failure through either disease or unfavorable conditions affecting all plants similarly. This has been countered by the practice of seeding nets with several strains, perhaps one selected for a dry, warm fall and another for a wet, cold fall. This "bet hedging" does not seem to cause any lowering of production (Inayoshi, personal communication). Kito (personal communication) and Waaland (personal communication) are establishing "gene banks" of conchocelis cultures to maintain the genetic diversity, which is being threatened by widespread use of only a few cultivars.

The matter of sexual reproduction in *Porphyra* has been an interesting enigma. Early workers (see Hawkes 1978 for extensive review) found a variety of structures and cytological phenomena purported to be evidence of sexual reproduction. Many of these involved fungal parasite structures. However, the Japanese and Chinese workers proceeded with their breeding work, naturally assuming (correctly) that sexual reproduction was the norm. Western scientists, either being unaware of earlier work or doubting its validity, balked at the idea of sexuality in *Porphyra* until recently (Hawkes 1978). From the work of Miura using color mutants (Miura and Kunifuji 1980; Miura 1985), it is now evident that sexual reproduction occurs in *Porphyra yezoensis* and inheritance follows Mendelian genetics. The impact of these major breakthroughs on nori production is shown in Figure 4-1A.

The details of the biological life history of the commercial species of *Porphyra* are shown in Figure 4-2. Though variations exist (Cole and Conway 1980), this is the life history that has been explored and exploited in the industry.

The complex series of biological events that must be carried out on an industrial scale are given in outline in Figure 4-3. The details for production are given in a large number of specialized manuals in either Japanese (Ueda 1973; Noda and Iwada 1978) or Chinese (IOEP 1976; IOESP 1978), although recent efforts in the United States have produced two manuals in English (Byce et al. 1984; Melvin, Mumford, and Byce 1986). A large number of review articles in English focus mainly on the biological aspects of *Porphyra* (Imai 1971; Okasaki 1971; Bardach, Ryther, and McLarney 1972; Suto 1974; Wildman 1974; Miura 1975; Umebayashi 1975; Korringa 1976; Kito 1977; Chapman and Chapman 1980; Miura 1980; Hansen, Packard, and Doyle 1981; Tseng 1981; Waaland 1981; Kramer, Chin, and Mayo 1982; Kafuku and Ikenoue 1983; Oohusa 1984; Waaland et al. 1986). Japanese literature is voluminous and is reviewed in Imai (1971) and Miura (1975).

C. Current methods

The modern practice in the cultivation of *P. tenera* and *P. yezoensis* commences with carpospores (Fig. 4-2), which are used to produce a stock of free-living conchocelis phase (diploid sporophyte) cultures. These are maintained in vitro by vegetative fragmentation. Either diploid carpospores from the blade or vegetative fragments from stock cultures of free-living conchocelis phase are used to start the large-scale culture of the conchocelis phase growing in a shell substrate. Inoculation of shells

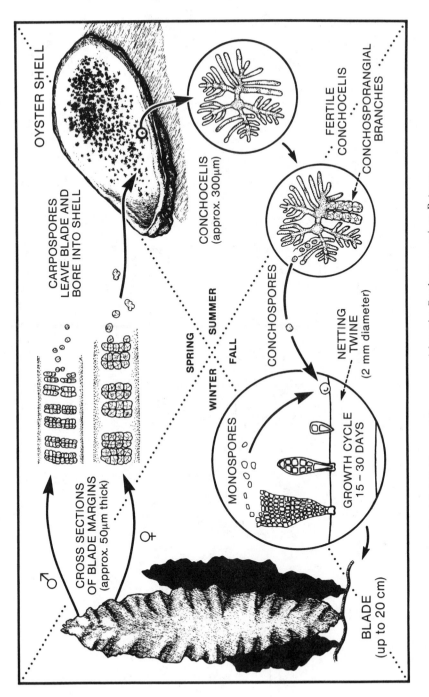

Fig. 4-2. Life history of the commercial species *Porphyra yezoensis* or *P. tenera*.

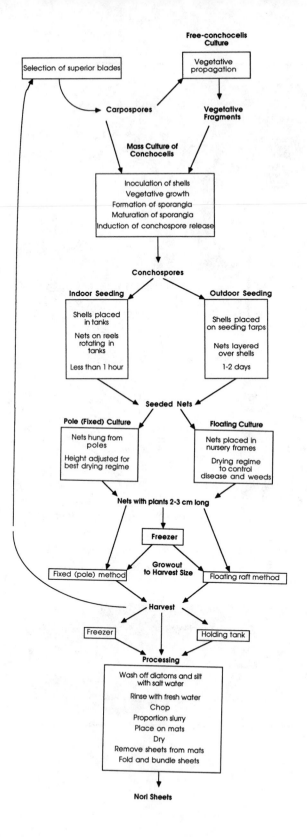

Free-conchocelis Culture

Selection of superior blades

Vegetative propagation

Carpospores

Vegetative Fragments

Mass Culture of Conchocelis

Inoculation of shells
Vegetative growth
Formation of sporangia
Maturation of sporangia
Induction of conchospore release

Conchospores

Indoor Seeding

Shells placed in tanks

Nets on reels rotating in tanks

Less than 1 hour

Outdoor Seeding

Shells placed on seeding tarps

Nets layered over shells

1-2 days

Seeded Nets

Pole (Fixed) Culture

Nets hung from poles

Height adjusted for best drying regime

Floating Culture

Nets placed in nursery frames

Drying regime to control disease and weeds

Nets with plants 2-3 cm long

Freezer

Growout to Harvest Size

Fixed (pole) method

Floating raft method

Harvest

Freezer

Holding tank

Processing

Wash off diatoms and silt with salt water

Rinse with fresh water
Chop
Proportion slurry
Place on mats
Dry
Remove sheets from mats
Fold and bundle sheets

Nori Sheets

Fig. 4-4. Tray culture of the conchocelis phase of *Porphyra*. Oyster shells inoculated with vegetative fragments are hung in pairs from rods. The cultures are grown for a 3–6-month period under controlled conditions for light, temperature, salinity, and nutrients. (Washington; photo by T. F. Mumford.)

usually takes place in the spring. The plants grow just beneath the shell surface and eventually cover the entire inside half of the shell. The use of shell substrate makes handling the cultures easy. The shells may be scrubbed of unwanted epiphytes, leaving the endolithic conchocelis phase untouched; hence, strictly unialgal cultures are not necessary. The arrangements of shells may be shinglelike in trays of various sizes or strung together and suspended in tanks (Fig. 4-4).

During the summer months temperatures reach 26–30°C in the conchocelis cultures. Light levels are reduced, and nutrient levels shifted to a higher P:N ratio (3:1) to induce the formation and maturation of sporangia. About five days before the desired time of conchospore release, several changes are made to induce mass release. Sporangium maturation is initiated by dropping the temperature 5°–10°C and raising light levels 18–27 $\mu E \cdot m^{-2} \cdot s^{-1}$, and/or by shortening daylength (to 8L:16D). After four to six days under these conditions massive simultaneous spore release can be achieved by exposure to full sun. The conchospores are released from the sporangia only during the first hours of daylight.

Fig. 4-3. Flow diagram for the production of hoshi-nori (sheets of *Porphyra*) as practiced in Japan.

Fig. 4-5. Indoor seeding of netting with *Porphyra* conchospores. A spore suspension is made in deep tanks, and the reels with netting are rotated through the suspension for a few minutes until a sufficient number of spores is attached. (Japan; photo by J. Merrill.)

Two general methods of getting the conchospores to attach to the netting are used. The indoor method involves placing the shells in deep tanks; when the spores release, a spore suspension is created. Nets are wound on large reels, and these are partially immersed in the tanks and rotated until sufficient spores adhere to the netting twine (Fig. 4-5). The second general method is outdoor seeding, which is done by placing the shells in a floating horizontal raft and spreading a large number of nets over the shells. When released, the spores float and adhere to the netting (Fig. 4-6). A successful seeding requires one to five days, after which the nets are moved to the field for nursery cultivation.

If the plants are to be grown in the traditional pole (Fig. 4-7) or semi-floating culture (Tseng 1981, figs. 20, 21, p. 715), the nets are lashed horizontally at a carefully chosen tidal height to achieve an optimum tidal exposure regime (Miura 1975). This regime is critical; too much drying damages the plants, and too little allows the growth of other algae. If the more recently developed type of raft culture is followed, the plants are placed in structures called *fujo ikada* or nursery frames (Fig. 4-8). These floating structures allow nets to be dried in a controlled regime. The nets can be raised and lowered at will.

When the plants on the nets have reached 2–3 cm in length, the nets are dried, rolled up, and frozen. Although *Porphyra* may be frozen at any size from spores on up, this is the optimal size. The unique ability to withstand freezing is not surprising, considering that the plants nor-

Fig. 4-6. Outdoor seeding of netting with *Porphyra* conchospores. The conchocelis-phase-bearing shells are placed on a semifloating tarp. Up to 50 nets are spread over the shells; the spores float up and attach to the netting. (Washington; photo by T. F. Mumford.)

mally are found living high in the temperate intertidal zone in the winter and must survive periodic freezing. Before the plants are frozen, they are "hardened" for the previous week by drying them more and more each day, and then drying them well before placing them at −20°C where they may be held for six months or more. This technique was first used commercially in Japan about 1965. It allows the farmer to freeze many more nets than he has growing areas and permits flexibility in disease or fouling control, or if environmental conditions become temporarily unsuitable.

Once the plants have reached a size of 2–3 cm they no longer need to be dried and may be placed in floating rafts of various configurations. There they grow to a size of 15–20 cm in a few weeks and are harvested.

Fig. 4-7. Traditional fixed-pole culture of *Porphyra*. Netting is tied between poles driven into the bottom of shallow bays. (Matsushima Bay, Japan; photo by T. F. Mumford.)

After the first net is grown to maturity and several harvests taken from it, another net is brought from the freezer and placed in the field. This process may be repeated three to four times until the growing season ends. Prior to the introduction of this technique only one net was grown and harvested, and the production season was shorter.

Harvesting, once a highly labor-intensive process done by hand or by hand-held rotating cutters on wands, now is more automated. In Japan,

Fig. 4-8. Nursery frames for *Porphyra* germling culture. Seeded nets are placed on these floating structures. The nets may be raised (as shown) and dried to control fouling and disease. (Washington; photo by T. F. Mumford.)

the usual harvester consists of a rotating reel with blades. Typically it is mounted in a boat that is run under the nets, which are pulled over the harvester sideways by two people (Figs. 4-9, 4-10). The cutter then cuts or tears the hanging blades from the net. Recently, several companies tested a boat-mounted harvester in which a single operator sits under a cage. The boat is then driven under the nets and harvesting takes place quickly. Another type recently introduced consists of a harvester mounted several feet under water in front of the boat. The harvester pushes the net under water, and the cut material is sucked up into the boat.

The harvested *Porphyra* is quickly brought ashore for processing, or it is stored in a saltwater holding tank or frozen until processed. Processing is an automated version of a centuries-old process that is akin to paper making. The plants are rinsed, chopped, made into a slurry, and poured out on mats, which are then run through a drier. The rate of drying is controlled by temperature and conveyor speed and is critical. The sheets are peeled from the mats and bundled. The entire process takes place in a large machine and is continuous (Fig. 4-11). Up to 4,500 sheets are produced per hour.

The traditional sheets in China and Japan are a standardized weight (3 g) and size (20–21 cm × 17–19 cm). Korean sheets tend to be somewhat lighter at 2–2.5 g. Circular sheets are produced in China.

Fig. 4-9. Floating culture method for *Porphyra*. (Japan; photo by A. Miura.)

Fig. 4-10. Harvesting from fixed-pole cultures of *Porphyra*. (Ariake Bay, Japan; photo by A. Miura.)

In Japan, the process to this point is usually carried out by a single farmer/family. The finished product (Japanese = *hoshi-nori*) is then transferred to a cooperative where the sheets are graded according to quality and sold at auction. The cooperative takes a percentage of the

Fig. 4-11. One-man processing machine for *Porphyra*. The machine takes a slurry of *Porphyra* and makes it into hoshi-nori. (Photo courtesy of Nichimo Co., Japan.)

sale price (10–15%) and returns the rest to the farmer. The sheets then enter the wholesale distribution and secondary processing sectors.

Secondary processing involves toasting the sheets to produce *yaki-nori* for the sushi trade, or toasting and seasoning the sheets, which are then often cut into smaller pieces and packaged into tins as *ajitske-nori* for the gift trade.

VI. Present and future trends

A. Production and trade

The principal nori-producing countries are Japan, Korea, and China. Production figures for Japan and Korea are given in Figure 4-1 (A, B). Total production from these three countries in 1984 was approximately 47,550 dry metric tons distributed as follows: Japan, 25,800 (@ 3 g per sheet); China, 12,000 (Fei, personal communication); Korea, 9,750 (@ 2.5 g per sheet). Chinese production was equivalent to 4 billion sheets.

Data showing the exports of *Porphyra* from Japan and Korea are shown in Tables 4-5 and 4-6. A large amount of *Porphyra* also is exported to Korean workers in Saudi Arabia, but data on exact quantities are not available.

Recent efforts by the Washington Department of Natural Resources

Table 4-5. *Porphyra products exported from Japan in 1984*

Country	Hoshi Nori		Yaki-nori + Ajisuke-Nori		Total value	%
	Quantity (1,000 × sheets)	Value (¥ × 1,000)	Quantity (1,000 × sheets)	Value (¥ × 1,000)	in $ ($ = ¥210)	
Taiwan	160,810	1,379,952	13,502	30,565	6,716,748	45
U.S.A.	25,961	326,255	101,661	955,207	6,102,200	41
Hong Kong	105	2,696	69,610	168,301	814,271	6
Singapore	195	3,932	22,068	69,739	350,814	2
Canada	591	10,294	3,080	24,471	165,548	1
Korea	1,610	19,548	1,505	1,057	98,119	1
Other	2,720	42,993	9,787	67,428	525,814	0
Total	191,992	1,785,670	221,213	1,316,768	14,773,514	

Table 4-6. *Porphyra products exported from Korea in 1984 (in US $)*

	Kim[a]	Tot[b] (processed)	Total
Saudi Arabia	Most, greater than U.S.		
United States	805,000	71,000	876,000
Taiwan	303,000	80,000	383,000
Japan	13,000	13,473,000	13,486,000
Total exports	US $3,373,000	$13,801,000	$17,174,000

Source: From Korean Trade Association, Los Angeles, 1985.
[a]"Kim" is the standard Korean sheet weighing 2.5 g each.
[b]"Tot" is both flavored and toasted *Porphyra*.

and the University of Washington have demonstrated the biological feasibility of growing *Porphyra* in Washington State (Mumford and Melvin 1983). Using Japanese and local species, researchers have grown *Porphyra* in commercially viable quality and quantities at several locations. One commercial company is in operation in 1988. An environmental impact statement has been prepared describing the impacts of nori farming and processing in Washington (Mumford and Hanson 1987).

B. Genetics, strain selection, and propagation

Recent scientific advances in genetics, physiology, and the biochemistry of food value are now being incorporated into the centuries of empirical knowledge gained by thousands of farmers. The results of this research are quickly being used in new production techniques.

With the advent of Miura and Kunifuji's work (1980) with color mutants (Miura 1985) and the discovery of the germinating conchospore as the site of meiosis (Ma and Miura 1984), great progress should be made in the near future in the genetics of *Porphyra*. At the present time, however, strain selection is the major area of progress. In the last 15 years strains have been selected for a long narrow shape, late maturation (Miura 1984), and monospore production. Narrow plants give a greater yield on netting. When plants become reproductive they stop rapid growth because margin erosion partially offsets growth. The secondary settlement of monospores on the netting can help overcome an otherwise insufficient initial seeding. In addition, nets can be seeded entirely with monospores, which lowers costs for conchocelis production per net. Monospores also grow and mature faster and may be used as a primary seed source (Li 1984).

The primary source of spores for seeding nets has been conchospores. This requires the culturing of the conchocelis phase on shells, a process that is relatively expensive. In addition, unwanted genetic diversity is introduced at the time of meiosis in the first four cell divisions of the germinating conchospore. Vegetative propagation would solve both of these problems. By eliminating the conchocelis phase, production costs would be lowered and genetic diversity eliminated.

Other complications include selecting for desirable traits only in the blade (gametophytic) phase. Desirable traits also occur in the conchocelis phase, but their linkage to blade characteristics is not clear. In addition, the conchocelis phase of hybrids may be infertile or of low viability (Kito, personal communication). By propagating vegetative cells from the blade in tissue culture, much greater control can be maintained over desirable plant genotypes. Protoplast production and fusion techniques (Polne-Fuller and Gibor 1984; Saga and Sakai 1984; Chen 1986; Fujita and Migata 1986) are now being studied actively in several laboratories.

Another technique is to produce the conchocelis in large amounts in free-living culture (Waaland et al. 1986; Kito, personal communication). Although this has been successful on a small scale, it does not seem economically feasible on a large scale for those lacking the specialized equipment.

Some of the new techniques will require, at least initially, specialized expertise and expensive equipment. Hence, farmers will have less control and higher costs because a middleman will be introduced in the process. At the present time, average cost of the conchospores per net is US $0.72 (Inayoshi, personal communication).

Disease remains one of the biggest threats to nori producers. Most of the diseases are fungal (Imai 1971; Andrews 1976) and are controlled by drying, freezing, and maintaining healthy plants by good culturing practices. In Japan, farmers have resisted using chemicals on this food crop. A green color mutant has been found to have high resistance to "red-rot," a *Pythium* sp. that attacks and quickly decimates *Porphyra*. This resistance may be due to the fact that strain selection has been for thin cell walls in *Porphyra*, and there seems to be an inverse relationship between susceptibility and cell wall thickness (Kito, personal communication). The green mutant has relatively thick cell walls. It is not clear if there is a relationship between the color and resistance.

A new disease has appeared in southern Japan. Called *suminori* or charcoal nori, it is caused by a bacterium. The plants appear normal, but after processing the sheets are gray and lusterless – a worthless product. No known control exists.

A potential problem with the use of vegetative cell propagation is that the fungal disease *Olpidiopsis* (white wasting disease) is not killed by freezing, and infected cells can be propagated. Very careful attention will have to be paid to assure disease-free stocks (Fujita, personal communication)

C. Production, processing, and quality control

The traditional and still typical Japanese nori farmer is largely self-sufficient. He grows his own conchocelis from plants selected from his own ground; he produces and then harvests and processes in his own compound. This is changing in Japan. A farmer may now buy conchocelis shells from a large firm that grows only conchocelis, or he may buy already seeded nets from another firm. He may sell or consign the raw *Porphyra* to a processor. In Korea, though almost all farms are independent and owner-operated, there are also a few large corporate-style farms. China's production is based on work cooperatives. Although benefits may arise from segmentation of the various production steps, it is not without its liabilities. For example, an owner-operator may have much more incentive to perform the small detailed steps required to produce a high-quality product, but this attitude may be lacking in hired help. Division of the steps also gives less personal control over the whole process.

Nori production in Japan increased steadily through the 1970s as new sites came into use, new production techniques became available, and strong markets existed. The record production of 10 billion sheets in 1974, however, greatly exceeded market demand for the product, and as a result the price declined markedly. Repeated overproduction led to voluntary production restraints. The problem seems to be an excess of low- and medium-quality nori because of the greater production from floating raft-style culture. Though this lower-quality product is often made into flavored nori or other products, or exported to less discriminating buyers, its presence in the marketplace exerts a downward pressure on all products.

The industry is moving toward planned production, which includes determining appropriate production levels from an examination of demand, improving the product line, cutting production of products not highly desired by consumers, and maintaining an appropriate price by increasing business efficiency. In Japan, great emphasis is now being placed on maximizing the utilization of the culture areas. Studies of the environmental parameters of the areas are aimed at lowering the quantity and raising the quality of the product.

The best-quality nori is produced in the Ariake Sea of southern Japan. The technique used there is the fixed-pole method, where periodic drying occurs throughout the growing season. Prices are high, and farmers tend to maintain control over production and processing. Demand for this high-quality nori is increasing. As mentioned above, when all other factors are equal, the floating method results in a somewhat lower-grade product overall. In the floating-raft sector of the Japanese industry, great emphasis is being placed on lowering the cost of production and making a uniform-quality product by mixing batches of nori during processing.

In the past 8–10 years, the appearance of a so-called one-man processing machine (Fig. 4-11) in Japan has led to an increase in productivity and higher quality control. It also led to high capital costs for the individual farmer. With overproduction forcing prices down, many farmers encountered financial problems, which led to the phenomenon of several farmers buying the expensive machinery together. This trend has also led in a few cases to the division of the processing steps among several different operating farms.

In efforts to increase quality, manuals were produced (Noda and Iwada 1978) pointing out the importance of quality control at every step of the long, complex process. Quality control is at the heart of every item discussed above. Oohusa (1984) published a series of papers on the chemical constituents affecting quality of nori and how these are influenced in processing and storage. Temperature, moisture content, and oxygen all degrade the finished products and must be carefully controlled.

VII. Impacts of nori cultivation in Japan

A. *Cooperatives and village structure*

Nori cultivation is performed by members of fisheries cooperatives found in every fishing village along the coast of Japan. The cooperatives are open only to fishermen who engage in the nori fishery more than 90 days a year. Most members are engaged in agriculture or other fisheries during the off season for nori.

In order to perform nori cultivation, a right of demarcated fishery is necessary. The national minister of agriculture, forestry, and fisheries has delegated authority to the prefectural governor to issue the right of a demarcated fishery. These rights are issued to cooperatives by the prefectural governor after consideration by the Sea-area Fishery Adjustment Commission. Competition between nori and other fisheries is resolved by this commission. The right of a demarcated fishery is not issued to any organization other than fisheries cooperatives; hence, nori farming companies do not exist in Japan. Terms of the rights are usually issued for five years, although in some instances one-year terms are authorized.

The cultivation areas are divided into several classes according to their productivity. Assignment of the cooperative members to certain areas is by lottery.

B. Impacts of environmental pollution

Oil spills and thermal effluent from power stations are two of the major pollutants affecting nori culture. By far the most serious problem is oil pollution. Oil spills damage the nori plants and contaminate nets, ropes, and floats. It is impossible to use fouled equipment again, and nori tainted with oil cannot be eaten.

To prevent damage in such areas as Tokyo Bay where the likelihood of pollution is high, continual airplane and boat surveillance is conducted during the growing season. When oil is discovered it is removed or the nori nets and gear are removed from the water.

The nori farmer is compensated for damage by the polluter if the latter can be identified. If no one can be identified, payment comes from the Foundation for Relief Fund for Fishery Damage by Pollution of the Sea by Oil. Money comes into this fund from the oil industry, the national government, the city of Tokyo, Hokkaido Prefecture, and all the districts of the Federation of Fisheries Cooperatives. The largest damage by oil was in excess of (¥ 800 million (approximately US $4 million) and occurred in Tokyo Bay near Chiba Prefecture.

In Japan the construction of nori cultivation facilities is not considered to spoil the beauty of the sea and does not compete with recreational uses. Japanese policy is that the existing sea areas should be used effectively for fisheries and that wealth should be acquired from the sea.

VIII. Acknowledgments

Funding to T. F. Mumford was provided by the Washington Department of Natural Resources, Project R/A-12, "Aquaculture of Seaweeds on Artificial Substrates," Grant no. NA81AA-00030 from the National Oceanic and Atmospheric Administration to the Washington Sea Grant Program, and by the Pacific Northwest Regional Commission, Grant no. 10090098, "Nori Cultivation Analysis and Planning."

T. F. Mumford wishes to thank Dr. Hitoshi Kito for many informative discussions about the nori industry in Japan, and Dr. James Craigie and Dr. J. E. Merrill for their critical reviews of the manuscript.

IX. References

Abbott, I. A., and E. H. Williamson. 1974. *Limu: An Ethnobotanical Study of Some Edible Hawaiian Seaweeds*. Pacific Tropical Botanical Gardens, Hawaii Publication. 20 pp.

Abe, S., and T. Kaneda. 1972. The effect of edible seaweeds on cholesterol metabolism in rats. *Proc. Int. Seaweed Symp.* 9, 562–5.

Andrews, J. H. 1976. The pathology of marine algae. *Biol. Rev.* 51, 211–53.

Arasaki, S., and T. Arasaki. 1983. *Low Calorie, High Nutrition Vegetable from the Sea to Help You Look and Feel Better.* Kodansha International Press, New York. 196 pp.

Bardach, J. F., J. H. Ryther, and W. O. McLarney. 1972. *Aquaculture: The Farming and Husbandry of Freshwater and Marine Organisms.* John Wiley, New York. 868 pp.

Bird, K. T. In press. *Macroalgal Cultivation for Food, Chemicals, and Energy.* Elsevier, Amsterdam.

Bold, H. C., and M. J. Wynne. 1985. *Introduction to the Algae: Structure and Reproduction.* 2nd ed. Prentice-Hall, Englewood Cliffs, N.J. 706 pp.

Byce, W. J., T. F. Mumford, Jr., M. Inayoshi, D. J. Melvin, and V. Bryant. 1984. *Equipment for Nori Farming in Washington State.* Vol. 1, *Outdoor Seeding.* Vol. 2, *Nursery Culture.* Washington Department of Natural Resources, Olympia. 154 pp.

Chapman, V. J., and D. J. Chapman. 1980. *Seaweeds and Their Uses.* 3rd ed. Chapman and Hall, London. 334 pp.

Chen, L. C. M. 1986. Cell development of *Porphyra miniata* (Rhodophyceae) under axenic culture. *Bot. Mar.* 29, 435–9.

Cole, K., and E. Conway. 1980. Studies in the Bangiaceae: reproductive modes. *Bot. Mar.* 23, 545–53.

Drew, K. B. 1949. *Conchocelis*-phase in the life history of *Porphyra umbilicalis* (L.) Kütz. *Nature* 164, 748.

Electric Power Research Institute. 1979. Comparative assessment of marine biomass materials. Technical Planning Study TPS-77-735.

Fujita, Y., and S. Migata. 1986. Isolation of protoplasts from leaves of red algae *Porphyra yezoensis. Jap. J. Phycol.* 34, 63.

Fujiwara-Arasaki, T., N. Mino, and M. Kuroda. 1984. The protein value in human nutrition of edible marine algae of Japan. *Hydrobiologia* 116/117, 513–16.

Glazer, A. N., V. T. Oi, and L. Stryer. 1985. Fluorescent immunoassay employing a phycobiliprotein labeled ligand or receptor: detection of a member. U.S. Patent no. 4520110, May 28, 1985.

Hansen, J. E., J. E. Packard, and W. T. Doyle. 1981. *Mariculture of Red Seaweeds.* Rep. no. T-CSGP-002. California Sea Grant College Publication, La Jolla. 42 pp.

Haughland, R. 1985. *Handbook of Fluorescent Probes and Research Chemicals.* Molecular Probes, Junction City, OR.

Hawkes, M. J. 1978. Sexual reproduction in *Porphyra gardneri* (Smith *et* Hollenberg) Hawkes (Bangiales, Rhodophyta). *Phycologia* 17, 329–53.

Imai, T. 1971. *Aquaculture in Shallow Seas.* Translation of *Senkai Kanzen Yosholi (Senkai Yoshoku no Shinpo)*, available from National Technical Information Service, Springfield, VA, Rep. TT-72-52021. 615 pp.

IOEP [Section of Experimental Phycoecology, Institute of Oecanology, Academica Sinica]. 1976. All-artificial spore-collecting cultivation of Tiaobanzicai (*Porphyra yezoensis* Ueda). *Sci. Sin.* 19, 253–9.

IOESP [Section of Experimental Phycoecology and Section of Systematic Phycology, Institute of Oceanology, Academia Sinica]. 1978. *Cultivation of Zicai (Porphyra yezoensis* Ueda). Science Press, Beijing. (In Chinese.)

Kafuku, T., and H. Ikenoue. 1983. *Modern Methods of Aquaculture in Japan*, pp. 192–200. Kodansha, Tokyo.

Kang, J. W., and N. P. Ko. 1977. *Seaweed Cultivation*. Tae-hwa, Busan, Korea. (In Korean.)

Kito, H. 1977. Recent problems of nori (*Porphyra* spp.) culture in Japan. In *Proceedings of the Sixth U.S.–Japan Meeting on Aquaculture*, ed. C. J. Sindermann, pp. 7–12. Santa Barbara, CA, Aug. 27–28, 1977. U.S. Department of Commerce Rep. NMFS, Circ. 442.

Korringa, P. 1976. *Farming Marine Organisms Low on the Food Chain*. Elsevier, Amsterdam. 264 pp.

Kramer, Chin, and Mayo, Inc. 1982. *Market Analysis and Preliminary Economic Analysis for Products of the Red Seaweed Porphyra*. Report to the Washington Department of Natural Resources, Olympia. 230 pp.

Kurogi, M. 1953a. Study on the life-history of *Porphyra*. I. The germination and development of carpospores. *Bull. Tohoku Reg. Fish. Res. Lab.* 2, 67–103. (In Japanese with English summary.)

Kurogi, M. 1953b. On the liberation of monospores from the filamentous thallus (*Conchocelis*-stage) of *P. tenera* Kjellm. *Bull. Tohoku Reg. Fish. Res. Lab.* 2, 104–8. (In Japanese with English summary.)

Li, S. Y. 1984. The ecological characteristics of monospores of *Porphyra yezoensis* Ueda and their use in cultivation. *Hydrobiologia* 116/117, 255–8.

Ma, J., and A. Miura. 1984. Observations on the nuclear division in conchospores and their germlings in *Porphyra yezoensis* Ueda. *Jap. J. Phycol.* 32, 373–8. (In Japanese with English summary.)

Matsuki, K. 1960. Studies on the absorption rate and availability of foods. *Bull. Med. Keio Univ.* 37, 203–24.

Melvin, D. J., T. F. Mumford, Jr., and W. J. Byce. 1986. *Techniques for Nori Farming in Washington*. Vol. 1, *Conchocelis Culture*. Washington Department of Natural Resources, Olympia. 67 pp. plus appendixes.

Michanek, G. 1975. *Seaweed Resources of the Ocean*. FAO Fisheries Technical Paper no. 138, Fisheries and Agriculture Organization, United Nations, Rome. 127 pp.

Migita, S. 1951. On the abnormal germination of *P. tenera* Kjellm. Lecture at the meeting of the Japan Society of Fisheries at Fukuoka. (Ref. in Kurogi 1953a.)

Miura, A. 1975. *Porphyra* cultivation in Japan. In *Advance in Phycology in Japan*, ed. J. Tokida and H. Hirose, pp. 23–304. Junk, The Hague.

Miura, A. 1979. Studies on genetic improvement of cultivated *Porphyra* (Laver). In *Proceedings of the Seventh Japan–Soviet Joint Symposium on Aquaculture*, pp. 161–8, Sept. 1978, Tokyo.

Miura, A. 1980. Seaweed cultivation: present practices and potentials. In *Ocean Yearbook 2*, ed. E. M. Borgese and N. Ginsberg, pp. 57–68. University of Chicago Press, Chicago.

Miura, A. 1984. A new variety and a new form of *Porphyra* (Bangiales, Rhodophyta) from Japan: *Porphyra tenera* Kjellman var. *tamatsuensis* Miura var. nov. and *P. yezoensis* Ueda form. *narawaensis* Miura, form. nov. *J. Tokyo Univ. Fish.* 71, 1–37.

Miura, A. 1985. Genetic analysis of the variant color types of light red, light green, and light yellow phenotypes of *Porphyra yezoensis* (Rhodophyta, Bangiaceae). In *Origin and Evolution of Diversity in Plants and Plant Communities*, ed. H. Hara, pp. 270–84. Academia Scientific Books, Tokyo.

Miura, A., and Y. Kunifuji. 1980. Genetic analysis of the pigmentation types in the seaweed susabi-nori *(Porphyra yezoensis)*. *Iden* [Heredity] 34, 14–20. (In Japanese.)

Morrice, L. M., M. W. McLean, W. F. Long, and F. B. Williamson. 1984. Porphyran primary structure. *Hydrobiologia* 116/117, 572–5.

Mumford, T. J., Jr., and L. Hanson. 1987. *Programmatic Environmental Impact Statement: Nori Farming and Processing*. Washington Department of Natural Resources, Olympia. 33 pp. plus appendixes.

Mumford, T. F., Jr., and D. J. Melvin. 1983. Pilot-scale mariculture of seaweeds in Washington. In *Seaweed Raft and Farm Design in the United States and China*, ed. L. B. McKay, pp. 1-1–1-18, New York Sea Grant Publication, Albany.

Mumford, T. F., Jr., J. R. Waaland, L. G. Dickson, and D. J. Melvin. 1985. *Porphyra* aquaculture in the Pacific Northwest. *J. Phycol.* (Suppl.) 21, 7.

Noda, H., and S. Iwada. 1978. *Nori Seihin Kojo no Tebiki*. National Federation of Nori and Shellfisheries Cooperative Association, Tokyo. (In Japanese, "A guide to the improvement of nori products"; translation available from the Department of Natural Resources, Olympia, WA, 234 pp.)

Nori Times. 1986. In search of new measures: crisis serving as motivation. July 11 (#1131). Published by Zen-Nori, Tokyo.

Okasaki, A. 1971. *Seaweeds and Their Uses in Japan*. Tokai University Press, Tokyo. 165 pp.

Oohusa, T. 1984. Technical aspects of nori *(Porphyra)* cultivation and quality preservation of nori products in Japan today. *Hydrobiologia* 116/117, 95–114.

Polne-Fuller, M., and A. Gibor. 1984. Developmental studies in *Porphyra*. I. Blade differentiation in *Porphyra perforata* as expressed by morphology, enzymatic digestion, and protoplast regeneration. *J. Phycol.* 20, 609–16.

Ragan, M. A. 1981. Chemical constituents of seaweeds. In *The Biology of Seaweeds*, ed. C. S. Lobban and M. J. Wynne, pp. 589–626. Botanical Monographs, vol. 17. University of California Press, Berkeley.

Rees, D. A., and E. Conway. 1962. The structure and biosynthesis of porphyran: a comparison of some samples. *Biochem. J.* 84, 411–16.

Roland, W., and M. J. Coon. 1984. Postharvest recovery of beds of the edible red alga *Porphyra perforata*. *Can. J. Bot.* 62, 1968–70.

Saga, N., and Y. Sakai. 1984. Isolation of protoplasts from *Laminaria* and *Porphyra*. *Bull. Jap. Soc. Sci. Fish.* 50, 1085.

Sakagami, Y., T. Watanabe, A. Hisamitsu, Kamibayshi, K. Honma, and H. Manabe. 1982. Anti-ulcer substances from marine algae. In *Marine Algae in Pharmaceutical Science*, ed. H. A. Hoppe and T. Levring, pp. 99–108. de Gruyter, Berlin.

Suto, S. 1974. Mariculture of seaweeds and its problems in Japan. In *Proceedings of the First U.S.–Japan Meeting on Aquaculture*, Tokyo, Oct. 18–19, 1971. National Oceanic and Atmospheric Administration Tech. Rep. NMFS CIRC-388, pp. 7–16.

Tseng, C. K. 1981. Commercial cultivation. In *The Biology of Seaweeds*, ed. C. S.

Lobban and M. J. Wynne, pp. 680–725. Botanical Monographs, vol. 17. University of California Press, Berkeley.

Turner, N. C. 1973. The ethnobotany of the Bella Coola Indians of British Columbia. *Syesis* 6, 193–220.

Turner, N. C., and M. A. M. Bell. 1971. The ethnobotany of the Coast Salish Indians of Vancouver Island. *Econ. Bot.* 25, 63–104.

Turner, N. C., and M. A. M. Bell. 1973. The ethnobotany of the southern Indians of British Columbia. *Econ. Bot.* 27, 257–310.

Ueda, S. 1973. *Manual of Nori Cultivation.* Zen-Nori (Japan Federation of Nori Cultivation and Shellfish Farm Cooperatives of Shellfish and Laver Cooperatives), Tokyo.

Umebayashi, O. 1975. Seaweed culture in Japan. In *Culture of Marine Life: Textbook for Marine Fisheries Research Course.* Japan International Cooperation Agency, Tokyo. 164 pp.

Waaland, J. R. 1981. Commercial utlization. In *The Biology of Seaweeds,* ed. C. S. Lobban and M. J. Wynne, pp. 726–41. Botanical Monographs, vol. 17. University of California Press, Berkeley.

Waaland, J. R., L. G. Dickson, E. C. S. Duffield, and G. M. Burzyki. 1986. Research on *Porphyra* aquaculture. In *Algal Biomass Technologies: An Interdisciplinary Perspective,* ed. W. R. Barclay and R. P. McIntosh. *Nova Hedw.* 83, 124–31.

Wildman, R. 1974. Seaweed culture in Japan. In *Proceedings of the First U.S.–Japan Meeting on Aquaculture,* Tokyo, Oct. 18–19, 1971. National Oceanic and Atmospheric Administration Tech. Rep. NMFS CIRC-388, pp. 97–101.

Williams, M. D. 1979. The harvesting of "slukus" *(Porphyra perforata)* by the Straits Salish Indians of Vancouver Island, British Columbia. *Syesis* 12, 63–9.

Woessner, J. W. 1981. *The Domestication of Macroalgae.* Ph.D. diss., University of California, Santa Barbara. 287 pp.

Yamamoto, I., and H. Maruyama. 1984. Inhibitory effects of dietary seaweeds *(Undaria, Porphyra, Laminaria)* on the growth of spontaneous mammary carcinoma in C3H mice. Abstract. Proceedings of the Japanese Cancer Association, 43rd Annual Meeting, Fukuoka, Oct. 1984.

Yamamoto, I., and H. Maruyama. 1985. Effect of dietary seaweed preparations on 1-2-dimethyhydrazine-induced intestinal carcinogenesis in rats. *Cancer Lett.* 26, 241–51.

Yaphe, W. 1984. Properties of *Gracilaria* agars. *Hydrobiologia* 116/117, 171–86.

5. Cultivated edible kelp

LOUIS D. DRUEHL

Department of Biological Sciences, Simon Fraser University, Burnaby, B.C.,
Canada V5A 1S6

CONTENTS

I. Introduction

Edible kelps are species of brown algae (Phaeophyta) belonging to the order Laminariales. All members of this order are more or less edible, but some are considered more desirable than others. Table 5-1 lists the more commonly selected species. *Undaria* and *Laminaria,* the two most economically important edible kelp, are illustrated in Figure 5-1.

Kelps are large seaweeds, often several meters in length. Usually they consist of a blade, stipe, and holdfast (a rootlike organ). Some species have special blades (sporophylls) arranged along the stipe, and some have midribs running the length of the blade. The life cycle of kelp consists of a large spore-producing stage (the sporophyte) alternating with a microscopic gamete-producing stage (the gametophyte) (Fig. 5-2). Kelps are usually restricted to colder waters, where they usually dominate the lower intertidal and upper subtidal floras.

II. Edible kelp products

Generally, kelp is valued for its micronutrients (e.g., iodine) and to a lesser extent for its vitamin and amino acid content. Kelp carbohydrates are characterized by β-1,3- and 1,6-glucoside linkages, which are not normally digested by humans. Thus, the bulk of the tissue is roughage in the human diet. Generally, edible kelp is employed in two categories of cuisine.

A. Oriental cuisine

Laminaria, known as *kombu* in Japan and *haidai* in China, and *Undaria,* known as *wakame* in Japan and *qundai-cai* in China have long been a desirable part of the Asian diet. Asians use *Laminaria* species from the section Fasciatae, particularly *L. japonica,* in their cooking. These *Laminaria* are not found outside of Asia. *Laminaria groenlandica* and *L. saccharina* of the section Simplices, and *Cymathere triplicata* have been successfully substituted for *L. japonica* in North American cuisine. The substitute for *Undaria,* which is also restricted to Asian waters, is young *Alaria.*

Table 5-1. *Species of Laminariales presently considered edible kelp*

Family and species	Cultivation status	Location
Laminariaceae		
Laminaria section		
Fasciatae		
L. *japonica* Aresh.	Extensive	China, Japan, USSR, Korea
L. *angustata* Kjell.	Extensive	Japan
L. *ochotensis* Miyabe	Extensive	Japan
L. *religiosa* Miyabe	Extensive	Japan, Korea
Section Simplices		
L. *groenlandica* Rosenv.	Light	Canada
L. *saccharina* (L.) Lamour.	Light	Canada, Great Britain,[a] USA[a]
Cymathere triplicata (P & R) J. Ag.	Experimental	Canada
Alariaceae		
Alaria esculenta (L.) Grev.	Experimental	Great Britain[a]
Undaria pinnatifida (Harv.) Sur.	Extensive	China, Japan
Lessoniaceae		
Macrocystis integrifolia Bory	Experimental	Canada
M. *pyrifera* (L.) C. Ag.	Experimental	USA,[a] China[a]
Nereocystis luetkeana (Mert.) P. & R.	Nil	Canada, USA

[a]Present interest is in biogas or kelp chemicals.

Laminaria is used to prepare stock for numerous dishes. The stock is prepared from either powdered kelp or kelp strips and is utilized as a tea or as the base for various soups, broths, and marinades. *Laminaria* strips are often used between the flame and food to be barbecued or as a liner in pots used in the preparation of vegetables, seafoods, and meats. This kelp is not usually eaten; rather, it imparts its flavor to the prepared foods. Kelp strips are also used in rice and with simmered vegetables and seafoods, in which case it is eaten.

Undaria is almost always eaten as fragments; the sporophylls and mid-ribs are considered the choicest parts. The product is available dry, fresh, or salted.

Macrocystis is a popular substrate for the deposition of herring roe.

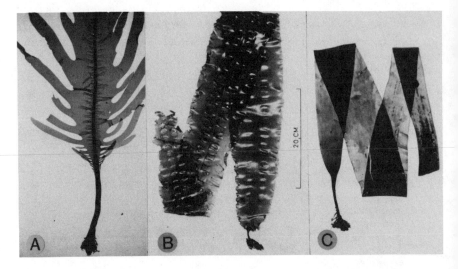

Fig. 5-1. Photographs of *Undaria pinnatifida* (A); *Laminaria cichorioides,* a section Simplices representative (B); and *Laminaria ochotensis,* a section Fasciatae representative (C).

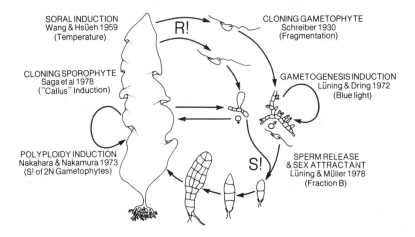

LIFE CYCLE MANIPULATIONS

Fig. 5-2. Generalized kelp life cycle indicating some manipulations of possible interest to mariculturists. R! = meiosis, S! = syngamy.

Presently, free (Alaska) and penned herring (British Columbia) spawn on kelp (and almost anything else). This special herring roe on a kelp product is highly prized by the Japanese.

For a full account of kelp in Japanese cuisine the reader is referred to Arasaki and Arasaki's *Vegetables from the Sea* (1983).

Table 5-2. *Iodine concentration as percentage of dry weight*

Species	Iodine (%)	Reference
Laminaria saccharina	0.10–0.50	M. Amat, pers. comm.
Laminaria digitata	0.13–1.20	Jensen & Haug 1956
Laminaria hyperborea	0.24–1.40	Jensen & Haug 1956
Macrocystis integrifolia	0.07–0.14	Rosell & Srivastava 1984
Nereocystis luetkeana	0.07–0.18	Rosell & Srivastava 1984

B. Health and novelty cuisines

Powdered kelp *(Laminaria, Macrocystis, Nereocystis)* from North America and Europe is marketed in pill and powder form for the health food markets. In some instances, the kelp powder is basically *Ascophyllum* (a fucoid) blended with *Laminaria* to achieve a higher iodine content.

Many of the novelty kelp foods are available for the health food market. These include kelp in canned salads (*Laminaria*, USSR), sweet pickled kelp (*Nereocystis*, United States), and candied kelp chips (*Laminaria* and *Cymathere*, Canada). Kelp and seasalt (the kelp supplements the iodine) and flaked kelp (reminiscent of parsley flakes) are also on the market.

III. Kelp composition

Kelp has long been famous for its high iodine content (Table 5-2). Tables 5-3 and 5-4 provide a brief summary of other components of *Laminaria* and *Undaria* (and cabbage for comparison) generally thought to be important in human nutrition. The values given in these tables should be used as rough indicators because they do not reflect seasonal or species variation. For example, the ascorbic acid content of *L. japonica* was found to be 3 mg·100 g dry wt^{-1}; *L. saccharina*, 7 mg·100 g^{-1}; *L. groenlandica*, 12 mg·100 g^{-1}; and *Cymathere triplicata*, 56 mg·100 g^{-1} (Druehl, unpublished data). Haug and Jensen (1954) and many others have found striking seasonal variation in many kelp components.

The tastiness of edible kelp is difficult to quantify. In part, the salty nature of the product appeals to some palates. Hosoda (1975) has quantified three substances thought to influence the tastiness of *L. longissima* (Table 5-5).

IV. Kelp production methods

A. History

Edible *Laminaria* was collected by the Ainu in northern Japan as early as the eighth century A.D. Soon the wild harvest was dominated by the

Table 5-3. *Concentration of some nutritionally interesting components of edible kelp and cabbage (units· 100 g dry wt^{-1})*

Nutrient	Unit	*Laminaria*	*Undaria*	Cabbage
Food energy	kcal	43	45	24
Protein	g	1.68	3.03	1.21
Lipid	g	0.56	0.64	0.18
Carbohydrate	g	9.57	9.14	5.37
Fiber	g	1.33	0.54	0.80
Ash	g	6.61	7.20	0.72
Calcium	mg	168	150	47
Iron	mg	2.9	2.2	0.56
Magnesium	mg	121	107	15
Phosphorus	mg	42	80	23
Sodium	mg	233	872	18
Potassium	mg	89	50	246
Zinc	mg	1.23	0.38	0.18
Copper	mg	0.13	0.28	0.023
Manganese	mg	0.20	1.40	0.159
Ascorbic acid	mg	3.00	ND	47.3
Alpha-tocopherol	mg	0.87	ND	1.67
Thiamin	mg	0.05	0.06	0.05
Riboflavin	mg	0.15	0.23	0.03
Niacin	mg	0.47	1.60	0.30
Folic acid	µg	180.00	ND	56.7
Vitamin A	IU	116	360	126
Saturated fatty acids	g	0.247	0.130	0.023
Monosaturated fatty acids	g	0.098	0.058	0.013
Polysaturated fatty acids	g	0.047	0.218	0.087
Cholesterol	g	0	0	0

Source: Anonymous 1984.

Japanese. Intially, only the privileged classes in Kyoto were allowed to eat *Laminaria;* later, samurai and then the public had access to this product (see Kawashima 1984 for a more detailed history). During this early period, edible kelp was exported from Japan to China (Tseng 1981).

Prior to the development of kelp cultivation, *Laminaria* and *Undaria* populations were enhanced by increasing available substrate and controlling competitors. Recently, dynamite blasting of colonized rocky shores to clear the existing biota has succeeded in allowing *Laminaria* to dominate the exposed shore for a few years (Hasegawa and Funano 1968). Other habitat manipulation techniques have involved placing stones and concrete blocks on the ocean floor to increase substrate availability for *Laminaria* growth (Hasegawa 1976).

Edible kelp cultivation was initiated in China and Japan in the early

Table 5-4. *Amino acid composition (g · 100 g dry wt⁻¹) of Laminaria, Undaria, and cabbage*

Amino acids	*Laminaria*	*Undaria*	Cabbage
Tryptophan	0.048	0.035	0.012
Threonine	0.055	0.165	0.042
Isoleucine	0.076	0.087	0.061
Leucine	0.083	0.257	0.063
Lysine	0.082	0.112	0.057
Methionine	0.025	0.063	0.012
Cystine	0.098	0.028	0.010
Phenylalanine	0.043	0.112	0.039
Tyrosine	0.026	0.049	0.021
Valine	0.072	0.209	0.052
Arginine	0.065	0.092	0.069
Histidine	0.024	0.015	0.025
Alanine	0.122	0.136	0.042
Aspartic acid	0.125	0.179	0.119
Glutamic acid	0.268	0.199	0.270
Glycine	0.100	0.112	0.027
Proline	0.073	0.092	0.238
Serine	0.098	0.078	0.071

Source: Anonymous 1984.

Table 5-5. *Concentrations of tasty substances in Laminaria longissima (g · 100 g dry wt⁻¹)*

	Mannitol	Glutamic acid	Succinic acid
Young plants	17.3	2.2	0.01
Mature plants	16.2	2.5	0.07
Sori	20.4	2.9	0.15

Source: Hosoda 1975.

1950s (Tseng 1981; Kawashima 1984). Cultivation has two major components: (*a*) seedstock production, which involves establishment of small sporophytes on string or other easily transported substrate starting from meiospores under controlled conditions; and (*b*) the raft/rope cultivation in the sea of the seedstock to maturity and harvesting.

B. Seedstock

The following methods for seedstock production have been successfully employed for *Laminaria, Cymathere,* and *Macrocystis* in my laboratory.

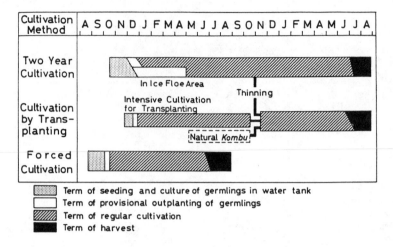

Fig. 5-3. Major seasonal events in the two-year cultivation method, the forced cultivation method, and the transplanted seedstock method of rearing edible kelp. (From Kawashima 1984.)

The methods evolved from numerous discussions with Drs. Y. Hasegawa, T. Kaneko, S. Kawashima, Y. Sanbonsuga, and several kombu fishermen.

Seedstock is produced from meiospores released from the sori of either wild or previously cultivated sporophytes. The sori are wiped vigorously with paper toweling and left in a cool (10–15°C), dark place for up to 24 hours. The sori are then placed in PES (Provasoli's Enriched Seawater: Provasoli 1968) and enriched with 80 μmol·L^{-1} of potassium iodide (KI) (Hsaio and Druehl 1973). Spore release usually occurs in one hour. The resulting suspension is poured through several layers of cheesecloth to remove foreign matter. The spore suspension is diluted with a larger volume of PES and allowed to inoculate the desired substrate (hydrophilic materials preferred) for 10 to 24 hours. The substrate is then rinsed off, immersed in fresh PES, and maintained at ca. 12°C under white fluorescent light (70 μE·m^{-2}·s^{-1}, 16 h light:8 h dark). Aeration is continuous, and the PES is changed every 10 to 14 days.

Seedstock, consisting of sporophytes 4–6 mm long is usually available after 45 to 50 days. These procedures, with variations in media and timing, and so on, are essentially employed universally.

Initially, *Laminaria* seedstock in Japan was produced in the late autumn when most sporophytes have sori. Seedstock produced in this way was available for outplanting from December to February, and the resulting crop would be ready for harvest in approximately 20 months. This system is referred to as the two-year cultivation method (Kawashima 1984; Fig. 5-3).

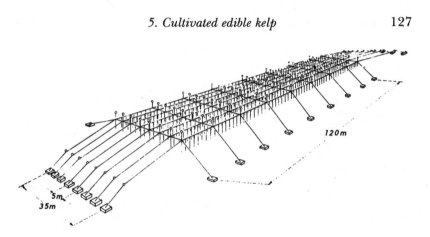

Fig. 5-4. General kelp farm arrangement. (From Kawashima 1984.)

Many edible *Laminaria* species require two years to obtain properties desirable to Asian cuisine. However, it is expensive and difficult to maintain a kelp crop in the ocean for such an extended period. Between 1966 and 1970 Hasegawa (1971) and his co-workers developed the forced cultivation method of seedstock production. The basis of this system is the production of seedstock from sori-bearing sporophytes in the summer (Fig. 5-3). These plants, which have spent three autumn months in the field prior to their second growth season, behave as second-year plants. Thus, plants cultivated in the field for 10 to 11 months have most of the qualities of plants subjected to the two-year cultivation method. The physiological basis of this response is not understood (Kawashima 1984).

The Chinese employ a modified forced cultivation method (Fei, personal communication). They produce seedstock in the summer and maintain this stock in elaborate refrigerated greenhouses until ambient seawater temperatures are sufficiently low to support kelp growth. By this time, the seedstock is 3–5 cm long and when outplanted will quickly take on second-year characteristics. Thus, the Chinese are able to exploit an abbreviated cool growth season in a climate with an extended warm period.

Seedstock often provides more than the desired number of sporophytes on the raft/rope system used for outplanting. The excess plants are frequently used to plant additional systems. Further, wild plants may be incorporated into the field culture system (Kawashima 1984).

C. Raft/rope cultivation

Commercial systems for rearing seedstock to maturity in the sea usually consist of an anchored horizontal grid of ropes buoyed 2–7 m below the surface (Fig. 5-4). In some instances, only the horizontal ropes are planted, or an elaborate system of vertical ropes is planted (Fig. 5-5).

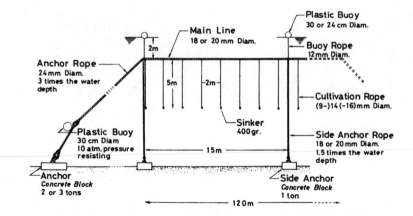

Fig. 5-5. Detail of kelp farm structure showing arrangement of vertical ropes. (From Kawashima 1984.)

The seedstock is placed in a relatively sheltered area of the sea for 7 to 10 days where the weaker plants are culled from the string by water motion. The string is then cut into small pieces and either inserted into the warp of the culture rope or attached by ribbon ties or staples. Regular maintenance includes thinning clusters of plants, removing trapped debris, and cleaning fouling organisms from the culture lines. The depth of cultivation may be altered to avoid adverse nutrient, thermal, or insolation conditions.

In contrast to other Laminariales, *Macrocystis* is cultivated either on the ocean bottom or at a constant distance above the bottom on buoyed and anchored substrates (Liu et al. 1984; Neushul and Harger 1985; Druehl and Wheeler 1986). In all other regards the cultivation of *Macrocystis* is similar to that of other edible kelp.

Usually, edible kelp plants are harvested once, at the end of the growth season. *Macrocystis* may be harvested several times over a few years, and the Chinese sometimes harvest the distal end of *Laminaria japonica* prior to the end-of-season major harvest.

V. Productivity

Productivity of rope-cultured plants is most easily compared on a wet weight per unit length of culture rope. Realized values range as high as 28 wet kg·m^{-1} rope (Table 5-6). Figure 5-6 illustrates a harvestable crop of *Laminaria saccharina* with an estimated weight of 8–10 wet kg·m^{-1} of culture rope.

Productivity comparisons on an areal basis are more difficult as different farm configurations result in more or less dense planting pat-

Table 5-6. *Production values per meter of culture rope for some edible kelp species*

Species	Locality	Culture duration (months)	Wet weight (kg·m⁻¹ rope)	Reference
Laminaria saccharina	Great Britain	6	4.2–28.4	Kain, per. comm.
Laminaria saccharina	B.C., Canada	8	3.0–8.0	Druehl, unpubl. data
Laminaria groenlandica	B.C., Canada	18	9.6–20.5	Druehl, unpubl. data
Laminaria religiosa	Korea	5	6–9	Chang & Geon 1970
Laminaria japonica	USSR	6	10	Buyankina 1977
Laminaria japonica	China	6	5–20	Fei, pers. comm.
Laminaria japonica	Japan	11	8.4–24.0	Kawashima, pers. comm.
Laminaria diabolica	Japan	20	10.2–15.9	Kawashima, pers. comm.
Alaria esculenta	Great Britain	3	7.2–11.9	Kain, pers. comm.
Undaria pinnatifida	Japan	5	<5–10	Akiyama & Kurogi 1982

Fig. 5-6. Photograph of *Laminaria saccharina* in British Columbia after eight months of cultivation. Estimated weight: 8–10 wet kg·m⁻¹ cultivation rope.

Table 5-7. *Production values per hectare kelp farm*

Species	Location	Wet weight (mt·ha^{-1})	Reference
Laminaria saccharina	B.C., Canada	26.4[a]	Druehl, unpubl. data
Laminaria groenlandica	B.C., Canada	33.0[a]	Druehl, unpubl. data
Laminaria japonica	USSR	50–60	Buyankina 1977
Laminaria japonica	Japan	49–139	Kawashima, pers. comm.
Laminaria diabolica	Japan	59–128	Kawashima, pers. comm.
Laminaria japonica	China	55	Fei, pers. comm.

Note: See Table 5-6 for duration of culture.
[a]Horizontal culture ropes only; other farms employed vertical ropes (extrapolated values).

terns. On the basis of limited data it appears that production in excess of 130 wet mt·ha^{-1} per crop may be expected (Table 5-7).

VI. World production of edible kelp

World production of edible kelp probably exceeds 300,000 dry mt·y^{-1} (Tseng 1981). China has cultivated as much as 250,000 dry mt·y^{-1} of *Laminaria japonica* (Tseng 1986). Japan relies on cultivated *Undaria,* but the cultivation of *Laminaria* is becoming increasingly important in meeting marketing requirements (Fig. 5-7). Thus, on a world scale, cultivation of kelp presently dominates over wild harvest. As competition for coastal waters increases and demand for edible kelp increases, a greater importance will be placed on extensive cultivation. Ultimately, offshore cultivation, presently being considered by the Chinese, will develop (Fei, personal communication).

VII. Domestication of edible kelp

For the cultivation of kelp to achieve the sophistication of terrestrial agronomy, genetic control and the ability to manipulate the life cycle are required. Genetic control is essential for product quality and uniformity, and life-cycle control is needed for seedstock production (Druehl and Boal 1981).

The Chinese have been successful in identifying varieties of *Laminaria*

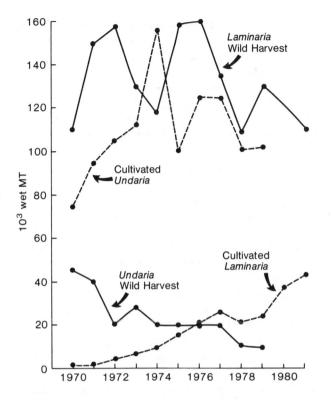

Fig. 5-7. Biomass of wild harvested and cultivated *Laminaria* and *Undaria* in Japan. (After Kawashima 1984; Akiyama and Kurogi 1982.)

japonica that are more tolerant of high temperatures, concentrate greater levels of iodine, and demonstrate more rapid growth than native stocks (Tseng 1981). These successes were achieved through inbreeding and induced mutation in combination with rigorous selection.

Hybridization of varieties, species, and even genera may result in unique and superior strains of kelp (Sanbonsuga and Neushul 1978). Further, polyploid series may be produced, and these may have special applications in kelp cultivation as they have had in terrestrial crops (Nakahara and Nakamura 1973). Further, sterile triploids may be introduced to foreign shores to assess potential ecological hazards of the introduction.

Initial studies indicate that the kelp life cycle is amenable to manipulation at several points (Fig. 5-2). Presently, asexual macroscopic gametophytes can be produced from single spores by withholding blue light (Lüning and Dring 1972). These gametophytes can be cloned by fragmentation (Schreiber 1930) to provide populations of genetically identical gametophytes, which, when crossed, produce genetically identical sporophytes. An additional advantage to this system is that superior strains may be maintained for long periods, providing uniform seed-stock year after year (Neushul 1978).

The ability to clone the sporophyte directly would represent an optimal system. Once a superior strain is identified, its uniform propagation would be assured. Initial studies indicate that this ability may be achieved in the not-too-distant future (Saga, Uchida, and Sakai 1978; Fries 1980; Fang 1984; Liu et al. 1984; Yan 1984).

VIII. Conclusions

The market for cultivated edible kelp will continue to grow. This will result from decreased access to wild plants, an increasing human population, the manufacture of new kelp products, and increased acceptance of established products. Much of the increased demand for kelp products will originate in the Western Hemisphere where there is a growing awareness of the value of sea vegetables by Occidentals and an increasing Oriental population.

The Chinese have demonstrated the feasibility of cultivating kelp in the Yellow Sea where hydrographic conditions were considered hostile to kelp growth (Tseng 1981). Further, they are considering offshore cultivation (Fei, personal communication). It seems reasonable that future development will occur in high-latitude waters where temperature, nutrient, ice, and topographic conditions favor kelp cultivation (e.g., New Zealand and British Columbia).

IX. Acknowledgments

My understanding of edible kelp cultivation has benefited greatly from discussions with Drs. X. Fei, Y. Hasegawa, T. Kaneko, S. Kawashima, and Y. Sanbonsuga.

X. References

Akiyama, K., and M. Kurogi. 1982. Cultivation of *Undaria pinnatifida* (Harvey) Suringar: the decrease in crops from natural plants following crop increase from cultivation. *Bull. Tohoku Reg. Fish. Res. Lab.* 44, 91–100.

Anonymous. 1984. *Composition of Foods.* Agriculture Handbook no. 8-11. U.S. Department of Agriculture, Human Nutrition Information Service. 502 pp.

Arasaki, S., and T. Arasaki. 1983. *Vegetables from the Sea.* Japan Publications, Tokyo. 196 pp.

Buyankina, S. K. 1977. Biotechniques of cultivation of *Laminaria japonica* off Primorye. *Vsesoiuznyi Nauchno: Issledovatelskit Institut Morskogo Rybnogo Kho-ziaistva i Okeanografii* 124, 52–6. (In Russian.)

Chang, J.-W., and S.-H. Geon. 1970. Studies on the culture of *Laminaria* (1) on the transplantation of tangle *Laminaria religiosa* Mijabe in temperate zone (the coast of Ilsan-Dong, Ulsan City). *Bull. Fish. Res. Dev. Agency* 5, 63–72. (In Korean.)

Druehl, L. D., and R. Boal. 1981. Manipulation of the laminarialean life-cycle and its consequences for Kombu mariculture. *Proc. Int. Seaweed Symp.* 10, 575–80.

Druehl, L. D., and W. Wheeler. 1986. Population biology of *Macrocystis integrifolia* from British Columbia, Canada. *Mar. Biol.* 90, 173–9.

Fang, T. C. 1984. Some genetic features revealed from culturing the haploid cells of kelps. *Hydrobiologia* 116/117, 317–18.

Fries, L. 1980. Axenic tissue cultures from the sporophytes of *Laminaria digitata* and *Laminaria hyperborea* (Phaeophyta). *J. Phycol.* 16, 475–7.

Hasegawa, Y. 1971. Forced cultivation of *Laminaria*. *Proc. Int. Seaweed Symp.* 7, 391–3.

Hasegawa, Y. 1976. Progress of *Laminaria* cultivation in Japan. *J. Fish. Res. Bd. Can.* 33, 1002–6.

Hasegawa, Y., and T. Funano. 1968. The cleaning effects by means of blasting. *Hokusuishi Geppo* 26, 916–20. (In Japanese.)

Haug, A., and A. Jensen. 1954. *Seasonal Variations in Chemical Composition of Alaria esculenta, Laminaria saccharina, Laminaria hyperborea, and Laminaria digitata from Northern Norway.* Norsk Institutt for Tang-og Tareforskning, Rep. no. 4, 14 pp.

Hosoda, K. 1975. Studies on the components of naga-kombu, *Laminaria longissima*-III: tasty substances. *Bull. Jap. Soc. Sci. Fish.* 41, 739–42. (In Japanese.)

Hsaio, S. I., and L. D. Druehl. 1973. Environmental control of gametogenesis in *Laminaria saccharina*. III. The effects of different iodine concentrations and chloride and iodide ratios. *Can. J. Bot.* 51, 989–97.

Jensen, A., and A. Haug. 1956. *Geographical and Seasonal Variation in Chemical Composition of Laminaria hyperborea and Laminaria digitata from the Norwegian Coast.* Norsk Institutt for Tang-og Tareforskning, Rep. no. 14, 8 pp.

Kawashima, S. 1984. Kombu cultivations in Japan for human foodstuff. *Jap. J. Phycol.* 32, 379–94.

Liu, T., R. Son, X. Liu, D. Hu, Z. Shi, G. Liu, Q. Zhon, S. Cao, S. Zhang, J. Cheng, and F. Wang. 1984. Studies on the artificial cultivation and propagation of giant kelp *(Macrocystis pyrifera)*. *Hydrobiologia* 116/117, 259–62.

Lui, W. S., Y. L. Tang, X. W. Liu, and T. C. Fang. 1984. Studies on the preparation and on the properties of sea snail enzymes. *Hydrobiologia* 116/117, 319–20.

Lüning, K., and M. J. Dring. 1972. Reproduction induced by blue light in female gametophytes of *Laminaria saccharina*. *Planta* 104, 252–6.

Lüning, K., and D. J. Müller. 1978. Chemical interaction in sexual reproduction of several Laminariales (Phaeophyceae): release and attraction of spermatozoids. *Z. Pflanzenphysiol.* 89, 333–41.

Nakahara, H., and Y. Nakamura. 1973. Parthenogenesis, apogamy, and apospory in *Alaria crassifolia* (Laminariales). *Mar. Biol.* 18, 327–32.

Neushul, M. 1978. The domestication of the giant kelp, *Macrocystis,* as a marine plant biomass producer. In *The Marine Plant Biomass of the Pacific Northwest Coast,* ed. R. Krauss, pp. 193–81. Oregon State University Press, Corvallis.

Neushul, M., and B. W. W. Harger. 1985. Studies of biomass yield from a near-shore macroalgal test farm. *J. Solar Energy Eng.* 107, 93–6.

Provasoli, L. 1968. Media and prospects for the cultivation of marine algae. In *Cultures and Collections of Algae: Proceedings of the U.S.–Japan Conference,* Hakone, ed. A. Watanabe and A. Hattori, pp. 63–75. Japanese Society of Plant Physiology.

Rosell, K. G., and L. M. Srivastava. 1984. Seasonal variation in chemical constituents of the brown algae *Macrocystis integrifolia* and *Nereocystis luetkeana. Can. J. Bot.* 62, 2229–36.

Saga, N., T. Uchida, and Y. Sakai. 1978. Clone *Laminaria* from single isolated cell. *Bull. Jap. Soc. Sci. Fish.* 44, 87.

Sanbonsuga, Y., and M. Neushul. 1978. Hybridization of *Macrocystis* with other float-bearing kelps. *J. Phycol.* 14, 214–24.

Schreiber, E. 1930. Untersuchungen über Parthenogenesis, Geschlechtsbestimmung, und Bastardierungsvermögen bei Laminarien. *Planta* 12, 331–53.

Tseng, C. K. 1981. Commercial cultivation. In *The Biology of Seaweeds,* ed. C. S. Lobban and M. J. Wynne, pp. 680–725. Blackwell Scientific Publications, London.

Tseng, C. K. 1986. Mariculture in China: an overview. In *Huntsman Marine Science Symposium, 1986.* St. Andrews-by-the-Sea, N.B. (Abstract.)

Wang, C., and L. Hsüeh. 1959. Observations of the sorus formation of mature *Laminaria* summering in low temperature culture room. *Acta Bot. Sin.* 8, 259–61. (In Chinese.)

Yan, Z. 1984. Studies on tissue culture of *Laminaria japonica* and *Undaria pinnatifida. Hydrobiologia* 116/117, 314–16.

6. Food and food products from seaweeds

ISABELLA A. ABBOTT

Department of Botany, University of Hawaii, Honolulu, HI 96822

CONTENTS

I. Introduction

Recent reviews of uses of seaweeds for food (Abbott and Cheney 1982; Abbott 1984) and industry (Jensen 1979), including those in this volume (Mumford and Miura, Chapter 4; Druehl, Chapter 5; Lewis, Stanley, and Guist, Chapter 9), and the effort to standardize binomials for the commercial species of *Eucheuma* (Doty and Norris 1985) reflect the interest in the uses of algae in general. The seaweeds and freshwater or terrestrial algae used for food can be divided into two major categories: those that are collected from the wild and used for food, traded, or sold by individuals; and those that are cultivated, harvested, and sold by companies, often conglomerates. This chapter primarily covers seaweeds collected from natural stands as whole pieces for food and provides some additional perspectives on their commercial uses and potential for aquaculture.

In Hawaii, thousands of people enjoying a day at the beach may make casual, small collections of just two species of edible *Gracilaria;* such collections can reduce the amount of these seaweeds available in the commercial marketplace for the next six to eight weeks. The seaweed so collected is not sold but is used strictly for home consumption. Thus, the sale of fresh seaweed as a vegetable and its use as an important addition to raw fish (Abbott 1984) are affected for some time afterward because "wild" stocks are the principal source of these Hawaiian sea vegetables.

One of the reasons for the unbalanced supply and demand in Hawaii is the occurrence in one relatively small geographical area of ethnic groups (Hawaiians, Chinese, Japanese, Koreans, Filipinos) who favor *Gracilaria* species (Figs. 6-1, 6-2) as a fresh food (Abbott 1978). If these collectors and consumers were distributed over a larger coastline such as that of California, the supply could easily meet the demand. In order to try to meet the demand for fresh *Gracilaria* in Hawaii, raceway-grown *Gracilaria* has been offered for sale, but results have been mixed because of inconsistent production.

Further examples of the pronounced impact on natural seaweed crops of individuals harvesting algae for their own use are found in

[136]

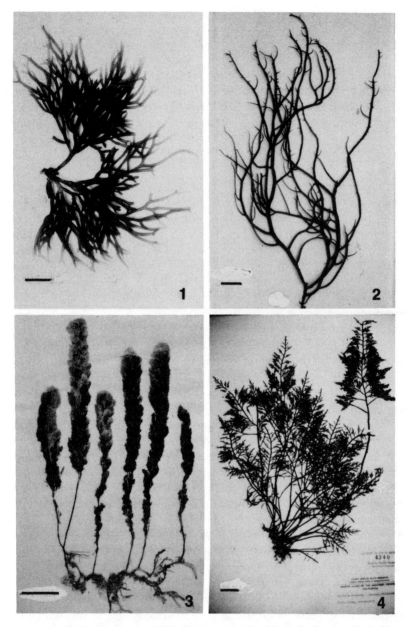

Figs. 6-1–6-4. **1:** *Gracilaria coronopifolia* (limu manauea), sold as a fresh vegetable in Honolulu markets. **2:** *Gracilaria parvispora* (limu ogo), sold as a fresh vegetable in Honolulu markets, usually mixed with raw fish. **3:** *Asparagopsis taxiformis* (limu kohu), before preparation as food, which commands about US$15–18 per pound fresh weight in Hawaiian markets. **4:** *Gelidium robustum,* the principal agarophyte from California and Pacific Mexico. Scale = 1 cm.

coastal China where agar-producing species are collected, in Japan where other agar-producing species are affected, and in coastal Chile where *Gracilaria* has been nearly eliminated (Santelices, personal communication, 1985) because of overharvesting. In all of these places some means of mitigating overharvesting and of finding economically viable ways to grow seaweeds for the "whole-piece" food or colloid market are obviously necessary. It is unusual to find an existing market for which the problems of supply seem to be without reasonable solution.

II. Whole pieces for food

Although the use of algae as food has been recorded by the Chinese for millennia (Tseng 1983) and the Chinese at present use more species in food preparation than any other people (Xia Bangmei, personal communication, 1985), it is the Japanese who have introduced their favorite recipes for algae to Westerners and thus expanded our awareness of algae as food. However, the snack known as dulse *(Palmaria [Rhodymenia] palmata)* from the North Atlantic has been eaten for a long time as well. The name *dulse* may have come from Ireland, the Gaelic word *duileasq* being its likely origin. Although phycologists tend to restrict common names to distinct taxa, in the days of the Scottish phycologist R. K. Greville (1830) the term *dulse* was also applied to algae that are now known as separate genera, *Dilsea* and *Iridaea*. Dulse is currently widely available in Atlantic Canada and New England. In 1986 it was selling for Can. $15–22 per kilogram in Halifax, Nova Scotia, supermarkets.

Several recent studies among non-Western cultures have shown that native names of edible seaweeds are frequently species-specific (Abbott 1984; Xia Bangmei, personal communication, 1985) and thus are among the best examples of accurate folk taxonomy. In reviewing the literature on native names, it appears that anthropologists who ask questions about food frequently do not have the botanical background that would allow accurate identification of the plants being discussed. The converse, when botanists ask questions, may fail because of lack of training in anthropological methods. In the case of Hawaiian, Chinese, and Japanese informants from whom the writer has sought information, being able to speak their primary language yielded at least 75% more information than did a discussion in English, even from persons who speak English rather well but as a second language. A substantial vacuum of information exists among peoples of the Pacific basin regarding the names they have for algae, the number of taxa eaten (or used for other purposes), and how these are prepared for use. For example, although the word *agar* or *agar-agar* occurs commonly in Western seaweed literature, it is Malay in origin and would imply a diverse use of seaweeds in Malaysia, but very few references on folk uses by Malays or neighboring Indonesians (Soegiarto 1979) are available.

In most cases the algae mentioned in the examples above are eaten in whole pieces. They are treated as vegetables and may be combined with other sources of protein such as beef, pork, fish, chicken, or soybean curd (tofu). *Amanori* or *Asakusanori* (dried *Porphyra* species from Japan) contain in general for each 100 grams about half the necessary daily basic protein requirement of an adult (Chapman and Chapman 1973; Abbott 1984). For other seaweeds (Chapman and Chapman 1980) 100 grams dry weight provide more than the necessary daily requirement of vitamin A, riboflavin, and vitamin B_{12} and about half the daily requirement of vitamin C. Extensive analyses of nitrogen and protein fractions including 17 or 18 amino acids are shown in Arasaki and Arasaki (1983) for many edible Japanese algae. Some are prepared as pickled vegetables by Japanese and Koreans, and perhaps these are the most commonly prepared dishes that are consumed daily in Japan and Korea. They also appear as *namasu* and *kimchee,* respectively, in Honolulu markets, selling as commonly as cucumber pickles are sold in American markets. Some seaweeds are used by Chileans in stews with other vegetables.

In Hawaii, the most expensive of the fresh seaweeds is prepared *Asparagopsis taxiformis* (Fig. 6-3) or *limu kohu* (Hawaiian), which sold for US $12–15 per pound in October 1985 (a fairly stable price), followed by fresh *Gracilaria parvispora* (Fig. 6-2) at $2.95–3.95 per pound (fresh weight), and *G. coronopifolia* (Fig. 6-1) at a usually cheaper price. *Codium* spp. (Fig. 6-5) and *Caulerpa racemosa* (Fig. 6-6) when available might sell for US $1.95–2.95 per pound in Honolulu markets, as would species of *Dictyopteris, Grateloupia filicina,* and *Halymenia formosa.* About a dozen species of algae are sold fresh in Honolulu markets. About six species of dried seaweeds from the Orient are also available. The value of dried sheets of nori (*Porphyra* spp.) in Japan in 1985 was estimated to be equivalent to US $679 million. In Honolulu during the holiday season preceding the New Year of 1986, specialty nori could be purchased for US $35 for eight ounces when "ordinary" sushi nori sold for $16 for the same weight.

In general, all ethnic groups favor *Gracilaria* species for which the state of Hawaii has reported sales of at least US $80,000 per year in recent years (at an average of $3 per pound, that's over 13 tons fresh weight per year!). Table 6-1 lists the various uses of selected algae as food by different peoples. Other food uses of seaweed are becoming rare, such as the somewhat gelatinous Welsh dish known as "laver bread," which combines *Porphyra* with cooked oatmeal, and *Porphyra* or *Ulva* mixed with milk in a kind of pudding produced by coastal Chileans.

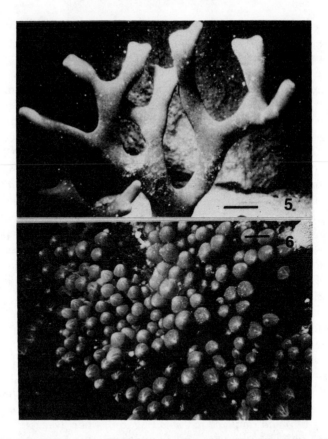

Figs. 6-5–6-6. **5:** *Codium reediae* (limu a'ala'ula), sold to Hawaiians, Filipinos, and Japanese in Honolulu markets. Scale = 1 cm. (Courtesy of W. H. Magruder.) **6:** *Caulerpa racemosa* as collected by Filipinos in Hawaii and the Philippines. Scale = 0.5 cm. (Courtesy of M. S. Foster.)

On certain Caribbean islands, a species of *Gracilaria* is sold in markets as "Irish moss" or "sea moss." It is reputed to have aphrodisiac properties. The moss is also used as a base for a nonalcoholic beverage by the local population (J. Craigie, personal communication, 1986).

III. By-products or extractives in food

The edible uses of dried, ground, and extracted algal products, on the other hand, are almost too numerous to name. The additives occur in an increasing number of prepared food products, and only because of the United States' federal label requirements are consumers able to discover the addition of seaweed colloids. These are principally carrageen-

Table 6-1. *Some algae used in large quantities as whole pieces for food, or whose rarity or cost makes them desirable as food*

Common name	Scientific name	How used
Amanori (Japanese)	*Porphyra tenera, P. yezoensis,* and other species	Sushi & other dishes
Zicai (Chinese)	*P. yezoensis, P. haitanensis, P. tenera*	Usually cooked
Kim (Korean)	*P. yezoensis, P. tenera*	Toasted
Laver (Welsh)	*P. umbilicalis*	"Bread"
Kombu (Japanese)	*Laminaria japonica*	Wide food usage
Hai dai (Chinese)	*L. japonica*	Wide food usage
Wakame (Japanese)	*Undaria pinnatifida*	Wide food usage
Qun dai cai (Chinese)	*U. pinnatifida*	Wide food usage
Miyeouk (Korean)	*U. pinnatifida*	Chiefly soup
Sui song (Chinese)	*Codium fragile*	Vegetable
Miru (Japanese)	*C. fragile*	Vinegared vegetable
? (Korean)	*C. fragile*	Winter vegetable
Popoklo (Filipino)	*C.* species	Fresh salad
Wawaeiole (Hawaiian)	*C. edule*	Salted relish
A'ala'ula (Hawaiian)	*C. reediae*	Salted relish
Ar-arucep (Ilocano)	*Caulerpa racemosa*	Fresh salad
Lato (Tagalog)	*C. racemosa*	Fresh salad
? (Fijian)	*C. racemosa*	With fresh grated coconut
Rimu (Tahitian)	*C. racemosa*	With fresh grated coconut
Tosakanori (Japanese)	*Meristotheca papulosa*	Seafood and salad
Dulse (North Atlantic; eastern U.S. and Canada)	*Palmaria palmata*	Salty snack
Cochajugo (Chilean)	*Durvillaea antarctica*	In stews
Limu kohu (Hawaiian)	*Asparagopsis taxiformis*	Salted relish
Limu manauea (Hawaiian)	*Gracilaria coronopifolia*	Salted relish
Limu ogo (Hawaiian)	*G. parvispora*	With raw fish
Tecuitlatl (Aztec)	*Spirulina* spp.	"Bread" or cakes
Dihe (Chad)	*Spirulina* spp.	Biscuits
Koxianmi (Chinese)	*Nostoc commune*	Dessert soup
Fat tsai (Chinese)	*N. flagelliforme*	New Year's dish

ans, alginates, and agars; their preparation and properties are described in Lewis, Stanley, and Guist, Chapter 9.

Agar is the most expensive of the seaweed extracts, owing to its relative scarcity and its specialized applications. By and large, the traditional species of agarophytes (principally *Gelidium* spp. [Fig. 6-4] and *Pterocla-*

dia spp.) grow slowly, do not readily respond to aquaculture conditions, and are not as easily harvestable as are carrageenophytes. However, a third genus that is increasingly relied upon for agar is *Gracilaria*. Its management as a crop (Santelices et al. 1984) in Chile has been well documented, but unfortunately the high price of this important export has stimulated overharvesting that will require several years for recovery. A similar "boom-and-bust" cycle characterized the early years of *Eucheuma* farming in the Philippines (Doty 1979). Species of *Gracilaria* are more plentiful and more widely distributed than are *Gelidium* and *Pterocladia* species, and they possess unique properties that make their agars especially desirable. For example, a high room temperature in a store or truck will not melt frosting prepared with *Gracilaria* agar. The price of this food-grade agar is at present stable, between US $10 and $12 per kg (Doty, personal communication, 1985). Agar used in microbiological media must conform to certain gelling and melting temperature ranges. For these purposes, the agar sells for US $18–25 per kg. Approximately three-quarters of all agar are used in microbiological media. Lewis, Stanley, and Quist (Chapter 9, Table 9-2) show that 36,000 dry metric tons of seaweed were used in 1980, and it is estimated that the market can support an 8–12% annual increase (Moss and Doty 1985). A relatively small chemical fraction of agar, agarose, is used for special gel electrophoresis methods where a nonionic solid is required. The current cost of agarose depends upon its purity and ranges from US $650 to $1,000 per kg.

The continued demand for agar blocks by Chinese and Japanese for their puddings is surprising in light of the high cost of agar resulting from the strong demand for it in world markets, a situation that places an increased cost on even the lowest grades. Several laboratories that I know "recycle" for a few times the agar they have used in culture media. Kappa-carrageenan, which has a lower price, can be found in commercial agar that is sold for media preparation (G. A. Santos, personal communication, 1985). The world shortage of the raw material for agar makes the investigation of its bioengineering an attractive option for solving the problem.

IV. Aquaculture of commercial algae

A. Carrageenophytes

Chondrus crispus or Irish moss has been used for generations on both sides of the Atlantic for the production of blancmange, a delicious vanilla-flavored pudding. Much of the Irish moss currently used this way is harvested by individuals for their personal use and does not appear

Table 6-2. *World use of carrageenans (in thousands of pounds)*

Product types	USA	Europe	Japan	Others
Ice cream	3,000–3,500	1,500–2,000	500	500
Chocolate milk	1,500–1,700	700–900	100	200
Milk dessert gels	600–800	1,500–1,800	—	200
Milk-base foods[a]	1,500–1,700	1,000–1,200	?	?
Protein-base foods[b]	1,100–1,500	1,000–1,200	200	400
Water-base desserts	600–700	1,000–1,100	800–900	400
Water-base liquids (i.e., fruit juices)	700–900	400–500	?	?
Other (mostly foods)[c]	—	3,500–7,500	—	—

[a]Evaporated milk stabilizer, coffee cream stabilizer, instant type breakfast drinks, artificial milk.
[b]Extruded artificial meats, sausage stabilizers, meat sauces, gravies.
[c]Cake mixes, biscuit mix, icings, candies, vegetable toppings.
Source: Modified from Moss and Doty 1985.

in the usual harvest data. Table 6-2 is modified from Moss and Doty (1985) and shows the principal food uses for carrageenans.

Commercial beds of *Chondrus crispus* are found in Nova Scotia, New Brunswick, and Prince Edward Island in Canada and in Maine in the United States and may be described as wild stock. Simpson et al. (1979) showed that the species can be grown under aquaculture conditions; however, the method is both capital- and energy-intensive and is not yet commercially developed. Most *Chondrus* carrageenan now available commercially is a kappa–lambda mixture. The carrageenans extracted from Chilean species of *Iridaea* (known in the trade as *Gigartina radula*) are also mixed in the dried weed that is delivered to the United States.

Doty (1979) has shown encouraging success in the Philippines for the growth of *Eucheuma* species (Figs. 6-7, 6-8) on small labor-intensive farms. This farmed *Eucheuma* now furnishes most of the world's supply of carrageenan. Different species of *Eucheuma* may be grown that provide only kappa or only iota carrageenans. The three main commercial species of *Eucheuma* that furnish carrageenans (Moss and Doty 1985) are *Eucheuma alvarezii* (*E. cottonii* of the trade) (Fig. 6-7), providing 30,000 dry tons per year from the Philippines for kappa-carrageenan; *E. denticulatum* ("*E. spinosum*" of the trade), furnishing ca. 6,000 tons per year from the Philippines and Indonesia for iota-carrageenan; and *E. gelatinae,* providing about 200 tons per year. Although the last-named species is at present imported only from Indonesia, its total tonnage is expected to increase owing to a high-quality β-carrageenan produced by this species. The recent reliability of Indonesian sources of *E.*

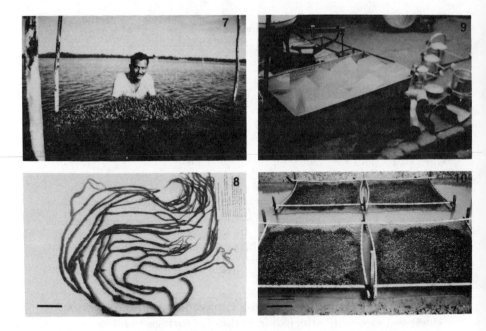

Figs. 6-7–6-10. **7:** *Eucheuma alvarezii*, one of the most common of the farmed species in the Philippines, being supported along the side of a 2.5 m wide net by Vicente Alvarez for whom the species was named. (Courtesy of M. S. Doty.) **8:** *Eucheuma alvarezii* var. *ajak-assi* from the Sulu Archipelago, southern Philippines. Scale = 2 cm. (Courtesy of M. S. Doty.) **9:** Blocks of agar (extracted from *Gelidium* species) offered for sale in a street market in Qingdao, China. **10:** Nets of *Gracilaria lemaneiformis* experimentally grown near Monterey, California. Scale = 0.25 m.

denticulatum has tended to stabilize the market price to about US $600 per dry ton, a higher price than the *cottonii* now at $350 per dry ton. According to Moss and Doty (1985), the 1982 world volume (for which there are adequate figures) of carrageenan production was 13,000 metric tons, with a value of about US $101 million. An additional $12 million can be added for carrageenan extracted by a different chemical method.

B. Agarophytes

World figures for the seaweeds that produce agar are provided in Lewis, Stanley, and Guist (Chapter 9, Table 9-2). Before World War II, Japan was the leader in the production and sale of agar from its own wild seaweed stocks. After overharvesting native materials, collection was expanded to some of the Japanese-owned island territories in Micronesia and Taiwan, but soon these seaweed sources also dwindled and were finally shut off because of the war. Not only did Japan sell its agarophytes abroad, but it also used perhaps one-third of its production for its own food consumption as *kanten*, the basis of sweetened jellylike puddings, and *tokoroten*, more suitable as an entree (Arasaki and Arasaki 1983). Along coastal China during the summer, a favorite food item found in street markets (Fig. 6-9) is blocks of agar extracted from dried *Gelidium* (also offered for sale separately) since the cool gels can be combined with other foods for light summer dishes (Xia Bangmei, personal communication, 1985). Gelidiaceous algae are not cultured in China, but the amount of wild seaweeds at present seems to be sufficient for both local laboratory and folk uses (Tseng et al. 1962). Not enough will be available for export, however, unless the technology for growing these rather "difficult" algae can be improved.

Before overharvesting of *Gracilaria* in Chile became apparent, a managed fishery from beds of wild stock appeared possible (Santelices et al. 1984) near Coquimbo, central Chile. The algae gatherers, numbering approximately 100 families, lived on or near the beaches and formed a fishery group that could be taught simple agronomic and ecologic practices that would maintain the wild stock in good condition. The pressures of the world market for agar destroyed the fishery before it had more than one year's start.

Experimental planting of *Gracilaria* species on *hibi* nets in central California (Fig. 6-10) showed biomass regrowth that exceeded the wild controls over the period of a year (Abbott and Hansen 1984). Preliminary reports of net-grown *Gracilaria* in St. Lucia Island in the Caribbean were given by Smith, Nichols, and MacLachlan (1984). No large-scale net planting of *Gracilaria* species from which crop statistics can be gathered is known at this time. It is surprising that, considering the price of the dried plants (upward of US $600 per dry ton) and the obvious shortage of agar in the world, no large-scale farming of *Gracilaria* has been undertaken.

V. Future of phycocolloids and algae as food

Hawaii and the Philippines are two areas where a shortage of the raw seaweeds for home consumption brings complaints from sellers and buyers alike. In Hawaii, price is not the overriding factor, but the lack of "ogo" or *Gracilaria parvispora* (Abbott 1985) is regarded as a serious problem. In the Philippines, *gulamon* or *Gracilaria* is infrequently seen in the markets and is scarce, as is the even greater favorite *Caulerpa racemosa*, desired for its peppery flavor.

The greatest difficulty, however, is in the inconsistent supply to the marketplace of the seaweeds that supply the giant colloid industry. Few long-range plans can be made by these manufacturers because their supplies may dry up for reasons beyond their control. In the colloid business, the use of carrageenan as a stabilizer cannot easily be shifted to the use of alginates, for example. Thus, if the supply of dried seaweeds stops for whatever reason, it may become not only financially prohibitive to replace the colloid but also impossible to obtain any of the raw materials. The rise of the use of xanthan gum, now made by microorganisms on a steady, reliable basis, will cause changes in the way seaweed supply and the cost of the colloids will be handled. Some stability in supplies, thus, will permit the expansion of the use of seaweed colloids in food and industry.

Excellent ideas for future uses of phycocolloids in food, such as softpackage aseptic pouches (instead of tin cans) whose contents need not be refrigerated (or can be) or preparation of convenience foods for microwave use, are given by Glicksman (1982). These forward-looking ideas can be implemented only if a suitable, dependable source of phycocolloids can be achieved.

VI. References

Abbott, I. A. 1978. The uses of seaweeds as food in Hawaii. *Econ. Bot.* 32, 409–12.

Abbott, I. A. 1984. *Limu: An Ethnobotanical Study of Hawaiian Seaweeds,* 3rd ed. Pacific Tropical Botanical Gardens, Lawai, HI. 35 pp.

Abbott, I. A. 1985. New species of *Gracilaria* Grev. (Gracilariaceae, Rhodophyta) from California and Hawaii. In *Taxonomy of Economic Seaweeds with Reference to Some Pacific and Caribbean Species,* ed. I. A. Abbott and J. N. Norris, pp. 115–21. California Sea Grant College Program, La Jolla.

Abbott, I. A., and D. P. Cheney. 1982. Commercial uses of algal products: Introduction and bibliography. In *Selected Papers in Phycology II,* ed. J. R. Rosowski and B. C. Parker, pp. 779–87. Phycological Society of America, Lawrence, KS.

Abbott, I. A., and J. E. Hansen. 1984. *Multiple Species Utilization of the Herring Eggs-on-Seaweed Fishery.* California Sea Grant Biennial Report for 1980–2. California Sea Grant College Program, La Jolla.

Arasaki, S., and T. Arasaki. 1983. *Vegetables from the Sea.* Japan Publications, Tokyo. 196 pp.

Chapman, V. J., and D. J. Chapman. 1973. *The Algae.* 2nd ed. St. Martin's Press, London. 497 pp.

Chapman, V. J., and D. J. Chapman. 1980. *Seaweeds and Their Uses.* 3rd ed. Chapman and Hall, London. 334 pp.

Doty, M. S. 1979. Status of marine agronomy, with special reference to the tropics. *Proc. Int. Seaweed Symp.* 9, 35–58.

Doty, M. S., and J. N. Norris. 1985. *Eucheuma* species (Solieriaceae, Rhodophyta) that are major sources of carrageenan. In *Taxonomy of Economic Seaweeds with Reference to Some Pacific and Caribbean Species,* ed. I. A. Abbott and J. N. Norris, pp. 47–61. California Sea Grant Program, La Jolla.

Glicksman, M. 1982. The hydrocolloids industry in the '80s: problems and opportunities. *Prog. Food Nutr. Sci.* 6, 299–321.

Greville, R. K. 1830. *Algae Brittanicae; or, Descriptions of the Marine and Other Inarticulated Plants of the British Islands, Belonging to the Order, Algae, With Plates Illustrative of the Genera.* MacLachlan and Stewart, Edinburgh. 237 pp.

Jensen, A. 1979. Industrial utilization of seaweeds in the past, present, and future. *Proc. Int. Seaweed Symp.* 9, 17–34.

Moss, J. R., and M. S. Doty. 1985. *Hawaii as a State in Which to Grow and Process Seaweeds into Commercially Valuable Products.* Report to Aquaculture Development Program, Hawaii. 125 pp. Department of Land and Natural Resources, Hawaii.

Santelices, B., J. Vasquez, U. Ohme, and E. Fonck. 1984. Managing wild crops of *Gracilaria* in central Chile. *Proc. Int. Seaweed Symp.* 11, 77–89.

Simpson, F. J., P. Shacklock, D. Robson, and A. C. Neish. 1979. Factors affecting cultivation of *Chondrus crispus* (Florideophyceae). *Proc. Int. Seaweed Symp.* 9, 509–13.

Smith, A. H., K. Nichols, and J. McLachlan. 1984. Cultivation of seamoss *(Gracilaria)* in St. Lucia, West Indies. *Proc. Int. Seaweed Symp.* 11, 249–51.

Soegiarto, A. 1979. Indonesian seaweed resources: their utilization and management. *Proc. Int. Seaweed Symp.* 9, 463–76.

Tseng, C. K. 1983. *Common Seaweeds of China.* Science Press, Beijing. 316 pp.

Tseng, C. K., T. J. Chang, C. F. Chang, X. Enzhan, X. Bangmei, D. Meiling, and Y. Zhongdiai, 1962. *Economic Marine Algae.* Science Press, Beijing. 198 pp. (In Chinese.)

7. *Spirulina:* a model for microalgae as human food

ALAN JASSBY

Institute of Ecology, University of California, Davis, CA 95616

CONTENTS

[149]

I. Introduction

Regional food shortages result from an interplay of many factors and do not necessarily represent natural constraints on agricultural production. War, misguided development and aid strategies, and unequal control over food-producing resources can play roles at least as important as physicial limitations on the carrying capacity of land for human populations. The most appropriate solutions thus may lie in the social, not the technological, domain. In the longer term, though, this planet still has to face a certain logic of numbers: the Malthusian confrontation between an exponentially growing population and a finite carrying capacity.

Current population growth rates obviously cannot continue unabated, but the earth's actual carrying capacity is widely disputed. What are the physicial limits, and when will the confrontation occur at current growth rates? A recent study by the United Nations Food and Agriculture Organization (FAO) concluded that the developing world as a whole is capable of producing sufficient conventional foods to sustain 150% of its year-2000 population, even under the most modest assumptions of technological and capital input (Higgins et al. 1983). As completely unrestricted movement of food within the developing world is unrealistic, however, a more detailed analysis was performed for individual countries. The results imply that 65 countries (of 117 studied) have insufficient resources to meet their year-2000 food needs; the "excess" population (i.e., the people entirely dependent on imported food) will number about 440 million. Even with the most optimistic assumptions about available technical and capital help, 19 countries will be unable to meet their food needs with conventional crops; the "excess" population in these countries will be about 50 million. If these people are to become independent of food imports, they must turn to unconventional food sources that make use of ignored resources or have higher yields than traditional crops.

Several industrialized countries recognized decades ago that microalgae might satisfy a need for unconventional crops. The earliest relevant research dates from the 1940s in Germany, the United States, and

[150]

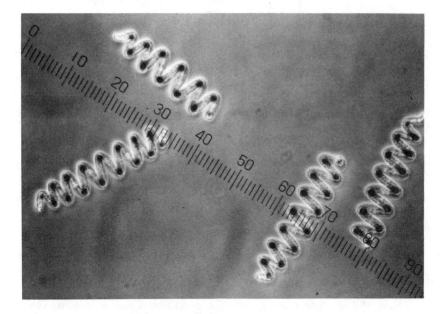

Fig. 7-1. Typical helical filaments of the blue-green alga *Spirulina platensis*. The smallest division in the ruled line corresponds to 7.6 μm. (Courtesy of Earthrise Farms.)

England and from the early 1950s in Israel, the USSR, and Japan (Burlew 1953; early USSR work mentioned by Gromov 1968). People worldwide have been exposed to enthusiastic predictions of the role microalgae can play in ameliorating resource scarcity. Is there any basis for these predictions, or are they just exuberant extrapolations from laboratory and pilot studies? We shall examine several topics that bear on this question. Recent publications deal in depth with various aspects of microalgal biology, cultivation, and product development (Shelef and Soeder 1980; Richmond 1986; Borowitzka and Borowitzka, in press); these reports need not be repeated here. Although some overlap is inevitable, I will confine this discussion to an evaluation of the food potential for microalgae using proven technology.

Commercial microalgal production for food began more than 20 years ago, thus enabling a realistic appraisal of the obstacles to further development. Unfortunately, most information from industrial-scale facilities remains unpublished, particularly data on actual yields, costs, and market conditions. As my industrial experience has centered on *Spirulina* (Fig. 7-1), production of this blue-green alga serves as a model for investigation here. Only one additional algal genus, namely *Chlorella,* currently is of commercial importance for human food use; other microalgae such as *Scenedesmus, Dunaliella,* and *Porphyridium* are the subject of intensive research for food applications, but commercial production, if any, is just beginning. In any case, most of the arguments regarding *Spirulina* apply to *Chlorella* as well as to microalgae not yet in full-scale cultivation.

II. Traditional use of inland algae

The algae of inland waters cannot match the cultural and economic importance of the larger marine algae. Seaweeds have been collected and consumed since prehistoric times and play a significant role in the food traditions of many countries, particularly China, Korea, and Japan (see Abbott, Chapter 6). The current wholesale market value of seaweed sold as food is approximately $1 billion annually (McConnaughey 1985).

Nevertheless, the use of nonseaweed algae is hardly a new idea. Numerous examples of traditional use of inland algae can be cited from more than 15 countries (Table 7-1). Although a majority of the cases are in South and East Asia, four different continents – Africa, Asia, and the Americas – as well as Oceania are represented.

Most of the algae used cannot be considered strictly microscopic, unlike the inland algae currently of commercial importance. *Nostoc,* which accounts for 16 of the examples, consists of gelatinous aggregations of filaments. Although individual filaments are microscopic, aggregations can occur as globules the size of hens' eggs (*N. pruniforme*) or as loose-lying continuous masses of considerably larger dimensions (*N. commune* var. *flagelliforme*). Similarly, *Phormidium* and *Aphanothece* appear as large aggregations visible to the naked eye and composed of numerous filaments or solitary cells, respectively, embedded in mucilage. The microscopic filaments of *Spirulina* do not form palmelloid aggregations but often mass into floating clumps that are pushed against the shore by wind.

Oedogonium and *Spirogyra,* both unbranched threadlike filaments, often occur as attached or free-floating silky masses. *Prasiola* and *Enteromorpha* (also distributed in marine habitats) appear in rapidly running streams and are easily visible thalloid types fixed to the substrate by rhizoids. The freshwater red alga *Lemanea,* usually found during the cold season attached to rocks in swift water, forms stiff olive-green or green-black bristles reaching 20 cm in length. The only unicellular alga among these examples is *Chroococcus turgidus,* and it is never collected alone but occurs in mixtures dominated by *Phormidium tenue* along with other algae and invertebrates.

These examples illustrate an important point: When wild populations of algae could be collected conveniently because individuals were macroscopic or capable of massing in large amounts, they played a culinary and therapeutic role similar to that of many higher plants. Although

Table 7-1. *"Wild" populations of inland algae that have been collected for food*

Country	Algae	Reference
Bolivia	*Nostoc commune*	de Lagerheim 1892
Burma	*Spirogyra*	Biswas 1953
Chad	*Arthrospira platensis*[a]	Dangeard 1940
China	*Nostoc commune* var. *flagelliforme*	Okamura 1913
	Nostoc edule[b]	de Lagerheim 1892[c]
	Prasiola yunnanica	Jao 1947
Ecuador	*Nostoc commune, N. ellipsosporum*	de Lagerheim 1892
Fiji	*Nostoc*	Wood 1965
"Himalayas"	*Prasiola*	Léonard & Compère 1967[d]
India	*Lemanea mamillosa*	Khan 1973
	Oedogonium, Spirogyra	Tiffany 1958
"Indochina"	*Spirogyra*	Léonard & Compère 1967
Indonesia (Java)	*Nostoc commune*	de Lagerheim 1892[e]
Japan	*Aphanothece sacrum, Nostoc commune, N. verrucosum, Prasiola japonica*	Watanabe 1970
Mexico	*Chroococcus turgidus, Nostoc commune, Phormidium tenue*	Ortega 1972
	"tecuitlatl"[f]	Deevey 1957; Farrar 1966
Mongolia	*Nostoc commune*	Elenkin 1931
	Nostoc edule[b]	Johnston 1970[g]
Okinawa	*Nostoc*	Johnston 1970
Peru	*Nostoc pruniforme*	Aldave-Pajares 1978
Thailand	*Nostoc verrucosum*[h]	Smith 1933
US (Hawaii)	*Enteromorpha*	Taft 1965
USSR (Siberia)	*Nostoc edule*[b]	Johnston 1970[g]

[a] *Arthrospira platensis* (Nordst.) Gom., called *Spirulina platensis* (Gom.) Geitl. by Léonard and Compère 1967.
[b] Synonymous with *Nostoc pruniforme*.   [c] Cites work by Rabenhorst.
[d] Cites personal communication from Bourrelly; countries not specified.
[e] Cites work by Wittrock and Nordstedt and by Cohn.
[f] Aztec word referring almost certainly to blue-green algae, perhaps to *Spirulina* (Johnston 1970).
[g] Cites oral presentation by Hooker
[h] Identification suggested by Johnston (1970), based on Smith's description.

inland algae may not have provided significant protein or energy in the diet, they were used in a variety of soups, spreads, and sauces and may have been important sources of vitamins and minerals. Traditional consumption of nonseaweed algae may have been limited only by difficulty of collection. Precedent thus exists for using microalgae or closely related species as food; no evidence exists for any deep or widespread cultural resistance to expanding their use through cultivation.

But why should we want to expand their food use? Are there any advantages from the viewpoint of nutrition, economics, or resource use compared to conventional foods?

III. Nutritional aspects

Certain microalgae have an unusual breadth of nutritional quality when compared to the higher plants in our diet. Among microalgae currently of commercial importance, *Spirulina* stands out. The chemical composition of *Spirulina* varies widely according to strain and culture conditions, but the most interesting features are almost always present. Average composition data for spray-dried *Spirulina platensis* powder produced at Earthrise Farms, the pioneer U.S. facility located in the Imperial Valley of southern California, are used here as a basis for discussion. Earthrise Farms began commercial operation in late 1982; the (unpublished) data presented here were collected during the 1983 and 1984 growing seasons (see also Jassby, Chapter 8).

A. Protein

Protein quality is estimated by feeding tests in which intakes and losses are measured under standardized conditions. These assays enable proteins to be ranked by their *net protein utilization* (NPU) values, a number that simultaneously reflects both the *digestibility* of a protein and its *biological value* once absorbed (Miller and Payne 1961). A completely utilized food with an ideal amino acid pattern would have an NPU of 100.

NPU measurements reported for raw *Spirulina* range from 46 to 63, depending on the algal source and experimental conditions (Clément, Giddey, and Menzi 1967; Bourges et al. 1971; Durand-Chastel and Clément 1975; Narashima et al. 1982). The protein quality of certain *Spirulina* strains thus compares favorably with the best plant sources such as soybeans (only brown rice is significantly higher). Animal proteins, however, generally exceed *Spirulina* in NPU value.

The ultimate measure of a protein source is *usable protein,* the product of total protein and NPU. When *Spirulina* powder is compared with the best conventional food (fresh weight basis) that each food group can offer, it leads in both total and usable protein (Table 7-2). On a dry weight basis – a more legitimate test – *Spirulina* does not fare as well. Poultry and fish are superior (>45% usable protein), whereas *Spirulina* falls into the same class as meat and dairy products (30–45%). *Spirulina* remains superior, however, to all other vegetable products (<30%).

NPU measurements to date probably underestimate the potential

Table 7-2. *Protein quality and quantity for Spirulina and common food groups on the basis of fresh weight (fw) and dry weight (dw)*

Food group	NPU (%)	Total protein (% fw)	Usable protein (% fw)	Total protein (% dw)	Usable protein (% dw)
Spirulina[a]	46–63	63	29–40	66	30–42
Dairy (Parmesan cheese)	70	36	25	51	36
Poultry (turkey, roasted slices)	70	31	22	81	57
Legumes (soybeans)	61	34	21	38	23
Seafoods (tuna, canned, drained)	80	24	19	62	50
Meat (pork loin, broiled, medium fat)	67	29	19	54	36
Nuts and seeds (pine[b])	50	32	16	34	17
Grains (hard red spring wheat)	60	14	8	16	10
Vegetables (broccoli[c])	<60	4	<3	38	<24

Note: Each food group is represented by the member with the highest usable protein content (% fw).
[a]Total protein (% fw) is the average for spray-dried powder from Earthrise Farms; seasonal range is 55–70%; powder averages 5% moisture content.
[b]Moisture content assumed equal to that of peanuts.
[c]NPU for broccoli unknown; 60 is maximum known for common plant foods other than legumes, nuts and seeds, and grains.
Sources: Unpublished data (*Spirulina* total protein); Lappé 1982; Adams 1975; see text for references on *Spirulina* NPU.

protein quality of *Spirulina*. The biological value of most microalgal proteins is limited primarily by low levels of the sulfur-containing amino acids (Florenzano 1975). *Spirulina* is no exception, as suggested by the amino acid pattern (Rao, Venkataraman, and Duggal 1981) or by methionine supplementation of *Spirulina* diets (Narashima et al. 1982). Yet amino acid composition varies significantly among wild strains (Garson, Maigrot, and Busson 1969), and high-methionine mutants have been isolated (Riccardi et al. 1981). Furthermore, Wouters (1969) showed that the essential amino acid pattern of a single strain responds to variations in the culture medium: Increased sulfate raises levels of sulfur-containing amino acids. Amino acid patterns of *Spirulina* also can be improved by enzyme modifications known as plastein reactions (Arai, Yamashita, and Fujimaki 1976). Even in the absence of any advances *Spirulina* must be considered an excellent protein source, at least from the purely nutritional viewpoint.

Table 7-3. *Vitamin and mineral content of spray-dried Spirulina powder and amounts required to provide the daily requirement in each case*

Nutrient	Content[a] (mg·kg^{-1})	Amount[b] (g)
Vitamins		
A	230,000 RE[c]·kg^{-1}	4.3
D	n.d.	—
E	40 TE[d]·kg^{-1}	250
C	200	300
Thiamine	37	41
Riboflavin	46	37
Niacin	130 NE[e]·kg^{-1}	150
B$_6$	7.0	310
Folic acid	0.51	780
B$_{12}$	1.7	1.8
K	n.d.	—
Biotin	0.046	2,200–4,300
Pantothenic acid	4.6	870–1500
Minerals		
Calcium	14,000	87
Phosphorus	6,700	180
Magnesium	2,900	140
Iron	1,500	12
Zinc	21	710
Iodine[f]	not detected (<5)	>30
Copper	8.0	250–380
Manganese	37	68–140
Fluorine[f]	n.d.	—
Chromium	2.4	21–83
Selenium[f]	not detected (<0.5)	>100
Molybdenum[f]	n.d.	—

[a]Unless otherwise indicated, content is average (mg·kg^{-1}) for spray-dried powder from Earthrise Farms; powder averages 5% moisture content.
[b]Amount of powder providing RDA for the age–sex group with the highest requirement or, if no RDA established (vitamin K, biotin, pantothenic acid, copper, manganese, flourine, chromium, selenium, molybdenum), amounts providing the range for estimated safe and adequate intake (ESAI).
[c]RE = retinol equivalents.   [d]TE = alpha-tocopherol equivalents.
[e]NE = niacin equivalents.   [f]See text for data from other sources.
Sources: Unpublished data; FNB 1980

B. Vitamins and minerals

The Food and Nutrition Board (FNB) of the U.S. National Research Council periodically publishes *recommended dietary allowances* (RDA) for various nutrients (FNB 1980), allowing estimation of the amount of *Spirulina* powder required to satisfy daily needs (Table 7-3). The data indicate unusually high levels of iron and vitamins A and B$_{12}$ in *Spirulina*.

1. Vitamin A. Preformed vitamin A is not present in plants, but *Spirulina* is rich in provitamin A carotenoid pigments, which are converted to true vitamin A in the intestines or when absorbed through intestinal walls. The vitamin A activity is due primarily to β-carotene, constituting 67% (Tanaka, Matsuguchi, and Katayama 1974) to 79% (Palla and Busson 1969) of the total carotenoids. Xanthophylls, particularly echinenone and cryptoxanthin, provide most of the remaining vitamin A activities.

β-Carotene levels in Earthrise Farms *Spirulina* powder range from 1,200 to 1,400 mg·kg^{-1}. The average level is equivalent to 230,000 retinol equivalents (RE)·kg^{-1}, so that only 4.3 g provides the RDA (Table 7-3), without counting the contributions from provitamin A xanthophylls. No conventional plant food can compare on the basis of fresh weight, although spinach and carrots attain 100,000–150,000 RE·kg^{-1} on a dry weight basis; dehydrated liver may contain up to 680,000 RE·kg^{-1} (Adams 1975). The undisputed leader, though, is the microalga *Dunaliella*, strains of which contain over 50 times the β-carotene in *Spirulina* (Ben-Amotz and Avron 1980).

2. Vitamin B_{12}. Vitamin B_{12} activity derives from cobalt-containing corrinoid substances known as cobalamins. Standard methods to determine B_{12} unfortunately do not distinguish between cobalamins and other corrinoids with no B_{12} activity for humans (Herbert et al. 1982). In one investigation of *Spirulina* tablets (Herbert and Drivas 1982), over 80% of the B_{12} determined by standard microbiological assay were analogues. The standard assay results are used for comparison here, however, as the division into true vitamin B_{12} and cobalamin analogues without B_{12} activity is unknown for common foods.

Vitamin B_{12} activity averages 1.7 mg·kg^{-1} for Earthrise Farms *Spirulina* (ranging from 1.1 to 2.4), similar to other commercial sources. Higher values have been reported (Materassi, Ricci, and Tofani 1974) but are not characteristic. Just 1.8 g, then, provides 100% of the RDA. Although B_{12} is concentrated in certain animal products (e.g., beef liver is similar to *Spirulina* on a dry weight basis), it occurs in negligible quantities in common plant foods (Orr 1969).

3. Iron. Like many other minerals, iron in *Spirulina* reflects physical and chemical conditions during growth. Most samples contain 500–1,000 mg·kg^{-1} (cf, Laquerbe, Busson, and Maigrot 1970; Santillán 1982), although culture conditions at Earthrise Farms result in an average of 1,500 mg·kg^{-1} (ranging from 1,000 to 1,900). Only 12 g supplies the RDA for iron. *Spirulina* contains an order of magnitude more iron than conventional foods, plant or animal, even when compared on a dry weight basis.

Availability of dietary iron is influenced by its chemical nature. Heme iron (i.e., iron complexed with certain porphyrin compounds like those in blood hemoglobin) is the most available. Nonheme iron, 60% of the iron in animal tissues and almost all in plant foods, is only 13–35% as absorbable (Monsen et al. 1978). Although blue-green algae have no significant heme, they do have a related linear tetrapyrrole called phycocyanobilin, also found among red and cryptomonad algae. Phycocyanobilin usually occurs attached to a protein, forming a compound known as phycocyanin. Studies at Dainippon Ink and Chemical in Japan (H. Shimamatsu, personal communication) suggest that phycocyanobilin can form soluble complexes with iron and other minerals under conditions resembling digestion. Although the mechanism is speculative, experiments by Johnson and Shubert (1986) and K. Takemoto of Saitama Medical College in Japan (H. Shimamatsu, personal communication) verify that the availability of *Spirulina* iron to animals is unusually high.

4. Other vitamins and minerals. For reasons to be discussed below, not more than 50 g of *Spirulina* should be consumed daily. Even with this limitation, *Spirulina* can be a major source of thiamin and riboflavin (Table 7-3). Similar or higher levels, however, are found in many common foods even on a fresh weight basis.

As with iron, other mineral levels respond to concentrations in culture medium and to harvest procedures; much variation can be found among samples from different sources. The calcium content of 1.4% (Table 7-3), for example, is unusually high (cf. Santillán 1982). Fluorine was not measured in Earthrise Farms material, but Boudène, Collas, and Jenkins (1976) reported a range of 76–630 $mg \cdot kg^{-1}$ in samples from different sources. A single gram of *Spirulina* would contain more than the estimated safe and adequate intake (ESAI; FNB 1980) at the higher concentration, but no accumulation was found by these workers in feeding tests with rats. The same samples yielded a range of 0.01–0.39 $mg \cdot kg^{-1}$ selenium. Laquerbe et al. (1970) reported a molybdenum level of 5.7 $mg \cdot kg^{-1}$ in two strains, so 26–88 g of powder could supply the ESAI for molybdenum.

Paoletti, Pelosi, and Filastò (1971) demonstrated a linear relation between iodine levels in *Spirulina* cells and culture medium, up to a maximum level in cells of 3.2 $mg \cdot kg^{-1}$ weight. This value is similar to the 4 $mg \cdot kg^{-1}$ actually found in powder from Siam Algae Company in Thailand (unpublished data). As the Siam Algae Company farm is situated near the ocean, high levels of iodine are expected, resulting from either atmospheric fallout or intrusion of seawater into groundwater supplies to the growth ponds. At 4 $mg \cdot kg^{-1}$, 38 g of powder provide the RDA for iodine.

These data demonstrate that 50 g of *Spirulina* can be a useful source of calcium, iodine, manganese, chromium, and molybdenum (and perhaps a source of excessive fluorine), but the variation among samples needs further study before we can generalize.

C. Safety for human consumption

Any food grown or processed under unsanitary conditions may be contaminated by toxic organisms or chemicals. Here, the discussion is confined to the safety of algae produced according to good manufacturing practices, that is, presumably uncontaminated and not containing toxic breakdown products. Other public health issues are reviewed elsewhere in this volume (see Jassby, Chapter 8; Gorham and Carmichael, Chapter 16).

1. Nucleic acid content. The high nucleic acid content of algae and other microbial organisms limits their use as food (Waslien 1975). All nucleic acids produce uric acid upon digestion, which can precipitate and form sodium urate crystal deposits in cartilage. Extreme sensitivity of the joints, a disease known as gout, can result. Kidney stones also can be deposited in the urinary tract.

The United Nations Protein-Calorie Advisory Group (PAG) established a guideline for maximum nucleic acid intake of 2 $g \cdot d^{-1}$ (PAG 1975). Some bacteria contain up to 18% nucleic acid, yeast and fungi usually 6–11%, and algae the least (Chen and Peppler 1978). *Spirulina* contains approximately 4% (Jassey, Berlot, and Baron 1971); a maximum of about 50 g *Spirulina* powder, then, is permissible under the PAG guideline.

Waslien and her co-workers (1970) provide supporting evidence; they found that uric acid levels did not increase in humans receiving up to 30 $g \cdot d^{-1}$ of *Chlorella* protein, equivalent to 50 g of algal powder. Similarly, Kofranyi (1978) determined that uric acid excretion fell within the normal range with consumption of 45 $g \cdot d^{-1}$ *Scenedesmus* powder. As *Spirulina* has a low ratio of nucleic acid to protein – the lowest among 13 algal strains examined by Paoletti, Materassi, and Balloni (1974) – 50 $g \cdot d^{-1}$ of *Spirulina* probably is safe as well. Of course, one must also account for other nucleic acid sources in the diet, especially organ meats, which can contain 2–3% nucleic acids.

2. Human studies. A daily energy intake of 2,000–2,500 kcal is recommended for the average adult doing light work (FNB 1980). Unless there is excessive fat in the diet, this caloric intake corresponds to roughly 500 g dry wt of food intake. The nucleic acid guideline thus implies that *Spirulina* can constitute up to about 10% of an adult's diet. Are there any other toxicity problems when *Spirulina* is consumed at this level?

Few clinical or field studies shed light on the safety of *Spirulina* for humans. *Spirulina* powder was fed to undernourished children (Ramos Galvan 1973) and adults (Sautier and Trémolières 1976) in a clinical setting, and no adverse physiological or chemical changes were noted. Algae constituted more than 10% of the diet in both cases, but each patient was studied for one to two weeks, enough time to assess only acute effects. More recently, Fica, Olteanu, and Oprescu (1984) fed *Spirulina* tablets to undernourished patients but made no explicit conclusions on toxicity aside from noting the absence of acute problems. Other human studies were conducted in Japan, but these involved small doses and were designed to examine specific therapeutic effects (see Section V).

As mentioned earlier, *Spirulina* has been used traditionally in Chad, where consumption is restricted to 9–13 g per meal and occurs at 10–60% of the meals (Delpeuch, Joseph, and Cavalier 1976). No attempt has been made to isolate *Spirulina* as a factor in the health of these people. Among the thousands of people in the United States, Japan, and Europe who use *Spirulina* as a nutrient supplement, acute problems – usually gastrointestinal upset – are reported rarely. The average dose, however, is only several grams per day. More information is available from animal studies.

3. Animal studies. Perhaps the most comprehensive animal study was sponsored by the United National Industrial Development Organization on rats and mice (Chamorro Cevallos 1980): No chronic or subchronic toxicity, mutagenicity, or teratogenicity and no adverse effects on reproduction or lactation were detected (feeding levels of 10–30% of the total diet were used). Published studies from independent laboratories around the world confirm the absence of toxic effects when *Spirulina* provides a significant amount of dietary protein (Bourges et al. 1971; Til and Willems 1971; Boudène et al. 1976; Février and Sève 1976; Tulliez et al. 1976; Contreras et al. 1979; Yoshino et al. 1980; Krishnakumari, Ramesh, and Venkataraman 1981; Yap et al. 1982). Many relevant investigations were conducted in Japan under private auspices but do not appear in the technical literature (H. Shimamatsu, personal communication). These include studies on acute toxicity, by K. Akatsuka and his colleagues at the Meiji College of Pharmacy in 1979 and by the Japan Scientific Feeds Association in 1982; chronic toxicity, by K. Takemoto of the Saitama Medical College in 1981 and 1982; and reproduction, by Y. Maeda and co-workers of the Animal Reproduction Research Laboratory in 1979. All of the cited studies used rats as the test animal; none showed any deleterious effects.

Spirulina has been used in numerous other animal feeding studies that do not address the question of toxicity. Their goal, like the human studies mentioned above, was assessment of weight gain and nitrogen balance. Some of these experiments detected retarded growth at feeding levels above 10% (e.g., Blum and Calet 1976; Bezares Sansores, Rossainz, and Avila Gonzales 1977 – with poultry in both cases), but the results were attributed to limitations in digestibility or net energy. Brune's (1982) work with poultry is an important exception. His data suggest the presence of growth-retarding factors in the fat fraction that can be compensated by vitamin A and methionine supplementation or eliminated by fat extraction. Earlier, Brune (1980) found evidence for a similar vitamin A activity-decreasing factor in *Scenedesmus*. *Spirulina* was used as the sole protein source in Brune's studies, so there is still no evidence for toxicity or growth retardation at the 10% level. Antinutritional factors also are present in many common foods and must be inactivated before consumption or the quantity of the food must be limited in the diet (Liener 1966).

D. Food applications

Assuming that a maximum of 50 $g \cdot d^{-1}$ *Spirulina* powder can be incorporated safely into the diet, what nutritional role can be envisioned for this alga? *Spirulina* certainly has little value as a calorie source. Typical samples containing about 65% protein, 15% carbohydrate, and 10% fat (the remainder being ash and moisture) contribute about 4 $kcal \cdot g^{-1}$, a daily maximum of only 200 kcal.

Spirulina is more interesting as a protein source: 50 g contain at least half the protein RDA (FNB 1980), itself a liberal quantity for most of the population. Furthermore, the amino acids of *Spirulina* and some conventional grains complement each other, at least in feeding studies with rats. Narashima et al. (1982), for example, demonstrated an improved NPU with a 1:1 mixture of *Spirulina* and barley. Similarly, Devi and Venkataraman (1983a) found that a mixture of *Spirulina* and rice or *Spirulina* and wheat had much higher protein efficiency ratios than any component alone.

Spirulina may be particularly suitable as a protein source among malnourished populations. The four most prevalent deficiency diseases of public health importance are protein–calorie malnutrition (PCM), nutritional anemia, xerophthalmia, and endemic goiter (Ramalingaswami 1977). *Spirulina*'s composition addresses each of these problems through its protein content (vs. PCM), iron and B_{12} (vs. anemia), provitamin A (vs. xerophthalmia), and iodine (vs. goiter). In fact, even small supplementary quantities of 10 $g \cdot d^{-1}$ could alleviate the latter three deficiency diseases.

Palatability is not an overwhelming problem when 10 $g \cdot d^{-1}$ or less is used. At least 3 g can be easily incorporated into a single portion of a drink or sauce. The Kanembu of Chad use an average of 10 g per serving in the form of sauces eaten either alone or with beans, meat, or fish (Delpeuch et al. 1976). As a last resort, this quantity can be consumed in the form of tablets.

Using larger quantities, especially as much as 50 $g \cdot d^{-1}$, poses additional problems. *Spirulina,* like most algae, has a strong color, odor, and taste that does not easily hide under or blend with conventional foods. A French study (Sautier and Trémolières 1976) found that chocolate and tomato were the most acceptable flavors, but none of the recipes containing significant quantities of unrefined *Spirulina* was popular. On the other hand, tolerance improved when the strong color and odor were decreased by an alcohol extraction, and hydrolyzed *Spirulina* powder, with no algal taste or odor, was completely accepted. Further processing of spray-dried *Spirulina* powder (e.g., Baron 1976) thus may be necessary before it can be consumed in quantity.

IV. Economic and environmental aspects

A. Current world production and use

1. Major producers. Ten commercial farms of significant size (>5 $mt \cdot y^{-1}$ dry wt) produced food-grade *Spirulina* in 1984. Three of these farms are located in temperate-zone deserts (Earthrise Farms in the United States and Ein Yahav Algae and H. K. Spirulina in Israel); five near the Tropic of Cancer (Japan Spirulina Company on Miyako Island and Nan Pao Resin, Sei-Shu Chlorella, En-Toh Microalgae, and Toh-Kai Chlorella in Taiwan); and two in the tropics (Siam Algae Company in Thailand and Sosa Texcoco in Mexico).

The total world production on a dry weight basis was just over 700 mt and the mean yield 20 $mt \cdot ha^{-1} \cdot y^{-1}$, or 5.5 $g \cdot m^{-2} \cdot d^{-1}$ averaged over the year. Mean daily productivity is far less than the 10–20 $g \cdot m^{-2} \cdot d^{-1}$ and higher typically predicted from laboratory or pilot studies. Even the current productivity leader, Siam Algae Company, lies at the low end of the range. What is the basis for this discrepancy?

Unresolved technical problems plague industrial facilities. Fast-growing pure cultures do not persist indefinitely. They decline in productivity or become contaminated with "weed" algae and must be discarded, resulting in lost production time. Many pilot studies do not last long enough to encounter these "aging" problems.

In addition, productivity in large commercial ponds (up to 5,000 m^2 at Earthrise Farms) usually is less than in small pilot ponds. This probably is due to differences in water motion, which can affect nutrient diffusion, gas exchange with the atmosphere, survival of grazing organ-

Table 7-4. *World commercial production of food-grade Spirulina (1984)*

Country	Farms	Approximate latitude (°N)	Area (ha)	Yield $(T \cdot y^{-1})$	Yield $(T \cdot ha^{-1} \cdot y^{-1})$	Yield[a] $(g \cdot m^{-2} \cdot d^{-1})$
Thailand	1	14	1.8	65	36	9.9
Mexico	1	19	10	330	33	9.0
Japan	1	25	1.3	30	23	6.3
Israel	2	30	1.7	25[b]	15	4.0
US	1	33	5.0	70[c]	14	3.8
Taiwan	4	24	16	200	12	3.3
World	10		36	720	20	5.5

Note: Farms producing more than $5T \cdot y^{-1}$ food-grade *Spirulina*. All yields are expressed on the basis of a spray-dried powder.
[a]Average based on 365 days.
[b]Yield for 1985.
[c]Yield for 1987.
Sources: Thailand, Japan, and Taiwan: H. Shimamatsu, personal communication; Mexico: Durand-Chastel and Santillán 1975, Santillán 1982; US: unpublished data; Israel: A. Vonshak, personal communication.

isms, and the light regime experienced by individual cells. The exact mechanism awaits elucidation.

Seasonality of production is a major limitation outside of the tropics. Higher insolation and a longer growing season due to more favorable temperatures usually result in a productivity increase with decreasing latitude. Taiwan is an exception to this pattern because of frequent summer storms that hamper production (Table 7-4). The ponds of Japan Spirulina Company on Miyako Island are protected by greenhouses, which explains why productivity is so much higher than in nearby Taiwan. At Earthrise Farms, suboptimal or completely unsuitable conditions for *Spriulina* culture occur during most of the year, and the growing season is only six months long (May to October), whereas 9–12 months of production are possible in Thailand and Mexico. Pilot studies often are conducted only during the seasons most favorable to growth, and extrapolation to annual yields may be biased outside the tropics.

The discrepancy between pilot and full-scale facilities thus can be explained by the effects of culture age and size and by seasonality. With the solution of outstanding technical questions and wider use of greenhouses and geothermal heat sources, dramatic productivity increases are possible.

2. Market conditions. Before 1981, the demand for food-grade *Spirulina* came primarily from the small health food market in Japan, amounting to about 20 mt·y^{-1} by 1980. Most of this material was produced by the Japan Spirulina Company (since 1977) and the Siam Algae Company (since 1978). An additional 150–200 mt·y^{-1} of Sosa Texcoco *Spirulina* were imported into Japan for use as a fish feed (*Spirulina* carotenoids enhance the colors of koi, the Japanese ornamental carp). Less than 20 mt·y^{-1} were consumed in the fledgling U.S. *Spirulina* market in 1980, and negligible quantities elsewhere. Ex factory prices for Sosa Texcoco *Spirulina* were about US $5,000 per mt at that time, compared to the prices of about US $20,000 for material from the farms in Japan and Thailand. Sosa Texcoco produces its *Spirulina* as a by-product of another process – evaporation of natural brines to extract baking and caustic soda – so that many investment, labor, and nutrient expenses can be discounted. Even so, Sosa Texcoco may have been selling below production cost during that initial period of low demand.

In 1981, *Spirulina* became known as a dieting aid, and demand sky-rocketed. A shortage developed, despite the fact that Sosa Texcoco, producing most of the world's *Spirulina,* raised prices to US $15,000–30,000 per mt. The new demand naturally stimulated additional production, eventually resulting in the establishment of seven other significant farms (Table 7-4) and innumerable pilot projects since 1981.

Fraudulent products purporting to contain *Spirulina,* concern over hygiene, and, presumably, the absence of miraculous effects on body weight all led to a decline in global *Spirulina* demand from the peak of 1981. U.S. consumption, for example, dropped from a peak of several hundred tons per year in 1982 to the current level of 75–100 mt·y^{-1}. Although the market for food-grade *Spirulina* in Japan has steadily increased, the growth rate has slowed, and current consumption is similar to that in the United States. Europe and the rest of the world probably account for less than an additional 100 mt·y^{-1}.

Despite the obvious oversupply, a 1985 survey of 58 buyers of bulk *Spirulina* powder in the United States (unpublished data) revealed an average price of US $20,000 per mt, ranging from $11,000 (for Sosa Texcoco material) to $28,000. There should be little, if any, profit margin included in these figures because of the competitive situation for producers. Why does *Spirulina* cost so much to produce? (The same question can be asked of *Chlorella*, which is sold at a similar price.)

B. Production costs

The technical literature contains many projections on construction and production costs of industrial microalgal facilities, including *Spirulina* farms. Few, if any, of these projections are based on experience above the pilot or even laboratory scale, and they typically suffer from severe inaccuracies. Problems encountered with full-scale systems producing food-grade material simply cannot be anticipated at the smaller scale in

Table 7-5. *Major categories of investment and production costs for a 5 ha Spirulina farm, with a harvest capacity of 40 kg·h⁻¹ dry wt, producing 60 T·y⁻¹ food-grade algae*

Item	Amount (US $1,000)	Amount (%)
Initial investment costs		
Land	—	—
Engineering & permits	88	5.6
Site preparation	50	3.2
Water and power networks	257	16.5
Laboratory, office, shop	79	5.1
Nutrient storage, supply	50	3.2
Cultivation system	493	31.6
Harvest system	541	34.7
Total	1,558	100
Annual production costs		
Labor	320	31.4
Fixed operating	163	16.0
Variable operating	103	10.1
Repair and maintenance	42	4.1
Administrative	78	7.7
Cost of capital/profit	156	15.3
Depreciation	156	15.3
Total	1,018	100

such a novel undertaking as microalgal production. The cost breakdown presented here (Table 7-5) is based on the construction and operating experience at Earthrise Farms. The analysis does not necessarily describe actual expenses; we now know that certain expenses are unnecessary or that design improvements merit additional expense.

Land costs are not included as they are so variable and can be considered a separate investment. Growing area is assumed to be 5 ha. The ponds are lined with a plastic membrane and the culture circulated by paddlewheels. The dewatered product is dried in a spray dryer. The harvest system is designed to handle 40 kg·h⁻¹ dry wt, or almost 1 mt·d⁻¹ if operated around the clock. This design capacity may seem excessive for a 5 ha farm in a temperate climate – 1 mt·d⁻¹ corresponds to 20 g·m⁻²·d⁻¹ – but investment costs are low for incremental increases in harvest capacity at this level of production and labor savings are significant when 24-hour harvest is unnecessary. After estimating the investment expenses, total production costs are determined by setting both cost of capital/profit and depreciation to 10% of construction costs (Table 7-5).

As mentioned previously, a temperate-zone farm can reliably produce

12 mt·ha^{-1}·y^{-1} of food-grade *Spirulina* powder, a total of 60 mt on 5 ha. According to Table 7-5, the ex factory price for *Spirulina* is then US $17,000 per mt, within the range of current world market prices. The yield probably can be doubled in temperate areas by increasing productivity with minor technical improvements and by lengthening the growing season with pond covers, which should not affect cultivation system costs significantly unless ponds are extremely large. Harvesting system costs also do not change as the design capacity is more than sufficient. Variable operating expenses – nutrients, drying, and packaging – double, and a 10% increase in labor costs might be necessary to handle the additional harvest plant operation. The ex factory price for 120 mt·y^{-1} of algae becomes US $9,600·mt^{-1}. Much lower costs at this scale of production would require a major breakthrough in production technology or strain selection, and even the present analysis makes unproven assumptions about the net effect of pond covers. Covers actually may lower production if the temperature improvement is more than offset by the decrease in solar radiation, and careful study should precede their installation.

Further decreases can be realized in areas with special local resources. For example, natural deposits of carbon dioxide lie underneath and adjacent to the Salton Sea near Earthrise Farms. As commercial carbon dioxide accounts for a few percentage points of the production cost by itself, exploitation of local deposits can yield significant savings. Similarly, geothermal heat is available at selected locations; Ein Yahav Algae in the Arava Valley of Israel is experimenting with the use of geothermal water combined with pond covers to extend the growing season. Creating alkaline water conditions with bicarbonate and carbonate is the major expense in preparing *Spirulina* culture medium and, at the Sosa Texcoco facility, the presence of naturally alkaline waters represents a major saving in operating expenses. The high production costs of small temperate-zone farms may force them to situate in locations with special resources (e.g., Wadi El Natroun in Egypt; El-Sherif and Clément 1982).

In tropical areas, a tripling of actual production to 36 mt·ha^{-1}·y^{-1} has been achieved (Table 7-4), even without pond covers. Average solar radiation is higher, production is possible all year long, and the equable climate favors a greater stability of mass cultures. Labor costs, the biggest single production expense, often are lower than for industrialized countries of the temperate zones. On the other hand, equipment and nutrient costs may be higher for imported items. Making the most simplistic assumption – that variable operating costs will triple and other expenses will experience no net change – yields an estimate of US $6,800 per mt for a 5 ha farm.

Numerous "economies of scale" are possible among both capital and operating costs. The harvesting system, for example, constitutes more than one-third of the investment for a 5 ha farm, but incremental costs for larger capacity decrease rapidly. The staff required to operate a food-grade farm is almost independent of farm size, as increasing the capacity of equipment does not require additional personnel. Similar comments apply to administrative and many fixed operating expenses. Tropical farms of much larger size thus represent the best direction for decreasing production expenses, in the absence of major technical advances. An accurate cost projection is not possible at this point.

C. Comparison with conventional foods

The least expensive sources of protein currently available in the United States are soybeans and wheat (Lappé 1982); both are appropriate for comparison with *Spirulina* because of a similar NPU. The production costs on a protein basis are roughly US \$1 per kg for soybeans and US \$2 per kg for wheat, so US \$1.50 per kg can be considered a target price for *Spirulina* protein. At a 63% average protein content, the spray-dried algae would have to be produced at approximately US \$1,000 per mt, a cost far below the capability of existing farms, regardless of latitude and ignoring any processing to remove color and odor. Although the high content of certain vitamins and minerals in *Spirulina* is an advantage over conventional protein sources, the RDA for these nutrients can be provided by other components of a well-balanced diet; even supplemental sources can provide the RDAs at less expense than *Spirulina* (e.g., the β-carotene in *Spirulina* increases its wholesale value less than \$0.50 per kg). Indeed, the current production costs for *Spirulina* protein resemble those for animal protein. *Spirulina* and other microalgae certainly cannot compete with conventional plant protein sources at current levels of production. On the other hand, competitive prices cannot be ruled out as the scale of production increases, even without major technological advances.

Apart from price, how do microalgae compare with conventional crops in utilization of natural resources? Perhaps the most striking difference is in land use. *Spirulina* can be grown on land completely unsuitable for terrestrial crops – as long as the terrain is flat – and requires much less area for a given protein production. Assuming 20 $mt \cdot ha^{-1} \cdot y^{-1}$ dry wt (global average: Table 7-4) and 63% average protein content, *Spirulina* protein yields of 13 $mt \cdot ha^{-1} \cdot y^{-1}$ are 20 times that of soybeans, its closest competitor. *Spirulina* production also may be unusually efficient in water or nutrient use, at least on a protein basis.

Although these features may be of interest in areas where land, water, or fertilizer availability limits agricultural production below local needs, they do not concern the poor who have no money to buy microalgae.

Regardless of efficient use of land and perhaps other resources, production costs must drop radically if commercial microalgae are to benefit the hungry (i.e., the poor). Unfortunately, the demand that could spur construction of large tropical farms does not yet exist, in large part because of current high prices. Are there other approaches to production that can relieve this impasse?

V. Therapeutic applications

Higher plants have provided many of our pharmaceutical needs for thousands of years. Microalgae have played an insignificant role, but not because of a lack of chemical diversity. As suggested earlier, the main limitation is difficulty of collection. Now that microalgae are produced in mass quantities and subjected to intensive chemical analyses, their richness in potential pharmaceutical applications is becoming apparent.

In the case of *Spirulina*, a number of possibilities are in various development stages (Table 7-6). Among published studies, the ability to lower serum cholesterol is well established and also has been observed with both *Chlorella* (Okuda et al. 1975) and *Scenedesmus* (Rolle and Pabst 1980). Also of interest is the study by Becker et al. (1986) supporting the use of *Spirulina* in weight control, the original foundation of its rise to popularity. In one of the more significant studies, Iijima and his coworkers demonstrated that oral administration of phycocyanin elevated spleen lymphocyte activity in mice, an observation that is now undergoing confirmation at Tohoku University School of Dentistry in Japan (H. Shimamatsu, personal communication). A Japanese patent for using phycocyanin as an antitumor agent already has been awarded on the basis of Iijima's work (DIC 1983). If phycocyanin is indeed a stimulant of certain immune system activities, it may find application against a wide range of diseases. Ongoing research at Harvard Medical School (Schwartz and Shklar 1986) also supports the use of *Spirulina* extracts in the treatment of certain cancers. All of these published and unpublished experiments require elaboration and confirmation, but the potential for diverse and significant therapeutic applications is undeniable and deserves more attention.

Applications can be predicated on the chemical composition of *Spirulina*, in addition to the possible therapeutic uses listed in Table 7-6. Hydrolyzed *Spirulina* seems to be completely palatable (Delpeuch et al. 1976) and may serve as a component of "elemental diets" (Bounous 1984) for patients undergoing and recovering from various traumas. The content of gamma-linolenic acid (GLA: 1% of dry wt) in *Spirulina* is unusually high and may give it a further role in treating certain degenerative diseases (Tudge 1981). Similarly, the β-carotene content is of interest because of epidemiological studies linking high dietary β-caro-

Table 7-6. *Published studies and unpublished experiments on potential therapeutic applications of Spirulina*

Application	Subject	Reference
External wounds	Humans	Clément, Rebeller, & Zarrouk 1967; Yoshida 1977
Infection (antibiotic action)	Microbial cells	Martinez Nadal 1970; Jorjani & Amirani 1978
Pesticide poisoning	Human cells	Matsueda et al. 1976
Hypothyroidism	Poultry	Babaev et al. 1979, 1980
Infection (immune stimulation)	Rabbits	Besednova et al. 1979
Hypercholesterolemia	Rats	Chen et al. 1981; Devi & Venkataraman 1983b; Kato et al. 1984
	Humans	Nakaya et al. 1986
Obesity	Humans	Becker et al. 1986
Oral cancer	Human cells	Schwartz and Shklar 1986
Hypochromic anemia	Humans	Takeuchi 1978[a]
Diabetes	Humans	Takeuchi 1979[a]
Hepatitis, cirrhosis	Humans	Takeuchi 1979; Miyairi 1982[a]
Pancreatitis	Humans	Tanaka 1980[a]
Cataracts	Humans	Yamazaki 1980[a]
Constipation	Humans	Sakai 1981[a]
Liver cancer (immune stimulation)	Mice	Iijima 1982[a]
Iron-deficient anemia	Rats	Takemoto 1982[a]
Stress ulcer	Rats	Takemoto 1982[a]
Allergy	Humans	Watanabe 1982[a]
Hypertension	Rats	Iwata 1984[a]

[a]H. Shimamatsu, personal communication; reports available from Drs. K. Takeuchi/Tokyo College of Medicine & Dentistry, M. Tanaka/Kyoto Medical College, Y. Yamazaki/Tokyo College of Medicine & Dentistry and Yamazaki Opthalmic Clinic, T. Sakai/University of Eastern Japan, N. Iijima/St. Maryanne College of Medicine, K. Takemoto/Saitama Medical College, K. Watanabe/Dohai Memorial Hospital, K. Iwata/Women's Nutritional College, all located in Japan.

tene intake with lower cancer rates (Shekelle et al. 1981). Both of these compounds are available from other sources at lower cost, but they do add to the value of *Spirulina* even if they cannot substantiate its price by themselves.

Almost any of the therapeutic applications of Table 7-6, if and when they are confirmed, can support the current production cost of *Spirulina*. Most of the unpublished Japanese studies with humans obtained positive results with less than 5 $g \cdot d^{-1}$, equivalent to an ex factory cost

of about US $0.10 and insignificant beside the cost of many question-
able medical treatments for the same problems. We earlier dismissed a
strictly nutritional role for *Spirulina* in industrialized countries because
of widespread availability of inexpensive alternatives. The information
of Table 7-6, however, does provide a separate basis for regular use of
Spirulina in small amounts.

Malnutrition results not only in the four major deficiency diseases
identified earlier but also in a suppression of certain activities of the
immune system. As a result, victims of malnourishment – whether it be
starvation or undernutrition – are more susceptible to infectious disease. If its immune stimulating activity is confirmed, then phycocyanin
can be added to the list of *Spirulina* constituents that seem so ideally
suited for renourishment programs.

How do these essentially therapeutic applications change the potential for making use of the strictly nutritional attributes of *Spirulina,* particularly its protein content? In certain cases where the pharmaceutically active ingredient has been identified, it can be extracted and used
independently. Phycocyanin, for example, which typically constitutes
10–20% of the dry weight, is water-soluble and easily extracted on a
large scale. The process has been developed primarily by Dainippon Ink
and Chemical of Japan (e.g., Shimamatsu and Kato 1979), which markets the blue pigment as a food coloring. If phycocyanin's medical
promise is realized, its price could completely offset the cost of production and extraction, yielding a high protein residue of low cost and less
intense color. Additional extractions perhaps could remove a fat-solu-
ble fraction (where most of the odor resides) with separate medical
applications. One then can envision a large tropical farm at which the
microalga is separated into several valuable extracts and a protein-rich
residue. The extracts are exported for foreign exchange, and the residue is made available locally at low cost. Because of the absence of
offensive color and odor, the residue is easily incorporated into traditional foods. All of the necessary steps have not been developed, but
this direction seems most promising for the future of current mass cultivation technology.

VI. Microalgae and hunger

A. Harvesting wild populations

Even if the above approach has value, its fruition will not be realized
soon. Can microalgae offer any respite from current problems of famine and undernutrition? Proposals have been made to take immediate

advantage of the large standing crops of phytoplankton, particularly *Spirulina,* present in the Rift Valley alkaline lakes of East Africa, which includes and is adjacent to areas that recently experienced famine. Despite the appeal of this suggestion – making use of a forgotten resource at a critical place and time – the following issues cannot be overlooked:

1. *Spirulina* density in these lakes is extremely variable in time and may fluctuate over more than an order of magnitude (Tuite 1981) for reasons not fully understood. Can such a patchy and unpredictable resource merit the construction of a large harvesting facility?

2. *Spirulina* concentrations usually are too low for efficient harvesting. The efficiency of harvesting devices drops rapidly as the algal concentration falls below 400 mg·L^{-1}. Most of these lakes have biomass levels below 100 mg·L^{-1} (Melack 1981, assuming chlorophyll *a* about 1% dry wt; Tuite 1981). Higher concentrations may occur as surface mats, but this material is undependable and often decomposed.

3. *Spirulina* harvest may eventually exhaust the nutrient reservoirs of certain lakes if natural nutrient inputs from rainfall, fallout, and watershed runoff are not sufficient to replenish nutrients removed by harvesting algae.

4. Large flamingo populations currently depend on natural *Spirulina* crops for their sustenance. In 1972–3, Lake Nakuru alone was host to a million lesser flamingos *(Phoeniconaias minor)* extracting about 60 mt·d^{-1} dry wt of *Spirulina* (Vareschi 1978). Half the world's flamingos may live in Rift Valley alkaline lakes, where they constitute an important tourist attraction and undoubtedly play a large role in local energy flow and materials cycles. How will large-scale algal harvest affect these birds and other animals that are directly or indirectly dependent on the *Spirulina* crop? (Vareschi and Jacobs [1985] provide a synopsis of the Nakuru ecosystem.)

5. Harvest of wild algal communities may pose a public health hazard because of toxic substances in the watershed or parasites of birds or other animals that can infect humans. Most of these lakes are not isolated from domestic or industrial pollution. Contamination by toxic algae also is a threat when harvesting wild communities. The U.S. Food and Drug Administration recently seized retail food supplements containing *Aphanizomenon flos-aquae,* a potentially toxic blue-green alga harvested from Klamath Lake, Oregon (Anonymous 1985).

These five points do not show that lake harvest is inevitably doomed to failure. They do point out, however, several technical, ecological, and public health issues that require serious evaluation before any large-scale venture is undertaken.

B. Village-scale production systems

The lake harvesting scheme and the industrial-scale farm are both large, mechanized, highly capitalized projects of the sort that has failed time and again in countries receiving development aid (Lappé, Collins, and Kinley 1980). Is there any way in which microalgal production can be incorporated into projects without these characteristics? Attempts are underway to bring small-scale microalgal production to the villages of developing countries. Demonstration projects in India, Togo, and Senegal currently are testing one scheme that addresses problems of poor sanitation and fuel scarcity as well as malnutrition (Fox 1984). The design consists of several integrated elements including community toilets, an anaerobic digester, and small *Spirulina* cultivation ponds. The digester processes sewage and other wastes, producing biogas for running community cooking facilities and a liquid effluent that is "pasteurized" in a solar heater and used to fertilize the algal ponds. *Spirulina* grows well in digester effluent diluted and enriched only with crude seasalt (Chaudhari, Krishnamoorthi, and Kotangale 1983). The algae are then harvested with woven cloth and dehydrated in a solar dryer. The system is designed to provide supplementary amounts of *Spirulina* – several grams each day – to the village inhabitants.

The goal of this design in undoubtedly ambitious. Problems with technology and training, public health hazards, and cultural obstacles to recycling waste all hinder progress. Simpler realizations of the general idea may be more likely to succeed soon, particularly designs oriented toward feed production (Nigam, Ramanatham, and Venkataraman 1981; Becker and Venkataraman 1984). Nonetheless, the system requires little investment, provides for community ownership and cooperation, and makes best use of existing resources with the aid of microalgae.

The record shows that large centralized undertakings fail to address the real causes of hunger, usually benefit only a minority, and even exacerbate difficulties for the rest of the population (IFDP 1979). Even if microalgal production does make better use of resources than conventional agriculture, no guarantee can be made that those truly in need will benefit. Agricultural planning often suffers from a narrow focus on productivity, as does the belief that microalgae can alleviate widespread malnutrition, but more efficient use of resources such as land is only one facet of the solution. Effective solutions are characterized by, or at least coexist with, more equable control over food-producing resources and an increase in local cooperative efforts, both of which are facilitated by these small integrated systems. Research and development of industrial-scale microalgal facilities is consuming huge financial resources;

one cannot help but wonder what benefits might arise if even a small fraction of this effort could be applied to integrated village systems.

VII. Acknowledgments

I am indebted to Bruce Carlsen, David Donnelly, Bob Henrikson, Yoshimichi Ohta, and Larry Switzer for their persevering support and good cheer during the development of Earthrise Farms. Additional thanks are due to Hidenori Shimamatsu of Dainippon Ink and Chemical in Tokyo for providing information on microalgal production and product development in East Asia.

VIII. References

Adams, C. F. 1975. *Nutritive Value of American Foods in Common Units.* U.S. Department of Agriculture Handbook no. 456. Superintendent of Documents, U.S. Government Printing Office, Washington, DC. 291 pp.

Aldave-Pajares, A. 1978. Cushuro: edible blue-green algae in the high mountain regions of Peru. *Biol. Soc. Bot. Liberated Trujillo* [Peru] 1, 5. (In Spanish.)

Anonymous. 1985. Algae eaters. *FDA Consumer* 19, 38–9.

Arai, S., M. Yamashita, and M. Fujimaki. 1976. Enzyme modifications for improving nutritional qualities and acceptability of proteins extracted from photosynthetic microorganisms *Spirulina maxima* and *Rhodopseudomonas capsulatus. J. Nutr. Sci. Vitaminol.* 22, 447–56.

Babaev, T., Y. Barashkina, M. Kuchkarova, A. Tulyaganov, G. Khodzhiakhmedov, and Y. Turakulov. 1979. Iodine-containing compounds of *Spirulina platensis* (Gom.) Geitl. *Uzb. Biol. Zh.* 5, 6–8. (In Russian.)

Babaev, T., K. Karimova, R. Sattyev, M. Kuchkarova, E. Zaripov, and Y. Turakulov. 1980. Thyroxine – active principle of *Spirulina. Uzb. Biol. Zh.* 1, 3–4. (In Russian.)

Baron, C. 1976. Study of *Spirulina* decoloration. *Ann. Nutr. Aliment.* 29, 615–22. (In French.)

Becker, E. W., B. Jakober, D. Luft, and R.-M. Schmülling. 1986. Clinical and biochemical evaluations of the alga *Spirulina* with regard to its application in the treatment of obesity: a double-blind cross-over study. *Nutr. Rep. Int.* 33, 565–74.

Becker, E. W., and L. V. Venkataraman. 1984. Production and utilization of the blue-green alga *Spirulina* in India. *Biomass* 4, 105–25.

Ben-Amotz, A., and M. Avron. 1980. Glycerol, β-carotene, and dry algal meal production by commercial cultivation of *Dunaliella.* In *Algae Biomass: Production and Use,* ed. G. Shelef and C. J. Soeder, pp. 603–10. Elsevier/North Holland Biomedical Press, Amsterdam.

Besednova, N., T. Smolina, L. Mikheiskaya, and R. Odova. 1979. Immunostimulating activity of lipopolysaccharides from blue-green algae. *Zh. Mikrobiol. Epidemiol. Immunobiol.* 56, 75–9. (In Russian.)

Bezares Sansores, A., M. Rossainz, and E. Avila Gonzales. 1977. Nutritional value of *Spirulina* algae *(Spirulina geitleri)* in diets for broiler chicks. *Tec. Pecu. Mex.* 32, 46–51. (In Spanish.)

Biswas, K. 1953. The algae as substitute food for human and animal consumption. *Sci. and Cult.* 19, 246–9.

Blum, J., and C. Calet. 1976. Nutritive value of *Spirulina* algae for the growth of broiler chicks. *Ann. Nutr. Aliment.* 29, 651–74. (In French.)

Borowitzka, M. A., and L. J. Borowitzka, (eds.). In press. *Microalgal Biotechnology.* Cambridge University Press, Cambridge.

Boudène, C., E. Collas, and C. Jenkins. 1976. Characterization and measurement of various mineral toxicants in *Spirulina* algae of different origins: evaluation of long term toxicity in rats of a sample of Mexican *Spirulina. Ann. Nutr. Aliment.* 29, 579–88. (In French.)

Bounous, G. 1984. A review of elemental diets during cancer therapy. *Anti-Cancer Res.* 3, 299–304.

Bourges, H., A. Sotomayor, E. Mendoza, and A. Chávez. 1971. Utilization of the alga *Spirulina* as a protein source. *Ann. Nutr. Aliment.* 29, 675–82. (In French.)

Brune, H. 1980. Microalgae for the nutrition of man and animals: results and problems. In *Prospects of Algae Research in Egypt: Proceedings of the Second Egyptian Algae Symposium, March 11–13, 1979, Cairo,* ed. M. M. El-Fouly, pp. 85–110. National Research Centre, Cairo.

Brune, H. 1982. Single cell algae *Spirulina maxima* and *Scenedesmus acutus* as a lone protein source for broilers. *Z. Tierphysiol. Tierernaehr Futtermittelkd* 48, 143–54. (In German.)

Burlew, J. S. (ed.). 1953. *Algal Culture from Laboratory to Pilot Plant.* Carnegie Institution, Publication 600. Washington, DC. 325 pp.

Chamorro Cevallos, G. 1980. *Toxicological Studies on* Spirulina *Alga: Sosa Texcoco, S.A., Pilot Plant for the Production of Protein from* Spirulina *Alga.* Rep. no. UF/MEX/78/048. United Nations Industrial Development Organization, Vienna, 177 pp. (In French.)

Chaudhari, P., K. Krishnamoorthi, and L. Kotangale. 1983. Growth potential of *Spirulina platensis,* a blue-green alga, in night soil digester effluent and saline water. *Indian J. Envir. Health* 25, 275–81.

Chen, L. C., J. S. Chen, and T. C. Tung. 1981. Effects of *Spirulina* on serum lipoproteins and its hypocholesterolemic activities. *Taiwan I Hseuh Hui Tsa Chih* 80, 934–42. (In Chinese.)

Chen, S., and H. Peppler. 1978. Single-cell proteins in food applications. *Dev. Indian Microbiol.* 19, 79–93.

Clément, G., C. Giddey, and R. Menzi. 1967. Amino acid composition and nutritive value of the alga *Spirulina maxima. J. Sci. Food Agric.* 18, 497–501.

Clément, G., M. Rebeller, and C. Zarrouk. 1967. Wound treating medicaments containing algae. French Patent no. 5279, assignee Institut Français du Pétrole, Sept. 11.

Contreras, A., D. Herbert, B. Grubbs, and I. Cameron. 1979. Blue-green alga *Spirulina* as the sole dietary source of protein in sexually maturing rats. *Nutr. Rep. Int.* 19, 749–64.

Dangeard, P. 1940. On a blue alga edible for man: *Arthrospira platensis* (Nordst.) Gomont. *Actes Soc. Linn. Bordeaux (Extr. Proc. Verb.)* 91, 39–41. (In French.)

Deevey, E. S., Jr. 1957. Limnological studies in Middle America. *Trans. Connecticut Acad. Arts Sci.* 39, 213–328.

de Lagerheim, M. G. 1892. La "Yuyucha." *La Nuovo Notarisia* 3, 137–8. (In Spanish.)

Delpeuch, F., A. Joseph, and C. Cavalier. 1976. Consumption as food and nutritional composition of blue-green algae. *(Oscillatoria platensis)* among several populations in the Kanem region of Chad. *Ann. Nutr. Aliment.* 29, 497–516.

Devi, M. A., and L. V. Venkataraman. 1983a. Supplementary value of the proteins of blue green algae *Spirulina platensis* to rice and wheat proteins. *Nutr. Rep. Int.* 28, 1029–36.

Devi, M. A., and L. V. Venkataraman. 1983b. Hypocholesterolemic effect of bluegreen algae *Spirulina platensis* in albino rats. *Nutr. Rep. Int.* 28, 519–30.

DIC [Dainippon Ink and Chemical]. 1983. Antitumoral agents containing phycobilin. Japanese Patent no. 58065216, assignees Dainippon Ink and Chemical and Tokyo Stress Foundation, Apr. 18, 6 pp.

Durand-Chastel, H., and G. Clément. 1975. *Spirulina* algae: food for tomorrow. In *Proceedings of the Ninth International Congress on Nutrition, Mexico, 1972,* vol. 3, pp. 85–90. S. Karger, Basel.

Durand-Chastel, H., and C. Santillán. 1975. Progress in industrialization of photosynthesis with the alga *Spirulina*. Presented at the first Chemical Congress of the North American Continent, Nov. 30–Dec. 5, Mexico City. (In Spanish.)

Elenkin, A. A. 1931. On some edible freshwater algae. *Priroda* 20, 964–91. (In Russian.)

El-Sherif, S., and G. Clément. 1982. Edible alga *Spirulina* sp.: discovery of natural lakes at Wadi El Natroun in the Arab Republic of Egypt. *Rev. Inst. Fr. Pet.* 37, 123–30. (In French.)

Farrar, W. V. 1966. Tecuitlatl: a glimpse of Aztec food technology. *Nature* 211, 341–2.

Février, C., and B. Sève. 1976. Incorporation of *Spirulina (Spirulina maxima)* into pig diets. *Ann. Nur. Aliment.* 29, 625–30. (In French.)

Fica, V., D. Olteanu, and S. Oprescu. 1984. Observations on the utilization of *Spirulina* as an adjuvant nutritive factor in treating some diseases accompanied by a nutritional deficiency: preliminary study. *Rev. Med. Interna. Neurol. Psihiatr. Neurochir. Dermato. Venerol. Ser. Med. Interna.* 36, 225–32. (In Rumanian.)

Florenzano, G. 1975. Microalgal protein as food. *Riv. Ital. Sostanze Grasse* 52, 11–25. (In Italian.)

FNB [Food and Nutrition Board, National Research Council]. 1980. *Recommended Dietary Allowances.* 9th rev. ed. National Academy of Sciences, Washington, DC. 185 pp.

Fox, R. 1984. *Algoculture.* Doctoral thesis. Louis Pasteur University, Strasbourg. 336 pp.

Garson, J., M. Maigrot, and F. Busson. 1969. Cyanophyceae utilization in human nutrition. *Med. Trop.* [Marseilles] 29, 536–9. (In French.)

Gromov, B. V. 1968. Main trends in experimental work with algal cultures in the U.S.S.R. In *Algae, Man, and the Environment*, ed. D. F. Jackson, pp. 249–78. Syracuse University Press, Syracuse, NY.

Herbert, V., and G. Drivas. 1982. *Spirulina* and vitamin B_{12}. *J. Amer. Med. Assoc. (Lett.)* 248, 3096–7.

Herbert, V., G. Drivas, R. Foscaldi, C. Manusselis, N. Colman, S. Kanazawa, K. Das, M. Gelernt, B. Herzlich, and J. Jennings. 1982. Multivitamin/multimineral food supplements containing vitamin B_{12} may also contain analogues of vitamin B_{12}. *J. Amer. Med. Assoc. (Lett.)*, 307, 255–6.

Higgins, G., A. Kassam, L. Naiken, G. Fischer, and M. Shah. 1983. *Potential Population Supporting Capacities of Lands in the Developing World.* Technical Report of Project FPA/INT/513. Food and Agriculture Organization of the United Nations, Rome. 139 pp.

IFDP [Institute for Food and Development Policy]. 1979. *Food First Resource Guide: Documentation on the Roots of World Hunger and Rural Poverty.* Institute for Food and Development Policy, San Francisco. 76 pp.

Jao, C.-C. 1947. *Prasiola yunnanica sp. nov. Bot. Bull. Acad. Sin.* 1, 110.

Jassey, Y., J.-P Berlot, and C. Baron. 1971. Comparative study of the nucleic acids of two *Spirulina* species: *Spirulina platensis* Geitler and *Spirulina maxima* Geitler. *C. R. Acad. Sci. Paris, Sér. D* 273, 2365–8. (In French.)

Johnson, P. E. and L. E. Shubert. 1986. Availability of iron to rats from *Spirulina*, a blue-green alga. *Nutr. Res.* 6, 85–94.

Johnston, H. W. 1970. The biological and economic importance of algae. III. Edible algae of fresh and brackish waters. *Tuatara* 18, 19–35.

Jorjani, G. H., and P. Amirani. 1978. Antimicrobial activities of *Spirulina platensis*. *Maj. Ilmy Puzshky Danishkadah Jundi Shapur* 1, 14–18.

Kato, T., K. Takemoto, H. Katayama, and Y. Kuwabara. 1984. Effect of *Spirulina (Spirulina platensis)* on dietary hypercholesterolemia in rats. *Jap. Nutr. Food Soc. J.* 37, 323–32. (In Japanese.)

Khan, M. 1973. On edible *Lemanea* Bory de St. Vincent: a fresh water red alga from India. *Hydrobiologia* 43, 171–5.

Kofranyi, E. 1978. The nutritional value of the green alga *Scenedesmus acutus* for humans. *Ergebn. Limnol.* 11, 150.

Krishnakumari, M., H. Ramesh, and L. Venkataraman. 1981. Food safety evaluation: acute oral and dermal effects of the algae *Scenedesmus actus* and *Spirulina platensis* on albino rats. *J. Food Prot.* 44, 934–5.

Lappé, F. M. 1982. *Diet for a Small Planet.* Ballentine, New York. 498 pp.

Lappé, F. M., J. Collins, and D. Kinley. 1980. *Aid as Obstacle: Twenty Questions about Our Foreign Aid and the Hungry.* Institute for Food and Development Policy, San Francisco. 191 pp.

Laquerbe, B., F. Busson, and M. Maigrot. 1970. On the mineral composition of two cyanophytes, *Spirulina platensis* (Gom.) Geitler and *Sp. geitleri* J. de Toni. *C. R. Acad. Sci. Paris. Sér. D*, 270, 2130–2. (In French.)

Léonard, J., and P. Compère. 1967. *Spirulina platensis*, a blue alga of great nutritive value due to its richness in protein. *Bull. Jard. Bot. Nat. Bruxelles* (1 Suppl.) 37, 1–23. (In French.)

Liener, I. E. 1966. In *World Protein Resources*, ed. R. F. Gould, p. 179. Advances in Chemistry Series no. 57. American Chemical Society, Washington, DC.

McConnaughey, E. 1985. Seaweed farming. *Appl. Phycol. Forum* 2(2), 1–3.

Martinez Nadal, N. G. 1970. Antimicrobial activity of *Spirulina maxima*. Presented at the Tenth International Congress of Microbiology, Aug., Mexico City.

Materassi, R., D. Ricci, and A. Tofani. 1974. Vitamin B_{12} production by *Spirulina* strains. *Riv. Ital. Sostanze Grasse* 51, 465–7. (In Italian.)

Matsueda, S., M. Nagaki, and K. Shimpo. 1976. Isolation and purification of the human erythrocyte-cholinesterase and the enzymoeffect of *Chlorella* components to the activity. *Sci. Rep. Hirosaki Univ.* 23, 17–23.

Melack, J. 1981. Photosynthetic activity of phytoplankton in tropical African soda lakes. *Hydrobiologia* 81, 71–85.

Miller, D. S., and P. R. Payne. 1961. Problems in the prediction of protein values of diets: the use of food composition tables. *J. Nutr.* 74, 413–19.

Monsen, E., L. Hallberg, M. Layrisse, D. Hegsted, J. Cook, W. Mertz, and C. Finch. 1978. Estimation of available dietary iron. *Amer. J. Clin. Nutr.* 31, 134–41.

Nakaya, N., Y. Honma, and Y. Goto. 1986. The effect of *Spirulina* on reduction of serum cholesterol. *Prog. Med.* 6, 3125–34. (In Japanese.)

Narashima, D., G. Venkataraman, S. Duggal, and B. Eggum. 1982. Nutritional quality of the blue-green alga *Spirulina platensis*. *J. Sci. Food Agric.* 33, 456–60.

Nigam, B. P., P. K. Ramanatham, and L. V. Venkataraman. 1981. Simplified production technology of blue-green alga *Spirulina platensis* for feed applications in India. *Biotechnol. Lett.* 3, 619–22.

Okamura, K. 1913. On Chinese edible *Nostoc* (Fahtsai) identified by Prof. Setchell as *Nostoc commune* var. *flagelliforme*. *Bot. Mag. Tokyo* 27, 177–83.

Okuda, M., T. Hasegawa, M. Sonoda, and Y. Tanaka. 1975. The effect of *Chlorella* on the levels of cholesterol in serum and liver. *Jap. J. Nutr.* 33, 3–8. (In Japanese.).

Orr, M. L. 1969. *Pantothenic Acid, Vitamin B_6, and Vitamin B_{12} in Foods*. U.S. Department of Agriculture Home Economics Research Rep. no. 36. Superintendent of Documents, U.S. Government Printing Office, Washington, DC. 53 pp.

Ortega, M. M. 1972. Study of the edible algae of the Valley of Mexico. *Bot. Mar.* 15, 162–6.

PAG [Protein-Calorie Advisory Group]. 1975. PAG ad hoc working group meeting on clinical evaluation and acceptable nucleic acid levels of SCP for human consumption. *PAG Bull.* 5(3), 17–26.

Palla, J.-C., and F. Busson, 1969. Study of the carotenoids of *Spirulina platensis* (Gom.) Geitler (Cyanophyceae). *C. R. Acad. Sci. Paris, Sér. D* 269, 1704–7. (In French.)

Paoletti, C., R. Materassi, and W. Balloni. 1974. Nucleic acid content of algae biomass. *Riv. Ital. Sostanze Grasse* 51, 432–4. (In Italian.)

Paoletti, E., E. Pelosi, and M. Narese Filastò. 1971. Accumulation of iodine in *Spirulina platensis*. *Agr. Ital.* [Pisa] 71, 319–26. (In Italian.)

Ramalingaswami, V. 1977. Knowledge and action in the control of vitamin A deficiency. *Ann. N.Y. Acad. Sci.* 300, 210–20.

Ramos Galvan, R. 1973. Clinical experimentation with *Spirulina*. Presented at the Colloquium on the Nutritional Value of the Alga *Spirulina,* May, Rueil-Malmaison, France. (In Spanish.)

Rao, D., G. Venkataraman, and S. Duggal. 1981. Amino acid composition and protein efficiency ratio of *Spirulina platensis. Proc. Indian Acad. Sci. Plant Sci.* 90, 451–6.

Riccardi, G., A. M. Sanangelantoni, D. Carbonera, A. Savi, and O. Cifferi. 1981. Characterization of mutants of *Spirulina platensis* resistant to amino acid analogues. *FEMS Microbiol. Lett.* 12, 333–6.

Richmond, A. (ed.). 1986. *Handbook of Microalgal Mass Cultivation.* CRC Press, Boca Raton, FL. 526 pp.

Rolle, I., and W. Pabst. 1980. On the cholesterol-lowering effect of single-cell green alga *Scenedesmus acutus* 276–3a. *Algensubstanz. Nutr. Metab.* 24, 291–301. (In German.)

Santillán, C. 1982. Mass production of *Spirulina. Experientia* 38, 40–3.

Sautier, C., and J. Trémolières. 1976. Food value of *Spirulina* algae in humans. *Ann. Nutr. Aliment.* 29, 517–34. (In French.)

Schwartz, J. L., and G. Shklar. 1986. Growth inhibition and destruction of oral cancer cells by extracts of *Spirulina. Proc. Amer. Acad. Oral Pathol.* 40, 23.

Shekelle, R., S. Liu, W. Raynor, M. Lepper, C. Muliza, and A. Rossof. 1981. Dietary vitamin A and risk of cancer in the Western Electric study. *Lancet* 11, 1185–9.

Shelef, G., and C. J. Soeder (eds.). 1980. *Algae Biomass: Production and Use.* Elsevier/North Holland Biomedical Press, Amsterdam. 852 pp.

Shimamatsu, H., and T. Kato. 1979. Blue pigments. Japanese Patent no. 54101833, assignee Dainippon Ink and Chemical, Aug. 10.

Smith, H. M. 1933. An edible mountain-stream alga. *J. Siam Soc. Nat. Hist. Suppl.* 9, 143.

Taft, C. E. 1965. *Water and Algae: World Problems.* Educational Publishers, Chicago. 236 pp.

Tanaka, Y., H. Matsuguchi, and T. Katayama. 1974. Comparative biochemistry of carotenoids in algae. IV. Carotenoids in cyanophyte, blue-green alga, *Spirulina platensis. Mem. Fac. Fish., Kagoshima Univ.* 23, 111–15.

Tiffany, L. H. 1958. *Algae, the Grass of Many Waters.* Charles C. Thomas, Springfield, IL. 199 pp.

Til, H. P., and M. Willems. 1971. *Subchronic (90-Day) Toxicity Study with Dried Algae (M₂) in Albino Rats.* Rep. no. R3352. Central Institute for Nutrition and Food Research, Zeist, Netherlands. 11 pp.

Tudge, C. 1981. Why we could all need the evening primrose. *New Scientist,* 19 Nov., 506.

Tuite, C. 1981. Standing crop and distribution of *Spirulina* and benthic diatoms in East African alkaline saline lakes. *Fresh Water Biol.* 11, 345–60.

Tulliez, J., G. Bories, C. Boudène, and C. Février. 1976. Hydrocarbons from *Spirulina* algae: their nature and fate of heptadecane in rats and pigs. *Ann. Nutr. Aliment.* 29, 563–72. (In French.)

Vareschi, E. 1978. The ecology of Lake Nakuru (Kenya). 1. Abundance and feeding of the lesser flamingo. *Oecologia* [Berlin] 32, 11–35.

Vareschi, E., and J. Jacobs. 1985. The ecology of Lake Nakuru (East Africa). 6. Synopsis of production and energy flow. *Oecologia* [Berlin.] 65, 412–24.

Waslien, C. 1975. Unusual sources of protein for man. *Crit. Rev. Food Sci. Nutr.* 6, 77–151.

Waslien, C., D. Calloway, S. Morgen, and F. Costa. 1970. Uric acid levels in men fed algae and yeast as protein sources. *J. Food Sci.* 35, 294–8.

Watanabe, A. 1970. Studies on application of Cyanophyta in Japan. *Schweiz. Z. Hydrol.* 32, 566–9.

Wood, E. J. F. 1965. *Marine Microbial Ecology.* Chapman and Hill, London. 243 pp.

Wouters, J. 1969. Influence of mineral surroundings on the weighable yield and the spectrum of amino acids of *Spirulina platensis* (Gom.) Geitl. *Ann. Physiol. Veg. Univ. Bruxelles* 14, 85–105. (In French.)

Yap, T., J. Wu, W. Pond, and L. Krook. 1982. Feasibility of feeding *Spirulina maxima, Arthrospira platensis,* or *Chlorella* sp. to pigs weaned to a dry diet at 4 to 8 days of age. *Nutr. Rep. Int.* 25, 543–52.

Yoshida, R. 1977. *Spirulina* hydrolyzates for cosmetic packs. Japanese Patent no. 52031836, assignee R. Yoshida, Mar. 10.

Yoshino, Y., Y. Hirai, H. Takahashi, N. Yamamoto, and N. Yamazaki. 1980. The chronic intoxication test on *Spirulina* produce fed to Wister-strain rats. *Eiyogaku Zasshi* 38, 221–5. (In Japanese.)

8. Some public health aspects of microalgal products

ALAN JASSBY

Institute of Ecology, University of California, Davis, CA 95616

CONTENTS

I. Introduction

Although several populations throughout the world gather wild algae from inland habitats for food, many people associate the word *algae* with decay and uncleanliness. In both developing and developed countries, incomplete waste treatment and inefficient fertilizer use lead to the foul odors and unsightly scums of hypereutrophic water. Most people first encounter microalgae in this negative context.

Commercial microalgal products, in particular *Chlorella* (a green alga) and *Spirulina* (a blue-green alga), have been accepted to a degree in industrialized countries. People are responsive to health-related information on food, and education on the unusual nutritional properties of certain microalgae often overcomes unconscious or uninformed prejudices. Nonetheless, inexperience with these unconventional foods breeds caution and may result in an overreaction to any public health problems that might arise. For example, *Salmonella* poisoning from spoiled meat products rarely is perceived as an indictment of meat as a nutrient source, but a similar incident with microalgae might be interpreted as a generic problem, not an isolated incident. Public health difficulties are of concern not only because they affect individual well-being but also because they might delay widespread acceptance of microalgal foods with all the attendant ecological and nutritional benefits.

Public health issues fall into two categories. First, we must ask whether substances indigenous to a microalgal strain are health hazards. Examples include neurotoxins of certain blue-green algae (Hughes, Gorham, and Zehnder 1958; Gorham and Carmichael, Chapter 16), allergens (McElhenney et al. 1962), and high nucleic acid content (Waslien et al. 1970). Hazardous strains must be detected by chemical investigations and feeding studies prior to any decision on commercial cultivation. Second, we must determine if production, harvesting, processing, and storage result in harmful properties, through either the decomposition of an indigenous substance or the introduction of contaminants. This second category forms our subject matter. Most issues of hygiene concerning conventional foods apply equally to microalgae,

[182]

Fig. 8-1. View of the main *Spirulina* culture ponds at Earthrise Farms in the low desert of southern California. Each pond is 5,000 m² in surface area and about 0.2 m deep. Photograph was taken from the top of the spray-drying tower. (Photo by author.)

and these issues are treated thoroughly in a number of food science publications (e.g., Banwart 1981 provides a comprehensive introduction). Here, I concentrate on practical problems unique to microalgal food production.

Many of the practical considerations dealt with in this chapter were uncovered during the operation of Earthrise Farms, a commercial *Spirulina* farm in the Colorado Desert of southern California (Fig. 8-1; see also Jassby, Chapter 7). The first phase of the Earthrise Farms project, consisting of 10 production ponds of 5,000 m² surface area each and a number of smaller seed and experimental ponds, began operation in the summer of 1982. Data presented in this chapter not published previously were collected by the staff of Earthrise Farms (see Section VIII, Acknowledgments) or by independent laboratories under contract to Earthrise Farms, unless otherwise indicated.

II. Pheophorbide

In 1977, 23 Tokyo residents contracted photosensitive dermatitis, a skin inflammation aggravated by exposure to light (Tamura et al. 1979). Investigation revealed that all patients recently had consumed the same brand of *Chlorella* tablets. The tablets contained large amounts of pheophorbide and its ester and provoked a similar condition in mice during subsequent laboratory experiments. These experiments also established a linear dose–response relationship between the dermatitis severity and the pheophorbide content of *Chlorella,* strongly suggesting that this phototoxic pigment caused the dermatitis outbreak (Miki et al. 1980; Seki et al. 1981).

Pheophorbides are naturally occurring degradation products of chlorophylls that form when a chlorophyll molecule loses its magnesium atom and phytol residue. All living plants capable of photosynthesis contain pheophorbide, and the amount usually increases when the plant dies. Pheophorbide formation depends on acidity, which removes the magnesium atom, and chlorophyllase activity, which catalyzes the loss of the phytol residue (Klein and Vishniac 1961). Pheophorbide is assimilated into the blood stream and travels throughout the body. When pheophorbide is exposed to light, after transfer to the skin or other organs near the body surface, oxygen is formed and cell membrane fatty acids are oxidized. The membranes rupture and capillary wall permeability increases, leading to rashes and itchiness (Matsuura et al. 1982).

In light of the dermatitis outbreak and subsequent findings, the Japanese Ministry of Public Welfare recommends a limit of 0.8 mg$\cdot$g^{-1} for existing pheophorbide in microalgae. In addition, total pheophorbide (the amount that potentially can form) is limited to 1.2 mg$\cdot$g^{-1}. Total pheophorbide is determined in the same manner as existing pheophorbide except for a preliminary three-hour incubation at $37°C$ to promote the conversion of chlorophyll to pheophorbide by the chlorophyllase present. The tablets responsible for the Tokyo outbreak contained up to 8.2 mg$\cdot$g^{-1} total pheophorbide.

Tamura et al. (1979) also established that ethanol used in granulation of *Chlorella* for tablet manufacture enhanced chlorophyllase activity. Acetone and methanol can play a similar role. The combination of high *Chlorella* chlorophyll content and contact with ethanol was responsible for the hazardous levels of pheophorbide. As a result, the Japanese ministry further recommended heating the dried powder for three minutes at $100°C$ prior to processing in order to inactivate chlorophyllase, as well as avoiding contact with alcohol and acetone.

Since the start of commercial *Spirulina* production, total pheophorbide has averaged 0.23 mg$\cdot$g^{-1} at Earthrise Farms. Aside from a single anomalous sample of 1.6 mg$\cdot$g^{-1}, all values are well within the Japanese ministry guideline, ranging from 0.053 to 0.88. A distinct seasonal pattern exists that bears a close relationship to temperature (Table 8-1). Ambient temperature is expected to influence pheophorbide levels via chlorophyllase activity, but other factors such as spray-drying conditions probably play a role. In general, pheophorbide may pose less of a prob-

Table 8-1. *Mean monthly ambient temperature, total pheophorbide (tPPB), standard plate count (SPC), and mold count for the main Spirulina growing season at Earthrise Farms in 1984.*

Month	Temperature (°C)	tPPB (mg·g^{-1})	SPC (10^5·g^{-1})	Mold (g^{-1})
Apr	19.0	n.d.	0.24	9.1
May	27.2	0.10	3.3[a]	16.5
Jun	28.8	0.26	4.6	12.0
Jul	33.2	0.48	6.5	12.6
Aug	33.2	0.48	6.0	18.7
Sep	31.6	0.33	1.9	28.0
Oct	20.9	0.10	0.77	26.0

[a]Three suspect values (out of 25 measurements) excluded from SPC average.

lem in *Spirulina* than in *Chlorella* because of the former's relatively low chlorophyll content (typically less than 1% dry wt, compared to 2–3% for commercial *Chlorella* products).

In contrast to the chlorophylls, no adverse effects associated with accessory pigments (i.e., carotenoids and phycobiliproteins) or their degradation products have been identified. Excessive consumption of carotenoids can alter skin color, but the coloration is harmless and disappears when intake is reduced (NRC 1980).

III. Microbial contaminants

A. Standard plate count

Microalgae usually are cultivated in large exposed basins allowing contact with animal life. Local ground or surface water is used in its natural condition as the volumes required are too large for economical treatment. Many opportunities exist in this situation for contamination of harvested algae by pathogenic microorganisms. The small size of microalgae prevents easy separation of product and microbial contaminants, especially as the mucilaginous sheaths of many algae provide a strong attachment for adhering microorganisms. Detailed examination of the microbial flora in dried product is required. The indicator microorganisms used for conventional foods are specific to the contamination source, not the food type, and these same methods usually are applicable for quality testing of microalgae.

The standard plate count (SPC) is an important exception to this principle. Outdoor microalgal cultures are relatively dense (100–1,000 mg·

L^{-1} on a dry weight basis) and have high levels of dissolved organic matter from extracellular products of photosynthesis and cell lysis. Water temperature often is high, especially for such desert types as *Spirulina* that grow optimally at around human body temperature. Bacteria proliferate under these conditions: SPC levels of 2×10^5 to $6 \times 10^5 \cdot mL^{-1}$ of culture reported for the Sosa Texcoco *Spirulina* farm in Mexico (Santillán 1974) also are typical of Earthrise Farms. *Spirulina* does have certain antimicrobial properties (Martinez Nadal 1971), but gram-negative bacteria were unaffected by *Spirulina* extracts in laboratory tests (Jorjani and Amirani 1978). Two reports identify *Pseudomonas* spp. as the predominant bacteria accompanying *Spirulina* in culture (David, Santillán, and Clément 1970; Kondrat'eva and Galkina 1979).

In general, SPC values of products from different sites will not be similar. Table 8-1 illustrates large seasonal changes even for a single site, depending on the ambient temperature. The variety of processing techniques, especially the presence or absence of a pasteurization step, introduces even greater discrepancies among SPC levels from different farms. Length of storage is an additional factor (Jacquet 1976). Accordingly, SPC values cannot serve as a universal indicator of the hygienic quality of microalgal powder.

Nonetheless, SPC analysis is of value in monitoring conditions within a particular site, where unusual SPC values often indicate abnormal pond or drying plant behavior. For example, when algae are subject to metabolic stress or parasite attack, cell lysis products provide a greater substrate for bacterial growth and result in an observable increase in SPC levels of dried algae. Undesirable culture conditions, however, are usually identified in more direct ways, including visual inspection, before SPC increase is apparent. SPC analysis is most useful in monitoring drying plant conditions. By bracketing each unit process with SPC analyses, conditions that favor breeding of microbes can be identified and particular pieces of equipment can be targeted for extra cleaning or replacement.

In general, SPC levels for *Spirulina* on a dry weight basis decrease two orders of magnitude during dewatering. Subsequent spray drying lowers SPC an additional order of magnitude to between 10^4 and $10^6 \cdot g^{-1}$. With the addition of a pasteurization step, SPC is reduced to between 10^2 and $10^4 \cdot g^{-1}$ at the Sosa Texcoco farm. The heat treatment is conducted at "65°C for 30 sec" (Gronek 1983).

B. Enteric indicators

Cultivation ponds in desert areas may be a source of water for animals, and the dense avian and rodent life in agricultural areas cannot be excluded completely from ponds. Contamination of the pond water with the intestinal flora of these visitors can occur, as is true for any conventional crop. The absence of sources of human fecal contamination does not rule out health hazards as man shares many pathogenic organisms with other animals, particularly *Salmonella* species in the case of birds and rodents. Unless a recent and massive contamination has occurred, pathogens in food or water exist in small numbers and often cannot be isolated. As a result, food products must be examined for "indicator" bacteria that inhabit the intestinal tract in large numbers but are not necessarily disease agents. The coliform group (FDA 1978) in particular is widely used as an indicator for food because of a long history of success in water analysis and the simplicity of detection.

The coliform test is only an indication of potential contamination and must be interpreted in context. In studies at Earthrise Farms, positive presumptive coliform tests are almost always followed by negative confirmed tests. Presumably, some of the noncoliform organisms present ferment nonlactose carbohydrates in the *Spirulina* samples. Studies on product from Chad, Algeria, and Mexico showed high coliform counts (Jacquet 1976), but current commercial systems generally produce coliform-free algal powder. Analysis of hundreds of *Spirulina* samples from commercial farms in Thailand, Japan, Taiwan, and Mexico, as well as from Earthrise Farms in the United States, shows a consistent picture: Whether pasteurization is used or not, coliforms are rarely present, and tests for *Escherichia coli* or fecal coliforms are almost always negative.

The apparent absence of enteric bacteria is surprising in view of the susceptibility of outdoor cultures to contamination. Perhaps the high osmolarity and pH of *Spirulina* ponds reduces survival of coliforms. Joshi, Parhad, and Rao (1973) and Parhad and Rao (1974), for example, have shown that a pH above 9 is lethal to coliforms and *Salmonella*. Fecal streptococci may be more appropriate indicators for *Spirulina* and other algae grown in environments of extreme temperature, pH, and salinity. Jacquet (1976) was able to find fecal streptococci in *Spirulina* powder where coliforms were not detected, but never the converse. *Streptococcus faecalis*, the principle species, has a wide temperature range (10–45°C) and can grow at a pH of 9 to 10 and in salinities of 6% to 7%. The fecal streptococci survive outside the intestine better than coliforms and tolerate spray drying and storage better than *E. coli* and coliforms (Banwart 1981).

C. Specific enteric pathogens

Unless the actual pathogens that coliforms or other indicators trace are equally susceptible to growing and processing conditions, the results to date are not sufficient evidence that all commercial algal products are free from enteric pathogens. Experiments with other foods that correlate the presence of coliforms with specific pathogens are not applicable to the unique conditions, especially the often extreme culture media, used in microalgal cultivation. Similar comments apply to the other common indicator groups: fecal coliform, *Escherichia coli,* and fecal streptococci. The value of applying traditional indicators to microalgal food products cannot be assumed blindly. In view of this uncertainty, analyses for specific pathogens must be conducted in more detail. Unfortunately, lack of facilities, trained analysts, and standardized methods typically makes routine examination for pathogens impossible at farms involved in commercial production. Pathogen tests are conducted sporadically, and a detailed examination is made only when a high SPC level, the presence of coliforms, or some other indicator sounds a warning signal.

The most common microbial pathogens in the water of industrialized countries with temperate climates are *Salmonella, Shigella,* enteropathogenic *E. coli, Leptospira,* and the enteric viruses (APHA 1971). Qualitative analysis for the first three groups is fairly standardized, and these can be examined on a regular basis. Testing of many commercial *Spirulina* products from around the world (unpublished data) has yet to reveal any occurrence of these organisms, and there appears to be no peculiar hazard of microalgal powder contamination by enteric bacteria.

Nevertheless, producers must remain alert for the unexpected, especially in managing huge outdoor cultures. Both enteropathogenic *E. coli* and *Leptospira* have been detected in a *Spirulina* growing system (David et al. 1970; Gonzáles Arroyo et al. 1976). Also, little effort has been made to screen for enteric viruses or *Clostridium perfringens* in quality-assurance tests. Along with *Salmonella* and *Shigella, C. perfringens* is an enteric pathogen most likely to be involved in food-borne illness; together, these three organisms account for more than 50% of food-borne disease cases (CDC 1977).

Systems using treated wastewater as a nutrient source (Oswald 1979) represent a completely different level of sanitary challenge. The possibilities of disease transmission (reviewed by Cooper 1962) are large and varied, and a great deal of study is required to determine the public health hazard. In certain small-scale systems integrating waste treatment and algal production (Fox 1983), a number of physical and chemical barriers limit pathogens from contaminating the algal product; but the final proof of a hazard-free dried microalgal product still awaits experimental verification.

D. Yeast and mold

Yeast and mold counts on potato dextrose agar are performed routinely for *Spirulina* powder at Earthrise Farms. Yeast is detected rarely, but three different mold types are observed regularly on plates. No attempt has been made yet to identify these types. Total mold counts, unlike SPC, do not follow seasonal weather patterns (Table 8-1).

Mold may develop in any food-processing plant that is not subjected to adequate cleaning, but algal cultures also can be a significant mold source. Mold counts generally are higher in ponds growing poorly for one reason or another, a phenomenon also observed by David et al. (1970), who found *Saprolegnia* associated with clumping of poorly mixed *Spirulina* cultures.

Because mold counts on potato dextrose agar do not appear to reflect sanitary conditions exclusively, it may be desirable to examine further for specific organisms such as "machinery mold" *(Geotrichum candidum)* when potato dextrose counts are high. *G. candidum* grows on uncleaned machinery surfaces and contaminates food, especially dairy products, during processing. Many microalgal and dairy products share certain biochemical features, particularly high protein levels, so *Geotrichum* may be a useful hygiene parameter for microalgal processing plants as well.

Few investigations of potential mycotoxin contamination in microalgae have been conducted. The only two analyses I am aware of showed no detectable aflatoxin B1, B2, G1, or G2 (5 ppb detection limit) in product from Thailand or from the pilot plant preceding Earthrise Farms in California (unpublished data).

E. Other pathogens

Staphylococcus is a common inhabitant of mucus membranes and human skin. The presence of large numbers of *Staphylococcus* indicates questionable sanitary practices, especially excessive human contact with the product. A specific health hazard due to staphylococcal enterotoxin also is indicated when *S. aureus* is present. Experiments show that ponds are unlikely to be a source of this pathogen. *S. aureus* disappears within 24 hours when inoculated into *Spirulina* cultures (Yanagimoto, Tahara, and Saitoh 1980). Direct skin contact with algae undergoing processing is unnecessary in a harvest plant although opportunity exists for such contact, especially with algae on a conveyor. Basic precautions on the part of plant workers can obviate any problems. In practice, I have seen no cases of coagulase-positive *Staphylococcus* in analyses of commercial microalgae from any source.

Occasionally, individuals report gastrointestinal upset after ingestion of microalgal powder or tablets. These cases are rare. Based on consumer correspondence and total volume of commercial material for one U.S. company manufacturing retail *Spirulina* products (Proteus Corporation), I estimate the frequency at approximately 10^{-4}. Examination of the returned product shows it to be free of identifiable pathogens. Do these incidents point to some unidentified pathogen associated with microalgal products? It is likely that these adverse reactions are allergic responses. Edema and elevated urinary corticosteroid excretion have been observed in experimental studies with *Scenedesmus* and *Chlorella* (reviewed by Waslien 1975), and gastrointestinal upset sometimes occurs even with autoclaved and, presumably, pathogen-free algae (Powell, Nevels, and McDowell 1961). As far as I am able to determine, the incidence of enteric problems with *Spirulina* is much less than with, for example, dairy products. Nevertheless, allergenicity to unusual proteins must be considered in plans to feed microalgae to malnourished individuals. For example, infant children in certain developing countries commonly suffer from lesions of the gastrointestinal tract and have increased immunological sensitivity (Miller 1968).

Whatever the role of allergic response, unusual pathogens cannot be ruled out at this stage of our experience with large-scale algal cultures. Cooper (1962) suggests that stagnant algal scums in hot weather may nourish growth of *Clostridium botulinum* Type C whose exotoxin is the causative agent of "duck sickness" (although generally not considered an agent of human botulism). Similar conditions can easily develop in improperly tended cultures, especially when mixing is inadequate or is stopped inadvertently.

F. Significance of moisture

Moisture is a critical factor affecting spoilage of "dried" algae, as microorganisms can maintain normal metabolic activities only within a certain range of water activity (A_w). Three regions of moisture content can be recognized in the sorption isotherm for a sample of Earthrise Farms *Spirulina* (Fig. 8-2), as is the case for most foods (Banwart 1981). Below about 2% moisture, water is in the form of a monomolecular layer, completely bound and unavailable for biological functions. Above about 12%, the algae contain free water that can support bacterial and fungal growth. Between 2% and 12%, moisture levels lie in an intermediate linear region where layers of water are added to the monomolecular film. Some water becomes available for mold and yeast growth in the upper half of this linear region, namely, between 7% and 12% in the case of this *Spirulina* sample.

Thus, a water content of 7% maximum should protect the algae from spoilage, although in practice this value has proved to be conservative. For example, mold decreased 67–85% in one month in samples containing 7.2–7.6% moisture, whereas mold increased 155% in a single week in a sample with 9.3% moisture. The lower value of 7% is preferred; it affords a margin of safety, particularly as sorption isotherms differ from sample to sample when particle size and ash content of the dried product vary. If moisture levels are too low, the algal powder creates dust easily and can irritate the lungs when inhaled.

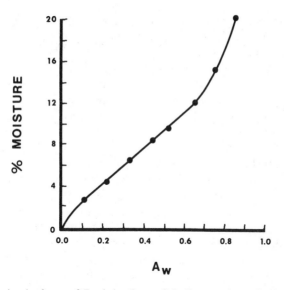

Fig. 8-2. Sorption isotherm of Earthrise Farms *Spirulina* powder at 25°C. A_w refers to water activity within the powder.

IV. Extraneous materials

A. Terminology

Extraneous material is the name given to "any foreign matter in product associated with objectionable conditions or practices in production, storage, or distribution" (AOAC 1980). If the objectionable matter is contributed by insects, rodents, birds, or other animal contamination, it is placed in a subcategory called *filth*. A variety of analytical methods exist depending on the food product: the U.S. Food and Drug Administration (FDA) has used methods for both *light filth* (AOAC 1980; sec. 44.056) and *sieved filth* (Quintero 1982) in examining *Spirulina*.

B. Invertebrates

Unidentified insect fragments form the bulk of foreign animal matter in microalgal products. Although insect fragments in conventional foods often indicate improper manufacturing or storage conditions, their presence in dried microalgae has an ambiguous interpretation. Aquatic invertebrates are indigenous to outdoor microalgal ponds, even to such extreme environments as a *Spirulina* culture. Waterboatmen (Corixidae), midges (Chironomidae), and brine flies (Ephydridae) commonly occur in *Spirulina* ponds where they depend on the algae for their sustenance. Aquatic beetles (Coleoptera) and mites (Hydracarina) occasionally are present. Adventitious guests such as thrips (Thysanoptera) and psocids (Psocoptera) are not unusual in agricultural areas where they occur as pests of local crops.

The presence of fragments from indigenous insects or arachnids in the dry powder represents no hazard to physical health, and their occurrence does not affect the aesthetic properties of the powder. Should these indigenous insect fragments be considered filth? That is, is the presence of this foreign animal matter objectionable in any way? These fragments can be interpreted as filth under existing regulations. A U.S. FDA training bulletin (Vazquez and Eisenberg 1977) cites the critical point: "Health can reasonably be defined as a state of complete physical, mental and social well-being, and not merely the absence of disease or infirmity. It is obvious that mental tranquility cannot coexist with the revolting feelings that insect- and rodent-defiled foods engender, even if, bacteriologically speaking, they are safe to consume." The FDA has issued alerts (1982) advising that imported *Spirulina* be examined for filth, and certain shipments have been refused entry on the basis of insect fragment levels.

A sensible standard for insect fragments in microalgae is difficult to develop. Each farm has a unique location so that susceptibility to aquatic and terrestrial insects varies enormously. In addition, certain processes used during harvesting can affect analytical results for filth by shattering fragments into particles too small to separate out for counting by the usual screening method. I have seen fragment levels ranging from 0 to 10^4 per 50 g sample from various farms around the world. Current FDA guidelines allow up to 150 unidentified insect fragments per 50 g of *Spirulina*. Identifiable fragments or whole insects or arachnids are another matter. Even a single specimen may indicate that a lot has been improperly processed or stored, depending on the nature of the contaminating organism.

Regardless of whether or not indigenous insect fragments are a health hazard, microalgal producers should make every effort to minimize aqautic insect levels in growth ponds: Insects consume the crop and are in direct competition with us for the pond harvest. Agricultural areas also harbor huge populations of terrestrial insects ready to infiltrate ponds and harvest plants. Although many measures can be taken to decrease contamination by both aquatic and terrestrial insects, no method has been found to completely eliminate the problem.

C. Hair and feather fragments

As previously mentioned, rodents and birds are attracted to ponds – at least partly by the rich insect life – and may carry microorganisms capable of infecting humans. Even if coliform results are negative, the presence of rodent hairs or feather fragments in dried microalgae is an indicator of potential contamination. Rodent contact can be decreased by destroying adjacent habitat, trapping, and protection of the ponds with appropriate fences. Birds are particularly difficult to dislodge from ponds once they adopt the territory and should be discouraged at the outset.

At Earthrise Farms, the source of rodent hair is probably the deer mouse *(Peromyscus maniculatus)*. Many birds could be a source of feather barbules: The farm lies along a major migratory route, and blackbirds, egrets, doves, and pigeons are resident in adjacent agricultural fields. However, the black-necked stilt *(Himantopus mexicanus)* and American avocet *(Recurvirostra americana)* have shown a particular interest in *Spirulina* ponds.

Current FDA guidelines permit 0.5 rodent hairs per 50 g of *Spirulina*. Hair from humans and indoor-dwelling rodents should be totally absent, but other rodent hairs and feather barbules are not completely avoidable in microalgal products. High winds, for example, can carry in rodent hairs from adjacent agricultural areas, as suggested by a correlation between wind speed and rodent-hair levels in harvested algae (unpublished data).

D. Other materials

Plant fragments are an inevitable contaminant in agricultural areas, but their presence usually is ignored in assessing extraneous material provided that they contribute insignificantly to the total product mass. Plant and insect fragments can be difficult to distinguish for an inexperienced analyst, often resulting in an overestimate of "unidentified insect fragments." At Earthrise Farms, the number of plant fragments usually exceeds the level of unidentified insect parts. Sand can be a seasonal problem, especially in windy desert areas, and grittiness in product may be caused also by nutrient precipitates. Occasionally, fibers of unknown origin are detected in a sample; these may originate from filter media used in dewatering.

V. Metals

A. Metal sources

Lead, mercury, cadmium, and arsenic are the metals most likely to adulterate microalgal products. Each is found as a trace contaminant in certain pesticides and fertilizers, so they are common in agricultural areas. They also are globally distributed as components of industrial waste. Nickel, copper, and zinc are common fertilizer components or contaminants, but they are substantially less toxic and have a narrow range of optimum concentrations for algae, at least in the case of *Chlorella* and *Spirulina* (Kallqvist and Meadows 1978; Gribovskaya et al. 1980; Pande, Sarkar, and Krishnamoorthi 1981; Kotangale, Sarkar, and Krishnamoorthi 1984). They are likely to terminate algae growth before being accumulated at levels toxic to humans. Tin, chromium, selenium, and aluminum are not such a universal threat, but local conditions must be appraised before they can be completely eliminated as a possible hazard. Aluminum frequently exceeds $100 \text{ mg} \cdot \text{kg}^{-1}$ in commercial microalgal products. The source is not known for certain but may originate in dust that blows into growth ponds.

Highest-grade nutrients, generally food- or feed-grade, should be used. Lower-grade phosphorus fertilizers in particular contain unacceptable levels of arsenic and fluorine, natural contaminants of many phosphate ores. Alkalinity and salinity sources also must be chosen carefully; the high concentrations normally used can introduce large quantities of heavy metal contaminants.

Certain plastic pond linings can be a heavy metal source as well. Lead frequently is used as a stabilizer in polyvinyl chloride linings, and mercury may be present to inhibit degradation by soil bacteria. In order to protect cultures from lead and mercury contamination, special double-layer liners can be custom-made by plastic membrane manufacturers. The portion in contact with the culture is stabilized by tin instead of lead, and only the bottom portion in contact with soil contains a microbial inhibitor.

B. Metal accumulation

Surface water can carry a significant heavy metal load, especially irrigation water receiving intentional or inadvertent drainage from croplands. Even amounts that appear insignificant in culture water can be concentrated many orders of magnitude within the cells of an algal crop. In one study with *Spirulina platensis*, mineral concentrations within fresh algae exceeded that in the medium for 20 of 21 elements tested, including nickel and lead (Laquerbe, Busson, and Maigrot 1970). Sodium, the only exception, was actively excluded from the cells, which is not too surprising for such a euryhaline organism.

Blue-green algae may be especially effective accumulators: Certain types excrete hydroxamate chelating agents (Murphy, Lean, and Nalewajko 1976) that can act as carrier molecules or increase the trace metal pool available near the cell surface. Experiments demonstrate that *Spirulina platensis* accumulates trace metals more effectively than *Chlorella vulgaris* (Gribovskaya et al. 1980), an advantage with regard to trace elements essential to humans but a liability if toxic metals are present.

Table 8-2. *Heavy metal guidelines relevant to microalgae*

Guideline	Lead	Mercury	Cadmium	Arsenic
Chlorella[a] ($\mu g \cdot g^{-1}$)	2.0	0.1	0.1	2.0
Spirulina[b] ($\mu g \cdot g^{-1}$)		20.0 total as lead		2.0
Single-cell protein[c] ($\mu g \cdot g^{-1}$)	5.0	0.1	1.0	2.0
Tolerable intake[d] ($\mu g \cdot d^{-1}$)	500	50	66–83	—

[a]Japan *Chlorella* manufacturers (H. Shimamatsu, personal communication).
[b]Japan Health Food Association (H. Shimamatsu, pers. comm.).
[c]Protein-Calorie Advisory Group (PAG 1974).
[d]Food and Agriculture Organization/World Health Organization (FAO/WHO 1972).

C. Metal levels in algae

No "characteristic" levels exist for toxic metals in microalgal products. Results are extremely variable, even for a single site, and some of this variation undoubtedly is a result of sample contamination or improper analytical technique. I have submitted identical *Spirulina* samples to independent commercial laboratories and received completely contradictory results, sometimes differing by a factor of 100. The need for quality control of independent laboratory analyses cannot be overstated.

At present no U.S. government standards exist for heavy metal content of microalgal products. The relative popularity of microalgae in Japan has led to the establishment of guidelines for *Chlorella* by a manufacturers' association and for *Spirulina* by a health food trade association. The *Chlorella* guideline is within the limits recommended for single-cell protein by the Protein-Calorie Advisory Group (PAG) of the United Nations (Table 8-2).

Atomic absorption (AA) results for Earthrise Farms *Spirulina* have satisfied the *Chlorella* guidelines consistently since the start of commercial production. However, all work was conducted at an independent commercial laboratory and is thus subject to the uncertainty described previously. Most publications on metals in microalgal foods also report low levels of no health concern (Boudène 1973; Santillán 1974; Chamorro Cevallos 1980; Becker 1984).

Higher levels appear occasionally. Boudène, Collas, and Jenkins (1976), for example, report concentrations of 1.78 $\mu g \cdot g^{-1}$ mercury and 8.5 $\mu g \cdot g^{-1}$ arsenic in certain *Spirulina* samples. Rat-feeding trials with the high-arsenic product demonstrated arsenic accumulation but no signs of toxicity. Lead levels up to 7.32 $\mu g \cdot g^{-1}$ have been determined by independent laboratories for certain Taiwanese products although most Taiwanese products are under the 3 $\mu g \cdot g^{-1}$ level. Even at 7.32 $\mu g \cdot g^{-1}$ lead, 68 g of dried algae daily would be required to exceed the provisional intake limits established by the Food and Agriculture Organization/World Health Organization (FAO/WHO) for a 70 kg person (Table 8-2).

In a recent study using an inductively coupled argon plasma-emission spectrometer (ICAP), mercury levels over 9 $\mu g \cdot g^{-1}$ were measured in some commercial and laboratory-grown *Spirulina* (Johnson and Shubert 1986). These results, however, are contradicted by numerous AA measurements by the Japan Food Analysis Center in which mercury never exceeded 1 $\mu g \cdot g^{-1}$ (personal communication, T. Kato, Dainippon Ink and Chemical) and by ICAP measurements that detected no mercury in similar commercial *Spirulina* products (personal communication, Y. Kasahara, Japan Jarrell Ash). The authors also were unable to replicate their results in follow-up tests (pers. comm., E. Shubert, University of North Dakota). In any case, a need clearly exists for conscientious management of potential metal sources and an adequate quality assurance program at all algal production facilities.

VI. Organic compounds

A. *Pesticides*

Pesticides can enter cultivation ponds either through fallout or with incoming water. The only complete protection against fallout is a pond enclosure, an expensive undertaking that also may produce undesirable temperature conditions in the heat of summer. Surface water is a constant threat, especially if the water has passed through agricultural areas, in which case it probably has been sprayed inadvertently or received cropland drainage containing pesticide residues.

Mircoalgal growing systems differ in their susceptibility to pesticide contamination. Although algae can accumulate a diversity of organic compounds, actual contamination levels will depend on the organism's chemical composition and pesticide persistence in the culture environment. The commercial microalgae *Spirulina* and *Chlorella* are sensitive to many pesticides (Bednarz 1981), so pesticide in cultivation ponds should be suspected during unexplained productivity declines. In the case of *Spirulina*, the medium's high pH discourages persistence of many pesticide compounds, but acute contamination episodes are possible under any culture conditions.

Daily analysis for pesticides is out of the question for most commercial growing facilities; efforts must be made at the start to avoid areas subject to pesticide contamination. Most farms using surface water have some on-site reservoir. The reservoir can be stocked with largemouth bass *(Micropterus salmoides)* or some other suitable fish population that is checked daily for signs of distress before water is released into the growth ponds. At least acute problems can be avoided in this manner.

Few data are available for pesticide levels in commercial microalgal products; in any case, I am aware of no published accounts that show detectable pesticide residue in commercially available *Spirulina* or other microalgal foods.

B. *Other compounds*

A major cost in microalgal production is the provision of carbon dioxide. The gas used is often a by-product of the natural gas or fermentation industry, but experiments have verified that excellent growth rates can be achieved with furnace gases as a source of carbon dioxide (Wachowicz and Zagrodzki 1975). The use of waste gas significantly reduces growing costs but adds a danger of contamination from carcinogenic or otherwise harmful hydrocarbons. The hydrocarbon 3, 4-benzopyrene often is used to indicate the presence of carcinogens of the aromatic polycyclic hydrocarbon series. Experiments with *Spirulina* demonstrated that 3,4-benzopyrene levels were independent of whether furnace or natural gas was used as the carbon dioxide source. The levels of $0.002-0.005$ $\mu g \cdot g^{-1}$ were lower than those of many common fruits and vegetables tested (Bories and Tulliez 1975).

Polyvinyl chloride pond liners contain toxic organic substances such as phthalic acid esters and unreacted vinyl chloride (Pim 1981), but no one has examined whether these substances are released into the water and accumulated by microalgae.

VII. Maintaining sanitary quality

In order to aid industry in complying with the Food, Drug, and Cosmetic Act of 1968, the U.S. Food and Drug Administration published regulations called Good Manufacturing Practices (GMPs: FDA 1985). GMP regulations cover all aspects of food processing, including personnel, techniques, and equipment. GMP amendments have been adopted specifying regulations for specific foods, although microalgae are not included at this time. The basic GMPs are completely applicable to microalgal production and harvest, and all producers should become familiar with them.

In addition to complying with GMPs, producers should include hygienic parameters as part of a consistent quality assurance program. The precise nature of this program must suit local conditions, especially

Table 8-3. *Suggested guidelines for quality of microalgae powder*

Microbial	
Coliform (or *E. coli*)	Negative
Salmonella	Negative
Shigella	Negative
Staphylococcus (coagulase +)	Negative
Mold	$<100 \text{ g}^{-1}$
Metals	
Lead	$<2.0 \ \mu\text{g}\cdot\text{g}^{-1}$
Mercury	$<0.1 \ \mu\text{g}\cdot\text{g}^{-1}$
Cadmium	$<0.1 \ \mu\text{g}\cdot\text{g}^{-1}$
Arsenic	$<2.0 \ \mu\text{g}\cdot\text{g}^{-1}$
Filth	
Unidentified insect fragments	<150 per 50 g
Rodent hairs	<0.5 per 50 g
Miscellany	
Moisture	$<7\%$
Pheophorbide (total)	$<1.2 \text{ mg}\cdot\text{g}^{-1}$
Appearance	Uniform powder
Color	Depends on species
Odor	Fresh, not decayed
Taste	Mild, not salty or bitter

as microbial and toxic chemical threats vary so greatly from region to region. In the absence of any site-specific information, however, one might start with the parameters and guidelines listed in Table 8-3. The rationale for these choices is presented throughout the text.

Microalgal producers are faced with negative attitudes for a number of reasons. The traditional association of algae with pollution and decay is firmly entrenched in many minds and, for others, microalgal foods have not fulfilled the claim made decades ago that they would avert famine in a densely populated world. More recently, extravagant medical claims made by unscrupulous retailers of microalgal products (Uretsky 1985) have destroyed the good will of many people. Nonetheless, anyone who becomes familiar with the vast diversity of microalgae, the many valuable attributes found in just the handful of strains studied to date, and the reality of dozens of commercial growing systems can only want to continue exploring the potential of these organisms. A sincere attention to health issues will prevent any further erosion of public confidence and allow this exploration to continue for the benefit of everyone.

VIII. Acknowledgments

I am grateful to my associates Bruce Carlsen, Ron Henson, Kathie Rutowski, Ted Sommer, and Allen Sonneville of Earthrise Farms, and Saburo Hamada, Yoshimichi Ohta, and Hidenori Shimamatsu of Dainippon Ink and Chemical for sharing their unique experience. Dr. Faouzi Senhaji of the Institut Agronomique et Vétérinaire Hassan II in Morocco kindly provided the sorption isotherm for *Spirulina*.

IX. References

AOAC [Association of Official Analytical Chemists]. 1980. *Official Methods of Analysis.* 13th ed. Washington, DC. 1,018 pp.

APHA [American Public Health Association]. 1971. *Standard Methods for the Examination of Water and Wastewater.* 13th ed. New York. 874 pp.

Banwart, G. J. 1981. *Basic Food Microbiology.* Abridged textbook ed. AVI Publishing, Westport, CT. 519 pp.

Becker, E. W. 1984. Biotechnology and exploitation of the green alga *Scenedesmus obliquus* in India. *Biomass* 4, 1–19.

Bednarz, T. 1981. The effect of pesticides on the growth of green and blue-green algal cultures. *Acta Hydrobiol.* 23, 155–72.

Bories, G., and J. Tulliez. 1975. Determination of 3,4-benzopyrene in the alga *Spirulina* produced and processed with different methods. *Ann. Nutr. Aliment* 29, 573–6. [In French.]

Boudène, C. 1973. Study of elements and toxic substances. Presented at the Colloquium on the Nutritional Value of the Alga *Spirulina*, Rueil-Malmaison, France, May. [In French.]

Boudène, C., E. Collas, and C. Jenkins. 1976. Characterization and measurement of various mineral toxicants in *Spirulina* algae of different origins: evaluation of long term toxicity in rats of a sample of Mexican *Spirulina*. *Ann. Nutr. Aliment.* 29, 579–88. [In French.]

CDC [Center for Disease Control]. 1977. *Foodborne and Waterborne Disease Outbreaks: Annual Summary 1976.* Rep. no. DHEW/PUB/CDC-78-8185. Center for Disease Control, Atlanta, GA. 88 pp.

Chamorro Cevallos, G. 1980. *Toxicological Studies on Spirulina Alga: Sosa Texcoco, S.A., Pilot Plant for the Production of Protein from Spirulina Alga.* Rep. no. UF/MEX/78/048. United Nations Industrial Development Organization, Vienna. 177 pp. [In French.]

Cooper, R. C. 1962. Some public health aspects of algal-bacterial nutrient recovery systems. *Amer. J. Pub. Health* 52, 252–7.

David, M., C. Santillán, and G. Clément. 1970. Contamination in outdoor cultures of the alga *Spirulina*. Presented at the Tenth International Congress of Microbiology, Mexico City, Aug. [In Spanish.]

FAO/WHO [Food and Agriculture Organization/World Health Organization]. 1972. *Evaluation of Certain Food Additives and of the Contaminants Mercury,*

Lead, and Cadmium: Sixteenth Report of the Joint FAO/WHO Expert Committee on Food Additives, Geneva, 4–12 April 1972. Geneva.

FDA [Food and Drug Administration]. 1978. *Bacteriological Analytical Manual.* 5th ed. Association of Official Analytical Chemists, Washington, DC.

FDA [Food and Drug Administration]. 1982. Import Alert no. 54-01, Feb. 10. Washington, DC. 1 p.

FDA [Food and Drug Administration]. 1985. Part 110: Current good manufacturing practice in manufacturing, processing, packing, or holding human food. In *Code of Federal Regulations. Title 21. Food and Drugs.*, pp. 91–8. Office of the Federal Register, General Services Administration, Washington, DC.

Fox, R. 1983. *Algoculture.* Doctoral thesis. Louis Pasteur University, Strasbourg. 336 pp.

Gonzáles Arroyo, S., R. M. Luna Lagunes, R. Hernández Velarde, P. Soriano Lucio, and V. J. Torres Gallardo 1976. Preliminary studies of bacterial contamination in seminatural cultures of *Spirulina. Salud Publica de Mexico* 5 (18), 705–10. [In Spanish.]

Gribovskaya, I. V., N. A. Yan, I. N. Trubachev, and G. K. Zinenko. 1980. Resistance of certain species of green and blue-green algae to an increased concentration of trace elements in the medium. In *Parametricheskoe Upr. Biosint. Mikrovodoroslei*, ed. F. Ya. Sid'ko and V. N. Belyanin, pp. 49–57. Izd. Nauka, Sib. Otd., Novosibirsk. [In Russian.]

Gronek, D. M. 1983. Citizen petition, Jul. 7. Food and Drug Administration, Rockville, MD. 3 pp.

Hughes, E. O., P. R. Gorham, and A. Zehnder. 1958. Toxicity of a unialgal culture of *Microcystis aeruginosa. Can. J. Microbiol.* 4, 225–36.

Jacquet, J. 1976. Microflora of *Spirulina* preparations. *Ann. Nutr. Aliment.* 29, 589–601. [In French.]

Johnson, P. E., and L. E. Shubert. 1986. Availability of iron to rats from *Spirulina*, a blue-green alga. *Nutr. Res.* 6, 85–94.

Jorjani, G. H., and P. Amirani. 1978. Antimicrobial activities of *Spirulina platensis. Maj. Ilmy Puzshky Danishkadah Jundi Shapur* 1, 14–18.

Joshi, S. R., N. M. Parhad, and N. U. Rao. 1973. Elimination of *Salmonella* in stabilization ponds. *Water Res.* 7, 1357–65.

Kallqvist, T., and B. S. Meadows. 1978. Toxic effect of copper on algae and rotifers from a soda lake (Lake Nakuru, East Africa). *Water Res.* 12, 771–5.

Klein, A. O., and W. Vishniac. 1961. Activity and partial purification of chlorophyllase in aqueous systems. *J. Biol. Chem.* 236, 2544–7.

Kondrat'eva, E. M., and T. B. Galkina. 1979. Formation of bacteriocenoses under different mineral nutrition conditions in the intensive cumulative culturing of algae. In *Rol Nizshikh Org. Krogovorote Veschchestu Zamknutykh Ekol. Sist., Mater. Vses. Soveshch. 10,* ed. V. A. Kordyum, pp. 257–63. Izd. Naukova Dumka, Kiev. [In Russian.]

Kotangale, L. R., R. Sarkar, and K. P. Kirshnamoorthi. 1984. Toxicity of mercury and zinc to *Spirulina platensis. Indian J. Envir. Health* 26, 41–6.

Laquerbe, B., F. Busson, and M. Maigrot. 1970. On the mineral element composition of two cyanophtes, *Spirulina platensis* (Gom.) Geitler and *Sp. geitleri* J. de Toni. *C. R. Acad. Sci. Paris, Sér. D,* 270, 2130–2. [In French.]

McElhenney, T. R., H. C. Bold, R. M. Brown, Jr., and J. P. McGovern. 1962. Algae, a cause of inhalant allergy in children. *Ann. Allergy* 20, 739–43.

Martinez Nadal, N. G. 1971. Sterols of *Spirulina maxima*. *Phytochemistry* 10, 2537–8.

Matsuura, E., Y. Saito, H. Ishida, Y. Seki, K. Miki, T. Fukimbara, H. Kawahara, and H. Ito. 1982. Toxic effects and changes of blood components by photo-dynamic action induced by pheophorbide a in rats. *Food Hyg. J.* [Japan] 23, 365–71. [In Japanese.]

Miki, K., O. Tajima, E. Matsuura, K. Yamada, and T. Fukimbara. 1980. Isolation and identification of a photodynamic agent of *Chlorella*. *Jap. Agric. Chem. Assoc. J.* 54, 721–6. [In Japanese.]

Miller, S. A. 1968. Nutritional factors in single-cell protein. In *Single-Cell Protein*, ed. R. I. Mateles and S. R. Tannenbaum, pp. 79–89. MIT Press, Cambridge, MA.

Murphy, T. P., D. R. S. Lean, and C. Nalewajko. 1976. Blue-green algae: their excretion of iron-selective chelators enables them to dominate other algae. *Science* 192, 900–2.

NRC [National Research Council]. 1980. *Recommended Dietary Allowances*. National Academy of Sciences, Washington, DC. 185 pp.

Oswald, W. J. 1979. Ecology of rural human settlements and recycling of natural resources. *Proc. GIAM* 5, 67–74.

PAG [Protein-Calorie Advisory Group]. 1974. PAG guideline (no. 15) on nutritional and safety aspects of novel protein sources for animal feeding. *PAG Bull.* 4(3), 11–17.

Pande, A. S., R. Sarkar, and K. P. Krishnamoorthi. 1981. Toxicity of copper sulfate to the alga *Spirulina platensis* and the ciliate *Tetrahymena pyriformis*. *Indian J. Exp. Biol.*, 19, 500–2.

Parhad, N. M., and N. U. Rao. 1974. Effect of pH on survival of *Escherichia coli. J. Water Poll. Contr. Fed.* 46, 980–6.

Pim, L. R. 1981. *Invisible Additives*. Doubleday Canada, Toronto. 267 pp.

Powell, R. C., E. M. Nevels, and M. E. McDowell. 1961. Algae feeding in humans. *J. Nutr.* 75, 7–12.

Quintero, J. A. 1982. *Rapid Method for the Analysis of Bulk Spirulina Powder for Light Filth*. Laboratory Information Bulletin no. 2636. Food and Drug Administration, Los Angeles. 1 p.

Santillán, C. 1974. Cultivation of the alga *Spirulina* for human consumption and for animal feed. In *Proceedings: Fourth International Congress of Food Science and Technology, Madrid, Spain, September 1974*. Instituto Nacional de Ciencia y Tecnologia de Alimentos, Consejo Superior de Investigaciones Científicas, Madrid.

Seki, Y., T. Fukimbara, E. Matsuura, and K. Yamada. 1981. Relation between photosensitization of rats by *Chlorella* powder and chlorophyllase activity in the powder. *Food Hyg. J.* [Japan] 22, 183–8. [In Japanese.]

Tamura, Y., T. Maki, Y. Shimamura, S. Nishigaki, and Y. Naoi. 1979. Causal substances of photosensitivity dermatitis due to *Chlorella* ingestion. *Food Hyg. J.* [Japan] 20, 173–80. [In Japanese.]

Uretsky, S. D. 1985. A pharmacist's guide to quack weight products. *Amer. Pharm.* 25, 24–9.

Vazquez, A. W., and W. V. Eisenberg. 1977. Introduction to analytical entomology. In *Training Manual for Analytical Entomology in the Food Industry*. FDA Tech. Bull. no. 2, ed. J. R. Gorham, pp. 1–3. Food and Drug Administration, Washington, DC.

Wachowicz, M., and S. Zagrodzki. 1975. Test of the use of furnace gases for cultivation of *Spirulina platensis*. *Przem. Spozyw*. 29, 397–400. [In Polish.]

Waslien, C. I. 1975. Unusual sources of protein for man. *Crit. Rev. Food Sci. Nutr*. 6, 77–151.

Waslien, C. I., D. H. Calloway, S. Morgen, and F. Costa. 1970. Uric acid levels in men fed algae and yeast as protein sources. *J. Food Sci*. 35, 294–8.

Yanagimoto, M., H. Tahara, and H. Saitoh. 1980. Behavior of coexisting microorganisms especially bacteria in *Spirulina* culture. *Rep. Nat. Food Res. Inst*. [Japan] 10, 84–90. [In Japanese.]

ALGAE IN INDUSTRY, ENVIRONMENTAL MANAGEMENT, AND AGRICULTURE

9. Commercial production and applications of algal hydrocolloids

JERRY G. LEWIS

Kelco Division of Merck and Co., Inc., P.O. Box 23076, San Diego, CA 92123-1718

NORMAN F. STANLEY

Marine Colloids Division, FMC Corporation, P.O. Box 308, Rockland, ME 04841

G. GORDON GUIST

Marine Colloids Division, FMC Corporation, P.O. Box 308, Rockland, ME 04841

CONTENTS

I. Introduction

Alginates, carrageenans, and agars are commercially extracted products from marine macroalgae of the divisions Phaeophyta (brown algae) and Rhodophyta (red algae) (Table 9-1). These extracts are commonly called hydrocolloids or phycocolloids because they display colloidal properties when dissolved in water and are extracted from seaweed (the Greek word for seaweed is *phykos*). They are composed of high-molecular-weight polymers of simple sugars and thus are polysaccharides. The basic structures of these sugars are shown in Figures 9-1–9-3.

Phycocolloids are primary constituents of brown and red algal cell walls. They serve a structural function analogous to, but differing from, that of cellulose in land plants. Whereas land plants require a rigid structure capable of withstanding the constant pull of gravity, marine plants must have a more flexible structure to accommodate the varying stresses of currents and wave motion. They have adapted accordingly by developing hydrophilic, gelatinous structural materials having the necessary flexibility. Alginates, carrageenans, and agars are not found in terrestrial plants, although according to Glicksman (1982) polymers with similar properties are extracted from trees (e.g., gum arabic), from seeds (e.g., guar and locust bean gum), and from products of bacterial fermentation (xanthan). All of these products compete in the food additive market which, in 1979, amounted to approximately 21,000 mt (460 million pounds) of gums in the United States alone. The phycolloids (alginates, carrageenans, and agars) accounted for only 3% of this total (Glicksman 1982).

In 1980, about 40% of the seaweeds used to extract phycocolloids came from developing countries (Table 9-2); however, these same countries produced only about 18% of the world's supply of the extracted products (Table 9-3), indicating that much of their seaweed is exported for processing.

Alginates, carrageenans, and agars possess unique functional properties. They thicken, gel, emulsify, and stabilize many food and industrial products (Tables 9-4–9-7). Thickening and gelling are their most important functions. The thickening capability of algal polysaccharides

[206]

Table 9-1. *Principal commercial sources of algal phycocolloids*

Agar producers	
Phylum: Rhodophyta	
Order: Nemaliales	
Gelidiella	1 sp.
Gelidiopsis	1 sp.
Gelidium	ca. 16 spp.
Pterocladia	4 spp.
Order: Gigartinales	
Gracilaria	ca. 6 spp.
Carrageenan producers	
Phylum: Rhodophyta	
Order: Gigartinales	
Chondrus	2 spp.
Eucheuma	5 spp.
Furcellaria	1 sp.
Gigartina	ca. 7 spp.
Gloiopeltis	1 sp.
Hypnea	2 spp.
Iridaea	3 spp.
Phyllophora	1 spp.
Alginate producers	
Phylum: Phaeophyta	
Order: Laminariales	
Ecklonia	2 spp.
Eisenia	1 sp.
Laminaria	ca. 8 spp.
Macrocystis	1 sp.
Order: Fucales	
Ascophyllum	1 sp.
Sargassum	ca. 4 spp.

Source: Gellenbeck and Chapman 1983.

Alginate

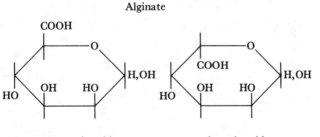

D-mannuronic acid   L-guluronic acid

Fig. 9-1. Monomer structure of alginate. Alginate polymers occur as three types, depending on the algae from which they are extracted. One segment is entirely D-mannuronic acid units; a second segment is entirely L-guluronic acid units; and the third segment alternates D-mannuronic acid and L-guluronic acid units.

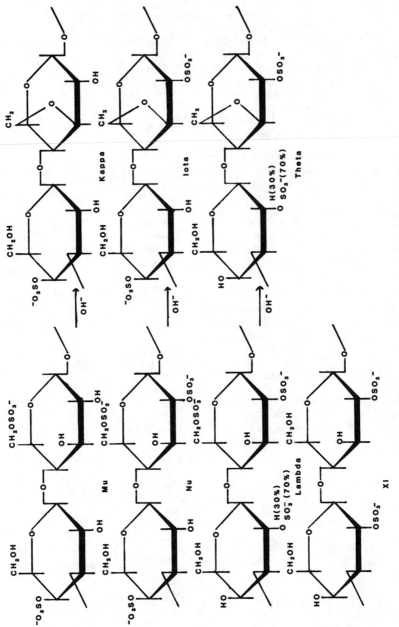

Fig. 9-2. Repeating units of limit carrageenans. (From Guisely et al. 1980.)

Agarobiose

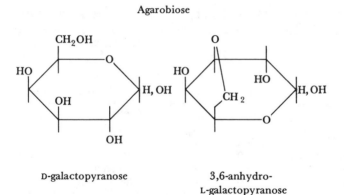

D-galactopyranose   3,6-anhydro-
                           L-galactopyranose

Fig. 9-3. Monomer structure of agarobiose, one of the components of agar.

Table 9-2. *Estimated world production of seaweeds used for colloids in 1980*

	Agar		Carrageenan and furcellaran		Alginate	
	Dry metric tons	% of total	Dry metric tons	% of total	Dry metric tons	% of total
Asia	18,400	51	17,300	44	5,900	6
Latin America	10,500	29	5,200	13	13,600	14
Europe	5,300	15	8,700	22	33,900	34
North America	Negligible		5,200	13	41,500	42
Other	1,800	5	3,100	8	5,100	5
Total	36,000	—	39,500	—	100,000	—

Source: Anonymous 1983.

is determined by measuring the thickness, or viscosity, of a solution of the polysaccharide in water. The phycocolloids are hydrophilic; when added to water they swell and thicken the solution, thus increasing its viscosity. Typically, viscosity measurements are made on solutions containing 0.5% to 2.0% of algal extract by weight. Because measurements of viscosity involve the flow of solutions, rheological terms are used to describe them. Viscosity is determined by viscometers, which measure the resistance of these solutions to flow when stressed by shear. A solution requiring a shear stress of one dyne per square centimeter to produce a shear rate of one reciprocal second is said to have a viscosity of

Table 9-3. *Estimated world production of seaweed colloids in 1980*

	Agar		Carrageenan and furcellaran		Alginate	
	Metric tons	% of total	Metric tons	% of total	Metric tons	% of total
Asia	3,600	52	500	4	1,800	8
Latin America	600	9	Negligible		400	2
Europe	1,600	23	7,100	59	13,200	60
North America	560	8	4,400	37	6,600	30
Other	600	9	Negligible		Negligible	
Total	6,960	—	12,000	—	22,000	—

Source: Anonymous 1983.

one poise (Anonymous 1980). Though the classical unit of viscosity is the poise, the international expression now in use for dynamic viscosity is the pascal second (Pa·s). One poise equals 0.1 Pa·s. Solutions of algal extracts can have viscosities from less than 10 mPa·s (10 centipoises) to more than 10 Pa·s (10,000 centipoises).

Solutions of algal polysaccharides that are dissolved in boiling water and allowed to cool may form gels. Gel formation is possible with as little as 0.2% polysaccharide by weight in solution. The mechanism of gelation for carrageenan and agar has been described as the aggregation of polymer chains into double helices as the solution cools. Gels of carrageenan and agar are thermally reversible; that is, the gels can be solubilized and reformed repeatedly by heating and cooling. Alginate gels are not thermally reversible. Typically, the gel temperatures of carrageenan and agar solutions are lower than their sol (melting) temperatures. The difference between the gel and sol temperatures, which can be as much as 50°C, is called hysteresis.

Gels of algal polysaccharides are generally made with 0.5% to 2.0% polymer by weight. Besides being characterized for gel and sol temperatures, these gels are tested for break force or gel strength and penetration. These parameters are measured on devices aptly called gel testers. These machines determine, by various means, the force necessary to break the surface of a gel (i.e., break force) and the amount of deformation of the surface of the gel at the break point (i.e., penetration). The units of gel break force are grams per square centimeter (or merely grams to describe gel strength). Penetration is measured in centimeters.

Solution viscosity, gel strength, and penetration are dependent on the type of algal polysaccharide, on the presence of metallic ions such as K^+

and Ca^{2+}, and on temperature and pH. The variations are almost limitless and make algal polysaccharides useful in many ways.

II. Alginates

A. History

The British pharmacist E. C. C. Stanford is credited with discovering alginates in the 1880s. However, commercial production of alginates did not begin until Kelco, now a division of Merck and Company was founded in 1929 in California (Anonymous 1976). Since then, the alginate industry has grown until at least a dozen countries have seaweed-processing plants, with major producers being the United States, the United Kingdom, Norway, Canada, France, Japan, and China. Such specialty chemical manufacturing facilities provide an important source of jobs and income for each of these countries.

Driving the expansion of manufacturing operations has been the increase in applications for alginates. Today there are literally hundreds of different uses, including pharmaceutical tableting, dental impression molding, food texturing, paper coating, textile printing, welding rod coating, and bacterial cell encapsulation (Tables 9-4–9-7).

Chemically, alginate polymers are composed of varying proportions of D-guluronic and L-mannuronic acid units (Fig. 9-1), which occur as three types of segments: individual groups of each of the two units and a third group of alternating guluronic and mannuronic acid units. Differences in proportions of these units and segments comprising the alginates from various species of brown algae account for differences in properties and functionality of the commercial products.

B. Production

Seaweed resources. Several genera of brown algae are utilized as raw materials by commercial alginate producers; these include *Macrocystis, Laminaria, Lessonia, Ascophyllum, Alaria, Ecklonia, Eisenia, Nereocystis, Sargassum, Cystoseira,* and *Fucus.* However, *Macrocystis pyrifera* and *Ascophyllum nodosum* are the principal sources of the world's alginate supply (McNeely and Pettitt 1973).

Alginate manufacturers locate their processing facilities close to the source of raw material to minimize seaweed transportation costs and to avoid the expense of drying large quantities of seaweed which are usually 80–90% water. However, several large producers purchase dried algae from around the world to supplement their native seaweed supply. Algae are harvested along the coasts of Australia, Iceland, India, Mexico, Argentina, Chile, and Africa, in addition to the major alginate-producing countries.

Fig. 9-4. *Macrocystis* kelp harvester. (Courtesy of Kelco, Division of Merck & Co., Inc., San Diego, CA.)

Annual differences in nutrient levels, water temperature, and weather affect the productivity of natural seaweed beds. Partially to overcome this problem and to extend the growing area, the Chinese have developed a system of artificial supports on which *Laminaria* is cultivated along the eastern coast of China (see also Druehl, Chapter 5). Other attempts at farming algae for alginate production have met with limited financial success.

Harvesting methods. Brown algae grow in ocean areas with rocky bottoms less than ca. 40 meters deep, the limit of useful sunlight penetration, and on rocky shores. Each plant usually has a holdfast anchoring it to a rock. Deepwater algae, such as *Macrocystis pyrifera*, have stipes up to 60 m long, which reach to the surface of the ocean and form a dense canopy buoyed by small float bladders connecting the leaflike fronds to the stipes.

Harvesting methods depend on the depth at which the particular species of seaweed grows. In some areas, gathering seaweed by hand is still practiced, but the major producers have fleets of modern boats designed for more efficient harvesting. For example, *Macrocystis* harvesters are equipped with underwater reciprocating blades, which mow the kelp about one meter below the surface (Fig. 9-4). The cut kelp is conveyed into a bin, which can hold up to several hundred metric tons before the ship returns to port.

It is advantageous for alginate producers to maintain a healthy, nat-

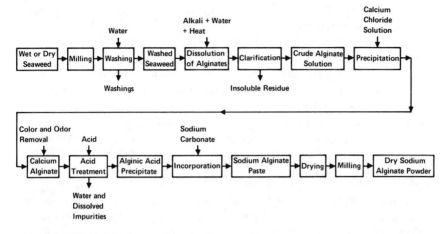

Fig. 9-5. Typical alginate manufacturing process. (From Anonymous 1976 [*Kelco Algin*].)

ural supply of algae along the coast. Therefore, harvesting is conducted at a frequency and depth that will protect the reproductive cycle of the particular species being harvested. Harvesting of a *Macrocystis* bed in California can be conducted up to three times a year. By California law, the kelp is cut no deeper than 1.25 m beneath the surface. Studies by several investigators have indicated that proper commercial harvesting results in no adverse impact on seaweed supply or nearby fish populations (Kimura 1980; Barilotti, Dayton, and McPeak 1985).

Extraction and processing. Once harvested and delivered to the factory, the seaweed is milled and washed to prepare it for alginate removal. Since alginates occur in the cell walls of brown algae as the insoluble mixed salts of sodium, calcium, magnesium, and potassium alginate, typical commercial processes for recovering purified alginate are based on ion exchange. The exact procedures are proprietary, but most are based on the patents of Green (U.S. Patent 2,036,934) or Le Gloahec and Herter (U.S. Patent 2,128,551) of about 50 years ago (Fig. 9-5).

For example, insoluble alginate salts in the seaweed may be converted with alkali to sodium alginate, which is readily soluble in water. Further dilution of the mixture with fresh water causes the algal cell structure to break down and release the sodium alginate into solution. Insoluble seaweed particles are then separated from the sodium alginate solution by standard solid–liquid clarification techniques such as sedimentation, screening, filtration, and centrifugation. Since sodium alginate is a linear polyelectrolyte of high molecular weight, even solutions at low concentrations have high viscosities. This thickening ability, which is a valuable property of the final product, makes the processing of alginates difficult and expensive.

Alginate is then recovered from the clarified sodium alginate solution by various techniques including precipitation with calcium chloride or sulfuric acid, electrolysis, or direct drying. Alginic acid is consequently neutralized with appropriate cation bases, then dried and milled to produce the final product.

Partially neutralized alginic acid may be further reacted with propylene oxide to produce propylene glycol alginate with various degrees of esterification. This product was developed for its improved acid stability, surface active properties, and resistance to precipitation by polyvalent metal ions (Anonymous 1976).

Commercially available alginates include the sodium, potassium, ammonium, and calcium salts in addition to alginic acid. Alginates are sold in various grades of molecular weight (viscosity), calcium content, particle size, purity, and cation form to provide specific functions in various applications. Prices in 1985 were US $4.50–18 per kilogram, depending on the type and purity of product.

As with many specialty chemicals, worldwide competition among producers of alginates continues to increase. Research and development of more efficient manufacturing processes and applications technology have created jobs for technical professionals such as chemists, biologists, and chemical engineers. An example of this research is the use of enzymes to modify the molecular structure of alginates. Researchers have isolated an epimerase enzyme that can be used to transform the mannuronic acid units into guluronic acid units, thereby changing the reactivity of the alginate product and improving its ability to form gels with calcium (Skjak-Braek and Larsen 1985).

Properties. Alginates are readily soluble in cold or hot water, developing viscosities up to 5,000 mPa·s for a 1% solution (Fig. 9-6). These solutions are pseudoplastic, becoming less viscous as more shear is applied (Fig. 9-7). An example of the importance of pseudoplastic rheology to modern society is in the application of paints. When paint is brushed onto a surface, a controlled rate of viscosity recovery is required to minimize brush drag, to promote leveling of the surface, and to prevent sagging and runoff (Fig. 9-7).

Physical variables that affect the flow characteristics of alginate solutions include temperature, shear rate, concentration, and polymer size. Chemical variables include pH, sequestrants, monovalent salts, and polyvalent cations. Alginates in solution are compatible with a wide variety of materials, including other thickeners, resins, sugars, oils, pigments, surfactants, preservatives, and enzymes.

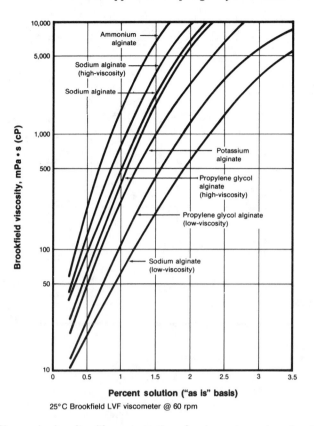

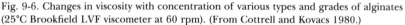

Percent solution ("as is" basis)

25°C Brookfield LVF viscometer @ 60 rpm

Fig. 9-6. Changes in viscosity with concentration of various types and grades of alginates (25°C Brookfield LVF viscometer at 60 rpm). (From Cottrell and Kovacs 1980.)

C. Applications

Alginates are widely used in foods such as frozen desserts, salad dressings, dairy products, bakery products, beverages, dessert gels, and fabricated foods (Tables 9-4 and 9-5). Numerous toxicological studies have demonstrated the high level of safety of alginates in food use (McNeely and Kovacs 1975). After an evaluation by the Select Committee on GRAS Substances of the Federation of American Societies for Experimental Biology, the FDA affirmed the "generally recognized as safe" (GRAS) status of alginates in 1982 and continued the food-additive status of propylene glycol alginate (Anonymous 1982b). Alginates are also permitted in foods under the European common market (EC) regulations and the Codex Alimentarius Commission of the United Nations Food and Agriculture Organization/World Health Organization.

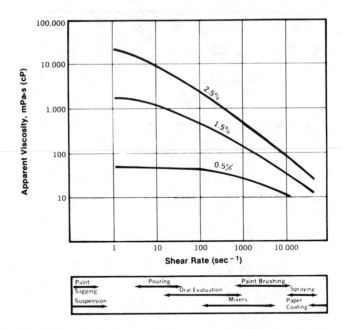

Fig. 9-7. Changes in apparent viscosity with shear rate of medium-viscosity sodium alginate solutions. (From Cottrell and Kovacs 1980.)

Table 9-4. *Food applications of seaweed hydrocolloid water systems*

	Agars	Alginates	Carrageenans
Frozen foods	X	X	X
Pastry fillings		X	X
Syrups		X	X
Bakery icings	X	X	X
Relishes		X	X
Cooked/instant puddings		X	X
Meringues	X	X	
Chiffons		X	X
Dessert gels	X	X	X
Candies	X	X	
Fruit juices	X	X	X
Jams and jellies		X	X
Sauces and gravies		X	X
Pimiento strips		X	X
Salad dressings		X	X

Table 9-5. *Food applications of seaweed hydrocolloid milk systems*

	Agars	Alginates	Carrageenans
Whipped toppings		X	X
Milk shakes		X	X
Skim milk			X
Evaporated milk			X
Chocolate milk			X
Cheeses	X	X	X
Cottage cheese			X
Infant formulations			X
Flans and custards		X	X
Yogurt	X		X
Instant breakfasts		X	X
Ice cream		X	X

Alginate solutions produce useful gels, fibers, and films when precipitated with certain polyvalent salts to form cross-linkages. Structured foods, from onion rings to imitation caviar, can be made from a mixture of vegetable or meat pieces, flavoring, and other ingredients encased in a firm alginate gel. Dietetic foods with improved palatability can be fabricated with less than 1% alginate. By the proper selection of gelling agent, gel structure and rigidity are controlled (King 1983).

Alginates have a wide range of industrial applications (Table 9-6), but the two major uses are in paper and textiles. For paper manufacture, alginates are used to improve ink holdout, to control the flow properties of coatings, and to improve smoothness of the sheet. In textiles, fiber reactive dye pastes are thickened with alginates to print sharp lines and to conserve dye.

In pharmaceuticals, alginates serve as protective coatings and suspending agents that increase the stability and bioavailability of the active ingredients (Table 9-7). Alginate gels are used as dental impression materials because of their excellent dimensional reproducibility, predictable setting time, and ability to set at mouth temperatures.

Although alginates have been commercially produced for over 50 years, the changing needs of society for controlling fluids and gels in foods, pharmaceuticals, and industry continue to expand the applications for alginates. In particular, their unique properties, natural source, and proven safety will increase the role of alginates in consumer products worldwide. Adequate raw materials and production capacity are already available throughout the world to support a significant increase in the demand for alginates.

Table 9-6. *Industrial applications of seaweed hydrocolloids*

	Agars	Alginates	Carrageenans
Paper sizings/coatings	X	X	
Adhesives	X	X	
Textile printing/dyeing	X	X	
Air freshener gels		X	X
Explosives		X	
Boiler compounds		X	
Polishes		X	
Tertiary oil treatment			X
Antifoams		X	
Ceramics		X	
Welding rods		X	
Cleaners		X	X
Castings and impressions	X	X	
Enzyme immobilization		X	X
Microtomy media	X		
Electrophoretic and chromatographic media	X		X
Conductivity bridges	X		

Table 9-7. *Medical and pharmaceutical applications of seaweed hydrocolloids*

	Agars	Alginates	Carrageenans
Laxatives	X		X
Bulking agents	X	X	X
Radiology suspending agents	X		
Capsules and tablets	X	X	X
Suppositories	X		
Anticoagulants	X		
Lotions and creams		X	X
Shampoos			X
Ulcer products		X	X
Toothpastes			X

III. Carrageenans

A. History

The Western world. In Europe and North America, the algae *Chondrus crispus* and *Gigartina stellata* of the Gigartinaceae, as sun-bleached whole plants, may have been used for centuries for making jellies and milk puddings (blancmange). Donovan (1837) wrote: "Carrageen. This is the

Irish name of the fucus crispus, commonly called Irish moss, introduced from Ireland as an article of food within the past ten years." The name *carrageen* frequently has been described as derived from an Irish coastal village of that name; however, Mitchell and Guiry (1983) find no evidence for this derivation. An alternative derivation is from the Irish word *carraigeen,* meaning "rock moss." Irish moss appears to have been imported into the United States early in the nineteenth century.

Domestic production was reported to have first been started by Dr. J. V. C. Smith, at one time a mayor of Boston. Smith is said to have recognized in 1835 that the plants growing off the Massachusetts coast were the same as the imported material (Smith 1905). As a result, moss harvesting promptly began at Scituate, Massachusetts, an area that remains a mossing center to this day (Humm 1951). Another account attributes the discovery to David Ward in 1847, with collection again said to have been in the Scituate area (Fraser 1942).

Stanford (1862) coined the name *carrageenin* for the gelatinous material extracted by water from *Chondrus crispus.* The present spelling, carrageenan, has been accepted within the past 25 years, this being consonant with the use of the *-an* suffix for the names of polysaccharides.

The modern carrageenan industry dates from the 1940s, receiving its impetus from the dairy industry where carrageenan was found to be the ideal stabilizer for the suspension of cocoa in chocolate milk. The pioneers in this industry were Jacques Wolf and Company, Krim-Ko Corporation, and Kraft Foods. Earl Jertson, an officer of Krim-Ko, later bought Krim-Ko's carrageenan operation at Scituate, Massachusetts, to found Seaplant Corporation. Following World War II, Algin Corporation of America in Rockland, Maine, started carrageenan production and, by aggressively developing improved processing methods and new seaweed sources, became a major producer by the midfifties. At this time, Algin Corporation participated in setting up the French company SATIA, which used Algin Corporation's technology under a licensing agreement. Today, SATIA is the second largest producer of carrageenans in Europe.

In 1959, Algin Corporation and Seaplant Corporation merged to form Marine Colloids, Inc., to become the world's largest manufacturer of carrageenans. Marine Colloids was acquired by FMC Corporation in 1977, becoming the Marine Colloids Division of FMC Corporation. In 1970, Stauffer Chemical Company purchased Kraft Foods' carrageenan operation at South Portland, Maine, but closed it down after a few years so that FMC Corporation is now the only carrageenan manufacturer operating in the United States. A Danish manufacturer of pectins, Copenhagen Pectin Factory, started to produce carrageenans in 1960 and soon established an enviable reputation for the high quality of its products. It was acquired by Hercules, Inc., in 1972 and is presently the second-largest producer of carrageenans worldwide. A smaller Danish manufacturer of carrageenans and agarose, Litex A/S, was sold to FMC Corporation in 1982, becoming a part of the Marine Colloids Division.

The Far East. Marine algae of the genus *Eucheuma* in the Solieriaceae family have a long history of use in southeastern Asia as articles of food, for their supposed medicinal properties, and in trades such as bookbinding where their mucilage is used as an adhesive. Although the Malay word *agar-agar* refers to *Eucheuma* species, it is now known that these yield carrageenans rather than agarose-type polysaccharides (Towle 1973). Eisses (1952, 1953) and Zaneveld (1955, 1959) have described a number of types of *Eucheuma* that have been harvested in southeast Asia, listing them by their botanical and native names. It was not until the *Eucheumas* were recognized in the 1950s as valuable carrageenophytes by the Western carrageenan industry that the present large-scale export of these species became established. The introduction of *Eucheuma* farming in the Philippines in 1971 greatly promoted this industry.

A recent development with a major impact on the carrageenan market was the advent in the late 1970s of semirefined carrageenan. This low-cost product from the Far East consists of whole *E. cottonii*, which has been chemically treated to enhance its gelling properties. It has proved to be an acceptable substitute for the refined extractive in a limited number of large-volume applications.

The origin of this product is obscure, but it appears to have been first produced by one or more small agar manufacturers in the region. In 1980, FMC Corporation entered this field by founding Marine Colloids Philippine Islands (MCPI) with a factory at Cebu to produce the semirefined *E. cottonii*. The plant was later sold to the local management and continues to operate independently, though in close cooperation with the parent company.

B. Chemistry

Structure. Carrageenans are linear polysaccharides with a repeating structure of alternating 1,3-linked β-D-galactopyranosyl and 1,4-linked α-D-galactopyranosyl units (Fig. 9-2). The 1,3-linked units may occur as the 2- and 4-sulfates or are occasionally unsulfated. The 1,4-linked units occur as the 2-sulfate, the 2,6-disulfate, the 3,6-anhydride, and the 3,6-anhydride 2-sulfate. Sulfation at C3 apparently never occurs.

Due largely to the investigations by Rees and his co-workers (Rees 1963; Dolan and Rees 1965; Anderson, Dolan, Lawson, et al. 1968; Anderson, Dolan, Penman et al. 1968; Anderson, Dolan, and Rees

1968; Lawson and Rees 1968; Anderson, Dolan, and Rees 1973; Lawson et al. 1973; Penman and Rees 1973a, b, c; Welti 1977), carrageenans are now defined in terms of chemical structure. Although a more or less continuous spectrum of carrageenans exists (Pernas et al. 1967), it is nevertheless possible to distinguish a small number of ideal or limit polysaccharides. The names mu, kappa, nu, iota, lambda, theta, and xi are presently applied to these limit carrageenans.

Native carrageenans from different algae may therefore be regarded as varying mixtures of the limit polysaccharides and intermediate hybrids ranging in degree of anhydridation and 2-sulfation of the 1,4-linked units.

The carrageenans may be divided into two general groups. One consists of mu, nu, kappa, iota, and their hybrids. Carrageenans in this group gel with potassium ions or can be made to gel by treatment with alkali. They are characterized by having their 1,3-linked units sulfated at C4. The other group consists of lambda, xi, and the products of their modification by alkali. These do not gel either before or after alkali treatment and characteristically have both their 1,4- and 1,3-linked units sulfated at C2 though occasionally the latter are unsulfated (Stancioff and Stanley 1969).

Commercial carrageenans comprise κ-, ι-, and λ-carrageenan. These extractives approximate the limit polysaccharides, the criteria of their value being functionality rather than strict chemical composition.

Properties. The chemical reactivity of carrageenans is primarily due to half-ester sulfate groups that are strongly anionic, being comparable to sulfuric acid in this respect. The free acid is unstable, and commercial carrageenans are available as stable sodium, potassium, and calcium salts or, most commonly, as a mixture of these. The associated cations together with the conformation of the sugar units in the polymer chain determine the physical properties of the carrageenans. For example, κ- and ι-carrageenans form gels in the presence of potassium or calcium ions, whereas λ-carrageenan does not. Reactivity with proteins is exhibited by both gelling and nongelling carrageenans, although regularity of the polymer is an important factor. In most if not all cases, ion–ion interactions between the sulfate groups of the carrageenan and the charged groups of the protein are involved. Reaction depends on protein/carrageenan net charge ratio and thus is a function of the isoelectric point of the protein, the pH of the system, and the weight ratio of carrageenan to protein (MacMullan and Girich 1963). The commercially important reaction of carrageenan with κ-casein in milk is specific for this protein and unique in that it can occur at pH levels above the isoelectric point of the casein.

C. Production

Seaweed resources. Carrageenophytes are now known to consist of numerous species from seven different families of red algae: Solieriaceae, Rhabdoniaceae, Hypneaceae, Phyllophoraceae, Gigartinaceae, Furcellariaceae, and most recently, Rhodophyllidaceae (Deslandes et al. 1985). Though not all of these have been, or are perhaps likely to be, exploited commercially, present-day sources of carrageenans go well beyond the original Irish moss. Seaweeds that have been utilized for carrageenan production include *Chondrus crispus, C. ocellatus, Gigartina stellata, G. acicularis, G. pistillata, G. canaliculata, G. radula, G. skottsbergii, Eucheuma cottonii, E. spinosum, E. gelatinae, Furcellaria fastigiata,* and *Hypnea musciformis.* Utilization of these not only has greatly extended the base and the geographical area from which the industry can draw raw materials but also has extended the range of properties of the extractives, as different species yield carrageenans of differing structure and properties.

Chondrus crispus is largely harvested in the Maritime Provinces of Canada, with smaller quantities collected along the coasts of Maine and Massachusetts. The difference in volume is due not so much to lack of abundance as to differences in the local economics. *C. crispus,* sometimes mixed with *Gigartina stellata,* is also harvested along the coasts of France and Spain.

Chondrus crispus, as well as other genera and species of Gigartinaceae, yields a mixture of κ- and λ-carrageenans from the commercially collected weed that comprises both gametophytic and tetrasporophytic plants. It is now known that κ-type carrageenans are produced by the gametophytes and λ-type carrageenans by the tetrasporophytes (McCandless, Craigie, and Walter 1973). A process has recently been developed for separating plants of the two reproductive stages (Carratech, Inc. 1986). This could become of great commercial value by permitting the production of pure λ-carrageenan.

Gigartina acicularis and *G. pistillata* occur and are harvested together along the coasts of southern France, northern Spain, Portugal, and Morocco. The composition of the stands gradually shifts with latitude. *Chondrus crispus* and *Gigartina stellata* predominate in the North with a shift to *G. acicularis* and *G. pistillata* in the South (Stancioff, personal communication). The latter two species are unique in that they yield a nongelling, predominantly lambda-type carrageenan.

Gigartina radula and *G. skottsbergii* are harvested in Chile and comprise a major resource for carrageenan production. Two varieties of *G. radula,* characterized as "narrow-leaf" and "broad-leaf" and yielding somewhat different extractives, are recognized.

Eucheuma cottonii and *E. spinosum* are now heavily used by carrageenan producers and are harvested in large quantities in Indonesia and the Philippines. *Eucheuma cottonii* is harvested to a lesser extent in southern Africa. Beds exist along the coast of Tanzania and undoubtedly elsewhere but remain to be exploited (Mshigeni 1984). *Eucheuma cottonii* and *E. spinosum* are remarkable in that these two speices, once thought to be varieties of a single species, yield quite different types of carrageenans. *Eucheuma cottonii*, which actually may comprise two very similar species, *E. cottonii* and *E. striatum* (Doty 1973), yields nearly ideal κ-carrageenan, whereas *E. spinosum* yields nearly ideal ι-carrageenan.

Furcellaria fastigiata yields furcellaran ("Danish agar"), often treated in the literature as a polysaccharide distinct from carrageenan but now considered, on the basis of chemical evidence (Lawson et al. 1973), to be a member of the carrageenan family of polysaccharides. Major quantities of *F. fastigiata* are harvested only in Denmark and in the Maritime Provinces of Canada. The Canadian weed is frequently a mixture of *Chondrus crispus* and *F. fastigiata* collected after being cast up on the beaches.

Harvesting methods. In contrast to the highly technical methods now employed for the extraction of carrageenans from the algae, harvesting remains a primitive, labor-intensive occupation. Much of the harvesting is done from small boats by hand raking and by drag rakes towed from inshore fishing boats. Collection of cast weed from beaches is also done to a large extent. Much research has been devoted to the development of mechanical harvesters, but a practical, cost-effective device has yet to be produced.

The only revolutionary advance to have occurred in carrageenophyte production has been the cultivation of *Eucheuma* species, now practiced on a large scale in the Philippines (Laite and Ricohermoso 1981). Due to the impact of cultivation, Philippine production increased dramatically between 1971 and 1984 whereas production from all other areas showed little change (Table 9-8).

Much of the credit for this impact must be given to the simplicity of the farming technique. It should be acknowledged, however, that about six years of research were required to develop exactly the right technique (Doty and Alvarez 1975; Doty 1979). Basically, a seaweed farm comprises networks of lines suspended from mangrove stakes and immersed a short distance below the low-tide level. Sprigs of *E. cottonii* seed stock are suspended from these lines, where they grow at a rate of about 3.5% per day to several times their original size (Doty 1979). They are then pruned back, with growth continuing from the remaining thallus. If senescence is apparent, the lines are reseeded with fresh stock.

Advantages of this farming technique include control and selection of fast-growing, hardy strains, enhanced photosynthesis due to

Table 9-8. *Production of carrageenan seaweed on
a world basis, in dry metric tons*

	1971	1979	1984
Canada	6,000	5,700	5,000
Phillippines	500	14,000	25,000
Chile	4,000	6,000	6,000
Indonesia	4,000	3,500	3,000
Others	5,500	4,500	4,500
Total	20,000	33,700	43,500

Source: Data Courtesy of FMC Corporation.

increased exposure to light, and relative immunity to attack by sea-floor grazers such as sea urchins. Problems include disease and senescence, manifested by bleaching and dying back of the thallus, a condition graphically described as "ice-ice." Farm site selection is important because a variety of biological, meteorological, and sociopolitical circumstances dictate whether or not a farm will succeed. *Eucheuma* farming is presently confined to small family-operated farms in the relatively storm-free southern Philippines (Parker 1974).

Attempts at cultivation of carrageenophytes other than *Eucheuma* have met with less success. Tank cultivation of *Chondrus crispus*, employing a superior strain (NRC T4), was studied extensively by researchers at the National Research Council of Canada in collaboration with the major carrageenan producers (Shacklock et al. 1973). However, tank cultivation is a capital- and energy-intensive process that has yet to be implemented commercially.

Extraction and processing. Seaweeds are dried, usually at the harvesting location, and baled for shipment to processing plants. In most collection areas, sun drying remains the most cost-effective technique although oil-fired mechanical dryers are used to a limited extent.

Specific details of extraction processes are loosely guarded as trade secrets by the several manufacturers but broadly follow a similar pattern. The dried weed may be washed to remove adhering salts, stones and marine organisms. Washed or unwashed weed, usually as a blend selected to achieve the desired properties in the extractive, is then digested with hot water under alkaline conditions to extract the carrageenan exhaustively. The alkali, usually calcium or sodium hydroxide, performs two functions. It promotes swelling and maceration of the weed to bring the carrageenan into solution and it effects removal of 6-sulfate groups from the carrageenan to generate 3,6-anhydro-D-galactose residues in the polysaccharide chain. These function to enhance the water gel strength and protein reactivity of the carrageenan (Stanley 1963). When the desired degree of conversion has been achieved, the solution of carrageenan is separated from weed solids by filtration or by centrifugation followed by filtration. Adjustment of pH and concentration of the filtrate by evaporation are done prior to recovery of the carrageenan from solution.

Several methods have been or are used to recover the carrageenan from solution. Direct drying of the concentrated filtrate on steam-heated rollers was used extensively in the past but presently is not used by any of the major producers. Products of much higher quality are obtained by precipitation of the carrageenan from solution by 2-propanol or other alcohols followed by further alcohol washes to dehydrate the coagulum. Residual alcohol is removed and recovered by drying the coagulum under vacuum. An interesting historical note is that perhaps the earliest process described for recovering carrageenan from Irish moss employed alcohol precipitation (Bourgade 1871).

Another process presently in use for the recovery of carrageenan from solution takes advantage of several properties of κ-carrageenan and its gels. First, κ-carrageenan solutions form gels in the presence of potassium ions. Second, these gels exude water by syneresis on standing, the more so when squeezed in a press. Third, much water separates from the gel when it is frozen and then allowed to thaw. This latter phenomenon is the same as that used for the production of agar. A limitation of the freeze–thaw process as applied to carrageenans is that it is applicable only to κ-carrageenan. Moreover, the requirement that potassium be present precludes making products where sodium is the major counterion.

Semirefined carrageenan from *Eucheuma cottonii* is prepared by a method that superficially resembles that for preparing French-fried potatoes. A basket of seaweed fronds is immersed and cooked in hot aqueous potassium hydroxide and then soaked in fresh water to extract most of the residual alkali. The product is dried and ground to produce a flour having many of the properties of conventional extracted carrageenans.

The economic advantage lies not in extracting the carrageenan from the seaweed but in performing the reaction that maximizes gel strength of the polymer within the plant structure. The product-to-water ratio is maximized, thereby reducing the cost of isolating the dry product (Lauterbach, pers. comm.).

D. Applications

Properties. The function of carrageenans in various applications depends largely on their rheological properties. Carrageenans are linear, water-soluble polymers that typically form highly viscous aqueous solutions due to their unbranched, linear macromolecular structure and poly-electrolytic nature. Commercial carrageenans are generally available in viscosities ranging from about 5 mPa·s to 800 mPa·s when measured at 1.5% concentration and 75°C.

Carrageenans specifically tailored for water-thickening applications are usually lambda types or the sodium salt of mixed lambda and kappa. They dissolve in either cold or hot water to form viscous solutions. Here high water viscosities are desirable, and the high molecular weight and hydrophilicity of lambda contribute to this. For gelling applications, a low viscosity in hot solution is usually desirable for ease of handling. Fortunately, high gel strength carrageenans (mixed calcium and potassium salts of kappa and iota) fulfill this requirement because of their lesser hydrophilicity and the effect of the calcium ions. The κ- and ι-carrageenans form gels on cooling of their hot solutions in the presence of certain cations, notably potassium. Heating is required to bring them into solution under these conditions. According to Rees (1972), carrageenans that form aqueous gels do so because of double helix formation followed by cation-induced aggregation. An alternative model of carrageenan gelation, based on cation-induced aggregation of single helices, has also been proposed (Paoletti, Smidsrød, and Grasdalen 1984).

Regardless of the mechanism, the occurrence of 1,4-linked 6-sulfated residues in the polymer chain of either κ- or ι-carrageenan appears to detract from the strength of their gels. This is ascribed to kinks, produced by these residues, in the chain, which inhibit the formation of double helices (Mueller and Rees 1968). Alkali modification of the carrageenan during processing increases the gel strength of the product by removal of these kinks through conversion of 6-sulfated residues to the 3,6-anhydride. Increased hydrophobicity from the added anhydride residues may also contribute to gelation.

All carrageenan products have the ability to form gels by cooling a solution of the carrageenan in hot milk. Even λ-carrageenan, which does not gel in water regardless of the cations present, will form a gel at levels of 0.2% by weight of the milk. This gelation is ascribed to the formation of carrageenan–casein bonds. With κ- and ι-carrageenan there is, in addition to the carrageenan–casein interaction, a water-gelling effect from the cations present in the carrageenan as well as the Ca^{2+} and the K^+ present in the milk.

Carrageenans are susceptible to depolymerization through acid-catalyzed hydrolysis. They can be used in acid systems, however, if not subjected to prolonged heating. The rate of hydrolysis at a given pH and temperature is markedly slower if the carrageenan is in the gel rather than the sol state.

Uses. Carrageenans are used to gel, thicken, or suspend; therefore, they are used in emulsion stabilization, for syneresis control, and for bodying, binding, and dispersion. Tables 9-4–9-7 list uses by end product. Major uses are in foods, particularly dairy products. Packaged milk gel products are direct descendants of the original blancmange.

Carrageenan is unique in its ability at very low concentrations (ca. 300 ppm) to suspend cocoa in chocolate milk; no other gum has been found to match it. A very delicate milk gel structure, undetectable on pouring or drinking the milk, is believed to hold the cocoa in suspension. A substantial differential between the concentration below which settling of cocoa occurs and that above which visible gelation is evident is required for practical stabilization. This is achieved by careful selection of weed species and molecular weight control during processing of the carrageenan.

The use of ι-carrageenan in dessert gel formulations produces gels with textures very similar to those of gelatin gels. They have an advantage over gelatin in that their melting point is higher so that they are well suited to tropical climates or where refrigeration is not available. A further advantage is that iota gels retain their tender structure on aging, whereas gelatin tends to toughen. This is important for ready-to-eat desserts popular in Europe.

In toothpaste, carrageenans function as a "binder" to impart the desired rheological properties to the paste and to provide the cosmetic quality of "sheen." Carrageenan suffers severe competition in the domestic toothpaste market from sodium carboxymethylcellulose, a much cheaper gum. Despite this, business has been retained and regained due to the superior quality and appearance carrageenan imparts to toothpaste. Overseas, carrageenan has maintained a strong position in this application due, among other factors, to its immunity to degradation by enzymes that attack cellulose gums.

Toxicological regulatory status. Carrageenan is listed by the U.S. Food and Drug Administration as "generally recognized as safe" (GRAS) (Federal Register, 44:40343–5 [1979]; 21CFR 172.620–6 [1981]). A review of the physiological effects of carrageenans has been published (Stancioff and Renn 1975).

World market. Sales of carrageenans in metric tons and dollars are shown in Table 9-9. Figures are shown for both extractive and semirefined carrageenan. The latter is used almost exclusively for the huge pet food market, and its explosive growth in the 1980–2 period, to the detriment of extractive sales, reflects the replacement of the latter by semirefined in pet food formulations. This changeover has now been accomplished, and steady growth is now projected for both types. The growth rate

Table 9-9. *World market for refined and semirefined carrageenans*

	Market size 1982		Compound volume growth (%)	
	Metric tons	US $ (millions)	1980–2	1983–8
Refined	10,800	88.3	−6.1	1.8
Semirefined	2,400	10.1	600.0	4.0
Total	13,200	98.4	3.6	2.2

Source: Data courtesy of FMC Corporation.

Table 9-10. *Distribution of carrageenan sales*

By end use	Percentage	By region	Percentage
Dairy	52	Europe	45
Water gel	16	North America	23
Other food	10	Latin America[a]	12
Nonfood	22	Asia	20

[a]Principally Chile, Argentina, and Mexico.
Source: Data courtesy of FMC Corporation.

reflects the maturity of the food-processing industry, which is the staple outlet for carrageenans. Table 9-10 shows the distribution of extractive sales by end use and geographical region.

E. Future prospects

After nearly 50 years of development, the carrageenan industry can now be said to have come of age and to be a mature industry. Its close ties to the food-processing industry, likewise in its maturity, suggest that future growth should be steady. New products and applications can be expected, but these may be slow in coming. An insight into the time scale can be gained from the observation that the last "new" application for carrageenan of any commercial significance was air freshener gels in the early 1970s.

A favorable factor has been the stabilization of seaweed supplies and prices due to the advent of *Eucheuma* farming. Barring political upheavals in the harvesting regions, the industry can remain assured of adequate supplies of good-quality seaweed at reasonable prices. Future progress now appears to lie in the areas of achieving cost reduction in processing and developing more versatile and better quality-controlled products.

IV. Agars

A. History

The red algal polysaccharide agar has been used for centuries (Tseng 1946). The name *agar* originated in Indonesia, and Indonesians still refer to some seaweeds and to jellies from seaweeds as "agar-agar." Indonesians probably learned about agar from Japanese and Chinese traders who brought *kanten* and *tungfen* to Indonesia from their home-lands. The Japanese refer to agar as *kanten,* meaning "cold weather" or "cold sky." *Tungfen* is Chinese for agar and means "frozen powder." Both names are references to processes used to prepare agar.

Europeans traveling to Indonesia learned about agar and introduced it to their native countries. In Europe, agar is known as Japanese isin-glass. *Isinglass* is Dutch for a gelatin prepared from the air bladders of sturgeons and used to clarify jellies.

The first reference to purified agar, according to Tseng (1946), was from seventeenth-century Japan. The story relates how, in 1658, an inn-keeper threw some agar jellies outside one cold evening. The next morning he found the remains of these jellies, which had presumably frozen in the cold night air and had then thawed and dried to flakes in the morning sun. The innkeeper rejuvenated the jellies by dissolving the flakes in hot water and allowing the solution to cool. He noticed the jellies from the rehydrated flakes had better clarity and texture than the original jellies.

Ironically, agar became best known not for its centuries-old use as a condiment but as a medium for culturing microbes. Frau Fanny Eil-shemius, the wife of a physician, discovered this application. Her hus-band introduced this use for agar to Robert Koch, the eminent bacte-riologist, who has since been credited for this discovery. Agar was so prized as a bacteriological medium that cessation of shipments of agar from Japan during World War II forced Allied countries to find alter-native sources. These efforts led to the discovery of many agar-bearing seaweeds and to the development of commercial industries for agar pro-duction in Europe, Africa, and North and South America.

B. Chemistry

Araki (1937) first identified the structural components of agar. He sep-arated two fractions, one of which was a neutral polymer, agarose, com-posed of repeating units of alternating 1,3-linked β-D-galactopyranose and 1,4-linked 3,6-anhydro-α-L-galactopyranose. The other fraction, called agaropectin, was highly ionic and contained polymers of galactose sulfate together with some D-glucuronic acid and pyruvic acid.

Later researchers (Duckworth and Yaphe 1971; Young, Duckworth, and Yaphe 1971) found that bacteriological-grade agar was more complex than an agarose and agaropectin mixture. They found a series of galactose polymers with ester sulfate, pyruvate, and other charged groups. Izumi (1971) concluded from his research on the structure of agar from *Gelidium amansii* that the term *agarose* still applied to the neutral galactose fraction, but the term *agaropectin* did not adequately define the variable group of charged moieties and therefore should not be used.

Currently, agarobiose (Fig. 9-3) is used to refer to the repeating β-D-galactopyranose and 3,6-anhydro-α-L-galactopyranose unit of agarose (Anonymous 1982a).

C. Production

Seaweed resources. Species of *Gelidium* and *Gracilaria* are the primary sources of commercial agar (Table 9-1). Agar has also been found in species of *Ceramium, Phyllophora, Pterocladia, Ahnfeltia, Campylaephora,* and *Acanthopeltis* (Percival and McDowell 1967).

In 1980, about 80% of the seaweed extracted commercially for agar came from Asia and Latin America (Table 9-2). Currently, most seaweed used for commercial agar production is harvested by hand from natural populations. Cultivation of agarophytes provides small amounts of raw materials. Current data on harvest and cultivation of agar-bearing algae are considered commercially confidential.

Extraction and processing. Modern commercial agar production starts with raw material selection because different species of agarophytes give agars with different properties. The seaweed is then pretreated mechanically and/or chemically to remove epiphytes, sand, salt, and pigments. The cleaned seaweed is dissolved in hot water to extract the water-soluble agar. This water-soluble fraction is separated from insoluble matter by centrifugation, filtration, or both. The supernatant/filtrate containing the agar may then be treated chemically to improve color. The solution is then gelled by cooling. Historically, the gel was subjected to a series of freeze–thaw cycles. Presently, hydraulic presses squeeze the gel. Water exuded by either process removes salts and other impurities from the gelled agar. The purified gel is then mechanically dried and packaged as sheets, flakes, or powder (Glicksman 1983).

D. Applications

Properties. The *Food Chemicals Codex* (Anonymous 1981) identifies agar as

a dried hydrophilic, colloidal polygalactoside extracted from *Gelidium cartilagineum* (L.) Gaillon (Fam. Gelidiaceae), *Gracilaria confervoides* (L.) Greville (Fam. Sphaerococcaceae), and related red algae (Class Rhodophyceae). It is commercially available in bundles consisting of thin, membranous agglutinated strips, or in cut, flaked, granulated, or powdered forms. It is white to pale yellow in color, is either odorless or has a slight characteristic odor, and has a mucilaginous taste. Agar is insoluble in cold water, but is soluble in boiling water.

Agar dissolved in hot water and permitted to cool will form thermally reversible gels (the gel will melt when heated and reform when cooled). Characteristically, solutions containing 1–2% agar by weight will gel at about 35°C and will melt at about 85°C. The hysteresis of the gel is of importance in commercial applications. Agar is used to stabilize bakery icings: The icing melts during baking, but the agar reforms to a gel-like glaze when cooled. Without agar, the icing would melt and run off. In addition, the hydrophilic nature of agar keeps baked goods moist during and after processing.

The 1–2% (w/w) agar gels are strong and brittle. Typically, a force of $500-1,000$ g·cm^{-2} is required to break these gels. The strength and brittleness of agar gels are proportional to the amount of 3,6-anhydro-α-L-galactopyranose in the agar. Confectionery manufacturers use the gel-forming characteristics of agar in their products so that they break cleanly and are not gummy or sticky.

The ionic nature of the agar molecule permits it to complex with proteins. The presence of proteins in wines, juices, and vinegar clouds these products. Manufacturers add agar during processing to bind with protein impurities. This facilitates removal of proteins by either filtration or centrifugation.

Products and applications. Tables 9-4–9-7 show the primary commerical applications for agar. The use of agar in baked goods, candies, juices, wines, and vinegar has already been mentioned. Agar provides body, mouthfeel, and texture to many products such as meringues, pie fillings, and cookies (Meer 1980).

Industrial applications for agar include use in adhesives and cosmetics, for electrophoretic and chromatographic separations, and as a casting base for dentistry and toolmaking. Agar has medicinal and pharmaceutical applications including use as a suspending agent for radiological solutions, as a bulk laxative, and as the formative ingredient for tablets and capsules to carry medications. In microbiology, agar is the medium of choice for culturing microbes on solid substrate. Since most microorganisms cannot metabolize agar, nutrients are added to agar media to support growth of cultured organisms (Meer 1980).

World market. Japan, Spain, Taiwan, Korea, Morocco, Chile, Portugal, and the United States were the major producers of commercial agar in 1982 according to the *Journal of Commerce.* Japan led the world with over 400 companies extracting agar. World consumption of agar in 1982 was 5,000 mt, about the same as it was in 1978 (Coyner et al.

1981). Imports of agar to the United States increased from 190 mt in 1955 to 700 mt in 1980. In 1979, the amount of agar used in food in the United States was estimated at 0.07% of the total amount of gums used in food (Glicksman 1982).

E. Future outlook

The current world market for agar is stable and mature. Production capacity for commercial agar exceeds world demand. Overutilization of natural populations of seaweeds containing agar may result in diminished supplies. Cultivation technology now being practiced should assure a continuing supply of high-quality raw material if cultivation can be carried out on a large scale. Genetic manipulation of agarophytes, in the developmental stages, promises to minimize seasonal variations in plant growth and agar quality.

The similarity in the properties among the gums used in foods means that agar faces stiff competition in these applications. The viability of the commercial agar business will depend on the development of new applications. Probably these new applications will require agar that demonstrates pH and temperature stability, salt tolerance, and cold solubility.

V. References

Anderson, N. S., T. C. S. Dolan, C. J. Lawson, A. Penman, and D. A. Rees. 1968. Carrageenans. V. The masked repeating structures of λ- and μ-carrageenans. *Carbohydr. Res.* 7, 468–73.

Anderson, N. S., T. C. S. Dolan, A. Penman, D. A. Rees, G. P. Mueller, D. J. Stancioff, and N. F. Stanley. 1968. Carrageenans. IV. Variations in the structure and gel properties of κ-carrageenan, and the characterisation of sulphate esters by infrared spectroscopy. *J. Chem. Soc.* (C), 602–6.

Anderson, N. S., T. C. S. Dolan, and D. A. Rees. 1968. Carrageenans. III. Oxidative hydrolysis of methylated κ-carrageenan and evidence for a masked repeating structure. *J. Chem. Soc.* (C), 596–601.

Anderson, N. S., T. C. S. Dolan, and D. A. Rees. 1973. Carrageenans. VII. Polysaccharides from *Eucheuma spinosum* and *Eucheuma cottonii:* the covalent structure of ι-carrageenan. *J. Chem. Soc. Perkin I,* 2173–6.

Anonymous. 1976. *Kelco Algin,* 2nd ed. Kelco Division of Merck & Co., San Diego, CA. 51 pp.

Anonymous. 1980. *Solutions to Sticky Problems.* Brookfield Engineering Laboratories, Stoughton, MA. 20 pp.

Anonymous. 1981. *Food Chemicals Codex,* pp. 11–12. 3rd ed. National Academy Press, Washington, DC.

Anonymous. 1982a. *The Agarose Monograph,* pp. 16–17. FMC Corporation, Rockland, ME.

Anonymous. 1982b. *J. Commer.,* 22B.

Anonymous. 1983. Seaweeds: products and markets. *Infofish* 4, 23–6.

Araki, C. 1937. Fractionation of agar-agar. *J. Chem. Soc. Jap.* 58, 1338.

Barilotti, D. C., P. K. Dayton, and R. H. McPeak. 1985. Experimental studies on the effects of commercial kelp harvesting in central and southern California *Macrocystis pyrifera* kelp beds. *Calif. Fish and Game* 71, 4–20.

Bourgade, G. 1871. Improvement in treating marine plants to obtain gelatine. U.S. Patent no. 112, 535.

Carratech, Inc. 1986. Charlottetown, P.E.I. (Patent pending.)

Cottrell, I. W., and P. Kovacs. 1977. Algin. In *Food Colloids*, ed. H. D. Graham, p. 443. AVI Publishing Co., Westport, CT.

Cottrell, I. W., and P. Kovacs. 1980. Alginates. In *Handbook of Water Soluble Gums and Resins*, ed. R. L. Davidson, pp. 2-1–2-43. McGraw-Hill, New York.

Coyner, E. C., J. Higuchi, O. Kamatari, and J. Bakker. 1981. *Water Soluble Resins.* SRI International, Menlo Park, CA.

Deslandes, E., J. Y. Floc'h, C. Bodeau-Bellion, D. Brault, and J. P. Braud. 1985. Evidence for *lambda*-carrageenans in *Solieria chordalis* (Solieriaceae) and *Callibepharis jubata, C. ciliata* and *Cystoclonium purpureum* (Rhodophyllidaceae). *Bot. Mar.* 28, 317–81.

Dolan, T. C. S., and D. A. Rees. 1965. The carrageenans. II. The positions of the glycosidic linkages and sulfate esters in λ-carrageenan. *J. Chem. Soc.* 3534–9.

Donovan, M. 1837. *Domestic Economy.* Vol. 2, p. 323. Longman, London.

Doty, M. S. 1973. Farming the red seaweed, *Eucheuma*, for carrageenans. *Micronesica* 9, 59–73.

Doty, M. S. 1979. Status of marine agronomy, with special reference to the tropics. *Proc. Int. Seaweed Symp.* 9, 35–58.

Doty, M. S., and V. B. Alvarez. 1975. Status, problems, advances, and economics of *Eucheuma* farms. *Mar. Techol. Soc. J.* 9, 30–5.

Duckworth, M., and W. Yaphe. 1971. The structure of agar. II. The use of bacterial agarose to elucidate structural features of the charged polysaccharides in agar. *Carbohydr. Res.* 16, 435–45.

Eisses, J. 1952. The research of gelatinous substances in Indonesian seaweeds at the Laboratory for Chemical Research, Bogor. *J. Sci. Res.* [Indonesia] 1, 44–9.

Eisses, J. 1953. Seaweeds in the Indonesian trade. *Indonesian J. Nat. Sci.* 1, 2, 3, 41–56.

Fraser, M. J. 1942. The Irish moss industry of Massachusetts. *Fish. Market News* 4, 24–8.

Gellenbeck, K. W., and D. J. Chapman. 1983. Seaweed uses: the outlook for mariculture. *Endeavour* 7, 31–7.

Glicksman, M. 1982. Comparative properties of hydrocolloids. In *Food Hydrocolloids*, vol. 1, ed. M. Glicksman, pp. 4–18. CRC Press, Boca Raton, FL.

Glicksman, M. 1983. Red seaweed extracts (agar, carrageenans, furcellaran). In *Food Hydrocolloids*, vol. 2, ed. M. Glicksman, pp. 73–113. CRC Press, Boca Raton, FL.

Guiseley, K. B., N. F. Stanley, and P. A. Whitehouse. 1980. Carrageenan. In *Handbook of Water Soluble Gums and Resins*, ed. R. L. Davidson, pp. 5-1–5-30. McGraw-Hill, New York.

Humm, H. J. 1951. The red algae of economic importance: agar and related phycocolloids. In *Marine Products of Commerce,* 2nd ed., ed. D. K. Tressler and J. McW. Lemon, pp. 47–93. Reinhold Publishing Corp., New York.

Izumi, K. 1971. Chemical heterogeneity of the agar from *Gelidium amansii. Carbohydr. Res.* 17, 227–30.

Kimura, R. S. 1980. The effects of harvesting *Macrocystis pyrifera* on algae in Carmel Bay. M.S. thesis, California State University, Fresno. 108 pp.

King, A. H. 1983. Brown seaweed extracts (alginates). In *Food Hydrocolloids,* vol. 2, ed. M. Glicksman, pp. 154–71. CRC Press, Boca Raton, FL.

Laite, P. S., and M. A. Ricohermoso. 1981. Revolutionary impact of *Eucheuma* cultivation in the South China Sea on the carrageenan industry. *Proc. Int. Seaweed Symp.* 10, 595–600.

Lawson, C. J., and D. A. Rees. 1968. Carrageenans. VI. Reinvestigation of acetolysis products of λ-carrageenan: revision of the structure of α-1,3-galactotriose, and a further example of the reverse specificities of glycoside hydrolysis and acetolysis. *J. Chem. Soc.* (C), 1301–4.

Lawson, C. J., D. A. Rees, D. J. Stancioff, and N. F. Stanley. 1973. Carrageenans. VIII. Repeating structures of galactan sulfates from *Furcellaria fastigiata, Gigartina canaliculata, Gigartina atropurpurea, Ahnfeltia durvillaei, Gymnogongrus furcellatus, Eucheuma cottonii, Eucheuma spinosum, Eucheuma isiforme, Eucheuma uncinatum, Aghardhiella tenera, Pachymenia hymantophora* and *Gloiopeltis cervicornis. J. Chem. Soc. Perkin I,* 2177–82.

McCandless, E. L., J. S. Craigie, and J. A. Walter. 1973. Carrageenans in the gametophytic and sporophytic states of *Chondrus crispus. Planta* [Berlin] 112, 201–12.

MacMullan, E. A., and F. R. Girich. 1963. The precipitation reaction of carrageenan with gelatin. *J. Colloid Sci.* 18, 526–37.

McNeeley, W. H., and P. Kovacs. 1975. The physiological effects of alginates and xanthan gum. In *Physiological Effects of Food Carbohydrates,* ed. A. Jeanes and J. Hodge, pp. 269–81. American Chemical Society, Washington, DC.

McNeely, W. H., and D. J. Pettitt. 1973. Algin. In *Industrial Gums: Polysaccharides and Their Derivatives,* 2nd ed., ed. R. L. Whistler and J. N. BeMiller, p. 50. Academic Press, New York.

Meer, W. 1980. Agar. *In Handbook of Water-soluble Gums and Resins,* ed. R. L. Davidson, pp. 7-1–7-19. McGraw-Hill, New York.

Mitchell, M. E., and M. D. Guiry. 1983. Carrageenan: a local habitat or a name? *J. Ethnopharmacol.* 9, 347–51.

Mshigeni, K. E. 1984. The red algal genus *Eucheuma* (Gigartinales, Solieriaceae) in East Africa: an underexploited resource. *Hydrobiologia.* 116/117, 347–50.

Mueller, G. P., and D. A. Rees. 1968. Current structural views of red seaweed polysaccharides. In *Drugs from the Sea,* ed. H. D. Freundenthal, pp. 241–55. Marine Technology Society, Washington, DC.

Paoletti, S., O. Smidsrød, and H. Grasdalen. 1984. Thermodynamic stability of the ordered conformations of carrageenan polyelectrolytes. *Biopolymers* 23, 1771–94.

Parker, H. S. 1974. The culture of the red algal genus *Eucheuma* in the Phillipines. *Aquaculture* 3, 425–39.

Penman, A., and D. A. Rees. 1973a. Carrageenans. IX. Methylation analysis of galactan sulphates from *Furcellaria fastigiata, Gigartina canaliculata, Gigartina chamissoi, Gigartina atropurpurea, Ahnfeltia durvillaei, Gymogongrus furcellatus, Eucheuma isiforme, Eucheuma uncinatum, Aghardhiella tenera, Pachymenia hymantophora,* and *Gloiopeltis cervicornis:* structure of ξ-carrageenan. *J. Chem. Soc. Perkin I,* 2182–7.

Penman, A., and D. A. Rees. 1973b. Carrageenans. X. Synthesis of 3,6-di-*O*-methyl-D-galactose, a new sugar from the methylation analysis of polysaccharides related to ξ-carrageenan. *J. Chem. Soc. Perkin I,* 2188–91.

Penman, A., and D. A. Rees. 1973c. Carrageenans. Part XI. Mild oxidative hydrolysis of κ- and λ-carrageenans and the characterisation of oligosaccharide sulphates. *J. Chem. Soc. Perkin I,* 2191–6.

Percival, E., and R. H. McDowell. 1967. *Chemistry and Enzymology of Marine Algal Polysaccharides.* Academic Press, London. 219 pp.

Pernas, A. J., O. Smidsrød, B. Larsen, and A. Haug. 1967. Chemical heterogeneity of carrageenans as shown by fractional precipitation with potassium chloride. *Acta Chem. Scand.* 21, 98–110.

Rees, D. A. 1963. The carrageenan system of polysaccharides. I. The relation between the κ- and λ-components. *J. Chem. Soc.* 1821–32.

Rees, D. A. 1972. Mechanism of gelation in polysaccharide systems. In *Gelation and Gelling Agents,* Symposium Proceedings no. 13, pp. 7–12. British Food Manufacturing Industries Research Association, London.

Shacklock, P. F., D. Robson, I. Forsyth, and A. C. Neish. 1973. *Further Experiments (1972) on the Vegetative Propagation of Chondrus crispus T4.* Tech. Rep. no. 18, Atlantic Regional Laboratory, National Research Council of Canada. 22 pp.

Skjak-Braek, G., and B. Larsen. 1985. Biosynthesis of alginate: purification and characterization of mannuronan C-5-epimerase from *Azotobacter vinelandii. Carbohydr. Res.* 139, 273–83.

Smith, H. M. 1905. *The Utilization of Seaweeds in the United States,* Bulletin, U.S. Bureau of Fisheries, vol. 24, 1904 [1905], pp. 169–71.

Stancioff, D. J., and D. W. Renn. 1975. Physiological effects of carrageenan. In *Physiological Effects of Food Carbohydrates,* ed. A. Jeanes and J. Hodge, pp. 282–95. American Chemical Society, Washington, DC.

Stancioff, D. J., and N. F. Stanley. 1969. Infrared and chemical studies on algal polysaccharides. *Proc. Int. Seaweed Symp.* 6, 595–609.

Stanford, E. C. C. 1862. On the economic applications of seaweed. *J. Soc. Arts* 10, 185–95.

Stanley, N. F. 1963. Process for treating a polysaccharide of seaweeds of the Gigartinaceae and Solieriaceae families. U.S. Patent no. 3,094,517.

Towle, G. A. 1973. Carrageenan. In *Industrial Gums: Polysaccharides and Their Derivatives,* 2nd ed., ed. R. L. Whistler and J. N. BeMiller, pp. 83–114. Academic Press, New York.

Tseng, C. F. 1946. Phycocolloids: useful seaweed polysaccharides. In *Colloid Chemistry, Theoretical and Applied,* ed. J. Alexander, pp. 629–734. Reinhold Publishing Corp., New York.

Welti, D. 1977. Carrageenans. XII. The 300 mHz proton magnetic resonance spectra of methyl β-D-galactopyranoside, agarose, *kappa*-carrageenan, and segments of *iota*-carrageenan and agarose sulphate. *J. Chem. Res.* (S), 312–13.

Young, K., M. Duckworth, and W. Yaphe. 1971. The structure of agar. III. Pyruvic acid, a common feature of agars from different agarophytes. *Carbohydr. Res.* 16, 446–8.

Zaneveld, J. S. 1955. *Economic Marine Algae of Tropical South and East Asia and Their Utilization.* Indo-Pacific Fisheries Council Special Publication, no. 3. IPFC Secretariat, FAO Regional Office for Asia and the Far East, Bangkok.

Zaneveld, J. S. 1959. The utilization of marine algae in tropical South and East Asia. *Econ. Bot.* 13, 89–131.

10. Lipids and polyols from microalgae

KENNETH G. SPENCER

Microbio Resources, Inc., 6150 Lusk Blvd., Suite B-105, San Diego, CA 92121

CONTENTS

I. Introduction

The lipids used by humans come from many sources: They are refined from petroleum, rendered from animal fat, squeezed from oil seeds, and extracted from many plant and animal sources. However, until recently virtually none of the useful lipids have come from the most productive plants on earth, microalgae. The possibility of using these algae as microscopic manufacturers of a variety of lipids will be addressed in this chapter. Polyols (polyhydric alcohols such as glycerol or mannitol) are also included because they represent a class of compounds that some microalgae are particularly proficient at making.

Certain algae are quite idiosyncratic in their production of high amounts of specific lipid metabolites. The lipids that could be produced commercially from algae are a heterogeneous mix of compounds that are grouped together only because of their solubility properties. Many of these compounds are inherently important to mankind because they are essential to human nutrition. In addition to nutrition, a much larger diversity of lipids is important in modern daily life as fuels, soaps, lubricants, pharmaceuticals, cosmetics, and pigments. Lipid compounds are also a fundamental chemical feedstock from which a host of other useful compounds are made. With several notable exceptions these lipids are not now produced commercially using algae. Whether any of the other possible compounds will be produced commercially will depend upon the inventiveness of biologists in inducing algae to make as much as possible of the desired product. At the same time biologists and engineers must cooperate in markedly reducing the cost of producing anything from microalgae.

For this review examples of several lipid classifications are discussed. Other potentially valuable lipid types, such as sterols, are not covered even though some compounds of these types are produced by some microalgae. The reader is referred to reviews by Dubinsky, Berner, and Aaronson (1978), Dubinsky and Aaronson (1982), Pohl (1983), and Cohen (1986) for listings of algal lipid products and further discussion of some of the topics not included here.

[238]

II. History of microalgal lipid production research

Much of the research on microalgal lipids received its impetus from industrial nations interested in alternative sources of food and fuel oils. During and after World War II, Germany and the United States initiated research efforts that resulted in some excellent physiological studies on the optimum production of storage lipids by microalgae (e.g., see Harder and von Witsch 1942; Spoehr and Milner 1949). By the mid-1950s most of the stress responses that prompt neutral lipid formation in various green algae and diatoms were described as a result of research efforts around the world. In the United States the most thorough investigation of the possibilities for microalgae production was conducted by the Carnegie Institution (Burlew 1953). Phycologists, physiologists, and engineers were brought together to address the problems of large-scale microalgal culture. Many questions were answered, but the economic forecasts that resulted were too equivocal to prompt the start of a new industry based on products from algae. Since that time, researchers in most industrial nations have continued to grow microalgae in the laboratory and in pilot plants with the hope that profitable mass cultivation technologies could be developed. Lipid formation by algae has been considered in these studies as an adjunct to wastewater treatment or as a by-product from algae cultivated as sources of single-cell protein. In the United States the Biomass Energy Technology program of the Department of Energy has supported basic research in this area since the mid-1970s. The goal of this research is to develop a technology to grow algae in the desert for the production of fuels. European research on growing microalgae for biomass was summarized in a recent volume (Palz and Pirrwitz 1983).

In recent years actual commercial production was initiated by a few companies in response to the high value of certain algal lipid products. During the 1960s the Grain Processing Corporation (Muscatine, Iowa) grew large quantities of the green alga *Neospongiococcum* for use as poultry feed because of its high xanthophyll content (Marusich and Bauernfeind 1981). In Belgium, Bioprex S.A. extracts terpenoids and other compounds from *Scenedesmus, Chlorella,* and other microalgae for use as components of cosmetics (E. Dujardin, personal communication). Work conducted in the USSR led to the industrial production of green algae for essential oils and pigments (see references in Dubinsky and Aaronson 1982). In the United States, Microbio Resources, Inc. (San Diego, California), in 1984 became the first company to successfully produce and market the β-carotene produced by *Dunaliella*. Other companies with an announced interest in producing high-value lipid products from microalgae currently are operating in Israel, Australia, and the United States. Manufacturing and marketing products from microalgae are currently an entrepreneurial enterprise. Some of the very high-value products such as β-carotene, cosmetics, and pharmaceuticals must be proven profitable before any of the commercially less valuable chemicals produced by algae can be considered for large-scale production.

The current state of the art in growing and harvesting microalgae in unialgal mass culture results in a production cost of at least US $10 per kg of dry whole algae. The final cost of producing an extracted lipid component depends on its relative percentage in the algal biomass and the cost of extracting it. So far most of the information on production costs of useful lipids from algae has come from the laboratory rather than the production plant.

III. Triacylglycerols

The vegetable oils, bulk commodities in today's marketplace, fill a multiplicity of needs. Most are used for human food preparation, but they are also important industrial feedstocks. Rather than being a single compound, each of the different oils represents a blend of triacylglycerols. Their composition depends upon the particular fatty acids being produced within the plant at the time. In most cases palmitic acid (a completely saturated 16-carbon compound) is the starting biochemical moiety from which other fatty acids are derived. Chain length modifications and desaturations that give rise to the other fatty acids occur to different degrees in the various plant tissues in which triacylglycerols accumulate. Studies show that the basic pathways in algae are directly analogous to those demonstrated in higher plant oil seeds. However, certain algal species produce long-chain or highly unsaturated fatty acids that are unique to those taxa. Variations in the component fatty acids of algal oil can be observed in response to growth conditions (Materassi et al. 1980). The fundamental difference between triacylglycerol formation in higher plants and microalgae is that in the case of algae the oil-accumulating cells are themselves fully photosynthetic. This means that the complete pathway from carbon dioxide fixation to triacylglycerol synthesis can be modulated by the cell. The microalgal cell has the potential for becoming a compact target for manipulation toward directed synthesis of a desired type of oil.

Conspicuous formation of lipids by microalgae is the result of the accumulation of storage triacylglycerols. Many genera of the Chlorophyceae, Bacillariophyceae, Chrysophyceae, and Xanthophyceae have oil storage as a diagnostic feature instead of or in addition to starch. Triacylglycerol accumulation has never been observed in the Cyanophyceae. The highest observed accumulations of lipid by microalgae are listed in Table 10-1. With the exception of *Botryococcus braunii*, all of the

Table 10-1. *Maximal lipid accumulation in microalgae*

Species	Freshwater or marine	Total lipid (% dry mass)	Mass yields $(g \cdot L^{-1})$	Reference
Botryococcus braunii	FW	86	—	Brown et al. 1969
Chlorella ellipsoidea	FW	84	0.8	Iwamoto et al. 1955
Nitzschia palea	FW	80	—	Opute 1974
Ankistrodesmus braunii	FW	73	—	Williams and McMillan 1961
Monallantus salina	M	72	—	Shifrin and Chisholm 1981
Chlorella pyrenoidosa	FW	70	—	Spoehr and Milner 1949
Chlorella vulgaris	FW	65	0.03	Piorreck and Pohl 1984
Dunaliella primolecta	M	54	—	Dubinsky et al. 1978
Ochromonas danica	FW	53	3.6	Aaronson 1973

algae listed primarily produced triacylglycerols. Lipid accumulations accounting for 50% of the dry cell mass are not uncommon. The large accumulations reported do not necessarily correspond to high production of lipids per unit time in outdoor cultures. High outdoor productivities of triacylglycerols have not yet been demonstrated.

The formation of oil in algal cells can be dramatically observed using oleophilic dyes such as Nile Blue A. Accumulation can first be seen as many small droplets in the cytoplasm. As the algae continue to photosynthesize under conditions that inhibit cell division, massive quantities of oil can be produced and can occupy the majority of the cell volume. Degradation of the chlorophylls and an increase in certain carotenoid pigments often accompany lipid synthesis (Lien and Spencer 1983). This synthesis provides the alga with a metabolic sink of almost fully reduced carbon molecules. Lipid synthesis is important at the time it occurs because it permits photosynthesis to continue. Moreover, the cells are later able to use this stored oil by metabolizing it in the dark or light for resynthesis of chlorophyll and for cell division once the stress condition has been removed (Spencer, unpublished data).

Triacylglycerol production in most algae is correlated to several types of stress conditions that can be easily envisioned as occurring in nature. Algae that produce large quantitites of lipid generally demonstrate the ability to subsist under the stress for a prolonged period. Numerous stress conditions have been documented to induce this lipid accumulation. Most research on this subject so far has been performed with green algae and diatoms. Nitrogen deprivation was the earliest condition observed to prompt lipid accumulation (Beijerinck 1904) and remains the most commonly observed. Other types of stress observed to precede lipid accumulation include sulfate limitation (Mandels 1943), phosphate limitation (Daniel 1956), desiccation (Spoehr and Milner 1949), and autoinhibitor formation (Scutt 1964). High CO_2 concentrations also were observed to promote lipid synthesis (Pratt and Johnson 1964). Under all of these stress conditions, lipid formation occurs as the photosynthetic rate is declining. Therefore, the best strategy for triacylglycerol production from these natural strains of oleaginous algae in mass cultures would be a two-stage process: rapid growth of cells under nonstress conditions followed by imposition of nitrogen deprivation or other stress in order to achieve maximum lipid content. Development of new algal strains that artificially produce storage lipids under conditions of maximal photosynthesis would be a breakthrough in the commercial development of microalgae.

Another important commercial objective is the development of algal oils with sharply defined and valuable compositions. Soybean and corn oils currently sell for less than US $1 per kg. A current estimate for the cost of a pure algal oil would be at least $10–20 per kg. Although most algal oils have fatty acid constitutions very much like the common vegetable oils, many unusual fatty acids can be synthesized by a variety of algae. Enhancing the production of a certain type of oil could be achieved in some cases by modification of growth conditions. To achieve improved strains, mutation and selection procedures could be employed to increase the activity of certain metabolic pathways. Future progress may result from insertion of the genes for formation of an unusual acyl lipid into an algal strain that can naturally direct a large amount of its photosynthate into this pathway.

IV. Hydrocarbons

The production of quantitites of hydrocarbons is a rather rare phenomenon in algae. Small quantities of C_{15}–C_{27} hydrocarbons have been found in various green (Gelpi et al. 1968) and blue-green algae (Gelpi et al. 1968; Goodloe and Light 1982). The presence of hydrocarbons is conspicuous only as the carotenoids in *Dunaliella* and as a mixture of unique C_{17}–C_{34} hydrocarbons in the green colonial alga *Botryococcus braunii*. Both of these organisms continue to receive attention as unique biological producers of these hydrocarbons.

Botryococcus has the capacity to divert a very large fraction of its photosynthate into several distinctive hydrocarbons, which have been named *botryococcenes*. They are acyclic olefin molecules. The molecule given the name botryococcene is $C_{34}H_{58}$. Other structural components of this oil are closely related. Their biosynthetic origin is just beginning to be examined; however, their branched chain structures suggest that they are products of a terpenoid pathway. Botryococcenes are produced internally and often are excreted to the outside of the cell where they can be observed as an oleaginous layer between and around the cells of this colonial organism.

Botryococcus has a documented role in several areas of petrochemistry (see also George, Chapter 13). Botryococcenes were reported to compose approximately 1% of a crude oil from Sumatra (Moldowan and Seifert 1980). An early stage in the production of such petroleum can be observed as deposits called coorongite (Cane 1969).

Botryococcus has been studied intensively as the alga most likely to succeed as a producer of renewable liquid fuels. Indeed, enough *Botryococcus* extract material was obtained by Wake and Hillen (1980) to confirm that the hydrocarbon mixture could be cracked into the same useful distillation products as petroleum. The factor most inhibiting to the mass cultivation of *Botryococcus* at the current time is the slow growth rate of the alga. Doubling times of more than one week are normal under laboratory conditions. However, the fact that blooms of the alga have been observed in nature (Wake and Hillen 1980) suggests that increased productivity is possible. Recent reports show that improvements can be made by changing culture conditions and by using certain strains of the organism. Casadevall et al. (1985) reported that doubling times of two to three days could be achieved in cultures mixed by means of airlift pumps. Wolf et al. (1985) observed 40-hour doubling times in laboratory cultures.

Botryococcus exists in several physiological states, which are characterized by varying rates of growth, carotenoid production, and botryococcene production. The current knowledge about these states was summarized by Wolf (1983). Recent *Botryococcus* research demonstrates the value of examining many natural strains of an alga to maximize productivity. Wolf, Nonomura, and Bassham (1985) described good productivity of botryococcenes in a strain selected from among 2,000 cloned isolates from a single natural collection. Variability in growth from several different strains was described by Berkaloff et al. (1984).

Hydrocarbons are an energy-intensive product. Thus, the mass accumulation of an alga synthesizing them in high amounts will naturally be less than that of a typical unicellular green alga. However, since high concentrations of the hydrocarbons can be produced, the possibilities for commercialization are greatly enhanced. Because products from *Botryococcus* will not be used for food, the alga is a potential candidate for growth on wastewater. The cost of producing large amounts of algal hydrocarbon would be dramatically reduced under this type of growth condition.

V. Algal solar energy conversion

The triacylglycerol and hydrocarbon producers just discussed are also algae that have been investigated as potential converters of solar energy. These oleaginous algae may be able to serve as energy-rich sources of biomass. Is the mass cultivation of microalgae for fuel a practical endeavor? There are two aspects to this question: (1) whether it is possible to harvest significantly more energy than is spent; and (2) whether any microalgal lipid fuel product could be sold for more than its production cost. So far most of the work on these questions has been theoretical with tests only at the small-scale culture level. No one has yet built any of the large-scale outdoor culture facilities required to get actual energy efficiency and financial numbers.

The balance sheet of the energetics of algal productivity must come out on the positive side if the culturing of these plants is to be deemed a renewable source of energy. Initially, this prospect seems quite favorable because photosynthesis is inherently quite efficient at converting absorbed light energy into organic chemical bond energy. The mechanism of photosynthesis as it is currently understood allows 28.5% of the energy in red light to be converted during the initial steps of photosynthesis. However, many factors reduce this value in the formation of lipids from sunlight in outdoor mass cultures. Green photosynthetic plants use only the visible portion (400–700 nm) of sunlight, which amounts to 48.5% of full sunlight. In addition, at least a small portion of this captured energy is rapidly lost to cell maintenance and respiration. If we take these factors into consideration, the maximum theoretical conversion of full sunlight by any green plant is approximately 12% (Radmer and Kok 1977). By optimizing conditions in the laboratory, physiologists have been able to approximate closely the maximum quantum efficiency of photosynthesis by using microalgae. Using laboratory measurements, Pirt et al. (1980) calculated even higher potential theoretical light conversion yields. However, their work requires a new interpretation of the mechanisms of photosynthesis and has yet to be repeated in other laboratories.

Microalgae have a distinct advantage over most land plants and macroalgae in achieving the theoretical maximum conversion rate. All of their cells can photosynthesize. The lack of structural components means that less energy needs to be spent on maintenance. In spite of these considerations, many factors prevent large-scale algal cultures from approaching the theoretical 12% conversion rate. Light reflection, photorespiration, suboptimal levels of nutrients or inorganic carbon, and competition with other organisms all have the potential to cause substantial reductions in conversion efficiency. Maximal full sunlight conversion rates measured in several pilot-scale cultures were found to be 0.9% (Toerien and Grobbelaar 1980), 2.8% (Shelef et al. 1980), and 5.2% (Tamiya 1953). These calculated efficiencies do not take into consideration the appreciable amount of energy that must be expended in order to maintain cultures, harvest the algae, and then extract or dry the resulting biomass to convert it to a usable fuel form. Indeed, algal mass culture systems must be designed very cleverly if any net energy is to be gained from the system. Many energy-saving steps no doubt can be taken to maximize efficiency, but the final result will be an efficiency that is markedly lower than that available from other forms of solar energy harvesting. Modern photovoltaic cells can turn more than 10% of the incident solar energy into electricity. Moreover, a simple flat-plate solar collector can use more than 50% of the sun's radiant energy for heating water (Meinel and Meinel 1976).

Although the outlook for effective solar energy conversion using microalgae is not optimistic, it is still likely that algae could be grown to produce specific fuel feedstocks and petroleum replacement products. Algal products have the potential for replacing the petroleum that will, one day, itself require more energy for exploration and delivery than it will yield. The important consideration at that point will become the financial cost of each form of fuel. The current production cost of $10 per dry kg from unialgal cultures precludes their use for energy production now. The challenge for anyone desiring to produce fuel from algae is therefore to devise cheaper production technologies. The economics of production of any algal product proposed as a fuel will always favor its production as a by-product of another process such as wastewater treatment (see Oswald, Chapter 11), protein production, or specialty chemical production. Another approach to the production of fuels from algae is the conversion of algal biomass by pyrolysis (i.e., subjecting the biomass to high temperature and pressure in a reducing atmosphere). Positive results using this procedure have been reported (Chin and Engel 1981; Goldman et al. 1981). Apparently, any algal residue can be converted to fuel feedstock by this method. The scale on which this has been tested so far, however, does not permit an economic analysis of the process.

VI. Carotenoids

Many terpenoid products formed by algae have potential commercial value if they could be produced in a highly concentrated form. The tetraterpene carotenoid pigments are emphasized here because β-carotene, the most common carotenoid pigment, became one of the first lipid products to be extracted from microalgae commercially. Beta-carotene has a well-defined role in human nutrition as a primary source of vitamin A. The molecule is a symmetrical 40-carbon molecule with an unsaturated six-carbon ring at each end. The presence of a long series of conjugated double bonds makes the molecule absorb visible light strongly. In the intestines each β-carotene molecule can be cleaved to form two molecules of retinol (vitamin A). In addition to its known role as a vitamin, the benefits of β-carotene as a natural antioxidant are being investigated. β-Carotene from algae is a mixture of several of the possible stereoisomers of the molecule, similar to those found in the leafy tissues of higher plants. Usually about half is the all-trans isomer, and the remainder exists as 9-cis and other cis isomers. All forms have vitamin A activity, with the all-trans form probably having the highest activity. The other possible natural roles of the various isomers are interesting research topics but difficult ones because of their instability in solution.

The market for β-carotene as a food supplement and food-coloring agent is increasing. Much of this market is currently supplied by the synthetic product. Synthetic all-trans β-carotene is manufactured using an organic synthesis of more than 10 steps. It is currently available from several large pharmaceutical companies at a price of about US $300 per kg for the pure compound. This relatively high price makes it easier for an algal product to be produced profitably compared to the other lipids previously discussed. The supply of β-carotene from other natural sources such as carrots has traditionally been limited, and the price has been high.

Although beta-carotene is found to some degree in almost all algae, the level in the green alga *Dunaliella salina* can be extremely high, making it a clear choice for production of this pigment. This species, like most species of *Dunaliella,* can grow in highly saline waters, but the other species lack the ability to synthesize large quantities of β-carotene. Natural collections of this alga have been made that contain as much as 14% of their dry mass as β-carotene (Aasen, Eimhjellen, and Liaanen-Jensen 1969). The control of its β-carotene synthesis was investigated by Loeblich (1982) and Ben-Amotz and Avron (1983b). The latter authors described *Dunaliella bardawil,* a strain with the same high carotene characteristics as *D. salina.* Under low light and nutrient-rich conditions these strains produce 1–5 pg of carotene per cell and are the same green color as most of the Chlorophyceae. Under conditions of excess light, β-carotene can rapidly accumulate to 30 pg per cell or more, and the cells are very orange in color. The carotene is concentrated in globules on the chloroplast membranes where it performs a photoprotective function (Burnett 1976). The exact mechanism of this protection and the physiological trigger for carotene synthesis in this organism are not currently understood.

The requirement for high light intensity means that outdoor cultivation is important for high yields. Open outdoor cultures are highly vulnerable to contamination, but the unique salinity tolerance of these *Dunaliella* strains reduces this problem. The ability to grow in a medium in which sodium chloride concentrations are close to saturation means that there are fewer types of competitive organisms. These biological factors, coupled with the financial incentive, explain why β-carotene has become one of the first compounds to be extracted and marketed from microalgae. Some of the conditions of growing *Dunaliella* at a pilot plant in western Australia were described by Borowitzka, Borowitzka, and Moulton (1984).

Beta-carotene is just one of many carotenoids found in high quantities in microalgae. Xanthophylls (oxygenated carotenoids) are prominent photosynthetic accessory pigments in many microalgae. Fucoxanthin is especially prominent in the Bacillariophyceae, Chrysophyceae, and Haptophyceae. Many other xanthophylls are produced by different genera and have been given chemotaxonomic significance. In the green algae, xanthophylls such as lutein, astaxanthin, and canthaxanthin often are observed to accumulate in response to the same stress conditions that prompt triacylglycerol accumulation. Physiological aspects of these pigments have not been examined in detail; however, they probably serve a photoprotective function (see Krinsky 1966).

Commercial use of the xanthophylls from *Neospongiococcum* already has been mentioned. The primary function of these carotenoids as a feed supplement is the imparting of color to animal products. The most prominent product in the xanthophyll market is currently lutein, which is used for pigmenting the skin and egg yolks of poultry. Although *Neospongiococcum, Coccomyxa,* and several other algae are recognized as good sources of this xanthophyll, they are not presently being grown commercially. Lutein currently is sold for approximately US $300 per kg as a component of feeds such as cornmeal, alfalfa meal, or powdered marigold petals (Marusich and Bauernfeind 1981).

Algae play a more significant role in the commercial feeds developed for shellfish and fish. Though many microalgae are cultivated as feed for invertebrates and small fish, only the role of carotenoids will be mentioned here. The bluish hues of many crustaceans result from carotenoproteins. The orange and red of crustaceans and of many fish are from xanthophylls. It is generally accepted that these animals do not synthesize carotenoids de novo but make specific changes in those they do ingest. Algae are therefore the most prevalent direct and indirect natural sources of carotenoid pigments in aquatic animals. Since the reddish pigmentation of many fish and seafoods is a very valuable commercial trait, the carotenoids must be supplied in some form when these animals are raised in aquaculture systems. In addition to meals from higher plants, the possible sources of carotenoids include crustacean extracts and pure carotenoid preparations. Since the pure carotenoids are currently rather expensive (e.g., canthaxanthin at approximately US $1,000 per kg), culture of organisms specifically for this purpose becomes feasible. The green alga *Haematococcus* (astaxanthin) and the blue-green alga *Spirulina* (zeaxanthin) have both been tried successfully. Some of the *Spirulina* currently produced in the United States is being sold for this purpose (Simpson, Katayama, and Chichester 1981).

VII. Polyols

Polyols represent yet another class of compounds known to be produced in large quantities by certain microalgae. Osmotic regulation is the physiological reason for the production of these polyols. Algae growing at salinity levels less than or equal to that of seawater generally maintain their osmotic balance with inorganic ions. For example, intracellular levels of potassium increase in response to increased salinity outside the cell. However, this strategy becomes difficult to maintain in response to highly saline environments. Many enzymes are inhibited by elevated levels of ions. Some organic molecules, including many polyols, are able to affect the osmotic activity of water without affecting the function of enzymes (for discussion, see Brown and Borowitzka 1979). Since the level of this metabolite becomes almost the only adaptation necessary for good osmotic control, the alga gains the capacity to exist and grow in a wide variety of salinities. Species of *Platymonas* (Prasinophyceae), *Pyramimonas* (Prasinophyceae), *Monallantus* (Eustigmatophyceae), and a number of brown algal macrophytes produce the six-carbon polyol, mannitol. The chrysophyte *Monochrysis lutheri* is a unique producer of cyclohexanetetrol. Microalgae capable of adapting to very high salinities are producers of glycerol. Since production of these metabolites is proportional to external molarity, these algae are the highest producers of polyols. Most species of *Dunaliella* as well as some species of *Chlamydomonas* (Chlorophyceae) and *Asteromonas* (Prasinophyceae) produce glycerol. In *Dunaliella* the intracellular molality of glycerol is approximately equal to the external concentration of sodium chloride with which it is in equilibirum (Brown and Borowitzka 1979; Ben-Amotz and Avron 1983a).

Dunaliella has the capacity for the production of very substantial amounts of glycerol. At very high salinities *Dunaliella* cells maintain glycerol levels that are approximately half of the cell mass. This alga has the capacity for rather rapid growth in mass culture, and the possible production of glycerol has been estimated to be approximately 10 g·

$m^{-2} \cdot d^{-1}$ or 37 $mt \cdot ha^{-1} \cdot y^{-1}$ (Chen and Chi 1981). There is an established market for pure glycerol as a bulk commodity at a current price of approximately US \$2 per kg. Glycerol is used in foods, cosmetics, and as an industrial chemical. It is currently derived from the saponification of natural fats as well as organic synthesis from propylene (D'Souza 1979). A small amount of the current production is now met from bacteria and yeasts. Natural glycerol commands approximately the same price as the synthetic glycerol. Mannitol is currently priced approximately three times as high as glycerol (*Chemical Marketing Reporter*, Jan. 13, 1986). However, since algal concentrations are less, the cost of producing mannitol from algae probably also would be higher. Although the concentration of polyols can be very high within algal cells, the intracellular volume is much less than 1% of the culture volume even in very dense algal suspensions. This means that the high costs of various concentrating techniques must be added to the costs of growing the algae. As with the other commodity lipids discussed previously, the production economics of these common polyols demands that they be a byproduct or a coproduct of an algal industry.

VIII. Summary

The concept of biosynthesizing lipid products using algae has not been explored extensively; hence, many of the products with exciting commercial potential are probably not mentioned here. As with other products from biotechnology, the technical capabilities and financial resources are just now coming together to enable real development. At this time the research commitment to algae is small in comparison to research with other microbes. There are two fundamental reasons for adopting algae as a means of producing a lipid: (1) the ability to produce the product photosynthetically; and (2) the unique natural capacity of algae to produce specific lipids. Photosynthesis eliminates the need for supplying an exogenous carbon source. Autotrophic media are inexpensive and somewhat less susceptible to contamination. However, since heterotrophic growth is often a reasonable alternative to photosynthetic growth, it seems likely that the second advantage will become the most compelling. The production of a unique lipid in even small quantities by an alga could make that particular alga a logical and fruitful starting point for strain enhancement.

There can be little doubt that at least a few of the lipids synthesized by microalgae have the potential to be exploited directly by humans. The unique lipids produced by algae have been explored mostly in basic research. Many have been noted as biochemical curiosities, and it is likely that many remain to be found. It is hoped that the commercial success of a few algal products will stimulate research with the multitude of species about which very little is known. Realization of their full potential will entail identifying useful algal compounds, learning how to induce the algae to produce them in sufficient quantities, and then learning to grow the algae rapidly and in large quantities. This process will require the cooperation of people with many types of expertise.

IX. Acknowledgments

The author wishes to thank Dr. A. M. Nonomura, Dr. R. K. Togasaki, and two anonymous reviewers for their critical reading of the manuscript.

X. References

Aaronson, S. 1973. Effect of incubation temperature on the macromolecular and lipid content of the phytoflagellate *Ochromonas danica*. *J. Phycol.* 9, 111–13.

Aasen, A. J., K. E. Eimhjellen, and S. Liaaen-Jensen. 1969. An extreme source of beta-carotene. *Acta Chem. Scand.* 23, 2544–5.

Beijerinck, M. W. 1904. Das Assimilationsprodukt der Kohlensaure in den Chromatorphoren der Diatomeen. *Rec. Trav. Bot. Neerl.* 1, 28–40.

Ben-Amotz, A., and M. Avron. 1983a. Accumulation of metabolites by halotolerant algae and its industrial potential. *Ann. Rev. Microbiol.* 37, 95–119.

Ben-Amotz, A., and M. Avron. 1983b. On the factors which determine massive beta-carotene accumulation in the halotolerant alga *Dunaliella bardawil*. *Plant Physiol.* 72, 593–7.

Berkaloff, C., B. Rousseau, A. Coute, E. Casadevall, P. Metzger, and C. Chirac. 1984. Variability of cell wall structure and hydrocarbon type in different strains of *Botryococcus braunii*. *J. Phycol.* 20, 377–89.

Borowitzka, L. J., M. A. Borowitzka, and T. P. Moulton. 1984. The mass culture of *Dunaliella salina* for fine chemicals: from laboratory to pilot plant. *Hydrobiologia* 116, 115–34.

Brown, A. C., B. A. Knights, and E. Conway. 1969. Hydrocarbon content and its relationship to physiological state in the green alga *Botryococcus braunii*. *Phytochemistry* 8, 543–7.

Brown, A. D., and L. J. Borowitzka. 1979. Halotolerance of *Dunaliella*. In *Biochemistry and Physiology of Protozoa*, 2nd ed., ed. M. Lvandowsky and S. H. Hutner, pp. 139–90. Academic Press, New York.

Burlew, J. (ed.). 1953. *Algal Culture from Laboratory to Pilot Plant*. Carnegie Institution, Publication 600. Washington, DC. 325 pp.

Burnett, J. H. 1976. Functions of carotenoids other than in photosynthesis. In *Chemistry and Biochemistry of Plant Pigments*, 2nd ed., vol. 1, ed. T. W. Goodwin, pp. 655–79. Academic Press, London.

Cane, R. F. 1969. Coorongite and the genesis of oil shale. *Geochim. Cosmochim. Acta* 33, 257–65.

Casadevall, E., D. Dif, C. Largeau, C. Gudin, D. Chaumont, and O. Desanti.

1985. Studies on batch and continuous cultures of *Botryococcus braunii:* hydrocarbon production in relation to physiological state, cell ultrastructure, and phosphate nutrition. *Biotech. Bioeng.* 27, 286–95.

Chen, B. J., and C. H. Chi. 1981. Process development and evaluation for algal glycerol production. *Biotech. Bioeng.* 23, 1267–87.

Chin, L.-Y., and A. J. Engel. 1981. Hydrocarbon feedstocks from algae hydrogenation. *Biotech. Bioeng. Symp.* 11, 171–86.

Cohen, Z. 1986. Products from microalgae. In *CRC Handbook of Microalgal Mass Culture,* ed. A. Richmond, pp. 421–54. CRC Press, Boca Raton, FL.

Daniel, A. L. 1956. Stoffwechsel und Mineralsalzernahrung einzelliger Grunalgen. III. Atmung and oxydative Assimilation von *Chlorella. Flora* 143, 31–66.

D'Souza, G. B. 1979. The importance of glycerol in the fatty acid industry. *J. Amer. Oil Chem. Soc.* 56, 812A–9A.

Dubinsky,Z., and S. Aaronson. 1982. Review of the potential uses of microalgae. In *Biosaline Research: A Look to the Future,* ed. A. San Pietro, pp. 181–206. Plenum Press, New York.

Dubinsky, Z., T. Berner, and S. Aaronson. 1978. Potential of large-scale algal culture for biomass and lipid production in arid lands. *Biotech Bioeng. Symp.* 8, 51–68.

Gelpi, E., J. Oro, H. J. Schneider, and E. O. Bennett. 1968. Olefins of high molecular weight in two microscopic algae. *Science* 161, 700–2.

Goldman, Y., N. Garti, Y. Sasson, B.-Z. Ginzburg, and M. R. Bloch. 1981. Conversion of halophilic algae into extractable oil. II. Pyrolysis of proteins. *Fuel* 60, 90–2.

Goodloe, R. S., and R. J. Light. 1982. Structure and composition of hydrocarbons and fatty acids from a marine blue-green alga, *Synechococcus* sp. *Biochim. Biophys. Acta* 710, 485–92.

Harder, R., and H. von Witsch. 1942. Bericht über Versuch zur Fettsynthese mittels autotropher Mikroorganismen. *Forschungsdienst* 16, 270–6.

Iwamoto, H., G. Yonekawa, and T. Asai. 1955. Fat synthesis in unicellular algae. I. Culture conditions for fat accumulation in *Chlorella* cells. *Bull. Agric. Chem. Soc. Jap.* 19, 240–52.

Krinsky, N. I. 1966. The role of carotenoid pigments as protective agents against photosensitized oxidations in chloroplasts. In *Biochemistry of Chloroplasts,* vol. 1, ed. T. W. Goodwin, pp. 423–30. Academic Press, New York.

Lien, S., and K. G. Spencer. 1983. Microalgal production of oils and lipids. In *Energy from Biomass and Wastes,* ed. D. L. Klass and H. H. Elliot, pp. 1107–21. Institute of Gas Technology, Chicago.

Loeblich, L. A. 1982. Photosyntheseis and pigments influenced by light intensity and salinity in the halophile *Dunaliella salina* (Chlorophyta). *J. Mar. Biol. Assoc. U.K.* 62, 493–508.

Mandels, G. R. 1943. A quantitative study of chlorosis in *Chlorella* under conditions of sulphur deficiency. *Plant Physiol.* 18, 449–62.

Marusich, W. L., and J. C. Bauernfeind. 1981. Oxycarotenoids in poultry feeds. In *Carotenoids as Colorants and Vitamin A Precursors,* ed. J. C. Bauernfeind, pp. 319–462. Academic Press, New York.

Materassi, R., C. Paoletti, W. Balloni, and G. Florenzano. 1980. Some considerations on the production of lipid substances by microalgae and cyanobacteria. In *Algal Biomass: Production and Use,* ed. G. Shelef and C. J. Soeder, pp. 619–26. Elsevier/North Holland Biomedical Press, Amsterdam.

Meinel, A. B., and M. P. Meinel. 1976. *Applied Solar Energy: An Introduction.* Addison-Wesley, Reading, MA. 651 pp.

Moldowan, J. M., and W. K. Seifert. 1980. First discovery of botryococcane in petroleum. *J. C. S. Chem. Comm.* 912–14.

Opute, F. I. 1974. Studies of fat accumulation in *Nitzschia palea. Ann. Bot.* 28, 889–902.

Palz, W., and D. Pirrwitz (eds.) 1983. *Energy from Biomass: Solar Energy R&D in the European Community,* ser. E, vol. 5, pp. 150–22. D. Reidel, Dordrecht.

Piorreck, M., and P. Pohl. 1984. Formation of biomass, total protein, chlorophylls, lipid and fatty acids in green and blue-green algae during one growth phase. *Phytochemistry* 23, 217–23.

Pirt, S. J., Y.-K. Lee, A. Richmond, and M. W. Pirt. 1980. The photosynthetic efficiency of *Chlorella* biomass growth with reference to solar energy utilization. *J. Chem. Tech. Biotech.* 30, 25–34.

Pohl, P. 1983. Lipids and fatty acids of microalgae. In *CRC Handbook of Biosolar Resources,* vol. 1, pt. 1, ed. O. R. Zaborsky, pp. 383–404. CRC Press, Boca Raton, Fl.

Pratt, R., and E. Johnson. 1964. Lipid content of *Chlorella* "aerated" with a CO_2-nitrogen vs. a CO_2-in air mixture. *J. Pharmacol. Sci.* 53, 1135–6.

Radmer, R., and B. Kok. 1977. Photosynthesis: limited yields, unlimited dreams. *BioScience* 27, 599–605.

Scutt, J. E. 1964. Autoinhibitor production by *Chlorella vulgaris. Amer. J. Bot.* 51, 581–4.

Shelef, G., Y. Azov, R. Moraine, and G. Oron. 1980. Algal mass production as an integral part of a wastewater treatment and reclamation system. In *Algal Biomass: Production and Use,* ed. G. Shelef and C. J. Soeder, pp. 63–189. Elsevier/North Holland Biomedical Press, Amsterdam.

Shifrin, N. S., and S. W. Chisholm. 1981. Phytoplankton lipids: interspecific differences and effects of nitrate, silicate, and light-dark cycles. *J. Phycol.* 17, 374–84.

Simpson, K. L., T. Katayama, and C. O. Chichester. 1981. Carotenoids in fish feeds. In *Carotenoids as Colorants and Vitamin A Precursors,* ed. J. C. Bauernfeind, pp. 463–538. Academic Press, New York.

Spoehr, H. A., and H. W. Milner. 1949. The chemical composition of *Chlorella:* effect of environmental conditions. *Plant Physiol.* 24, 120–149.

Tamiya, H. 1953. Mass culture of algae. *Ann. Rev. Plant Physiol.* 8, 309–34.

Toerien, D. F., and J. U. Grobbelaar. 1980. Algal mass cultivation experiments in South Africa. In *Algae Biomass: Production and Use,* ed. G. Shelef and C. J. Soeder, pp. 73–80. Elsevier/North Holland Biomedical Press, Amsterdam.

Wake, L. V., and L. W. Hillen. 1980. Study of a "bloom" of the oil-rich alga *Botryococcus braunii* in the Darwin River Reservoir. *Biotech. Bioeng.* 22, 1637–56.

Williams, V. R., and R. McMillan. 1961. Lipids of *Ankistrodesmus braunii. Science* 133, 459–60.

Wolf, F. R. 1983. *Botryococcus braunii:* an unusual hydrocarbon-producing alga. *Appl. Biochem. Biotech.* 8, 249–60.

Wolf, F. R., A. M. Nonomura, and J. A. Bassham. 1985. Growth and branched hydrocarbon production in a strain of *Botryococcus braunii* (Chlorophyta). *J. Phycol.* 21, 388–96.

11. The role of microalgae in liquid waste treatment and reclamation

WILLIAM J. OSWALD

Department of Civil Engineering and Public Health, 659 Davis Hall, University of California, Berkeley, CA 94720

CONTENTS

I. Introduction

This chapter reviews the applications of waste-grown microalgae, not only to treat wastewater and render liquid organic wastes less harmful to man and the environment but also to conserve energy and reclaim some of the water, nutrients, and energy from such wastes.

A. Water use

The development of large, dependable supplies of high-quality, potable water and the distribution of this water to residences, industries, and commercial establishments have been a boon to mankind in efficiency, convenience, and sanitation. Regrettably, these developments have also created the vexing problem of pollution of streams, lakes (see Stoermer, Chapter 3), estuaries, and even portions of the oceans. Although relatively small amounts of water are used for personal purposes such as drinking, cooking, and bathing, daily municipal water usage in the United States usually amounts to more than 100 gallons for each person. Most of the water is used to flush objectionable, infectious, and toxic wastes away from the users' immediate environment and, quite often, into the nearest natural body of water. In industry the weight of water used and discharged usually exceeds the weight of product by from 10-fold to 100-fold.

B. Wastewater

Wastewater, usually termed sewage, includes industrial wastes and storm drainage as well as human wastes. Disposal may be through underground sewers. Some areas have piped water supplies but lack sewers; in such cases, wastes are conveyed in open ditches or central and street-side gutters. In some places (e.g., China), concentrated suspensions of human wastes called night soil are conveyed by bucket, tank wagon, truck, and even boat, to be applied to crops as fertilizer. With increasing frequency, night soil is rejected in favor of commercial fertilizers but, used or unused, it ultimately contributes to water pollution through dumping or irrigation return flows.

[257]

C. Wastewater treatment

The removal of pollutants from wastewater for safe and nuisance-free disposal is the province of wastewater treatment. Wastewater treatment can be divided into five generic stages (Table 11-1), their order reflecting the most feasible and economical progression in removing unsightly, infectious, and biologically disruptive materials. Waste treatment methods are physical, chemical, biological, or combinations of all three. Conventional wastewater treatment is outlined in Table 11-1 and is reviewed in a number of texts (e.g., Metcalf and Eddy 1979; Gaudy and Gaudy 1980). In this chapter evidence will be presented that, where climate permits, algal-bacterial systems can equal or exceed conventional wastewater treatment and do so less expensively.

II. Biological waste treatment systems

Although primary treatment consists of essentially physical processes, biological processes are the only economical way to remove a significant fraction of dissolved organics from wastewater (secondary treatment).

A. Dissolved organic removal

Because more than half of the biochemical oxygen demand (BOD) in sewage is associated with the dissolved organic fraction, methods of growing and removing aerobic bacteria and fungi are normally applied in secondary treatment to convert soluble organics to insoluble, and hence removable, biomass. Vigorous aeration of concentrated suspensions of aerobic microbes in an aeration chamber, followed by their sedimentation or flotation in special tanks, is one of the more intensive methods of removing BOD from wastewater. Large portions of the settled materials are recycled to build up the microbial population in the aeration chamber, and the balance is introduced to an anaerobic tank where methane fermentation is fostered. By this method, termed *activated sludge,* up to 90% of the carbonaceous BOD can be removed from sewage in from six to eight hours.

Another somewhat less intensive method of BOD removal is to use surface films of bacteria, fungi, and protozoa on various types of solid media and intermittently dose them with sewage in the presence of air or pure oxygen. These films adsorb organic matter from the wastewater and, as they grow, age, and thicken, tend to fall away from the medium; they can then be removed from the wastewater by settling or flotation. The solids removed, containing much of the sewage organic matter, are introduced to anaerobic tanks designed to foster methane fermentation. By this method, termed *biofiltration,* up to 85% of the carbonaceous BOD can be removed from sewage in five or six hours.

Table 11-1. *Intensive (nonalgal) waste treatment*

Stage of treatment	What should be removed	Removal method	Relative cost (approx.)	Acceptable disposal of residuals	Acceptable treated water disposal	Main problem
Primary, 1	Grit and gravel floatables Settleable solids Absorbed colloids Pathogens	Settling Flotation Sedimentation Sedimentation Disinfection	1	Methane fermentation Dewatering Disinfection Burial	To deep ocean To fast streams Nonfood or feed irrigation	Day-by-day sludge handling Flies Odors
Secondary, 2	Colloidal and dissolved organics	Biological absorption & oxidation Activated sludge Biofiltration Disinfection	3	Methane fermentation Dewatering Disinfection Burial	Stabilization ponds Irrigation of crops not eaten raw Waste flushing Deep ponds Fast streams	Sludge bulking Sludge disposal Flies and odors
Tertiary, 3	Carbonates Ammonium Nitrates Phosphate	Precipitation Air stripping Ion exchange Nitrification Denitrification High pH precipitation Reduction	5	Fertilizer To air Brines for discharge To fast streams To air as N_2 Soil conditioner $Ca_3(PO_4)_2$ fertilizer Conc. to water	Nutrient-sensitive bodies of water Waste flushing Recreational waters Coastal estuaries	Temperature sensitivity Low reliability Brine disposal Air pollution
Quaternary, 4	Refractory organics Organic toxicants Herbicides/pesticides	Absorption on activated carbon	7	Drying and high-temp. incineration Toxic waste dump Toxic waste dump	Coastal estuaries Slow streams Stock water	Air pollution Low reliability
Quinary, 5	Heavy metals Soluble minerals Na^+, K^+, Mg^{2+}, Ca^{2+}	High pH precipitation Ion exchange Reverse osmosis Electrodialysis Distillation	15	Metal recovery Brine evaporation In ponds In ponds	Boiler feed water Recycled water supplies Greenhouse irrigation	Salt buildup on land Wildlife poisoning Brine disposal

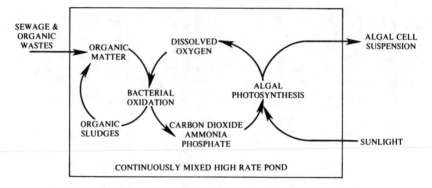

Fig. 11-1. Cycle for photosynthetic oxygenation of wastewater.

By extending residence time in their reactors and adding special reactors and flow patterns, both activated sludge and biofiltration systems can be modified to partially remove nutrients such as nitrogen and phosphorus as well as organic carbon.

A third method of achieving removal of soluble BOD is maintenance of an active algal-bacterial culture in continuous suspension in the presence of sufficient light, thus encouraging growth and dissolved oxygen release by algae and conversion of soluble organics to insoluble biomass by aerobic microbes. This process, termed *photosynthetic oxygenation* (Fig. 11-1), is carried out in special reactors termed *high-rate ponds* (Oswald and Gotaas 1957). Photosynthetic oxygenation is a more complete process than activated sludge or biofiltration because it involves both biological oxidation and photosynthetic reduction of oxidized nutrients, accomplishing a degree of tertiary treatment along with BOD removal. Photosynthetic oxygenation is often referred to as high-rate ponding. In this chapter the terms are used synonymously. Properly designed and operated high-rate ponds are capable of removing more than 90% of the carbonaceous BOD and up to 80% of the nitrogen and phosphorus. The time required, however, is measured in days rather than hours.

B. Nutrient removal

Intensive methods of tertiary waste treatment involve ammonium removal by ion exchange or air stripping, nitrate removal by anoxic reduction to nitrogen gas, and phosphate removal by lime precipitation or by anoxic reduction to phosphine (Table 11-1). Together, these intensive processes are several times more expensive than photosynthetic oxygenation in that they require extended residence times in expensive reactors and/or special chemicals, whereas in photosynthetic oxygenation nutrient removal is achieved economically in the same high-rate pond as biological oxidation (Table 11-2).

Table 11-2. *Waste treatment process in integrated microalgal-bacterial
systems*

Treatment process	Criterion of effectiveness	How accomplished in integrated ponds
Preaeration grit and screenables Removal and plain sedimentation	Odor removal Suspended solids removal	Recirculation of aerobic effluent from high-rate pond. Wastes are inserted at bottom of fermentation pits, where most settleable solids tend to remain until their organic fraction is decomposed by anaerobic bacteria. The inert fraction remaining has a small volume, and many years are required to fill the volume provided for inert accumulations.
Flotation	Grease and floatables removed	Inert floatables collect on specially designed and located scum ramps, which can be cleaned either manually or by mechanical equipment. Grease often becomes saponified at high pH and becomes biodegradable.
Methane fermentation	Combustible gas production	An anaerobic environment is created in the bottom of the ponds, permitting methane fermentation to occur. Protection from mixing and overturning is effected by elevated berms. If needed for energy purposes, special gas collection equipment can be installed at modest additional cost.
Biological oxidation	BOD removal	Oxygen in excess of BOD is produced by photosynthetic microalgae that are maintained in suspension by gentle flow-mixing in a high-rate pond. During algal growth the pH level in the pond increases to well above 9.
Algae removal	Percentage of suspended solids removed	Following gentle mixing in a high-rate pond, algae tend to flocculate and settle in the settling ponds. Algae accumulated in the settling ponds can be removed for use or can remain in the pond for several years. Since two settling ponds are always provided, removal can be by simple decanting and drying.
Nutrient removal	N, P removed	Organic nitrogen is converted to ammonium or to N_2 gas. N_2 gas is produced through heterotrophic nitrification and denitrification in the

Table 11-2. *(cont.)*

Treatment process	Criterion of effectiveness	How accomplished in integrated ponds
Nutrient removal *(cont.)*		anaerobic bottom of the primary pond and escapes to the atmosphere along with methane. Ammonium is taken up by algae during growth. Surplus ammonium is converted to NH_3, which escapes to the air during gentle mixing at high pH. Phosphorus is taken up by algae or precipitated at high pH as a calcium phosphate.
Disinfection	MPN, pathogen removal	Parasite ova remain in pits. Bacterial numbers are decreased by time, temperature, and high pH. A pH greater than 9.2 for 24 h is lethal to coliform and other enteric bacteria. pH levels above 9 occur daily in high-rate ponds. Virus die-away results from long detention periods.
Heavy metal removal	Percentage of removal	Algae have a high negative surface charge and, therefore, an affinity for heavy metals, which usually have a strong positive charge. By recirculation of high-rate pond effluent containing algae, heavy metals absorb to algae and settle out in the facultative pond. A high pH precipitates residual heavy metals in high-rate ponds.
Total dissolved solids removal	Percentage of removal	Calcium phosphate and calcium and magnesium carbonate precipitate at high pH. Potassium and magnesium are incorporated into algal cells. Sodium is not removed to any significant extent by microalgae and may increase slightly in long-detention ponds.
Refractory organic	Percentage of removal	Long detention times permit time for organic refractory degrading microorganisms to develop. High oxygen levels in high-rate ponds permit development of actinomycetes, which degrade compounds refractory to facultative heterotrophs.

C. Refractory and toxicant removal

During the fourth stage of treatment by conventional methods (Table 11-1), refractory substances cannot be oxidized at the low levels of dissolved oxygen required to achieve an economical rate of transfer of oxygen from air to water. Some toxicants and refractory organics may be either chemically converted to inert substances by strong oxidizing agents such as ozone, peroxide, or chlorine or adsorbed on activated carbon and certain resins. In high-rate ponds (Table 11-2), dissolved oxygen levels are sufficiently high to support the growth of lignin- and cellulose-oxidizing aquatic actinomycetes, and some toxic organics may be decomposed or inactivated in a similar manner. Detailed studies and proof of this function of high-rate ponds, however, remain to be made.

D. Demineralization

Few of the monovalent inorganic salts such as sodium chloride, divalent ions such as calcium, magnesium, sulfate, and carbonate, and toxic polyvalent metallic ions such as chromium are removed in intensive processes like activated sludge and biofiltration. Accordingly, if during intensive treatment their removal is necessary, special processes are required such as alkaline precipitation for removal of heavy metals, chemical precipitation or ion exchange for removal of calcium and magnesium, and reverse osmosis, ion exchange, or distillation for removal of such salts as sodium chloride.

In high-rate ponds, the high pH precipitation of calcium and magnesium is a common occurrence. Due to the fact that rapidly growing algae have a high affinity for multivalent cations, a significant degree of toxic metal removal also may occur as such algae grow and settle or float (Ramani 1974).

Although no detectable sodium chloride removal occurs in high-rate ponds, a short residence time of two to six days minimizes concentration of salts due to evaporation, compared with longer-residence-time waste impoundments (stabilization ponds) (Middlebrooks et al. 1979). Additionally, since they may be designed to soften water by precipitation of calcium and magnesium, high-rate ponds may be an economical method of preconditioning waters for distillation or reverse osmosis by precipitation of calcium carbonate and calcium sulfate. Precipitates of calcium sulfate and calcium carbonate interfere with heat transfer in boilers and foul or clog reverse-osmosis membranes, and their removal is required for successful reverse osmosis or distillation.

E. Treatment costs

Each degree of treatment in an intensive waste treatment system (Table 11-1) adds greatly to the total cost. Secondary treatment is often twice as costly as primary treatment, and complete tertiary treatment may cost three to five times as much as primary treatment (Dames and Moore 1978). Oxidation of toxic organic substances with ozone, chlorine, or hydrogen peroxide, and/or filtration through activated carbon or carbonated polyurethane foams, may cost up to 10 times as much as primary treatment. Finally, demineralization or desalinization of wastewater may cost 10 to 20 times as much as primary treatment. Complete removal of all added impurities in wastewater thus may cost about 30 times as much as primary treatment alone.

The cost of treating a unit of wastewater declines somewhat with an increase in size of the waste treatment plant. Thus, oxidation of a pound of BOD in a 1 million gallon per day (MGD) plant (0.45 kg in 3.7×10^6 L) may cost twice as much as oxidation of the same pound in a 10 MGD plant (Smith 1973; Dames and Moore 1978). This apparent cost savings resulted in the construction of large regional intensive treatment plants in the United States. Unfortunately, the practice has created more difficult disposal problems by concentrating so much wastewater in one location.

F. Regulation

In spite of projected higher unit costs, government agencies responding to real or suspected public health concerns are imposing more and more restrictive discharge requirements which, as implemented, require higher and higher degrees of waste treatment prior to discharge. For example, Public Law 92-500 requires treatment to the second degree for all wastes discharged to "navigable or contiguous waters." Subsequent amendments deal with nutrients and toxicants and sometimes prescribe either their prevention from entering or their removal from waste streams (USEPA 1984).

Many states have legislation limiting phosphate in detergents. Borated and sulfonated detergents are no longer used because of restrictive laws. The use of chlorinated hydrocarbons and tetraethyl lead is now strictly limited, and intentional discharge of these substances to public sewers violates federal law (USEPA 1984). Thus, directly or indirectly, the cost of liquid waste management and, indeed, management of all wastes – liquid, solid, and gaseous – has continued to rise. Because of these restrictions and costs, the use of impoundments to retain wastes has increased rapidly, but this solution has brought on a new wave of concern regarding air pollution and groundwater contamination by impounded wastes. New laws such as PL 99-499 require covers and lining of all impoundments known or suspected of containing toxic wastes (Sun 1986).

National, state, and local governments worldwide struggling with allocation of limited funds to a multitude of problems related to transportation, housing, water and food supplies, and waste collection can rarely find funds for even the first stage of intensive waste treatment. Without adequate treatment, the consequences, as any traveler to a developing country can attest, are the appalling pollution of streams and estuaries and the inexorable degradation of public health, safety, and welfare. Advanced integrated algal-bacterial systems, widely applied in developing areas of the world, would help to reverse this trend by providing economical and reliable waste treatment.

III. Design considerations

A. Algal concentrations

Because organic animal wastes typically contain all the nutrients required by plants, they can support growth of large concentrations of algae. It must be emphasized that among the major elements carbon is most likely to become limiting to algal growth because its ambient atmospheric concentration (0.033%) is far below optimum for absorption by growing cultures. Algae, growing in open waste ponds, are therefore almost entirely dependent on commensal bacteria for their carbon supply and rarely reach biomass levels greater than 300 mg·L^{-1}. Studies of algae grown in closed ecological life support systems (Shelef, Oswald, and McGauhey et al. 1970) with a CO_2-enriched atmosphere and continuous illumination at daylight intensities, and on human wastes 10 or more times as concentrated as domestic sewage, indicate that biomass levels as high as 6,000 mg·L^{-1} and primary productivity of up to 40 g·m^{-2}·day^{-1} can be attained.

In order to maximize algal productivity in outdoor ponds, sunlight, rather than nutrients, must be the factor limiting growth. If carbon is not limiting and a residence time of two to three days is provided, the sunlight-limited concentration of fast-growing, naturally occurring algae will be approximately

$$C_c = 6,000 \ d^{-1} \tag{1}$$

in which C_c is the algal dry weight concentration in mg·L^{-1}, d is the pond depth in cm, and 6,000 is an experimentally determined constant related to light absorption by algal-bacterial cultures (Oswald 1978). According to equation 1, if neither nutrients nor time are limiting, one can grow algae to any desired concentration by merely adjusting the depth or distance from the light source. In Shelef's studies (Shelef et al. 1970), a culture depth of 1 cm was used and, as predicted in equation 1, his cultures attained concentrations of about 6,000 mg·L^{-1}. In domestic sewage, since carbon tends to become limiting when algal dry weight concentrations reach 200–300 mg·L^{-1}, depths must be maintained at or about 20–30 cm to use sunlight efficiently.

A general rule of thumb regarding waste nutrient concentrations is that fast-growing algae can attain a dry weight concentration of 2.5 times C, 12 times N, and 100 times P, if no factors are limiting other than concentrations of available forms of C, N, and P (Oswald 1978). In sewage it is most likely that other elements required by algae such as potassium, magnesium, sulfur, and iron are present in excess. Domestic sewage usually contains sufficient available carbon (approx. 80 mg·L^{-1}) to yield green algal growths of at least 200 mg·L^{-1}; sufficient available nitrogen (approx. 20 mg·L^{-1}) to yield 250 mg·L^{-1} of biomass; and sufficient available phosphorus (approx. 5 mg·L^{-1}) to yield 500 mg·L^{-1} of biomass on a dry weight basis. From these values it is clear that carbon is usually the limiting nutrient since it will support only 200 mg·L^{-1} of algae compared with 250 for N and 500 for P. The availability of these elements is, of course, dependent on sewage bacteria and fungi releasing them from their organic forms – a step that, in turn, requires oxygen released through algal photosynthesis.

Time is an important factor in outdoor algal growth for two main reasons. First, the larger harvestable algae such as *Scenedesmus* sp. or *Microactinium* sp. have doubling times near one day, so residence time for continuous cultures must exceed one day to prevent them from being diluted away. The second, and usually more limiting, reason is that residence time must be provided to integrate enough solar energy to grow the algal concentration needed to release sufficient oxygen to satisfy the BOD (Oswald and Gotaas 1957). Oxidation of the BOD by microbes also requires time, but oxygen production by algae is a dominant constraint on the rate at which wastes are oxidized in high-rate ponds.

B. Algal productivity

Studies of algal productivities in sewage treatment systems have been made in several locations in the world and are reported elsewhere (Oswald and Eisenberg 1981). Some of the results from these studies are reproduced in Figure 11-2. Below a certain level of solar energy, termed S_o, net productivity is negligible, but above this "threshold" productivity becomes a linear function of solar energy. That is, the efficiency of total sunlight energy utilization for incremental values of S above S_o tends to be constant (near 5%) as long as nutrients are not limiting. In the Richmond studies (Fig. 11-2(D)), cultures were pushed to the point where carbon was, indeed, limiting to growth, and consequently productivity and efficiency declined above that value of S. However, within the positive linear portion, productivity was directly pro-

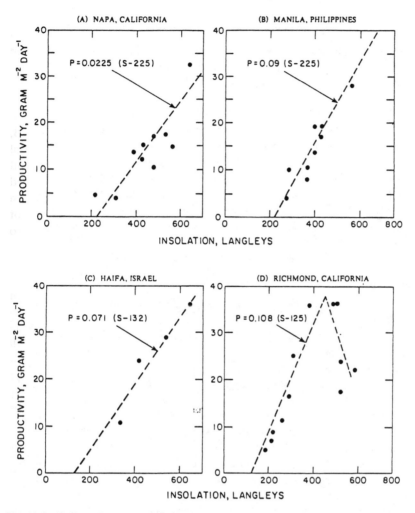

Fig. 11-2. Algal productivity in high-rate waste treatment ponds in various geographical locations.

portional to increases in solar energy input above S_o, and the efficiency was almost 6% of the residual value of S; that is, $S - S_o$.

From an engineering standpoint, productivities in Figure 11-2 lower than 3 $g \cdot m^{-2} \cdot day$ are not of a magnitude sufficient to contribute economically to production of algae or oxygen. Similarly, high productivities such as 30 $g \cdot m^{-2} \cdot day^{-1}$ cannnot be sustained due to climatic variations and hence, from a practical standpoint, may be disregarded in favor of medium productivities of 15–20 $g \cdot m^{-2} \cdot day^{-1}$. Pilot plants should be established at the site to determine such design characteristics.

IV. Operation of integrated algal-bacterial systems

Despite fears of groundwater contamination, (USEPA 1984; Sun 1986), the use of properly designed ponds as both treatment and disposal systems has much in its favor.

A. Advanced, integrated ponding systems

The advanced, integrated algal-bacterial systems developed at the University of California, Berkeley and shown schematically in Figure 11-3, go far beyond the mere use of algae as oxygen generators for bacterial oxidation of wastes. Rather, these systems involve ponds for settling and methane fermentation of suspended solids, oxidation of dissolved organics, removal of ammonium, nitrate, and phosphate, precipitation of calcium, magnesium, and metals, and disinfection. To achieve all of these requires at least four distinct ponds in a series, each designed to promote specific physical, chemical, and microbiological activities.

B. Fermentation pits

The primary (facultative) pond is equipped with one or more 3–4 m deep pits designed to prevent the intrusion of waters containing dissolved oxygen (Fig. 11-3). This is absolutely necessary to promote the desired methane fermentation in the pits (Parkin and Owen 1986). All wastewater is directed through these pits, where sedimentation of suspended solids and their conversion to methane occurs. Because the solid:liquid ratio in domestic sewage is usually less than 1:1,000 by weight, a liquid residence time in the pits of one or two days provides sufficient volume for solids to remain in the pits for hundreds of days. During this extended time, all fermentable solids that enter are converted to soluble organics, methane, carbon dioxide, nitrogen gas, soluble inorganics, and stable residues. There is also growing evidence that these intensely anoxic environments also favor biodegradation of some refractory toxic chemicals such as halogenated aliphatic compounds (Jewell 1987). Our studies at St. Helena, California, indicate that stable residues are all that permanently remains in the pits when fermentation is complete. The apparent rate of accumulation of stable residues is only about three liters per person per year (Meron, Zabat, and Oswald 1970).

C. Odor control

The fluid portion of the waste emerging from the pits is retained in a larger part of the primary pond above the pits for an additional 5 to 10 days. Here, if light and temperature permit, microalgae grow in profusion and produce sufficient dissolved oxygen to oxidize malodorous compounds such as hydrogen sulfide that may be in gases emerging

KEY TO NUMBERS ON FIGURE BELOW

1. SCREENING & GRIT REMOVAL	7. PADDLE WHEEL MIXER	13. ALGAE HARVEST
2. DISTRIBUTOR	8. HIGH RATE POND (SECONDARY)	14. LOW LEVEL TRANSFER
3. FERMENTATION PITS	9. HIGH LEVEL TRANSFER	15. MATURATION POND
4. FACULATIVE POND (PRIMARY)	10. ALGAE SUBSIDENCE CHAMBER	16. HIGH LEVEL TRANSFER
5. OXYGENATED WATER RETURN	11. ALGAE SETTLING PONDS	17. WATER REUSE
6. LOW LEVEL TRANSFERS	12. SETTLED ALGAE RETURN	

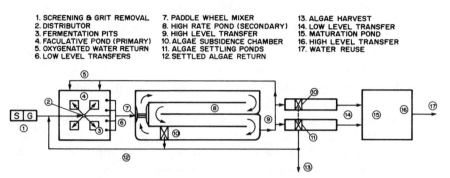

Fig. 11-3. Schematic diagram of an advanced, integrated algal-bacterial system for liquid waste treatment and oxygen, water, and nutrient recovery or reuse.

from the anoxic bottom. This feature thus controls the odors often associated with improperly designed and operated sewage ponds. The upper portion of the primary pond is equipped with an accessible beach located where wind-blown floatable solids can accumulate and be collected for disposal.

D. Sludge removal

At least two primary ponds usually are provided so one can be isolated for necessary maintenance. The author's experience with domestic sewage at St. Helena indicates that inert sludge removal from pits within primary ponds will not be required for at least 20 years. Thus, a major cost of primary and secondary sewage treatment, namely, day-by-day sludge handling, is virtually eliminated.

E. Seepage control

Although regulations may demand that these primary ponds be lined with clay or plastic, evidence in the literature and the author's experience indicate this to be an unnecessary expense. As originally demonstrated by Winneberger and McGauhey (1965) and reemphasized by Warshall (1979), continuously submerged soils saturated with anoxic sewage quickly become clogged with fine particles resulting from the combination of sulfide ions with iron, magnesium, and other metals. These fine particles lodge in soil voids and soon render the bottom impermeable. Only if dried and aerated do such soils again become permeable – events that should never occur in a properly designed pond. Accordingly, only a compacted earth lining is really required to protect groundwaters from intrusion of liquid from such anoxic pits. Even with linings, however, such pits are a highly economical fermentation system.

F. Methane capture

Currently, studies are underway to perfect methods to capture valuable methane as it emerges from such pits (Von Hippel and Oswald 1985).

G. Secondary (high-rate) pond

The exit ports for water from the primary facultative ponds are submerged so that floatables cannot enter the secondary pond. If land configuration permits, this secondary pond may be used as a sight buffer by constructing it to surround the primary ponds. This is the high-rate pond in which an algal-bacterial culture is grown to oxidize the remaining soluble organics and provide surplus dissolved oxygen and high pH. When needed, warm high-rate pond water, rich in dissolved oxygen and desirable green algae, is pumped to the upwind side of the primary ponds where it can spread over the surface to form a thermocline and provide a cap of oxygenated water. This oxygen will autocatalytically oxidize most odorous reduced compounds and provide sufficient dissolved oxygen to support surface-dwelling mosquito fish (*Gambusia* sp.) for control of mosquito and other insect larvae (Middlebrooks et al. 1979).

The high-rate pond is designed to grow a controlled culture of either settleable or floatable algae. All that is required is a shallow depth (20–40 cm) and continuous flow mixing, completely and gently, at a velocity of 15–20 cm·sec^{-1}. Most authorities today agree (see, e.g., Grobbelaar, Soeder, and Toerien 1981) that such mixing is most efficiently applied in endless channels using paddlewheels. The walls of shallow channels must be lined with concrete or thick plastic to prevent weed growth, but channel bottoms seal with biological slimes and need not be lined with such expensive materials. Furthermore, such slimes reduce hydraulic friction to a very low value (Meron et al. 1970). The net result is that more than a mile (1.61 km) of channel 40 cm deep $\times$ 12 m wide can be mixed with one properly designed and installed 6 m long $\times$ 2 m diam. paddlewheel powered by a 2.5 hp electric motor with reversible hydraulic drive and speed control (Oswald 1985). Although a 2.5 hp motor is provided, less than 1 hp is required to maintain a flow velocity of 15 cm·sec^{-1} in the channel.

H. Energy and cost savings

The energy input for well-designed paddlewheel-mixed high-rate ponds is about five kilowatt hours per hectare per day (kwh·ha^{-1}·day^{-1}). At latitude 35° minimal oxygen production on a cloudy winter day (e.g., 200 Langley's·day^{-1}, 12°C, 10 h daylight) will approach 50 kg·ha^{-1}·day^{-1} or about 10 kg·kwh^{-1}. On a warm summer day (e.g., 600 Langley's·day^{-1}, 30°C, 15 h daylight), oxygen production may reach 540 kg·

ha^{-1}·day^{-1} or about 108 kg·kwh^{-1}. This oxygenation efficiency can be contrasted with mechanical aeration, which rarely transfers oxygen from air to water at more than one kg·kwh^{-1}. Thus, depending on temperature and light, photosynthetic oxygenation, in which most of the energy comes from the sun, is from 10 to 100 times as efficient as electromechanical oxygenation systems. The use of high-rate ponds essentially exchanges land cost for energy savings. With current energy prices, one can afford to spend more than US $50,000 per hectare for land to build high-rate ponds in preference to operating electromechanical aeration systems.

Aside from their energy efficiency, high-rate ponds cost approximately 50 cents per gallon of daily capacity whereas mechanically aerated systems cost from two to five dollars per gallon of daily capacity, depending on their size and complexity. For example, in 1980 a 1 MGD high-rate pond cost about US $500,000 whereas a 1 MGD activated sludge unit cost about US $2.5 million (Dames and Moore 1978; Middlebrooks et al. 1979).

I. Algal harvest

Harvesting or removal of microalgae is a major cost in use of algae for oxygenation, particularly where a limit on suspended solids is imposed by law. Ponds that are not mixed tend to grow algae that do not settle because they are either small (e.g., *Chlorella* sp.), motile (e.g., *Euglena* sp. or *Chlamydomonas* sp.), or have buoyancy (e.g., *Oscillatoria* sp.). Aside from their failure to settle, none of these genera is desirable in ponds. *Chlorella* is periodically grazed, virtually to extinction, by rotifers, leaving the pond without dissolved oxygen. *Euglena* is mainly seasonal and moves up and down in ponds, producing net oxygen only when very near the water surface. *Chlamydomonas* appears and disappears sporadically, and *Oscillatoria* and other blue-green algae tend to form thick malodorous mats near pond edges. Ponds that are flow-mixed, on the other hand, tend to grow algal species that settle if not mixed or if mixed only intermittently (Nurdogan 1984). In well-oxygenated flow-mixed systems, small fast-growing algae are soon grazed to extinction by predators, leaving slower-growing but larger, not so readily grazed algae to predominate. These climax cultures include *Scenedesmus* sp. protected by spines and colonies of four or more cells and *Micractinium* sp. with setae too long for most rotifers to engulf (Nurdogan 1985). Neither *Euglena* nor filamentous blue-greens grow rapidly enough to compete with *Scenedesmus* sp. or *Micractinium* sp. for nutrients and light (Oswald 1978). Although many other algal genera occupy waste ponds from time to time (see, e.g., Silva and Papenfuss 1953; Sing and Saxena 1969; Palmer 1977), one of the latter two algal genera tends to predominate at all times in continuously mixed waste ponds.

Both *Scenedesmus* sp. and *Micractinium* sp. are relatively fast-growing algae capable of attaining production rates of 36 g dry wt·m^{-2}·day^{-1} (360 kg·ha^{-1}·day^{-1}) under ideal growth conditions (600 Langley's· day^{-1}, 30°C, 15 h light) in outdoor ponds. Because oxygen release amounts to about 1.5 unit weight O_2 per unit dry weight of algae synthesized, the rate of O_2 production can reach 540 kg·ha^{-1}·day^{-1}. However, in the engineering design of such systems, much lower productivities are assumed in order to compensate for cloudy days and provide a safety factor against the overloads that typically occur in sewage systems. In warm climates and in properly designed high-rate ponds, oxygen production averages 225 kg·ha^{-1}·day^{-1}.

Algal-bacterial cultures in continuously mixed high-rate ponds tend to form commensal or symbiotic flocs of algae and bacteria 0.1 mm or more in diameter. If permitted to stand quiescent, these flocs form even larger flocs that settle readily (Eisenberg et al. 1979). More recently, Nurdogan (1985) demonstrated that these flocs can be separated from effluents by dissolved air flotaton (DAF), often without added chemicals. The DAF process involves pressurizing some cell-free effluent with air at about 60 lb·in^{-2} (4 atmos.) and commingling it with algae-bearing water at one atmosphere. Air bubbles that form in the mixture attach to the algal-bacterial flocs and rise to the surface in a skimmable foam. Advanced DAF systems developed by the Krofta Corporation (Wang and Krofta 1985) can separate algae in as little as five minutes residence time in the DAF reactor, compared with 12 hours or more in a sedimentation tank or pond.

Algae from oxidation ponds (not high-rate ponds) are harvested by coagulation and DAF at Sunnyvale, California (25 MGD), and by coagulation and sedimentation at Napa, California (10 MGD) (Oswald 1985). In both cases, algae are removed to control suspended solids in effluents discharged to natural waters. At Napa, algal concentrate is wasted to the ponds, whereas at Sunnyvale, although algal concentrates are also returned to the ponds, studies are underway to ferment the algae in existing conventional sludge digesters to produce methane (Golueke, Oswald, and Gotaas 1957). This would constitute an important energy saving since the pumps are driven by methane-fueled diesel engines that sometimes require supplementary commercial gas (Ms. Helen Farnham, City of Sunnyvale, personal communication).

V. Current applications

A. Hollister and St. Helena

System designs of the type illustrated in Figure 11-3 have been applied in St. Helena (O.5 MGD) and Hollister, California (2.0 MGD) (Fig. 11-

4), and remarkably economical degrees of treatment have been achieved.

The operation of both systems is very similar. A major portion of the settleable solids in influent sewage is removed in fermentation pits of primary facultative ponds. Overlain by oxygenated water, the primary ponds emit virtually no odors. Because of submerged discharge pipes, floatables remain in the primary pond and can be removed on scum ramps. Water from each primary pond is discharged to the high-rate pond where most of the photosynthetic oxygenation occurs. The green water from the high-rate pond either is recirculated to cap the primary ponds or goes to a settling pond for one day's algal settling, then to maturation and disposal ponds for several weeks to several months where remaining algae can be consumed by predators. Remaining treated water percolates into the ground from which it is later pumped for irrigation. Complete primary and some quaternary treatment is accomplished in the primary facultative pond (Table 11-2). Secondary, tertiary, and some quaternary treatment is accomplished in the high-rate pond. Remaining suspended solids are removed and final disinfection is accomplished by residence time in the settling ponds and maturation ponds. The latter often contain large fish, which could be used for food after suitable depuration and evisceration. These fish graze on populations of *Daphnia* sp. and other small invertebrates that characterize maturation ponds (Chung and Lee 1980; Edwards 1985). In percolation of water from maturation ponds through the soil, all remaining solids and practically all possible pathogens are removed. Thus, high degrees of treatment are achieved.

The St. Helena system is more than 20 years old; the Hollister system is six years old. Each cost less than half as much to build as comparable conventional systems, which cost US $2–5 per gallon of capacity. Less than one-third as much is required to operate the advanced ponds. These examples indicate the dependability, economy, and effectiveness of properly designed integrated algal-bacterial ponding systems for domestic sewage treatment.

B. Reclamation and savings opportunities

From a futuristic standpoint two important steps were omitted in the Hollister and St. Helena systems: recovery of algae from the settling system and methane from the fermentation pits for power generation. Both of these processes are currently under study so that future systems can incorporate recovery of these potentially reclaimable, valuable commodities (Nurdogan 1985; Von Hippel and Oswald 1985). Systems that recover the algae and/or methane should further diminish and possibly eliminate the net cost of sewage treatment in sunbelt communities. For example, at Hollister, algae production in the 6 ha high-rate pond could

Fig. 11-4. Aerial photograph of the city of Hollister, California, advanced integrated waste treatment and disposal system constructed 1979–80.

amount to about 500,000 kg dry wt per year. If captured, methane from the facultative pond would amount to about 120,000 m^3 per year, enough to produce 240,000 kw-h of electrictiy. If a kilowatt hour is valued at 10 cents (US) and algae could be sold for 25 cents per kilogram, the gross income could approach $150,000 per year. In addition, although the city has no way to collect this value, the high-quality effluent water being reused for irrigation is worth at least $25,000 per year.

Concerning costs, Hollister's subsidized share of the $1.8 million cost of this plant was about $180,000 total or approximately $25,000 per year. Its annual cost of operations and maintenance is about $75,000. Overall costs then should be about $100,000 per year. A gross income of $150,000 per year would provide a surplus of $50,000 per year to help pay for the addition of algal harvest and heat–power generation systems. Even without the additions suggested above, Hollister's operating costs now are less than one-third those of comparably sized, less reliable, conventional intensive systems in nearby areas.

C. Further applications

One may ask, if these newer pond systems are so efficient and economical, why are not more being installed?

In the United States, Canada, western Europe, Japan, South Africa, Australia, and New Zealand equipment manufacturers advocate as conventional, and prefer to design, more complex and expensive systems for boreal and austral climates. Additionally, they tend to base their criticism of algal-bacterial systems on ponds that are archaic in design. This permits them to emphasize such negative aspects as odors, insect breeding, high land use, and discharge of suspended solids in the form of algae.

St. Helena was the first advanced ponding system ever built, and it resulted in such a high degree of waste treatment economy, and lack of nuisance, that all familiar with it were astonished, including its designers (Meron et al. 1970). Until the Hollister facility was built and proved equal to St. Helena in economy and performance, the general applicability of such systems was not recognized. Thus, few engineers are aware of advanced integrated ponds and their capabilities. It also should be recalled that systems of this type, unmodified, are limited in application to the sunbelt areas of the world (latitude 35°; minimums of 200 Langley's $\cdot$ day^{-1} and 10°C).

In developing countries very few sewage treatment plants of any kind exist, and the few mechanical plants that have been introduced work for only a few months, then fail. Similarly, poorly designed conventional ponds 1.5 m deep prove unsatisfactory due to odors, flies, organic carbon accumulation, and general lack of maintenance. Although the stage is set for more widespread adoption of advanced integrated ponds and several large systems are proposed for warmer regions of the world, only pilot systems are presently under active study (Becker and Ventkataraman 1982; Lee and Adan 1983; Edwards 1985).

VI. Future applications

As advanced integrated ponds are used more widely, development of a hierarchy of uses of waste-grown microalgae can be envisioned, with pharmaceutical and human food as the highest-quality, most valuable use, and fertilizer and energy production as the lowest-quality, least valuable use. The source of water and nutrients would determine the quality of the product (Oswald 1978; see also Spencer, Chapter 10).

A. Sewage-grown algae

Oxygen generation and energy production are likely to be the most feasible use of algae grown on domestic sewage because sewage always contains infectious organisms and often contains toxins, precluding its direct use to produce algae for human food or domestic animal feed. Such algae would at least be usable for energy generation.

B. Energy systems

The concept that waste-grown microalgae may be a key to producing economical electrical energy via their fermentation to methane has long been proposed (Golueke et al. 1957; Oswald and Golueke 1960). This is likely to be economical only if it involves microalgae grown on wastewater in high-rate ponds. Large-scale systems could probably be made economical by retaining and recycling water, carbon, and nitrogen, thus permitting the algal energy-fixing system to grow to a size that could contribute significantly to a community's energy needs. If grown, harvested, fermented to methane, and converted to electrical energy in a heat–power generator with proven technologies, one kilogram of waste-grown algae can be converted to one kilowatt hour of electrical energy (Eisenberg et al. 1979). The maximum dependable productivity of microalgae is about 360 kg·ha^{-1}·day^{-1}, so maximum capacity would be about 15 kw·ha^{-1}. Assured capacity would be only half this amount or about 7.5 kw·ha^{-1} or 180 kwh·ha^{-1}·day^{-1}, the average daily domestic consumption of electrical energy of about 30 persons in the United States. Clearly, the economic use of such a system would depend on the price of electrical energy from competitive sources (Hill et al. 1984; Laws 1984).

The author's current estimates are that solar energy via microalgal-bacterial systems would cost on the order of 25–30 cents (US) per kwh. Although this cost seems high, compared with current market costs of 10–12 cents per kwh, it would be a dependable, safe source of energy that could be employed in any sunbelt community having low-cost land, adequate water, and a sewage collection system.

Land use does not seem to be a major problem. Currently, in the United States, land devoted to food production amounts to about 162 million hectares, or about 0.64 ha per person. To generate the country's domestic electrical energy via algal-bacterial systems would require about 8 milliion sunbelt hectares or about an additional 0.033 ha per capita. In boreal or austral climes summer production would be possible also. Because unproductive agricultural land could be used for solar-energy-capturing algae ponds, sufficient land should be available in many locations.

C. Toxin removal

In California's Westlands Water District of the San Joaquin Valley, sub-surface tile drains are used extensively to remove water containing sodium chloride and other salts from agricultural soils. As a result drainage water contains several hundred parts per billion (ppb) of sele-nium and other objectionable minerals, a situation that has caused a crisis of major proportions (Weres et al. 1985). Concentrated in evaporation ponds, these minerals reached levels sufficient to cause overt teratogenic effects on coot and duck hatchlings. Because of wildlife and other interests, Kesterson Reservoir, the evaporation ponding system of the district's irrigation complex, was closed and the drains sealed.

Because microalgae grow vigorously in tile drainage water, it was proposed that they might be used in some way to remove selenium (Edwin W. Lee, U.S. Bureau of Reclamation, personal communication). Subsequent laboratory studies demonstrated that only about 5 mg of selenium were taken up by a kg (dry wt) of algae, an insignificant amount. On the other hand, each liter of drainage water supported growth of about 200 mg of microalgae, which, when concentrated and fermented, produced about 75 ml of pure methane and sufficient reduced organic residuals to convert soluble selenate to insoluble selenite (Oswald 1985). The methane, when burned in a heat–power system, should produce 5 to 10 times as much electrical energy as needed to operate the system. Carbon dioxide resulting from methane combustion and residual nitrogen and phosphorus from the digester can be returned to the algal growth system to sustain productivity. Thus, a microalgal-bacterial system may aid in removing selenium and other objectionable substances from subsurface tile drainage waters and might permit irrigated agriculture to continue in the western San Joaquin Valley and perhaps in other selenium-impacted areas.

D. Integrated feedlots

A further application of microalgal-bacterial systems that is likely to be highly economical is in integrated feedlots (Dugan, Golueke, and Oswald 1972). In these systems, animal manures, including both urine and feces, are removed from specially designed pens by flushing with water. After removal, the concentrated manures are washed and screened. Materials retained on screens may be refed, converted to methane in digesters, dried for use as bedding, or composted for use as soil conditioners. The washwater containing dissolved and suspended material passing the screen is introduced into a high-rate pond where masses of algae and bacteria grow commensally on the waste. The algal-bacterial biomass is then harvested by flotation or other techniques (Ayoub and Koopman 1986) and the water recycled to convey and dilute more waste. Concentrated algae are fed to animals as a protein supplement. In this way, as much as two-thirds of the fixed nitrogen in feeds can be converted to milk, meat, or eggs, compared to less than one-third now converted in conventional feedlots. Not only is fixed nitrogen recycled, but waste disposal problems are significantly mitigated. This system was demonstrated to be technically and economically feasible in 1972 (Dugan et al.) and has been studied and reported extensively since then (Shelef and Soeder 1980; Chung and Lee 1980;

Edwards 1985; and others). However, because of the sustained depressed farm economy worldwide since 1972 it has never been implemented in its entirety by any farm, fishery, or dairy. Nevertheless, as fixed nitrogen and other nutrients become less available and more expensive, future adaptation of integrated feedlots for poultry, fish, and animals seems inevitable.

E. Clean wastes

Many organic wastes, especially from food-processing plants, may be handled to avoid contamination with toxins and infectious materials. The production of human food, animal feed, or pharmaceutical-grade algae from such wastes is a clear opportunity that remains to be exploited.

F. Life support systems

The use of waste-grown microalgae in life support systems has been noted previously (Shelef et al. 1970) and is discussed in further detail in Wharton, Smernoff, and Averner (Chapter 19).

VII. Summary

Where climate permits their use, properly designed microalgal-bacterial systems are the most economical and reliable systems available for domestic sewage treatment and for treatment of industrial and agricultural organic liquid wastes. Applied to animal feedlots, such systems could double the amount of fixed nitrogen recovered as product from animal feeds and, in addition, produce a large surplus of electrical energy from the residues. Electrical power generation from nutrient-integrating algal-bacterial systems could provide a significant fraction of a community's electrical power needs in safe, ecologically sound systems. The use of waste-grown algae as a base commodity for production of many chemical derivatives such as paper, methane, films, pharmaceuticals, and fertilizers (Metting and Pyne 1986) is a potential of great promise that, as evidenced in other chapters of this book, needs to be, and is being, pursued in many laboratories throughout the world. Although both *Dunaliella* and *Spirulina* are now being grown commercially on inorganic media, the growth of selected species in wastewaters for specific products other than oxygen and biomass is unlikely. Nevertheless, the high productivities attainable and low or negative cost for water and nutrients will make waste-grown algal-bacterial biomass and oxygen production more and more attractive in the foreseeable future.

VIII. References

Anonymous. 1975. California State Department of Water Resources, Sacramento.

Ayoub, G. B., and B. Koopman. 1986. Algal separation by the lime–seawater process. *J. Water Poll. Contr. Fed.* 58, 924–31.

Becker, W., and L. V. Ventkataraman. 1982. *Biotechnology and Exploitation of Algae: The Indian Approach.* Agency for Technical Cooperation (GTZ), D-6236 Eschlorm, Federal Republic of Germany. 216 pp.

Chung, P., and R. C. T. Lee. 1980. *Animal waste treatment and utilization.* In *Proceedings of the International Symposium on Biogas, Microalgae, and Livestock Wastes*, pp. 415–89. Council for Agricultural Planning and Development, 37 Nonhai Rd., Taipei, Republic of China.

Dames and Moore, Engrs. 1978. *Construction Costs for Municipal Wastewater Treatment Plants, 1972–1977.* Tech. Rep. EPA 430/9-77-013 MCD 37, U.S. Environmental Protection Agency, Office of Water Program Operations, Washington, DC. 120 pp.

Dugan, G. L., C. G. Golueke, and W. J. Oswald. 1972. Recycling system for poultry wastes. *J. Water Poll. Contr. Fed.* 3, 432–40.

Edwards, P. 1985. *Integrated Resource Recovery, Aquaculture: A Component of Low Cost Sanitation Technology.* World Bank Technical Paper no. 36, UNDP Project Management Rep. no. 3. World Bank, Washington, DC. 45 pp.

Eisenberg, D. G., W. J. Oswald, J. R. Benemann, R. D. Goebel, T. T. Tiburzi. 1979. Methane fermentation of microalgae. In *Proceedings of the First International Symposium on Anaerobic Digestion*, pp. 123–35. University College Press, Cardiff, U.K.

Gaudy, A. F., Jr., and E. T. Gaudy. 1980. *Microbiology for Environmental Scientists and Engineers.* McGraw-Hill, New York. 736 pp.

Golueke, C. G., W. J. Oswald, and H. B. Gotaas. 1957. Methane fermentation of algae. *Appl. Microbiol.* 5, 47–55.

Grobbelaar, J. U., C. J. Soeder, and D. F. Toerien (eds.). 1981. *Wastewater for Aquaculture.* Proceedings of a Workshop on Biological Production Systems and Waste Treatment, University of O.F.S., Publications Series C, no. 3, Bloemfontein, S.A. 215 pp.

Hill, A., D. Feinberg, R. McIntosh, B. Neenan, and K. Terry. 1984. *Fuels from Microalgae: Technology Status, Potential, and Research Issues.* SERI/SP-231-2550. Solar Energy Research Institute, Golden, CO. 208 pp.

Jewell, W. J. 1987. Anaerobic sewage treatment. *Envir. Sci. Technol.* 21, 14–21.

Laws, E. A. 1984. *Research and Development of Shallow Algal Mass Culture Systems for the Production of Oils.* SERI/STR 231-2496. Solar Energy Research Institute, Golden, CO. 47 pp.

Lee, E. W., and B. L. Adan. 1983. Water quality management of Laguna de Bay. *J. Envir. Eng.* 109, 886–99.

Meron, A., M. Zabat, and W. J. Oswald. 1970. Designing waste ponds to meet water quality criteria. In *Proceedings of the Second International Symposium for Waste Treatment Lagoons*, pp. 186–94. Kansas State University, Manhattan.

Metcalf and Eddy, Inc. 1979. *Wastewater Engineering: Treatment, Disposal and Reuse.* 2nd ed. McGraw-Hill, New York. 920 pp.

Metting, B., and J. W. Pyne. 1986. Biologically active compounds from microalgae. *Enzyme Microbial Technol.* 8, 386–94.

Middlebrooks, E. J., N. B. Jones, J. H. Reynolds, M. F. Tordy, and R. Bishop.

1979. *Lagoon Information Source Book.* Ann Arbor Science Publishers, Ann Arbor, MI. 213 pp.

Nurdogan, Y. 1984. Shallow depth clarification of high-rate oxidation pond effluent. *University of California SEEHRL Lab. News Quarterly,* 32(1), 1–3.

Nurdogan, Y. 1985. Phosphorus removal by enhanced autoflocculation in high rate oxidation ponds. *University of California SEEHRL Lab. News Quarterly,* 35(4), 1–3.

Oswald, W. J. 1978. Engineering aspects of microalgae. In *Handbook of Microbiology,* vol. 2, pp. 519–52. CRC Press, Boca Raton, FL.

Oswald, W. J. 1985. *Potential for Treatment of Agricultural Drain Water with Microalgal-Bacterial Systems.* Contract no. 5-PG 20-06820 Final Report. U.S. Department of the Interior, Bureau of Reclamation, Mid-Pacific Region, Sacramento, CA. 74 pp.

Oswald, W. J., and D. M. Eisenberg. 1981. Energy from microalgae. In *Energy in the Man-Built Environment.* ASCE Speciality Conference Proceedings, ed. R. L. Anglin, Jr., pp. 244–54. American Society of Civil Engineers, New York.

Oswald, W. J., and H. B. Gotaas. 1957. Photosynthesis in sewage treatment. *Trans. Amer. Soc. Civ. Eng.* 122, 73–105.

Oswald, W. J., and C. G. Golueke. 1960. Biological transformation of solar energy. In *Advances in Applied Microbiology,* vol. 2, ed. W. W. Umbreit, pp. 223–62. Academic Press, New York.

Palmer, C. M. 1977. *Algae and Water Pollution.* EPA-600/9-77-036. Municipal Environmental Research Laboratory, U.S. Environmental Protection Agency, Cincinnati, OH. 124 pp.

Parkin, G. F., and W. F. Owen, 1986. Fundamentals of anaerobic digestion of wastewater sludges. *J. Envir. Eng.* 112, 867–920.

Ramani, A. R. 1974. *Factors Influencing Separation of Algal Cells from Waste Pond Effluents by Chemical Flocculation and Dissolved Air Flotation.* Ph.D. diss., University of California, Berkeley. 312 pp.

Shelef, G., W. J. Oswald, and P. H. McGauhey. 1970. Algal reactor for life support systems. *J. San. Eng., Div. Amer. Soc. Civ. Eng.* 96 (SAL), 91–110.

Shelef, G., and C. J. Soeder (eds.). 1980. *Algae Biomass: Production and Use.* Elsevier/North Holland Biomedical Press, Amsterdam. 852 pp.

Silva, P. C., and G. F. Papenfuss. 1953. *A Systematic Study of the Algae of Sewage Oxidation Ponds.* California State Water Pollution Control Board, Publication no. 7. 35 pp.

Sing, V. P., and P. N. Saxena. 1969. Preliminary studies on algal succession in raw and stabilized sewage. *Hydrobiolgoia* 34, 503–12.

Smith, R. 1973. *Electrical Power Consumption for Municipal Wastewater Treatment.* EPA-R2-73-281. National Environmental Research Center, U.S. Environmental Protection Agency, Cincinnati, OH. 89 pp.

Sun, M. 1986. Ground water ills: many diagnoses, few remedies. *Science* 232, 1490–3.

USEPA [U.S. Environmental Protection Agency]. 1984. *Toxic Substances Control Act (TSCA).* Report to Congress, U.S. Environmental Protection Agency, Office of Toxic Substances, Washington, DC. 57 pp.

Von Hippel, D. F., and W. J. Oswald, 1985. *Collection of Methane Evolved from Waste Treatment Ponds.* UCB/SEEHRL Rep. 85-11, U.C. Applied Technology

Program, Sanitary and Environmental Health Research Laboratory, University of California, Berkeley. 32 pp.

Wang, L., and M. Krofta. 1985. Application of dissolved air flotation to Lenox, Massachusetts, water supply: water purification by flotation. *J. N. Eng. Water Works Assoc.*, pp. 249–64, 265–84.

Warshall, P. 1979. *Septic Tank Practices.* LCCCN 77-76288. Anchor Press/Doubleday, New York. 177 pp.

Weres, O., A. F. White, H. A. Wollenberg, and A. Yee. 1985. *Geochemistry of Selenium in the Kesterson Reservoir and Possible Remedial Measures.* LBID 1014 Geochem GP Earth Sciences Div., Lawrence Berkeley Laboratory, University of California, Berkeley. 13 pp.

Winneberger, J. T., and P. H. McGauhey. 1965. *A Study of Methods of Preventing Failure of Septic Tank Percolation Systems.* SERL Rep. 65-17. Sanitary Engineering Research Laboratory, University of California, Berkeley. 33 pp.

12. Hydrogen production by algal water splitting

ELIAS GREENBAUM

Chemical Technology Division, Oak Ridge National Laboratory, Oak Ridge, TN 37831

CONTENTS

I. Introduction and overview

Photosynthesis is the oldest and most reliable method of converting solar energy into stored chemical energy. As the process occurs in nature, energy-rich molecules such as carbohydrates, proteins, lipids, and nucleic acids are synthesized from relatively simple renewable inorganic substrates including mineral salts, carbon dioxide, and water. Figure 12-1 is a schematic illustration of the overall photosynthetic reaction.

Biomolecular synthesis in the photosynthetic process is both energy-storing and entropically unfavorable. The material produced is energy-dense and represents a highly complex structured reordering of the inorganic substrate molecules. The overall process is driven by the energy and entropy densities of the photosynthetically active region of the solar-emission radiation field. The availability of energy-rich material that can be transported and consumed in a temporally convenient way has provided food, fuel, and materials for the development of human civilization.

Fossil fuels are the end products of very old photosynthesis transformed by appropriate geological conditions and forces. As this fuel supply continually diminishes and becomes less available because of economic, environmental, and political considerations, it is interesting to consider whether human intellect and ingenuity can rise to the challenge of synthesizing fuel from renewable inorganic resources rapidly enough to satisfy current and future needs. One obvious approach would simply involve the production and consumption of biomass itself. Another possibility is, however, a variant of photosynthesis in which the energy-rich product is molecular hydrogen, the primary substrate consumed is water, and oxygen is generated as a by-product. This process in its *in vitro* and *in vivo* forms has been referred to as biophotolysis of water (Hollaender et al. 1972; Gibbs et al. 1973; Lien and San Pietro, n.d.).

In 1984, the volume of molecular hydrogen produced in the United States exceeded 1 trillion cubic feet (*Chem. Eng. News* 1985) and had a total value of US $14 billion. Virtually all of this hydrogen was produced

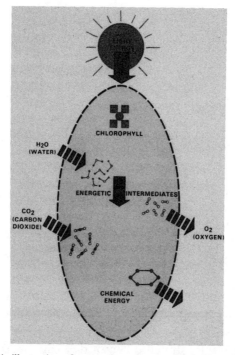

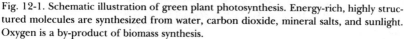

Fig. 12-1. Schematic illustration of green plant photosynthesis. Energy-rich, highly structured molecules are synthesized from water, carbon dioxide, mineral salts, and sunlight. Oxygen is a by-product of biomass synthesis.

by cracking and reforming fossil fuels at an energy investment of 1 to 2 quads (1 quad $\cong 10^{18}$ J). The most important uses of hydrogen are as a chemical feedstock for the synthesis of bulk commodity chemicals (e.g., ammonia and methanol) and as a reducing agent in the various hydro-treating processes used for petroleum refining. Liquid hydrogen (and oxygen) also serves as a rocket fuel in the U.S. space program.

II. Role of algae in hydrogen production

Gaffron and Rubin (1942) first discovered that the green alga *Scenedesmus* could evolve molecular hydrogen in a nitrogen atmosphere. Hydrogen was produced under both dark and light conditions; however, the rate of hydrogen evolution in the light was about 10 times greater than that in the dark.

Photosynthetic hydrogen and oxygen production by algae is unique among the bioconversion strategies in that biomass is not necessarily consumed in the act of energy production and utilization. In this respect, it is conceptually identical to the nonbiological photoelectrochemical (Nozik 1978; Wrighton 1979; Heller 1984) and photochemical (Bolton 1978; Kiwi, Kalyanasundaram, and Grätzel, 1982; Fendler 1985) approaches to the water-splitting problem. Biophotolysis of water is defined as the simultaneous photoevolution of hydrogen and oxygen. It is a variant of normal photosynthesis in which water serves as the primary substrate for source of reductant and the molecular hydrogen that is formed becomes the energy-rich reduced product.

The early controversy associated with the source of reductant in algal hydrogen evolution was authoritatively reviewed by Bishop, Frick, and Jones (1977). In the first experimental demonstration of hydrogen and oxygen production by algae, which was made by Spruit (1954, 1958), a novel polarographic technique was used to measure the gases, although technical limitations allowed only brief transients to be observed. Bishop and Gaffron (1963) demonstrated that 3-(3′,4′-dichlorophenyl)-1,1-dimethylurea (DCMU), a specific inhibitor of photosynthetic oxygen evolution, inhibited hydrogen production. Subsequent pioneering research by Gaffron, Stuart, and Kaltwasser (Kaltwasser, Stuart, and Gaffron 1969; Stuart and Kaltwasser 1970; Stuart 1971; Stuart and Gaffron 1971, 1972a–c; Stuart, Herold, and Gaffron 1972) laid the early groundwork for fundamental knowledge on the kinetics, mechanisms, and light response of photosynthetic hydrogen evolution in green algae. Healey (1970a) demonstrated different molecular mechanisms for the dark- and light-mediated synthesis of hydrogen in *Chlamydomonas moewusii*.

Due to the efforts of a group of forward-thinking scientists (Hollaender et al. 1972), the year 1972 marked a turning point for biofuels synthesis in general and algal hydrogen production in particular. Subsequent work in this field has frequently been perceived, seriously by some and cynically by others, in the context of fuel synthesis from renewable inorganic resources. Benemann and Weare (1974) demonstrated hydrogen evolution by heterocystous nitrogen-fixing *Anabaena cylindrica* cultures. Apart from green algae and cyanobacteria, the only other biologically based hydrogen-producing system based on the photosynthetic water-splitting system is the reconstituted in vitro system composed of isolated chloroplasts, ferredoxin, and hydrogenase (CFH) (Arnon, Mitsui, and Paneque 1961; Benemann et al. 1973; Krampitz 1975; Greenbaum 1980) or substituted analogues of this system (Krasna 1977; Greenbaum 1985c). All three of the only known biophotolytic systems – cyanobacteria, green algae, and the reconstituted in vitro CFH system – have been shown to be capable of the sustained simultaneous photoevolution of hydrogen and oxygen (Jones and Bishop 1976; Miyamoto, Hallenbeck, and Benemann 1979; Greenbaum 1980; Rosenkrans and Krasna 1984).

An authoritative review of the photobiological production of hydrogen, including hydrogen production by photosynthetic bacteria, has been given by Weaver, Lien, and Seibert (1980). Since photosynthetic bacteria are incapable of splitting water and grow only on energy-rich substrates, they are not *net* energy producers from the viewpoint of biological hydrogen production.

III. Principles of photosynthetic hydrogen production

The decomposition of liquid water to form gaseous hydrogen and oxygen is an energy-intensive chemical reaction (Latimer 1952):

$$H_2O(l) \rightarrow H_2(g) + 1/2\ O_2(g)$$
$$\Delta H^\circ = 68.32 \text{ kcal/mol} = 286 \text{ kJ/mol}$$
$$\Delta G^\circ = 56.69 \text{ kcal/mol} = 237 \text{ kJ/mol}$$
$$U^\circ = \Delta G^\circ/nF = 1.23 \text{ V}$$

where ΔH° is the enthalpy change, ΔG° is the Gibbs free energy change, n is the number of electrons transferred ($n = 2$ for the reaction as written above), F is the Faraday constant, and U° is the standard electrochemical potential. The physical significance of these parameters in this context is easy to understand. The enthalpy change is the energy released when the hydrogen and oxygen are burned. If hydrogen and oxygen are consumed in a fuel cell (a two-electrode device that can convert chemical energy directly to electrical energy), the standard electrochemical potential is the maximum possible voltage difference between the two electrodes. The Gibbs free energy is the maximum work that can be extracted from a mechanical device, such as a motor powered by the fuel cell. These parameters define the absolute minimum energetic requirements of any water-splitting system. Irrespective of the particular *device* that is used to split water (i.e., photobiological, photochemical, photoelectrochemical, thermochemical), the laws of thermodynamics define a common feature that transcends the specific characteristics of that device.

The potential difference required to split water in photosynthesis is generated in the photosynthetic reaction centers of Photosystems I and II (PS I and PS II), as shown in Figure 12-2. The light reactions are embedded in thylakoid membranes in such a manner that oxidizing and reducing equivalents are separated by the membrane. In analogy to electrolysis, the required potential difference appears across the thylakoid membrane. In accordance with the vectorial nature of photosynthesis, the lumen side of the thylakoid becomes positive with respect to the stroma side. Oxygen is evolved in the intrathylakoid space (lumen), and hydrogen is evolved in the extrathylakoid space (stroma).

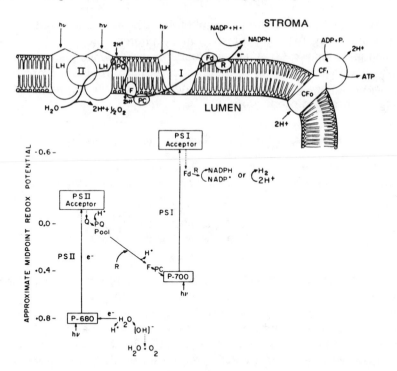

Fig. 12-2. Schematic illustration of a photosynthetic thylakoid membrane with light reaction centers and electron transport components. The Z-scheme, which summarizes the energetics of the overall electron transport process, is shown with the light reactions juxtaposed with their counterparts in the membrane. (Redrawn from Bogorad 1981. Copyright 1982 by the Rockefeller University Press. Used by permission.)

A. Hydrogenase

The Z-scheme of Figure 12-2 indicates two alternative pathways for electrons emerging from Photosystem I: (1) the enzymatic reduction of $NADP^+$ to NADPH and (2) the hydrogenase-catalyzed reduction of hydrogen ions to molecular hydrogen, $2H^+ + 2e^- \rightarrow H_2$. In normal aerobic photosynthesis, NADPH transports reducing equivalents into the Calvin cycle for the enzymatic reduction of atmospheric carbon dioxide. However, when certain algae such as *Scenedesmus* or *Chlamydomonas* are placed in a carbon-dioxide-free anaerobic atmosphere, they are capable, as first shown by Gaffron and Rubin (1942), of synthesizing the enzyme hydrogenase and evolving molecular hydrogen. By comparison with normal photosynthesis, photoproduced molecular hydrogen corresponds to carbon dioxide fixation products, and hydrogenase is analogous to the complement of Calvin cycle enzymes.

As reviewed by Kessler (1974), hydrogenase activity has been detected in about 50% of the eukaryotic algae examined. The group of hydrogenase-containing algae includes both unicellular and multicellular types. An example in the latter category is the seaweed *Ulva*. There is, however, an important difference between unicellular and multicellular algae with respect to hydrogenase activity. Unicellular algae are capable of catalyzing a variety of hydrogen-related reactions, including those that evolve and consume hydrogen or exchange hydrogen isotopes. Indeed, it is frequently assumed that algal hydrogenase activity, as demonstrated by any one of the hydrogenase-indicating reactions, implies the presence of all the rest. Although this is an experimentally verified fact for all unicellular algae tested, it is not the case for multicellular algae (Greenbaum and Ramus 1983). In contrast to unicellular algae, there is not a single known example of a macroscopic alga for which the *photo*evolution of hydrogen has been observed, in spite of the fact that hydrogenase activity has been reported for at least nine macroscopic algal species.

Although purified hydrogenase has been prepared from a large number of nonphotosynthetic bacteria (Schlegel and Schneider 1978), the corresponding work for algae is much more limited. The first algal cell-free hydrogenase preparation was reported by Abeles (1964). Erbes, King, and Gibbs (1979) partially purified hydrogenase from *Chlamydomonas reinhardtii* in order to examine the kinetics of oxygen inactivation. The most comprehensive work on the purification of an algal hydrogenase is that by Roessler and Lien (1984a), who also studied the effects of anions on hydrogenase activity (Roessler and Lien 1982) and of electron mediator charge properties on the reaction kinetics of hydrogenase from *Chlamydomonas* (Roessler and Lien 1984b). According to Roessler and Lien (1984a), the specific activity of pure *Chlamydomonas* hydrogenase is 1,800 μmol H_2 produced per mg protein per min with a molecular weight of $\sim 4.6 \times 10^4$ daltons. The enzyme has many acidic side groups, contains iron, and has an activation energy of 55.1 kJ/mol for hydrogen evolution.

Unlike hydrogen production in green algae, which is mediated by hydrogenase, hydrogen production in nitrogen-fixing cyanobacteria (blue-green algae) is mediated by nitrogenase and is an ATP-requiring process. According to Benemann et al. (1977), hydrogen evolution by the filamentous cyanobacterium *Anabaena* occurs in the heterocysts (Fig. 12-3). The source of reductant is photosynthate, which is generated in the vegetative cells and then transported to the heterocysts. Nonheterocystous cyanobacteria are also capable of hydrogen evolution. According to Mitsui et al. (1983), nitrogenase-catalyzed hydrogen evolution in these organisms occurs in the vegetative cells (Fig. 12-4).

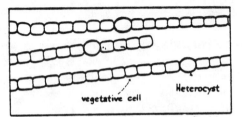

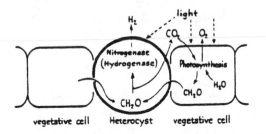

Microscopic View of *Anabaena cylindrica*

Fig. 12-3. Mechanism of hydrogen evolution in the heterocystous cyanobacterium *Anabaena cylindrica*. Photosynthate produced in the vegetative cells migrates to the heterocysts where it serves as the substrate for nitrogenase-catalyzed hydrogen evolution. (From Benemann et al. 1977.)

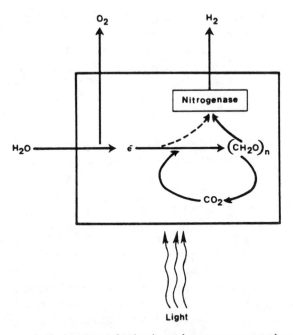

Fig. 12-4. Pathway for hydrogen production in nonheterocystous cyanobacteria. (From Mitsui et al. 1983. Copyright 1983, New York Academy of Sciences. Used by permission.)

Comprehensive review articles on the physiology and biochemistry of hydrogen metabolism in cyanobacteria and heterocysts have been provided by Houchins (1984) and Haselkorn (1978).

B. Light reactions

The electrochemical potential required to split water into molecular oxygen and hydrogen is driven by the absorption of light quanta in the photosynthetic reaction centers. The two-photosystem, or Z-scheme, model of photosynthesis allows an estimate of the maximum thermodynamic conversion efficiency obtainable in the photosynthetic process. The vertical scale of the Z-scheme in Figure 12-2 represents the midpoint oxidation/reduction potential at pH = 7 versus the standard hydrogen electrode (SHE). For the half-reaction

$$2H_2O \rightleftarrows O_2 + 4H^+ + 4e^-,$$

the midpoint versus SHE is $+0.8$ V at pH = 7. Since this number is the reversible thermodynamic value, the actual molecular species that is responsible for the oxidation of water must have an effective midpoint potential of at least $+0.8$ V. (This assumes that the internal pH of the cell is near 7.) In practice, the actual effective midpoint potential will be even greater (i.e., more oxidizing). Studies in which artificial electron donors and acceptors of known midpoint potentials are added to intact cells and isolated chloroplasts have shown that both Photosystem I (PS I) and Photosystem II (PS II) are capable of spanning potential differences of about 1 V each. It can be seen from Figure 12-2 that the movement of electrons on the oxidizing side of PS II (P_{680}) to the reduction level of ferredoxin spans a potential difference of at least 1.2 V. That is, the thermodynamically uphill movement of electrons through the photosynthetic electron transport chain results in an increase of energy of ~ 1.2 electron volts (eV) per electron transferred (27.7 kcal/mol of electrons transferred). This process is driven by the absorption of visible light quanta in the photosynthetic reaction centers.

The primary process of photosynthesis is such that the absorption of one quantum in a photosynthetic reaction center results in the transfer of one electron. Since there are two photoreactions in series, the absorption of two quanta is required to move one electron through the electron-transport chain of photosynthesis. Photosynthetically active radiation is in the wavelength range 400–700 nm (3.1–1.8 eV per photon), which contains about 47% of the power in the solar-emission spectrum.

The primary electron transfer of photosynthesis is a quantum conversion process that, presumably, takes place from the lowest excited singlet state of reaction-center chlorophyll (about 1.8 eV above ground state). The maximum theoretical efficiency of energy conversion by the photosynthetic apparatus is, therefore,

$$\frac{1.2 \text{ eV}}{2 \times 1.8 \text{ eV}} \times 47\% \simeq 13\%.$$

This efficiency is not as high as the maximum theoretical efficiency of electricity generation by silicon photovoltaic cells. It is, however, important to bear in mind that the end products of photosynthesis are storable, energy-rich molecules as compared with the electric power of photovoltaic cells. In addition, the process works quite well in a relatively impure environment.

The ability of higher plants to use water as the source of reductant for growth represents, from a thermodynamic point of view, the ultimate triumph of biology over an inorganic world. In spite of this importance, the actual molecular structure and nature of operation of the water-splitting complex on the oxidizing side of PS II are still unknown, although significant progress is being made. A phenomenological model advanced by Kok and co-workers (Kok, Forbush, and McGloin 1970; Forbush, Kok, and McGloin 1971) is frequently employed to describe the redox properties of the water-splitting complex. According to Kok's model, an electron carrier, S, is located on the oxidizing side of PS II and can exist in multiple metastable oxidation/reduction states. Each photon-driven redox state is more oxidizing than its predecessor by one oxidizing equivalent. When four oxidizing equivalents have been accumulated by the complex, it cycles back to its initial state with the corresponding oxidation of water and evolution of oxygen. As reviewed by Amesz (1983), manganese plays an important role in photosynthetic water splitting.

C. Electron transport chain and role of ATP

As indicated in Figure 12-2, the electron transport chain linking the two light reactions of photosynthesis acts as a molecular wire for the flow of electrons from water to an appropriate electron acceptor on the reducing side of PS I. Coupled with the electron transport is the translocation of protons from the thylakoid stroma to the thylakoid lumen. According to Mitchell (1961, 1966), the source of Gibbs free energy for the synthesis of ATP is the hydrogen ion gradient that is established as a result of this proton translocation.

The requirement for ATP is a key factor distinguishing hydrogen production by hydrogenase catalysis in green algae from nitrogenase catalysis in cyanobacteria. The photoproduction of hydrogen in green algae does not require ATP. Moreover, as shown by Kaltwasser et al. (1969), carbonyl cyanide m-chlorophenylhydrazone (Cl-CCCP), an uncoupler of phosphorylation, maximized rates of hydrogen photoproduction in

Scenedesmus. In contrast, Spiller et al. (1978) have shown that in the cyanobacterium *Nostoc muscorum* Cl-CCCP inhibited hydrogen photoproduction by a factor of 20 compared with the control. The additional requirement for ATP in nitrogenase-catalyzed hydrogen evolution decreases the maximum theoretical hydrogen-evolving efficiency in cyanobacteria to ~10%. This may be a significant factor in the development of a practical system.

D. Sources of reductant

There are two known sources of reductant for photosynthetic hydrogen evolution. The first is reductant generated by PS II and the water-splitting reaction. The second is the oxidation of endogenous reductants by components of the photosynthetic electron transport (PET) chain. Klein and Betz (1978) have demonstrated that starch is one such endogenous reductant. Gfeller and Gibbs (1985) have presented evidence supporting a chloroplastic respiratory pathway in the green alga *Chlamydomonas reinhardtii* whereby reducing equivalents generated during starch degradation enter the PET chain at the plastoquinone site. These reducing equivalents may then serve as electron donors for PS-I-mediated evolution of hydrogen. The two sources of reductant can be demonstrated explicitly in the light-saturated ratio of the simultaneous evolution of hydrogen and oxygen (Greenbaum 1984). Miura et al. (1986) have isolated and characterized a unicellular marine green alga exhibiting high activity in dark hydrogen production. However, irrespective of the immediate source of reductant, all hydrogen produced from photoautotrophically grown algae on a mineral salt solution is ultimately derived from water splitting.

IV. Biophotolysis of water

As indicated above, all three of the known biophotolytic systems have now been shown to be capable of simultaneous photoevolution of hydrogen and oxygen: green algae, cyanobacteria, and the in vitro chloroplast-ferredoxin-hydrogenase system. Seawater can also serve as the substrate for microalgal hydrogen evolution (Healy 1970b; Mitsui and Kumazawa 1977; Greenbaum, Guillard, and Sunda 1983). Because the biophotolysis of water is a variant of photosynthesis in which the energy-rich product is molecular hydrogen, the usual quantitative figures of merit that are used to characterize normal photosynthesis can be applied to biophotolytic systems. These quantitative figures are: light saturation response, turnover times, photosynthetic unit size, stress and endurance capability, and energy conversion efficiency.

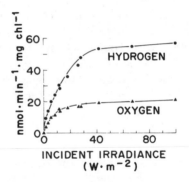

Fig. 12-5. Light saturation curves of the simultaneous photoproduction of hydrogen and oxygen by anaerobically adapted *Scenedesmus* D$_3$. (From Greenbaum 1984. Copyright 1984, Elsevier Science Publishers, B.V. Used by permission.)

A. Light saturation curves

Photosynthesis is linear at low light intensities and saturates at higher intensities. This same pattern is observed in the simultaneous photo-evolution of hydrogen and oxygen by anaerobically adapted green algae. The light saturation curves for the biophotolysis of water by *Scenedemus* D$_3$ are indicated in Figure 12-5. The following information has been deduced from light saturation studies (Greenbaum 1984): (1) The stoichiometric ratio of hydrogen and oxygen is light-dependent and, in general, not equal to 2; (2) the ratios of the light-saturated rates were greater than 2 for all of the systems studied; (3) far-red (≥700-nm) light greatly increased the ratio of hydrogen to oxygen; and (4) the light saturation curves for the simultaneous photoproduction of hydrogen and oxygen do not have the same analytical shapes; a higher light intensity is required to saturate the hydrogen curve than is required for the oxygen curve.

The biophotolysis of water is a strict analogue of normal photosynthesis in all respects except one. The light-saturated rate of biophotolytic oxygen evolution is only ~1% of that for photosynthetic oxygen evolution. The rate of photosynthesis is a product of two factors: the number of functional photosynthetic units and the turnover rate of each unit. The limiting aspect of biophotolytic hydrogen and oxygen production is the number of apparent functional photosynthetic units, a value that can be determined by using pulsed illumination to measure turnover times and photosynthetic unit sizes.

B. Turnover times

Photosynthesis saturates because, with increasing light intensity, the rate of quantum excitation of the reaction centers can exceed the

Table 12-1. *Summary of turnover times for light-driven photosynthetic hydrogen production at 20°C based on the repetitive double-flash technique*

Alga	Turnover time (ms)[a]
Chlamydomonas reinhardtii	0.3
Chlorella vulgaris	0.1
Scenedesmus obliquus	3.0
C. reinhardtii (TAP)	1.0

[a]The turnover time is defined as the time between two flashes for which there is a 50% increase between the initial steady-state flash-pair yield and the final steady-state flash-pair yield. (From Greenbaum 1979. Copyright 1979, Pergamon Press, Ltd.)

kinetic rate of thermally activated electron transport of the non-light-dependent biochemical reactions that serially link the light reactions. The turnover time of photosynthesis was first measured by Emerson and Arnold (1932a), who demonstrated that it was <40 ms at 25°C. A measurement of turnover time of photosynthetic hydrogen production and its comparison with the turnover time of normal photosynthesis are important considerations in the development of a practical engineering system.

As discussed by Diner and Mauzerall (1973), the turnover time of photosynthesis can be measured in two ways. The first, used by Emerson and Arnold, is by steady-state repetitive flash illumination and variation of the frequency of flashing. The second is the repetitive double-flash method in which the frequency of flash-pair spacing is kept constant and the delay between the flash pair is varied. The turnover times derived from these techniques are different because different physical phenomena are being measured. The turnover time of photosynthetic hydrogen evolution has been determined with each technique.

Measurement of the turnover time of photosynthetic hydrogen production using the repetitive double-flash method has provided data on the kinetics of thermally activated biochemistry on the reducing side of PS I. Table 12-1 summarized the turnover times measured by using this technique for selected algae. A turnover time of 6 msec for photosynthetic hydrogen evolution using the repetitive single-flash technique was determined for *Chlamydomonas reinhardtii* (Greenbaum 1982). This value is analogous to the turnover time of photosynthesis as measured by Emerson and Arnold.

These experimental data indicate that the turnover times of photosynthetic hydrogen evolution are comparable to those of normal photosynthesis. The results suggest that the low values of light-saturated hydrogen evolution by anaerobically adapted algae are not due to inherently slow turnover kinetics of hydrogen-evolving biochemistry. Instead, the limitation is primarily associated with the number of apparent functional photosynthetic units.

C. Photosynthetic unit of hydrogen evolution

According to Emerson and Arnold (1932b), there are, in the green alga *Chlorella vulgaris,* approximately 2,400 chlorophyll molecules per evolved molecule of oxygen evolved per single-turnover saturating flash of light. This number is an absolute figure of merit, which can be used to compare a variety of photosynthetic systems and their light-driven products. Two approaches are used to determine the size of the photosynthetic unit for hydrogen evolution. The first consists of measuring individually resolved bursts of hydrogen following single-turnover saturating light flashes at a low flash repetition rate (0.1 Hz) (Greenbaum 1977). The second, which is analogous to the Emerson and Arnold method, involves measuring the continuous steady-state rate of hydrogen (and oxygen) production under faster and varying repetitive flash illumination.

Theoretically, the value of the photosynthetic unit size for hydrogen evolution (the ratio of chlorophyll to hydrogen per single-turnover flash) should be predictable from a knowledge of the Emerson and Arnold photosynthetic unit size for oxygen evolution and the Z-scheme model of photosynthesis. If all the reducing equivalents that are expressed as molecular hydrogen are derived from the mainstream of the electron transport chain of photosynthesis, the unit size value for hydrogen evolution should be half that for oxygen evolution. For *Chlamydomonas reinhardtii* and some other algae, using the individual flash yield technique, this is fairly close to what is actually observed (Greenbaum 1977; Reeves and Greenbaum, unpublished data). At higher flash rates under steady-state hydrogen evolution, the yield of hydrogen per chlorophyll molecule per flash falls by a factor of ~100. The low light-saturated steady-state rates of hydrogen and oxygen evolution can, to a large extent, be explained almost entirely by a loss in the number of apparent functional photosynthetic units at higher equivalent light intensities. A fundamental understanding of this phenomenon would be an important first step in the development of a practical gaseous fuel-producing system based on algal water splitting. As discussed previously (Greenbaum 1984), part of the explanation may be associated with the heterogeneous partitioning of PS I and PS II between the stroma lamellae and grana stacks of photosynthetic thylakoid membranes.

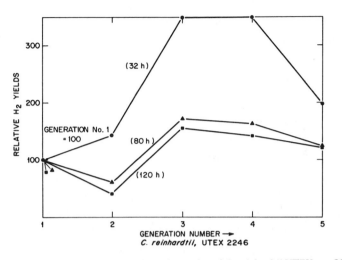

Fig. 12-6. Relative yields of hydrogen from five cycles of *C. reinhardtii* UTEX no. 2246 at: ●, 32; ▲, 80; and ■, 120 h of anaerobiosis. (From Reeves and Greenbaum 1985. Copyright 1985, Butterworth & Co., Ltd. Used by permission.)

D. Stress and endurance studies

The simultaneous photoproduction of hydrogen and oxygen represents a physiological stress for algae. However, eukaryotic green algae are very rugged microorganisms with respect to the biophotolysis problem. They can survive hundreds of hours of anaerobiosis and irradiation, a procedure that selects for strains of algae with enhanced properties for fuel production (Reeves and Greenbaum 1985; Ward, Reeves, and Greenbaum 1985).

Figure 12-6 illustrates the relative yields of hydrogen from five cycles of stressed and regreened *Chlamydomonas reinhardtii* UTEX 2246. In this experiment, the algal culture was subjected to repeated cycles of anaerobiosis, carbon dioxide deprivation, and irradiation as a means of testing the long-term stability of hydrogen and oxygen photoproduction and the effectiveness of these conditions as selection pressures for increasing hydrogen and/or oxygen yields. Simultaneous hydrogen and oxygen photoproduction yields were monitored for 160 hours. The cells were then removed from the reaction chamber and used to inoculate fresh growth medium to produce the culture for the next experiment. This cycle was repeated five times. Yields of hydrogen and oxygen improved after three cycles but declined in the fourth and fifth; unlike the second and third cycles, extended periods of aerobic growth were used for the fourth and fifth cycles. The stability of hydrogen and oxygen photoproduction was greater in the fifth cycle than in any of the

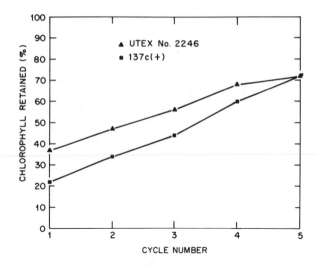

Fig. 12-7. Percentage of chlorophyll retained after each cycle of anaerobiosis and irradiation through five cycles. ▲, UTEX no. 2246; ■, 137c(+). (From Reeves and Greenbaum 1985. Copyright 1985, Butterworth and Co., Ltd. Used by permission.)

previous cycles. These subpopulations had hydrogen and oxygen production rates, at 160 hours, that were nearly equal to the rates at the beginning of the fifth-cycle experiments. In Figure 12-6, time profiles of the relative hydrogen yields from each of the five cycles, prepared at 32, 80, and 120 hours, show that the relative yield in each varies with the point in time at which the profile was taken. As indicated in Figure 12-7, chlorophyll retention increased with each successive cycle, indicating selection or adaptation for a more durable population of cells with respect to the light-harvesting component of the photosynthetic apparatus.

E. Energy conversion efficiencies

An important consideration for solar fuel synthesis is the conversion efficiency of light energy into chemical energy. This efficiency, η, is defined as

$$\eta = \frac{\text{rate of hydrogen production}}{\text{rate of total light energy input}}.$$

Approximately half of the power radiated in the solar-emission spectrum lies in the 400–700 nm wavelength interval. Efficiencies determined from absorbed photosynthetically active radiation can be con-

EFFICIENCY MEASUREMENTS

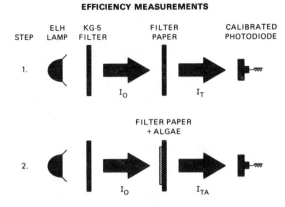

Fig. 12-8. Technique of measuring absolute conversion of light energy into hydrogen Gibbs free energy based on net absolute absorbed light energy. The net light absorbed by the algae, ΔI, is computed as the difference between transmitted light without and with algae entrapped on the filter paper. Efficiency is computed from the rate of energy output as expressed by molecular hydrogen divided by the rate of light energy input.

verted to approximate equivalent solar energy conversion efficiencies by dividing by two.

A new method for measuring absolute conversion efficiencies of light energy into chemical energy, as represented by the Gibbs free energy of molecular hydrogen, has recently been developed. This method is based on preparing well-defined geometric films of algae by entrapment on filter paper (Greenbaum 1985a) and measuring the absolute rate of hydrogen production for a given value of absorbed photosynthetically active radiation (Greenbaum 1985b). The principle of the method is illustrated in Figure 12-8. As shown in the figure, a two-step measurement is employed to determine the net amount of light absorbed by the algae. By measuring the net rate of hydrogen production and the net rate of absorbed radiation, equivalent solar conversion efficiencies can be computed. In the low-intensity, linear portion of the light saturation curve, conversion efficiencies of 10% have been measured for *Chlamydomonas reinhardtii, C. moewusii,* and *Scenedesmus* D_3. As expected, however, conversion efficiencies with these algae decrease with increasing light intensity. Kok (1973), however, has suggested a solution to this problem: a search for an alga with a small number of antenna chlorophylls per trap. Full sunlight excites a chlorophyll molecule 10 times per second. With 20 chlorophylls per trap (instead of the commonly found 200), an excitation rate of 200 per second would provide a closer match to the turnover-time rate (refer to Section IV, B, above) and yield higher efficiency in sunlight.

V. Conclusion

The key ideas of this chapter may be summarized as follows. First, algal photosynthesis is a self-replicating process that is capable of synthesizing energy-rich molecules from abundant inorganic materials, including atmospheric carbon dioxide as the *sole* carbon source. Second, it has been shown that, under appropriate physiological conditions, certain freshwater and marine green algae are capable of splitting water to molecular hydrogen and oxygen in a sustained steady-state reaction. In these algae, the gaseous-fuel-producing reaction can be driven by light throughout the visible portion of the solar-emission spectrum, including the long-wavelength (red) 700 nm region. No external energy sources are required. These two facts, combined with previous and recent experimental results, indicate that photosynthetic water splitting for hydrogen production is a promising research area for the production of gaseous fuels from inorganic resources.

The present availability of fossil fuels and the methods for producing and marketing them may remain the same for the next several decades. Renewed interest in the production of renewable fuels is likely to increase if disruptions of supplies occur or if environmental problems associated with burning and processing fossil fuels become dominant items on the social and political agenda.

VI. Acknowledgments

The author thanks C. M. Morrissey and J. E. Thompson for technical support and D. J. Weaver for secretarial support. He also thanks D. A. Graves, R. R. L. Guillard, L. W. Jones, M. E. Reeves, J. Ramus, and J. A. Solomon for criticism of and comments on the manuscript. This work was supported by the Gas Research Institute under contract GRI 5083-260-0880 with Martin Marietta Energy Systems, Inc., and by the Office of Basic Energy Sciences, U.S. Department of Energy. Oak Ridge National Laboratory is operated by Martin Marietta Energy Systems, Inc., under contract DE-ACO5-84OR21400 with the U.S. Department of Energy.

VII. References

Abeles, F. B. 1964. Cell-free hydrogenase from *Chlamydomonas*. *Plant Physiol.* 39, 169–76.

Amesz, J. 1983. The role of manganese in photosynthetic oxygen evolution. *Biochim. Biophys. Acta* 726, 1–12.

Arnon, D. I., A. Mitsui, and A. Paneque. 1961. Photoproduction of hydrogen gas coupled with photosynthetic phosphorylation. *Science* 134, 1425.

Benemann, J. R., J. A. Berenson, N. O. Kaplan, and M. D. Kamen. 1973. Hydrogen evolution by a chloroplast-ferredoxin-hydrogenase system. *Proc. Nat. Acad. Sci. USA* 70, 2317–20.

Benemann, J. R., P. C. Hallenbeck, J. C. Weissman, L. V. Kochian, P. Kostel, and W. J. Oswald. 1977. *Solar Energy Conversion with Hydrogen Producing Algae.* Final report to the Energy Research and Development Administration under contract no. E(04-3)-34.

Benemann, J. R., and N. M. Weare. 1974. Hydrogen evolution by nitrogen-fixing *Anabaena cylindrica* cultures. *Science* 184, 174–5.

Bishop, N. I., M. Frick, and L. Jones. 1977. Photohydrogen production in green algae: water serves as the primary substrate for hydrogen and oxygen production. In *Biological Solar Energy Conversion,* ed. A. Mitsui, S. Miyachi, A. San Pietro, and S. Tamura, pp. 3–22. Academic Press, New York.

Bishop, N. I., and H. Gaffron. 1963. On the interrelation of the mechanisms for oxygen and hydrogen evolution in adapted algae. In *Photosynthetic Mechanisms of Green Plants,* ed. B. Kok and A. Jagendorf, pp. 441–54. Publication 1145, National Academy of Sciences – National Research Council, Washington, DC.

Bogorad, L. 1981. Chloroplasts. *J. Cell. Biol.* 91, 256s–70s.

Bolton, J. R. 1978. Solar fuels. *Science* 202, 705–11.

Chem. Eng. News. 1985. 63, 11.

Diner, B., and D. Mauzerall. 1973. The turnover times of photosynthesis and redox properties of the pool of electron carriers between the photosystems. *Biochim. Biophys. Acta* 305, 353–63.

Emerson, R., and W. Arnold. 1932a. A separation of the reactions in photosynthesis by means of intermittent light. *J. Gen. Physiol.* 15, 391–421.

Emerson, R., and W. Arnold. 1932b. The photochemical reaction in photosynthesis. *J. Gen. Physiol.* 16, 191–205.

Erbes, D. L., D. King, and M. Gibbs. 1979. Inactivation of hydrogenase in cell-free extracts and whole cells of *Chlamydomonas reinhardtii* by oxygen. *Plant Physiol.* 63, 1138–42.

Fendler, J. H. 1985. Photochemical solar energy conversion: an assessment of scientific accomplishments. *J. Phys. Chem.* 89, 2730–40.

Forbush, B., B. Kok, and M. P. McGloin. 1971. Cooperation of charges in photosynthetic O_2 evolution. II. Damping of flash yield, deactivation. *Photochem. Photobiol.* 14, 307–21.

Gaffron, H., and J. Rubin. 1942. Fermentative and photochemical production of hydrogen in algae. *J. Gen. Physiol.* 26, 219–40.

Gfeller, R. P., and M. Gibbs. 1985. Fermentative metabolism of *Chlamydomonas reinhardtii. Plant Physiol.* 77, 509–11.

Gibbs, M., A. Hollaender, B. Kok, L. O. Krampitz, and A. San Pietro. 1973. *Proceedings of the Workshop of Bio-solar Conversion.* Indiana University, Bloomington.

Greenbaum, E. 1977. The photosynthetic unit of hydrogen evolution. *Science* 196, 878–9.

Greenbaum, E. 1979. The turnover times and pool sizes of photosynthetic hydrogen production by green algae. *Solar Energy* 23, 315–20.

Greenbaum, E. 1980. Simultaneous photoproduction of hydrogen and oxygen by photosynthesis. *Biotech. Bioeng. Symp.* 10, 1–13.

Greenbaum, E. 1982. Photosynthetic hydrogen and oxygen production: kinetic studies. *Science* 215, 291–3.

Greenbaum, E. 1984. Biophotolysis of water: the light saturation curves. *Photobiochem. Photobiophys.* 8, 323–32.

Greenbaum, E. 1985a. Hydrogen and oxygen production by photosynthetic water splitting: energy and quantum conversion efficiencies. In *Proceedings of the Second International Symposium on Hydrogen Produced from Renewable Energy,* ed. O. G. Hancock and K. G. Sheinkopf, pp. 145–53. Florida Solar Energy Center, Cape Canaveral.

Greenbaum, E. 1985b. Photosynthetic water splitting. June–Aug. 1985 Quarterly Report. Gas Research Institute, Chicago, IL.

Greenbaum, E. 1985c. Platinized chloroplasts: a novel photocatalytic material. *Science* 230, 1373–5.

Greenbaum, E., R. R. L. Guillard, and W. G. Sunda. 1983. Hydrogen and oxygen photoproduction by marine algae. *Photochem. Photobiol.* 37, 649–55.

Greenbaum, E., and J. Ramus. 1983. Survey of selected seaweeds for simultaneous photoproduction of hydrogen and oxygen. *J. Phycol.* 19, 53–7.

Haselkorn, R. 1978. Heterocysts. *Ann. Rev. Plant Physiol.* 29, 319–44.

Healey, F. P. 1970a. The mechanism of hydrogen evolution by *Chlamydomonas moewusii. Plant Physiol.* 45, 153–9.

Healy, F. P. 1970b. Hydrogen evolution by several algae. *Planta* 91, 220–6.

Heller, A. 1984. Hydrogen evolving solar cells. *Science* 223, 1141–8.

Hollaender, A., K. J. Monty, R. M. Pearlstein, F. Schmidt-Bleek, W. T. Snyder, and E. Volkin. 1972. *An Inquiry into Biological Energy Conversion.* University of Tennessee Press, Knoxville.

Houchins, J. P. 1984. The physiology and biochemistry of hydrogen metabolism in cyanobacteria. *Biochim. Biophys. Acta* 768, 227–55.

Jones, L. W., and N. I. Bishop. 1976. Simultaneous measurement of oxygen and hydrogen exchange from the blue-green alga *Anabaena. Plant Physiol.* 57, 659–65.

Kaltwasser, H., T. S. Stuart, and H. Gaffron. 1969. Light dependent hydrogen evolution by *Scenedesmus. Planta* 89, 309–22.

Kessler, E. 1974. Hydrogenase, photoreduction, and anaerobic growth. In *Algal Physiology and Biochemistry,* ed. W. D. P. Stewart, pp. 456–73. University of California Press, Berkeley.

Kiwi, J., K. Kalyanasundaram, and M. Grätzel. 1982. Visible light induced cleavage of water into hydrogen and oxygen in colloidal and microheterogeneous sytems. In *Structure and Bonding 49: Solar Energy Materials,* pp. 37–125. Springer-Verlag, Berlin.

Klein, V., and Betz, A. 1978. Fermentative metàbolism of hydrogen evolving *Chlamydomonas moewusii. Plant Physiol.* 61, 373–7.

Kok, B. 1973. Photosynthesis. In *Proceedings of the Workshop on Bio-solar Conversion,* ed. M. Gibbs, A. Hollaender, B. Kok, L. O. Krampitz, and A. San Pietro, pp. 22–30. Indiana University, Bloomington.

Kok, B., B. Forbush, and M. McGloin. 1970. Cooperation of charges in photosynthetic O_2 evolution. I. A linear four step mechanism. *Photochem. Photobiol.* 11, 457–75.

Krampitz, L. O. 1975. Biophotolysis of water to hydrogen and oxygen. In *Proceedings of the 169th National Meeting of the American Chemical Society*, pp. 29–32. Philadelphia.

Krasna, A. I. 1977. Catalytic and structural properties of the enzyme hydrogenase and its role in biophotolysis of water. In *Biological Solar Energy Conversion*, ed. A. Mitsui, S. Miyachi, A. San Pietro, and S. Tamura, pp. 53–60. Academic Press, New York.

Latimer, W. M. 1952. *Oxidation Potentials*, pp. 29–50. Prentice-Hall, Engelwood Cliffs, NJ.

Lien, S., and San Pietro, A. No date. *An Inquiry into Biophotolysis of Water to Produce Hydrogen*. Indiana University, Bloomington.

Mitchell, P. 1961. Coupling of phosphorylation to electron and hydrogen transfer by a chemiosmotic type of mechanism. *Nature* 19, 144–8.

Mitchell, P. 1966. Chemiosmotic coupling in oxidative and photosynthetic phosphorylation. *Biol. Rev.* 41, 445–502.

Mitsui, A., and S. Kumazawa. 1977. Hydrogen production by marine photosynthetic organisms as a potential energy resource. In *Biological Solar Energy Conversion*, ed. A. Mitsui, S. Miyachi, A. San Pietro, and S. Tamura, pp. 23–51. Academic Press, New York.

Mitsui, A., E. J. Philips, S. Kumazawa, K. J. Reddy, S. Ramachandran, T. Matsunaga, L. Haynes, and H. Ikemoto. 1983. Progress in research toward outdoor biological hydrogen production using solar energy, sea water, and marine photosynthetic microorganisms. *Ann. N.Y. Acad. Sci.* 413, 514–30.

Miura, Y., S. Ohta, M. Mano, and K. Miyamoto. 1986. Isolation and characterization of a unicellular marine green alga exhibiting high activity in dark hydrogen production. *Agric. Biol. Chem.* 50, 2837–44.

Miyamoto, K., P. C. Hallenbeck, and J. R. Benemann. 1979. Solar energy conversion by nitrogen-limited cultures of *Anabaena cylindrica*. *J. Ferment. Technol.* 57, 287–93.

Nozik, A. J. 1978. Photoelectrochemistry: applications to solar energy conversion. *Ann. Rev. Phys. Chem.* 29, 189–222.

Reeves, M., and E. Greenbaum. 1985. Long-term endurance and selection studies in hydrogen and oxygen photoproduction by *Chlamydomonas reinhardtii*. *Enzyme Microbial Technol.* 7, 169–74.

Roessler, P., and S. Lien. 1982. Anionic modulation of the catalytic activity of hydrogenase from *Chlamydomonas reinhardtii*. *Arch. Biochem. Biophys.* 213, 37–44.

Roessler, P., and S. Lien. 1984a. Purification of hydrogenase from *Chlamydomonas reinhardtii*. *Plant Physiol.* 75, 705–9.

Roessler, P., and S. Lien. 1984b. Effects of electron mediator charge properties on the reaction kinetics of hydrogenase from *Chlamydomonas*. *Arch. Biochem. Biophys.* 230, 103–9.

Rosenkrans, A. M., and A. I. Krasna. 1984. Stimulation of hydrogen photoproduction in algae by removal of oxygen by reagents that combine reversibly with oxygen. *Biotech. Bioeng.* 26, 1334–42.

Schlegel, H. G., and K. Schneider (eds.). 1978. *Hydrogenases: Their Catalytic Activity, Structure, and Function*. Erich Goltze KG, Göttingen.

Spiller, H., A. Ernst, W. Kerfin, and P. Boger. 1978. Increase and stabilization of photoproduction of hydrogen in *Nostoc muscorum* by photosynthetic electron transport inhibitors. *Z. Naturforsch.* 33C, 541–7.

Spruit, C. J. P. 1954. Photoproduction of hydrogen and oxygen in *Chlorella*. In *Proceedings of the First International Congress on Photobiology*, pp. 323–7.

Spruit, C. J. P. 1958. Simultaneous photoproduction of hydrogen and oxygen by *Chlorella*. *Meded. van de Landbou. Wageningen* 58, 1–17.

Stuart, T. S. 1971. Hydrogen production by photosystem I of *Scenedesmus:* effect of heat and salicylaldoxime on electron transport and photophosphorylation. *Planta* 96, 81–92.

Stuart, T. S., and H. Gaffron. 1971. The kinetics of hydrogen photoproduction by adapted *Scenedesmus*. *Planta* 100, 228–43.

Stuart, T. S., and H. Gaffron. 1972a. The gas exchange of hydrogen-adapted algae as followed by mass spectrometry. *Plant Physiol.* 50, 136–40.

Stuart, T. S., and H. Gaffron. 1972b. The mechanism of hydrogen photoproduction by several algae. I. The effect of inhibitors of photophosphorylation. *Planta* 106, 91–100.

Stuart, T. S., and H. Gaffron. 1972c. The mechanism of hydrogen photoproduction by several algae. II. The contribution of photosystem II. *Planta* 106, 101–12.

Stuart, T. S., E. W. Herold, Jr., and H. Gaffron. 1972. A simple combination mass spectrometer inlet and oxygen electrode chamber for sampling gases dissolved in liquids. *Anal. Biochem.* 46, 91–100.

Stuart, T. S., and H. Kaltwasser. 1970. Photoproduction of hydrogen by photosystem I of *Scenedesmus*. *Planta* 91, 302–13.

Ward, H. B., M. E. Reeves, and E. Greenbaum. 1985. Stress-selected *Chlamydomonas reinhardtii* for photoproduction of hydrogen. *Biotech. Bioeng. Symp.* 15, 501–7.

Weaver, P. F., S. Lien, and M. Seibert. 1980. Photobiological production of hydrogen. *Solar Energy* 24, 3–45.

Wrighton, M. S. 1979. Photochemistry. *Chem. Eng. News* 57, 29–47.

13. Products from fossil algae

RICHARD W. GEORGE

Manville Corporation, P.O. Box 287, La Grande, OR 97850

CONTENTS

I. Introduction

Fossil algae are used in the manufacture of a multitude of products utilized by industrial society. Product diversity may be exceeded only by the sheer numbers of algae themselves. An estimated 10,000 or more fossil and recent diatom species have been described (Schrader and Schuette 1981). Certain Cretaceous chalks in Kansas may contain 500,000 to 6 million coccoliths per mm^3 (Fig. 13-1), and some less detrital rocks are estimated to hold 5.5 million discoasters (an extinct taxon of calcareous nannoplankton) per mm^3 (Tappan 1980).

The intricately perforated frustules of diatoms (Fig. 13-2) are used to filter beer, swimming pool water, and municipal water supplies and may soon aid in the breakdown of toxic wastes. They are used in products ranging from cigars to paints and are added to embalming compounds and dental polish. Limestone, a thick accumulation of calcareous algae (Johnson 1961), has been used as construction stone for thousands of years. These calcareous rocks are important in the manufacture of cement, metallurgical products, plastics, glass, and chemicals. Countries suffering from the acknowledged effects of acid deposition use large quantities of limestone to combat atmospheric and aquatic acidity.

The fossil record of algae may extend beyond 3 billion years (deWit et al. 1982). Algae are credited with the Precambrian conversion of a reducing atmosphere to an oxygenic one (Tappan 1968; see also Harlin and Darley, Chapter 1). Thought to be linked to oceanic transgressions and regressions, phytoplankton population explosions and die-offs resulted in widespread accumulations of siliceous, calcareous, and organic algal remains (Tappan 1968; Tappan and Loeblich 1970; Tissot 1984). Upwelling may have been important in maintaining the continuous supply of nutrients necessary for prolonged population maximums (Calvert 1966a, b).

Many fossil lacustrine algal deposits are tied to past pluvial periods. These epochs of increased precipitation seem to have coincided with periods of volcanic maximums. Injection of volcanic ash into the stratosphere, through explosive eruption, caused a decrease in the earth's surface temperature, which in turn influenced the earth's climate (Axel-

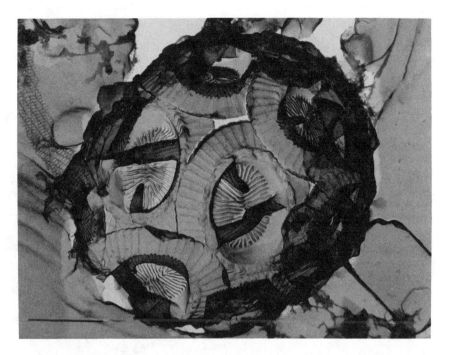

Fig. 13-1. SEM of the Atlantic coccolithophorid *Gephyrocapsa oceanica*. Each circular shield is a coccolith composed of the carbonate mineral calcite. Sedimentation usually separates the coccoliths from the coccolithophorid. Scale = 10 μm. (Courtesy of A. McIntyre-Lamont-Doherty Geologic Observatory.)

rod 1981). Increased precipitation coincident with lowered evaporation rates may have resulted in the formation of vast inland seas where algae could proliferate.

The following discussion separates products from fossil algae into (1) diatomite products, (2) calcareous algal products, and (3) hydrocarbon products. Diatomite is composed almost exclusively of diatom frustules (Bacillariophyceae). Other algae, including members of the blue-green algae (Cyanophyta), red algae (Rhodophyta), and coccolithophorids and related groups (Prymnesiophyta), may have been deposited as thick layers of calcareous ooze. Subsequent alteration (diagenesis) including compaction, recrystallization, and cementation may have transformed the ooze into the carbonate rock, limestone. Hydrocarbons may have formed from a variety of algal taxa. Other organisms, including bacteria (Williams 1984), zooplankton, and terrestrial flora and fauna, also may have been important contributors.

Additional applications of fossil algae deserve mention but will not be covered in this review. Fossil algal deposits may host (Zielinski 1982) or

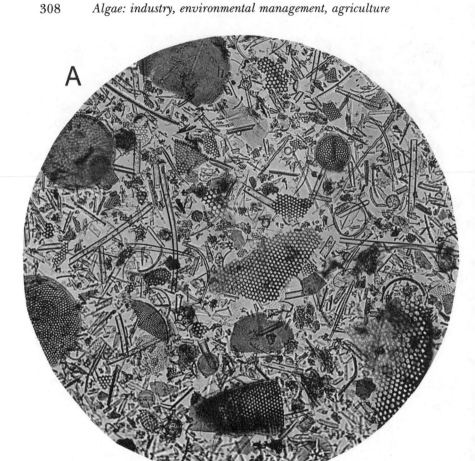

Fig. 13-2. Light micrographs of Lompoc, California, marine diatomite (A) and Soysambo, Kenya, lacustrine diatomite (B). Note the diversity of frustule structure and size and the distinction between marine and lacustrine diatom assemblages. Scale = 50 μm. (Courtesy of Manville Corporation.)

play a role in the formation of (Oehler and Logan 1977; deWit et al. 1982; Longheed 1983) precious and base metal deposits. Oxygen isotope ratios of diatom frustules are used to estimate aquatic paleotemperatures (Labeyrie 1974). Fossil algae yield information about evolution of the biota (Tappan 1980), past oceanic circulation, and water depth and chemistry (Lohman 1957; Bradbury 1975; Williams et al. 1979; Haq 1980). Diatom frustules have even been used to determine the provenance of pottery vessels and other prehistoric artworks (Matiskainen and Alhonen 1984).

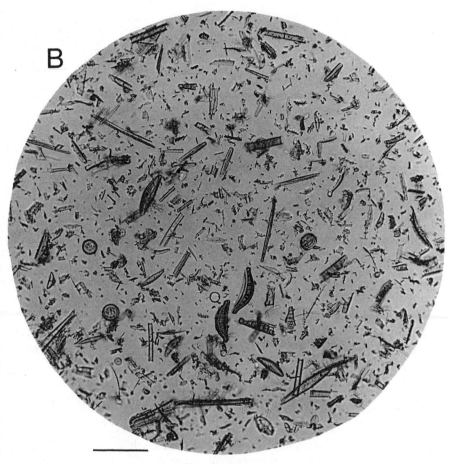

Fig. 13-2. *(cont.)*

II. Diatomite – industrial mineral

Diatomite, or diatomaceous earth, is a sedimentary rock formed primarily of the fossilized frustules of diatoms. Diatomite is the accepted term for those deposits that have potential for commercial exploitation. Certain specific criteria, including diatom assemblage, purity, and deposit size and minability, determine this potential. Accumulations of the microscopic siliceous frustules may reach thicknesses in excess of 150 m, and both marine and lacustrine deposits have potential for commerical production.

Diatomaceous silica, the mineral of which the diatom frustule is composed, is distinct from other silica minerals primarily because it is biogenic. Intricate frustule structure and diversity of form result in unique physical properties. Combined, chemistry and physical structure result in product characteristics unparalleled for many industrial applications. Classified as an industrial mineral, diatomite has a major use as filter aids, fillers, and absorbents.

Though diatomite use in construction materials dates back to 2,000 years before the present (B.P.), commercial diatomite production originated in Germany in the mid-1850s. Initial use was as a polish abrasive. Its insulative efficiency and filtration capability were recognized shortly thereafter in both Great Britain and Germany (Cummins 1975). Initial production in the United States in the late 1800s was for metal polish and, when small blocks were soaked in petroleum, for firelighters (Cummins 1973).

Diatomite is produced in over 30 countries (Meisinger 1985a). World production of diatomite in 1984 was about 1.7 million short tons (st) (Meisinger 1985b). The United States accounted for 38% of world production in 1984 with the USSR, France, and Denmark responsible for most of the remainder (Meisinger 1985b). Domestic production is estimated to be about 650,000 st in 1985 (Bureau of Mines 1985). The majority of domestic diatomite production comes from a large marine deposit in California, three lacustrine deposits in Nevada, and a lacustrine deposit in Washington. Two additional lacustrine deposits, one in northern California and the other in eastern Oregon, were brought on line in 1986.

A. Unique frustule structure

Industrial applications for diatomite take advantage of the rock's inherent physical and chemical properties. Low density, high porosity, large surface area, and low abrasion are a direct result of the structure of the diatom frustule.

The structures and perforations of the frustule are equally valuable in filtration and filler applications. Perforations through the basal siliceous layer, such as the areola and velum (Anonymous 1975), play an important role in many industrial uses. The smallest pores may be less than 1 μm in diameter, and up to 10–30% of the frustule surface area may be perforated (Tappan 1980). The combined volume of interfrustule and intrafrustule void space can cover up to 85–90% of the product surface area. In a diatomite filter aid these voids function to sieve particles, whereas in a filler they provide absorptive properties and large surface area.

Chemical properties can be as important as the physical characteristics of the diatom frustule. X-ray diffraction reveals that good commercial diatomite is composed primarily of amorphous or microamorphous hydrous silica: $SiO_2 \cdot nH_2O$ (Cummins 1960). Diatomaceous silica, containing traces of iron, alumina, and other minor constituents, is relatively inert and has a melting point around 1,400°C. Most commerical diatomites contain a minimum of nondiatomaceous material such as sand, ash, clay, and carbonate. They generally range from 86% to 94% silica and contain 1.5–5% alumina, 0.2–2% iron, and small amounts of other elements (Kadey 1983). High brightness, high resistivity, low compaction, and particle integrity are important characteristics of good natural earths.

B. Mining and processing

Current domestic diatomite mining is by the open pit method. Although open pit mining is dominant, some underground mining occurs in European and other countries (Meisinger 1985a). Open pit mining utilizes bulldozers, belt loaders, front-end loaders, power shovels, and trucks to remove and haul crude ore. The ore is segregated and stockpiled, allowing for subsequent blending of crude types. Figure 13-3 illustrates the unique process at Manville Corporation's Lompoc, California, plant, where 13 subsurface storage facilities, or "glory holes," feed an underground rail system that delivers crude to the mill.

Ore is moved through the mill via pneumatic processes. After gentle hammermill crushing the ore is milled and dried. It is passed in a stream of hot gases through cyclones, classifiers, and separators, resulting in further milling and classification. The fine diatom particles are collected in the bag house, and the coarser fractions are collected in the cyclone. A reduction in moisture content from as high as 60% to less than 5% also results. At this point the product may be bagged and marketed as "natural powders" (Fig. 13.4A).

Further processing involves calcining the natural powders in a rotary kiln. This results in partial removal of the thinner siliceous structures such as the velum and in agglomeration of fines. Increased particle size and frustule pore volume, which affect density, flow rate, and other properties, can thus be achieved. The calcined material is then classified in cyclones and separators, as above. For many products a flux, usually soda ash, is added prior to calcination. Additional adjustment of particle size and pore volume results (Fig. 13-4B).

Cyclones and classifiers separate all products into various-sized fractions, from fine to coarse, based on particle size, shape, and density. The unique properties of diatomite are not created by milling and processing. Rather, these inherent characteristics are selectively adjusted to maximize their utility in industrial product applications.

C. Filter aids

Filtration is the process of separating particles from a fluid as it passes through a permeable material. The photomicrographs in Figure 13-4

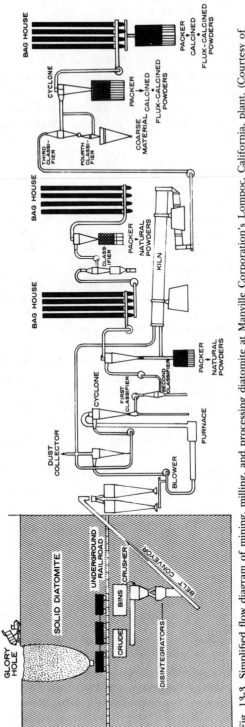

Fig. 13-3. Simplified flow diagram of mining, milling, and processing diatomite at Manville Corporation's Lompoc, California, plant. (Courtesy of Manville Corporation.)

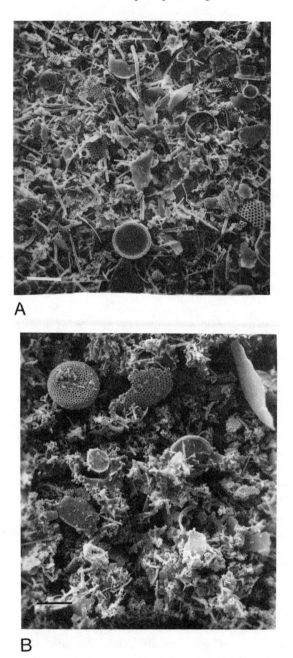

A

B

Fig. 13-4. SEM of "natural powder" filter aid product (A) and flux-calcined product (B). Note particle agglomeration in B. Scale = 50 μm. (Courtesy of Manville Corporation.)

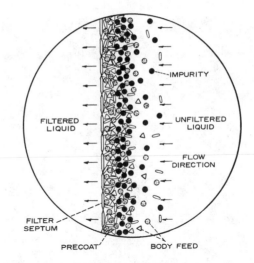

Fig. 13-5. Schematic representation of precoat and body-feed application in standard vacuum filter process.

illustrate the diverse void and particle structure of diatomite, revealing its suitability for filtration. Diatomite was first used in Great Britain in 1876 to filter sewage (Cummins 1975). Large-scale industrial use in the United States began in 1914 when diatomite was judged superior for filtration of cane sugar liquors (Cummins 1973). The development of calcined filter aids and of applicable filters further propelled diatomite into industrial markets. Fossil algal filter products have replaced many of the old sand filters for water and bag filters for sugar. At present, filter aids account for 67% of total annual domestic production (Meisinger 1985b).

Filtration begins with the application of a diatomite precoat to the filter septum by circulating clear liquid and filter aid under pressure or vacuum. Unfiltered liquid is then pumped through the precoated filter, and small amounts of filter aid (body feed) are periodically added. As shown in Figure 13-5, body feed results in the continuous formation of a relatively clean filtering surface. For liquids with very slimy solids, or a high volume of solids, the rotary vacuum precoat filter (RVPF) was developed (Smith 1976). A thick precoat is formed under vacuum on the RVPF, which is then partially submerged in the liquid to be filtered. As the drum rotates and filtration proceeds, a doctoring blade slices away a thin layer of precoat and filtered solids. The blade slowly advances toward the drum, continuously exposing a clean filter aid surface (Fig. 13-6).

Natural powders, with mean pore sizes of about 2 μm, pass fluids much more slowly than flux-calcined powders, which have mean pore

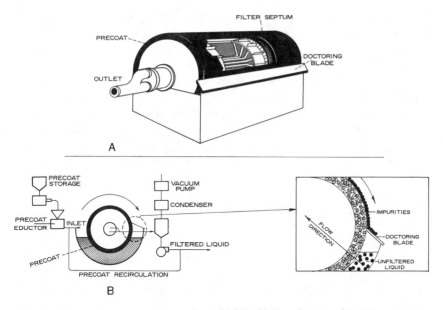

Fig. 13-6. (A) Rotary vacuum precoat filter (RVPF). (B) Flow diagram of RVPF operation. Enlarged view shows doctoring blade removing thin layer of impurities and diatomite.

sizes of up to 20 μm. These products are used in such familiar applications as swimming pool and municipal water supply filtration. Much larger quantities, however, are used in the filtration of beer, wine, sewage, fruit juices, sugar syrups, and animal fats and oils. In a brewery, fossil diatoms may filter raw water, hot or cold wort, and surplus yeast from culture. They also provide initial clarity and final "polish" to beer prior to packaging. Prices for these filter aids average US $200 to $300 per st.

Diatomite filter aids generally exhibit a negative surface charge as do most liquids to be filtered. Synthesis of calcium and magnesium silicates from diatomaceous silica reverses surface charge, adding adsorptive properties to filter aids. Gas chromatography columns, laboratory filtrations, and production of some pharmaceutical products may require such powders. Due to special treatment and handling, cost may be as much as US $150 per kg.

D. Mineral fillers

A mineral filler is an inert material added to a compound to produce a desired effect (Severinghaus 1983). Diatomite is added to embalming compounds as partial replacement for a more expensive ingredient and to paint for control of gloss and sheen. The latter is termed a functional

filler application because addition of diatomite produces a specific physical effect. Alfred Nobel used the first commercial diatomite filler in
1866 when he discovered that it stabilized nitroglycerine in the production of dynamite (Cummins 1975).

Again, it is the physical and chemical structure of the diatom frustule
that is responsible for diatomite's unique filler properties. High particle
and powder porosity imparts tremendous absorptive capability and
large surface area. Diatomite fillers can absorb 1.5 times their weight
and still maintain dry powder properties. High bulking value is possible
without significant increases in weight. Relative inertness, brightness,
refractive index, loose weight, and particle size are important filler
properties.

Most of the dollar return on diatomite fillers comes from its use in
the paint industry. The military stimulated this application during
World War II when it began using large quantities in the production of
camouflage paints and other low-luster coatings (Cummins 1973).
When added to paints and varnishes, the frustule fragments roughen
the coated surface on a microscopic level, creating a flattening effect by
diffusing reflected light. Many flat, semigloss, and other low-luster coatings contain diatomite powders that have been carefully classified for
particle-size distribution. Special calcium silicate powders additionally
function as pigment extenders.

The paper and paperboard industries use diatom frustules to control
pitch particles, increase bulk, and improve finish and opacity of the finished product. Diatomite raises the melting point and increases the durability of asphalt and is a processing aid for rubber goods. By protrud-
ing from blown polyethylene film, the microscopic frustule particles
prevent the surfaces from touching and sticking while still hot.

Diatomaceous silica is hard enough to scratch metal surfaces. Yet the
diatom frustules are so delicate that they break under pressure, resulting in a highly effective polishing capability. This abrasive property has
been utilized for centuries in metal polish compounds and more
recently in toothpastes, automobile polishes, and glass cleaners.

In matchheads, cigars, and acetylene tanks diatomite controls burn
and friction and is an anticaking agent in fertilizers and in the production of dry food products. Diatomaceous calcium silicates, capable of
absorbing three to four times their weight, increase the safety of handling hazardous liquids and serve as catalyst carriers. Lower product
grades are used in grease and oil absorbents and in the manufacture of
pet litter.

Most filler powders represent the baghouse fraction of milling which,
until the 1930s and 1940s, was generally regarded as waste and dis-
posed. Product costs range from less than US $50 per st for low-grade

absorbents to nearly $1,000 per st for premium functional filler products.

E. Biogenic silica and miscellaneous products

Diatomite is used in several industrial applications not easily categorized as filters or fillers. The production of sawed and natural diatomite bricks for use in construction and insulation was popular in the United States prior to 1950. Construction bricks are still popular in some European markets and probably in the USSR (Cummins 1975).

Diatomaceous silica reacts readily with a base to form silicate compounds. Thus, as a pozzolan (a functional cement additive), the hydrous SiO_2 reacts with the alkaline hydrolytic components of cement. This application was recognized in Denmark in the late 1890s as a means of increasing the durability of cements exposed to seawater (Cummins 1975). As mentioned earlier, diatomaceous silica is used in production of synthetic calcium and magnesium silicates. Initiated by Manville Corporation (then Johns-Manville Corporation) in 1956, the process involves hydrothermal reaction of the silica with lime or magnesia bases. Increased sorptive properties and adjusted chemical reactivity are the desired end results.

Möhler earth, an *in situ* mixture of marine diatomite and up to 30% clays, is produced in Denmark as a filler, insulation ingredient, pozzolan, and for various other applications. Over 125,000 st of Möhler (*Mo* from the Danish Island of Mörs and *ler* meaning "clay") is produced annually (Meisinger 1985b).

A new and intriguing application for diatomite as a "biocatalyst carrier" was recently introduced by Manville Corporation. Diatomite's complex void structures provide a microenvironment for living cells or enzymes (Fig. 13-7). Median pore diameter from as low as 200 nm to about 25 μm, is tailor-made to "fit" the cell or enzyme. Potential uses include the production of food products requiring enzymatic reactions or biosynthesis. Microbial breakdown of toxic wastes such as polychlorinated phenol (PCP) and biphenol (PCB) may be possible as the waste is passed through a diatomite–microbe medium. Reduction in toxic concentrations from 100 parts per million (ppm) to less than 12 ppm may be feasible.

III. Calcareous algae

Calcareous algae are important builders of marine and lacustrine carbonate rocks. Calcareous nannoplankton, including the asteroliths, coccolithophorids and other haptophytes (Tappan 1980) and certain green, red, and blue-green algae (Johnson 1961) are recognized as major con-

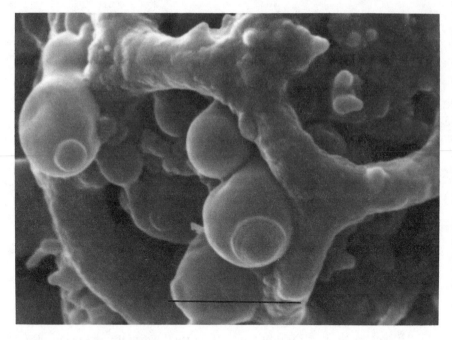

Fig. 13-7. SEM of yeast cells immobilized on diatomite "biocatalyst carrier." Scale = 5 μm. (Courtesy of Manville Corporation.)

tributors to carbonate deposition. Other organisms, including certain members of the Foraminiferida, Crustacea, and Mollusca also deposit carbonate. Calcareous algae, however, were the major biotic carbonate contributors to many carbonate rock deposits (Bathurst 1975; Haq 1978; Tappan 1980). Their role in carbonate rock formation has been increasingly recognized primarily through compositional studies of recent calcareous sediments (Carr, personal communication). It should also be noted that not all limestones are biogenic. Some are inorganically precipitated.

The coccolithophorids contribute rhombohedral and hexagonal calcite in the form of ornate shields called coccoliths whereas the asteroliths (discoasters – extinct since late Pliocene) deposited single tabular crystals in tubular form (Haq 1978). Various calcitic and aragonitic crystals are formed by members of the Chrysophyta, Rhodophyta, and Cyanophyta. Diagenesis of calcareous sediments may result in formation of limestone, composed primarily of calcite ($CaCO_3$). Subsequent alteration by hypersaline brines may then convert the limestone to dolomite ($CaCO_3 \cdot MgCO_3$) (Carr and Rooney 1983).

Ancient Egyptians used limestone to construct the great pyramids from 6,000 to 4,000 years B.P. (Power 1985). Presently limestones, dolo-

Table 13-1. *Limestone and dolomite sold or used by producers in the United States in 1983*

Use/type	Quantity[a] (1,000 st)	Value (US $1,000)
Concrete aggregate	93,406	364,863
Bituminous aggregate	72,008	312,790
Road stone and coverings	152,889	489,020
Riprap and railroad ballast	20,759	71,119
Other construction uses	34,288	122,520
Cement manufacture	78,604	233,017
Agricultural uses	20,914	108,941
Lime manufacture	22,422	91,340
Other uses	140,974	565,247
Total	624,500	2,329,000
Limestone (dimension stone)	245.084	19,122
Marble (dimension stone)	30.753	18,602

Note: Totals may not add up due to independent rounding.
[a]Crushed and broken.
Source: From Taylor 1984; Tepordei 1985.

mites, and lime are used worldwide in a myriad of industrial processes and applications (Table 13-1). An estimated 2.9 billion st of crushed rock, mostly carbonate rock, were produced worldwide in 1983 (Tepordei 1985). Approximately 630 million st of carbonate rock were produced domestically in 1983 comprising about 75% of all the stone quarried or mined in the United States (Taylor 1984; Tepordei 1985). It has been stated that lime (CaO) may be the most utilized of any chemical or mineral commodity (Boynton and Gutschick 1975).

A. Limestone and dolomite – industrial rocks

Carbonate rocks are found on all continents but vary substantially in chemical and physical characteristics. Though limestone and dolomite can be used interchangeably, their unique chemical compositions allow each to be utilized for specific applications. Mining methods include both underground and the more prevalent open pit method. Blasting is used to fracture the hard rock, and the pit is usually benched to facilitate mining and hauling (Carr and Rooney 1983).

Depositional environment determines deposit size, shape, and purity (Carr and Rooney 1983). A dense reefal limestone may produce excellent aggregate products, whereas a poorly cemented coccolith chalk may exhibit high reactivity and good absorptive properties, important in chemical use (Power 1985). Impurities such as clay, chert, and organics are detrimental for certain chemical uses yet, by adding color and texture, may be important for construction stone (Carr and Rooney 1983).

Limestone and dolomite are used for rough and dressed dimension and other building stone and for sculptural material. This requires rock of high physical strength and often includes the metamorphic variety marble. When crushed to small particle size, the whiter varieties are used as roofing granules and exterior housing ornamentation (Power 1985).

Large quantities of crushed or broken carbonate rock are used for aggregate, riprap, railroad ballast, drain fields, and fill (Table 13-1). Impact toughness, apparent specific gravity, and compressive strength are important properties (Carr and Rooney 1983). Low porosity and relative chemical purity contribute to overall load strength when carbonate rock is used as a concrete or roadbed aggregate. Limestone's use in seawall defenses has been important for some time. A large market has historically been provided by use as railroad ballast (Power 1985).

Pulverized limestone is used as a nutritional additive in animal feedstuffs and in some flours for human consumption. Calcium carbonate will see increased utilization as a neutralizer to combat the environmental degradation caused by acid rain. Currently it is sprayed in powder form in lakes and streams to raise pH values. Used for "scrubbing," carbonate lowers the acidity of industrial plant stack effluents (Power 1985). This use may require 49 million st annually by 1990 (Ozol 1984).

Calcium carbonate powders are economically important as industrial fillers. These products often require high brightness and relative chemical purity and are important in the plastics and rubber industries and in paper manufacture. In the paint industry, hydrophilic properties make carbonate powders especially appropriate for water-base paints (Smith 1984).

Carbonate fillers are also used in ceramics, fertilizers, and polishes (Power 1985). Filler product costs range from $14 per st to over $160 per st in the United States. Estimates of the overall domestic filler market in 1982 totaled 1.975 million st of carbonate valued at US $135 million (Smith 1984).

B. Lime – industrial mineral

Large amounts of carbonate rock are utilized for the production of lime and cement. The USSR is the world's leading producer of lime (Pressler and Pelham 1985). In 1984 about 12% of the crushed rock produced in the United States was used in lime and cement manufacture (Bureau of Mines 1985). In 1984 the United States produced about 16.1 million st of lime valued at an estimated US $853 million (Pressler and Pelham

1985). Lime is now utilized mainly by the chemical and metallurgical industries, reversing the historically construction-dominated market (Boynton et al. 1983).

Lime (CaO) is the general term for calcined limestone or dolomite, commonly known as quicklime or, when hydrated, as slaked lime. Calcination of these fossil algae results in a product with distinct chemical properties. Of utmost importance is the chemical purity of lime: its CaO and MgO content and the amount of such impurities as silica, alumina, and iron. Additional properties such as particle size, slaking rate, and plasticity are important for specific applications (Boynton et al. 1983).

The production of lime involves preparation of the stone, calcination, and sometimes hydration. Calcination involves heating the material in a kiln to about 1,000–1,300°C for the reaction:

$$CaCO_3 + heat \rightleftharpoons CaO \text{ (quicklime)} + CO_2\uparrow.$$

Since quicklime is highly reactive (reaction is reversible) it is often combined with water:

$$CaO + H_2O \rightleftharpoons Ca(OH)_2$$

to form the more stable hydrated lime. The original oxide may be regained by heating (Boynton et al. 1983). In the production of cement, carbonate calcination is accomplished in the presence of silica, alumina, and iron sources. This results in formation of a "clinker" composed of complex calcium silicates and aluminates, which is then ground to produce cement (Ames and Cutcliffe 1983). Coccolith chalks are traditionally used in the United Kingdom for cement manufacture (Power 1985).

A substantial tonnage of lime is used by the metallurgical industry in the production of steel, copper, aluminum, gold, silver, and uranium. Over 80% of domestic production is used by the steel industry where lime acts to flux out such impurities as alumina and silica. This may require 100–200 pounds of lime per ton of steel produced (Boynton and Gutschick 1975). Lime also plays a major role in the production of pig iron, where its calcium and magnesium oxides combine with silica and alumina to form a slag. Sulfur is removed by the slag as it floats to the top of the molten iron (Power 1985).

Lime is utilized in some pollution abatement processes, including mine waste neutralization, pH control in sewage treatment, and removal of acidic gases from stack effluents. A most valuable use is for "scrubbing" or flue gas disulfurization of coal-fired electric utility generating stations. Lime removes sulfur dioxide and other acidic constituents from flue gases, a process that could consume in excess of 3.5 million tons annually by 1990 (Boynton et al. 1983).

In the building products and ceramics industries, lime is used in the manufacture of calcium silicate materials and glass, respectively. Bottle glass plants use it extensively as a flux raw material. Large tonnages are used annually by the construction industries for road base stabilization, lime cement mortars, asphalt, and for production of cement (Boynton et al. 1983).

Fossil calcareous algal products are used in conjunction with products from their siliceous counterparts, the diatoms, in several industrial applications. Where diatomite filter aids clarify municipal water sources, lime often softens the water and, by raising pH, kills bacteria. About 1,400 pounds of quicklime are used to produce a ton of soda ash which, as discussed previously, is used as flux during calcination of diatomite. Diatomite increases quality control in the pulp and paper industry, and lime serves to recycle sodium carbonate for the regeneration of sodium hydroxide (Cummins 1973; Boynton et al. 1983). In the sugar industry, quicklime is essential for precipitation of phosphatic and organic acid compounds from sucrose juices. These compounds are then removed by diatomite filtration (Elsenbast and Morris 1942).

Recently, lime production seems to be experiencing an upswing initiated by the chemical and metallurgical industries. This trend is led primarily by steel industry applications, but increasing environmental needs, principally for "scrubbing," should also play a key role (Boynton et al. 1983).

IV. Algal organics and biostratigraphy

Previous discussion focused on the accumulation of the inorganic "skeletal" parts of fossil algae. The organic portion of the cell, including carbohydrates, lipids, and proteins, also may be preserved. Recent investigations show phytoplankton to be important contributors to fossil fuel accumulations (Hunt 1979; Tissot and Welte 1984). This conclusion is supported by the large fraction of total biomass represented by algae in most aquatic ecosystems (Hunt 1979; Tappan 1980; Tissot 1984; Tissot and Welte 1984).

Some hydrocarbon deposits, such as boghead coal, are formed almost entirely of a single algal type (Traverse 1955). Certain petroleum reservoirs and many oil shales contain cell remnants or molecular geochemical fossils (biological markers) that point to at least a partial algal origin (Tissot and Welte 1984). Geochemical fossils can now be isolated from hydrocarbons and can often be linked directly to source rock organic compounds. These molecular fossils were derived from precursor compounds and serve to represent and document the successive steps in the conversion of cellular compounds to petroleum (Welte 1965; Tissot 1984; Tissot and Welte 1984).

The Green River oil shales of Wyoming and Utah (Bradley 1931; Hunt 1979), the Brazilian Irati oil shale (Hunt 1979), the Kimmeridge clay oil shales of England (Gallois 1976) and the Hungarian Várpalota oil shales (Solti 1980) all show contributions from algae. The Sheep Pass formation, a suspected source rock for the Nevada Eagle Springs oil field, is described by Bortz and Murray (1979) as a "bioclastic, algal and lithographic lacustrine limestone." Gallois (1976) suggested that much of the North Sea oil may be attributed to organics provided by cocco-lithophorid blooms.

Hydrocarbon accumulations are found from the surface to depths of about 10 km and range in age from very recent to about 700 million years old (Melton and Giardini 1984). Approximately 77 billion st of oil were produced worldwide through 1984, and estimates of total world resources (including past production) range from 288 billion to 462 billion st (Tissot and Welte 1984; McMillan, personal communication). Production from oil shale has been inconsistent except in China and the USSR (Tissot and Welte 1984). An estimated 7,000 chemical products are manufactured using petroleum, but most oil production is consumed by the energy market (Hunt 1979).

The oil industry often utilizes fossil algae in the pursuit of additional resources. Certain calcareous nannoplankton and diatoms are used to date and correlate sedimentary strata. Biostratigraphy employs fossils or fossil assemblages to define discrete units of time within a sedimentary rock group (Berggren 1978).

A. Algal kerogen in petroleum and coal

Kerogen is a complex hydrocarbon formed from simpler organic molecules of living organisms. Cellular lipid compounds are probably the most important components (Tissot and Welte 1984). Average phytoplankton lipid content ranges from 4% to 28% on a dry weight basis (Hunt 1979). Under environmental stress, some algae may produce up to 90% lipids (Cane and Albion 1971; see also Spencer, Chapter 10), and diatoms may produce up to 70% lipids (Tissot and Welte 1984) on a dry weight basis. Chlorophylls alter to form the isoprenoids phytane and pristane, commonly found in petroleum and organic-rich sediments (Hunt 1979).

Conversion of cell material to kerogen and then petroleum is visualized as a series of maturation steps (Melton and Giardini 1984) (Fig. 13-8). Organics are first preserved in the sediments under anaerobic conditions and with bacterial action (Tissot 1984). As sedimentation proceeds and organics are deeply buried, diagenetic processes cause significant geochemical alteration. This includes polycondensation and insolubilization of the biologic polymers and results in formation of increasingly insoluble, complex hydrocarbons such as humin and kerogen. (Hunt 1979; Tissot and Welte 1984). Biogenic methane may also be produced (Melton and Giardini 1984), and lignites and subbitumi-

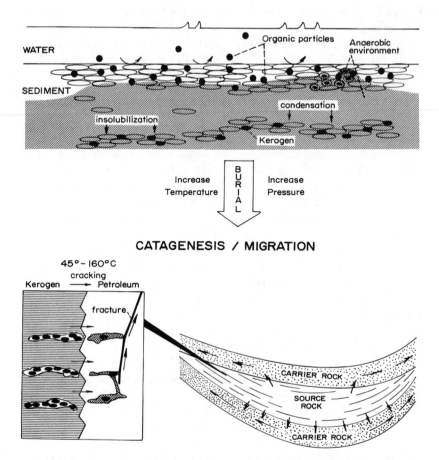

SEDIMENTATION / DIAGENESIS

WATER

Organic particles

Anaerobic environment

SEDIMENT

insolubilization

condensation

Kerogen

Increase Temperature

B U R I A L

Increase Pressure

CATAGENESIS / MIGRATION

45° – 160°C
cracking
Kerogen ⟶ Petroleum

fracture

CARRIER ROCK

SOURCE ROCK

CARRIER ROCK

Fig. 13-8. Physical evolution of petroleum from algal organics. Conversion of cell material to kerogen *(top)* probably involves anaerobic conditions and bacterial action. Increased temperature and pressure from burial convert kerogen to petroleum *(bottom)*. Migration through source and carrier rock may then form a petroleum reservoir. (Modified after Tissot and Welte 1984.)

nous coals may form from organic-rich sediments (Tissot and Welte 1984). Catagenesis, the second maturation step, results in severe changes in the organic compounds. High temperature and pressure, resulting from deep burial, cracks the kerogen to produce liquid petroleum (Tissot and Welte 1984). Hydrocarbons are thus formed from the lipids, proteins, and carbohydrates of once living algal cells.

Generally, three categories of kerogen are recognized. Type I is formed primarily of cellular lipid material, is often amorphous, and shows a direct link to algal organics. Some of the rich oil shales, such as the Green River shales, and the algal coals are examples. Type II kerogen consists of autochthonous (of local origin) organics and may have an algal contribution or origin. Terrestrial plant materials generally make up Type III kerogen (Tissot 1984). Kerogen is thought to be a natural precursor to in situ petroleum (Hunt 1979).

Most petroleum is not found in the sediments where it formed. Rather, it has migrated from these source beds (primary migration) through highly porous carrier beds (secondary migration) until trapped by an impermeable stratum (Welte 1965) (Fig. 13-8). Strata that host petroleum pools, mostly sandstones and carbonates, are termed *reservoir rocks* (Hunt 1979).

Exceptions to the above processes include the formation of oil shales and algal coals. Oil shales are shallow, fine-grained sediments that contain no oil and little bitumen. Type I and Type II kerogens are the dominant hydrocarbons. To extract oil, the shales must be heated to about 500°C, cracking the kerogen. This process, termed pyrolysis, is analogous to thermochemical conversion of kerogen during catagenesis (Tissot and Welte 1984). Most oil shales show algal contributions to kerogen formation (Gallois 1976; Hetényi and Sirokmán 1978; Hunt 1979; Solti 1980; Tissot and Welte 1984).

Certain lacustrine and brackish-water algae may produce rich hydrocarbon deposits termed *boghead coals.* (Tappan 1980). Most of these consist almost entirely of the chlorophyte *Botryococcus* sp., a form similar or identical to the extant *B. braunii* (Traverse 1955). Environmental stress has been shown to induce production in *Botryococcus* of up to 76% lipids (Hunt 1979) and possibly up to 90% (Cane and Albion 1971; Spencer, Chapter 10) on a dry weight basis.

The prolific occurrence of *Botryococcus* sp. cells in lacustrine coal deposits led to one of the first algae–hydrocarbon correlations in the late 1800s (Traverse 1955). *Botryococcus* is known from the Precambrian (Tappan 1980). It is found in many Cenozoic and older rocks (Tissot and Welte 1984) and is abundant in Pleistocene sediments in Europe, Australia, and the United States (Vimal 1953; Traverse 1955; Nagappa 1957; Rao 1957; Ramanujam 1966).

B. Biostratigraphy

Fossil algae are used increasingly in the field of micropaleontology to date and correlate sedimentary strata. The process is termed *biostratigraphy,* the biochronologic description, correlation, and division of strata (Berggren 1978). Biostratigraphy is based on the precept that species evolve and become extinct during specific time periods and can therefore be used to subdivide strata (Burckle 1978). By utilizing taxa with geologically short life spans (Fig. 13-9, 10), paleoecologic events are correlated with dated rocks to define an age span. Subsequent fossil identification allows a chronologic age to be applied to geographically removed strata. Wide geographic distribution of many biostratigraphic planktonic algae allows worldwide correlation of sediments (Wornardt 1969).

Biostratigraphy originally utilized the foraminiferal faunas (see Berggren 1978, pp. 2–5). However, calcareous nannoplankton including discoasters, coccolithophorids and related forms, and diatoms are currently some of the most useful biostratigraphic microfossils (Tappan 1980). Once such a fossil is identified, its upper and lower chronologic range is determined and then correlated to age dates obtained by such methods as radiometric paleomagnetic and K-Ar dating (Berggren 1978). Biozones, or groups of strata representing discrete units of time, are constructed. Zones are defined and correlated based on a distinctive natural assemblage or on a unique member of that assemblage (Berggren 1978). Exemplifying the latter, the marine diatom *Annellus californicus* is described by Wornardt (1979) as a "dream fossil." With a lifespan of only about 1 million years, *A. californicus* has been used to correlate and date early-middle Miocene strata throughout the Circum-Pacific (Wornardt 1979).

Nonmarine diatoms have significant biostratigraphic potential, but a lack of research data, thin strata, and other problems have hindered their utilization. Lohman (1961) showed that applicable nonmarine fossils with appropriately short age spans exist and can be dated. *Actinocyclus* and *Stephanodiscus* are the primary freshwater biostratigraphic genera. Strata may be dated using radiometric methods on the volcanic rocks that are often associated with lacustrine diatomites (Krebs and Bradbury 1984).

Mesozoic and Cenozoic zones are the most reliable to detect using microfossil biostratigraphy. Accuracy is about ±1 million years for late Cretaceous and Tertiary sediments, exceeding that of radioactive decay methods (Berggren 1978). Biostratigraphy is used extensively within the oil industry in the exploration for, and correlation of, petroleum source and reservoir rocks. It is also used to a lesser extent in the exploration for metallic and nonmetallic nonfuel minerals (Krebs and Bradbury 1984).

V. Conclusions

Diatoms, the most intensively utilized fossil algae, have penetrated a tremendous number of industrial and household product markets. The highly competitive diatomite industry requires producers to maintain innovative research and development programs and marketing schemes.

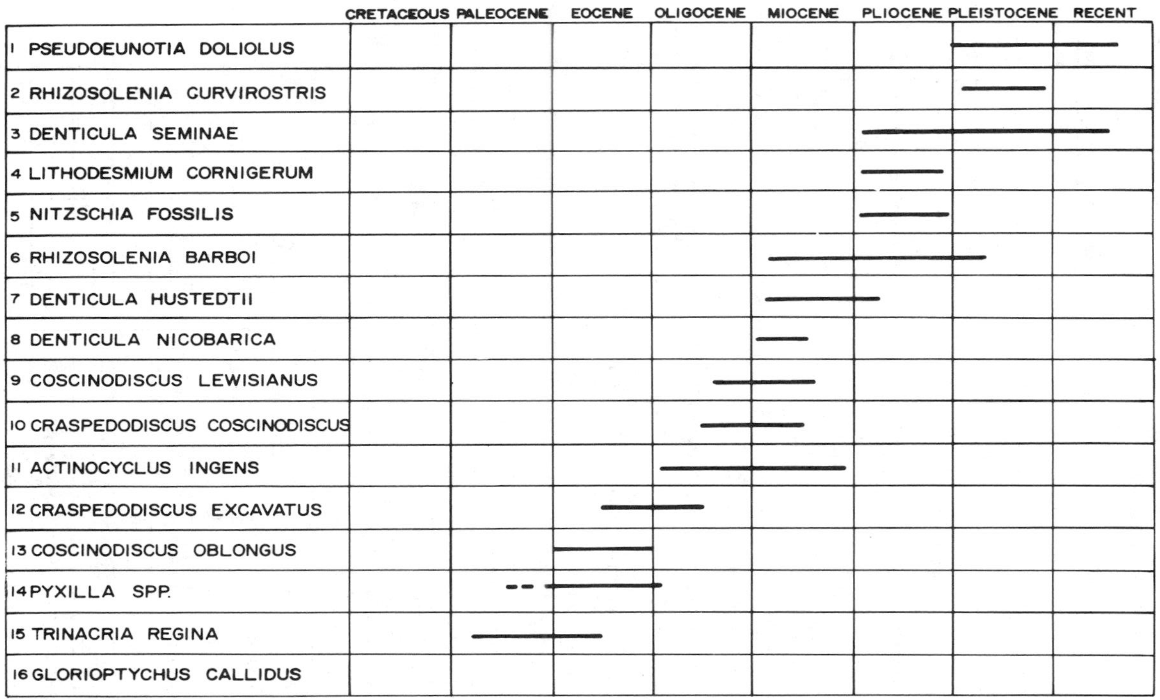

Fig. 13-9. Coccolith zonation for period from about 23 million years B.P. to 0.0 million years B.P. Bars represent lifespan (range) of species. (Reprinted by permission from Berger 1976.)

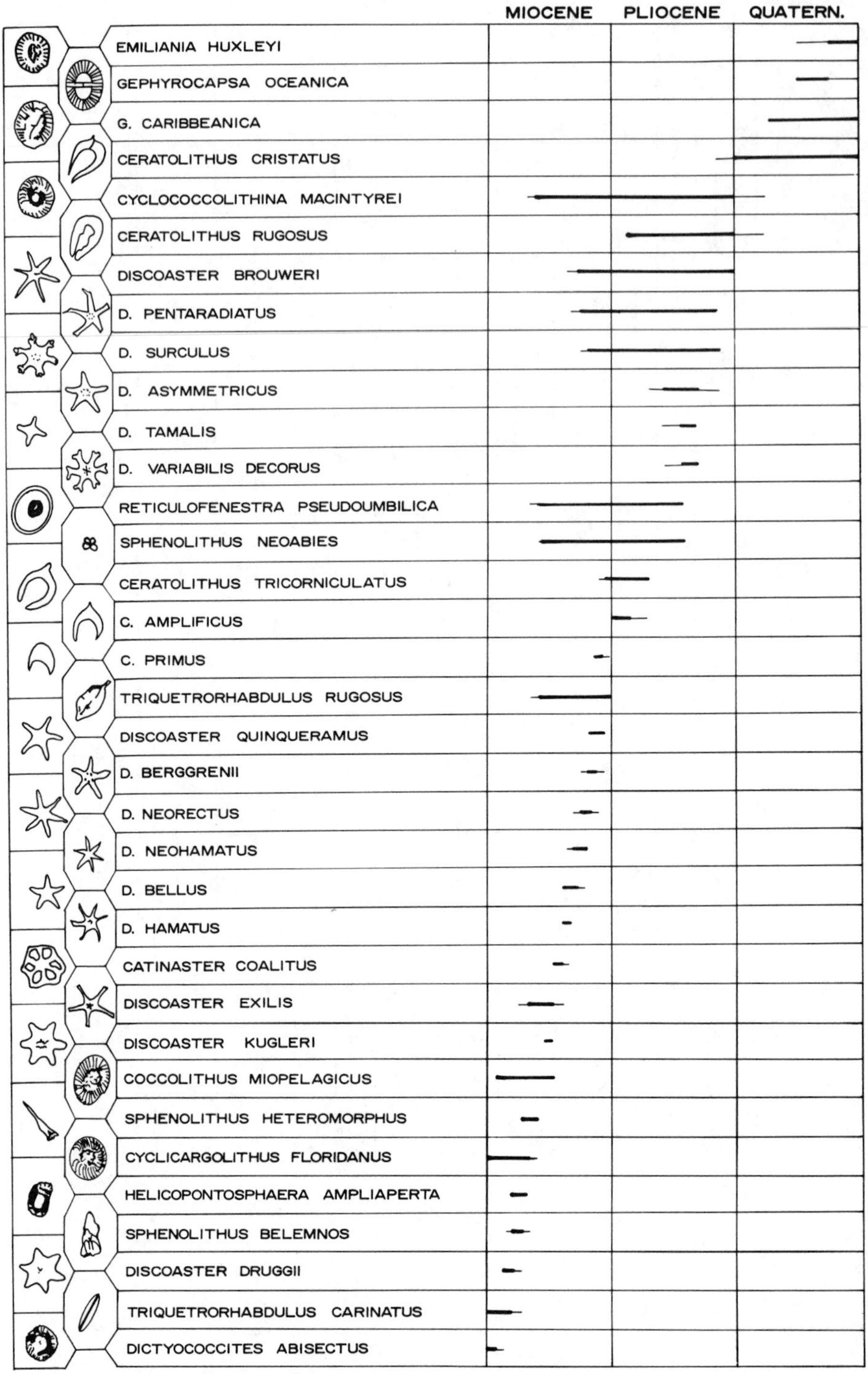

	MIOCENE	PLIOCENE	QUATERN.
EMILIANIA HUXLEYI			—
GEPHYROCAPSA OCEANICA			—
G. CARIBBEANICA			—
CERATOLITHUS CRISTATUS			—
CYCLOCOCCOLITHINA MACINTYREI	—	—	
CERATOLITHUS RUGOSUS		—	—
DISCOASTER BROUWERI		—	
D. PENTARADIATUS	—	—	
D. SURCULUS		—	
D. ASYMMETRICUS		—	
D. TAMALIS		—	
D. VARIABILIS DECORUS		—	
RETICULOFENESTRA PSEUDOUMBILICA	—	—	
SPHENOLITHUS NEOABIES	—	—	
CERATOLITHUS TRICORNICULATUS	—		
C. AMPLIFICUS	—		
C. PRIMUS	—		
TRIQUETRORHABDULUS RUGOSUS	—		
DISCOASTER QUINQUERAMUS	—		
D. BERGGRENII	—		
D. NEORECTUS	—		
D. NEOHAMATUS	—		
D. BELLUS	—		
D. HAMATUS	—		
CATINASTER COALITUS	—		
DISCOASTER EXILIS	—		
DISCOASTER KUGLERI	—		
COCCOLITHUS MIOPELAGICUS	—		
SPHENOLITHUS HETEROMORPHUS	—		
CYCLICARGOLITHUS FLORIDANUS	—		
HELICOPONTOSPHAERA AMPLIAPERTA	—		
SPHENOLITHUS BELEMNOS	—		
DISCOASTER DRUGGII	—		
TRIQUETRORHABDULUS CARINATUS	—		
DICTYOCOCCITES ABISECTUS	—		

Fig. 13-10. Lifespan of some stratigraphically useful marine diatoms. (Reprinted by permission from Berger 1976.)

As reserves of high-grade crudes are depleted, improved processing techniques, conservation, and aggressive exploration programs will be increasingly important. Academic research could benefit diatomite exploration programs by focusing on lacustrine paleobasin geology. Freshwater diatom silica sources and their relation to paleobasin geology also are important areas of research.

The importance of biotic carbonate deposition is becoming increasingly clear, though its parameters remain undefined. Taxonomic difficulties add to the problems inherent in assigning depositional and diagenetic pathways to metamorphic rocks (see Bathurst 1975, pp. 217–30, 278–93; Williams 1984). The steel industry and environmental use are predicted to dominate carbonate rock consumption trends in the near future (Boynton et al. 1983; Carr and Rooney 1983). Recognition of high-purity crudes, conservation, and recycling will be important in long-range utilization of world carbonate reserves (Carr and Rooney 1983).

The future rapid consumption of known world petroleum reserves will lead to more sophisticated exploration techniques. Conservation, innovative exploration techniques, and enhanced oil recovery and mining techniques will be important in utilization of reserves. Exploration success will necessitate recognition of optimum reservoirs and the ability to relate them to source rocks and migration processes (Zieglar and Spotts 1976). Biostratigraphic use of fossil algae will continue to be an important exploration tool. Enhanced recovery techniques of nonconventional crudes (i.e., heavy oil, oil shales) may become increasingly feasible economically as dwindling conventional reserves are consumed (Hunt 1979; Tissot and Welte 1984).

Fossil algae are of great importance to industrial society. Their siliceous, calcareous, and organic components have been used by humans for thousands of years. Coincident with the development of end uses for fossil algal products has been the recognition that these materials are biogenic. As technologies become more sophisticated and world resources are depleted, the biogenic pathways and ages of formation of fossil algal occurrences will be better understood. Anticipation and excitement will stimulate the research of fossil algae for many years to come.

VI. Acknowledgments

The Manville Corporation supported the preparation of this chapter, and the author appreciates its contributions and assistance. Sincere thanks is expressed to E. L. Mann and Lynn Le Roy of the Manville Corporation, Norman Cimon, and senior editor of the text, Dr. Carole

A. Lembi, for careful editing and helpful discussion. Shelley Cimon prepared the figures and contributed to their design. This chapter would not have been possible without the editing, word processing, and support of Lynn M. George.

VII. References

Ames, J. A., and W. E. Cutcliffe. 1983. Cement and cement raw materials. In *Industrial Minerals and Rocks*, 5th ed., vol. 1, ed. S. J. Lefond, pp. 133–59. Society of Mining Engineers, New York.

Anonymous. 1975. Proposals for a standardization of diatom terminology and diagnoses. In *Third Symposium on Recent and Fossil Marine Diatoms Kiel, Royal Octavo, VIII*, ed. Reimer Simonsen. Germany.

Axelrod, D. I. 1981. *Role of Volcanism in Climate and Evolution*. Geological Society of America, Boulder, CO. 59 pp.

Bathurst, R. G. C. 1975. *Carbonate Sediments and Their Diagenesis*. Elsevier Scientific Publishing Co., New York. 568 pp.

Berger, W. H. 1976. Biogenous deep sea sediments: production, preservation, and interpretation. In *Chemical Oceanography*, ed. J. P. Riley and R. Chester, pp. 265–388. Academic Press, New York.

Berggren, W. A. 1978. Marine micropaleontology: an introduction. In *Introduction to Marine Micropaleontology*, ed. B. U. Haq and A. Boersma, pp. 1–17. Elsevier/North Holland, New York.

Bortz, L. C., and D. K. Murray. 1979. Eagle Springs Oil Field, Nye County, Nevada. In *1979 Basin and Range Symposium*, ed. G. W. Newman and H. D. Goode, pp. 441–53. Rocky Mountain Association of Geologists and Utah Geological Association, Denver, CO.

Boynton, R. S., and K. A. Gutschick. 1975. Lime. In *Industrial Minerals and Rocks*, 4th ed., ed. S. J. Lefond, pp. 737–56. Society of Mining Engineers, New York.

Boyton, R. S., and K. A. Gutschick, R. C. Freas, and J. L. Thompson. 1983. Lime. In *Industrial Minerals and Rocks*, 5th ed., vol. 2, ed. S. J. Lefond, pp. 809–31. Society of Mining Engineers, New York.

Bradbury, J. P. 1975. *Diatom Stratigraphy and Human Settlement in Minnesota*. Geol. Soc. Amer. Special Paper 171. 74 pp.

Bradley, W. H. 1931. *Origin and Microfossils of the Oil Shale of the Green River Formation of Colorado and Utah*. U.S. Geol. Survey Prof. Paper 168. U.S. Government Printing Office, Washington, DC. 58 pp.

Burckle, L. H. 1978. Marine diatoms. In *Introduction to Marine Micropaleontology*, ed. B. U. Haq and A. Boersma, pp. 245–66. Elsevier/North Holland, New York.

Bureau of Mines. 1985. *Mineral Commodity Summaries*. U.S. Department of the Interior, Washington, DC. 185 pp.

Calvert, S. E. 1966a. Accumulation of diatomaceous silica in the sediments of the Gulf of California. *Geol. Soc. Amer. Bull.* 77, 569–96.

Calvert, S. E. 1966b. Origin of diatom-rich, varved sediments from the Gulf of California. *J. Geol.* 74, 546–65.

Cane, R. F., and P. R. Albion. 1971. The phytochemical history of torbanites. *J. and Proc. Roy. Soc. New South Wales* 104, 31–7.

Carr, D. D., and L. F. Rooney. 1983. Limestone and dolomite. In *Industrial Minerals and Rocks,* 5th ed., vol. 2, ed. S. J. Lefond, pp. 833–68. Society of Mining Engineers, New York.

Cummins, A. B. 1960. Diatomite. In *Industrial Minerals and Rocks,* ed. J. L. Gillson, pp. 303–14. American Institute of Mining, Metallurgical, and Petroleum Engineers, New York.

Cummins, A. B. 1973. *The Celite Enterprise: An Informal Account of Johns-Manville's Diatomite Business.* Johns-Manville Products Corp. Internal Document, Warren, NJ.

Cummins, A. B. 1975. *Terra Diatomacea.* Johns-Manville Corp. Internal Document, Warren, NJ.

deWit, M. J., R. Hart, A. Martin, and P. Abbott. 1982. Archean abiogenic and probable biogenic structures associated with mineralized hydrothermal vent systems and regional metasomatism, with implications for Greenstone Belt studies. *Econ. Geol.* 77, 1783–802.

Elsenbast, A. S., and D. C. Morris. 1942. Diatomaceous silica filter-aid clarification. *Indust. Eng. Chem.* 34, 412–18.

Gallois, R. W. 1976. Coccolith blooms in the Kimmeridge clay and origin of North Sea oil. *Nature* 259, 473–5.

Haq, B. U. 1978. Calcareous nannoplankton. In *Introduction to Marine Micropaleonotology,* ed. B. U. Haq and A. Boersma, pp. 79–107. Elsevier/North Holland, New York.

Haq, B. U. 1980. Biogeographic history of Miocene calcareous nannoplankton and paleoceanography of the Atlantic Ocean. *Micropaleontology* 26, 414–43.

Hetényi, M., and K. Sirokmán. 1978. Structural information on the kerogen of the Hungarian oil shale. *Acta Mineralogica-Petrographica* 23, 211–22.

Hunt, J. M. 1979. *Petroleum Geochemistry and Geology.* W. H. Freeman and Co., San Francisco. 617 pp.

Johnson, J. H. 1961. *Limestone-building Algae and Algal Limestones.* Johnson Publishing Co., Boulder, CO. 297 pp.

Kadey, F. L., Jr. 1983. Diatomite. In *Industrial Minerals and Rocks,* 5th ed., vol. 1, ed. S. J. Lefond, pp. 677–708. Society of Mining Engineers, New York.

Krebs, W. N., and J. P. Bradbury. 1984. Neogene and Quaternary non-marine diatom biostratigraphy. In *Geological Uses of Diatoms: Short Course.* National Geological Society of America Meeting, Reno, NV.

Labeyrie, L. 1974. New approach to surface seawater paleotemperatures using $^{18}O/^{16}O$ ratios in silica of diatom frustules. *Nature* 248, 40–2.

Lohman, K. E. 1957. *Cenozoic Nonmarine Diatoms from the Great Basin.* Ph.D. thesis, California Institute of Technology, Pasadena.

Lohman, K. E. 1961. *Geologic Ranges of Cenozoic Nonmarine Diatoms,* U.S. Geol. Survey Prof. Paper 424D, pp. 234–6.

Longheed, M. S. 1983. Origin of Precambrian iron-formations in the Lake Superior Region. *Geol. Soc. Amer. Bull.* 94, 325–40.

Matiskainen, H., and P. Alhonen. 1984. Diatoms as indicators of provenance in Finnish sub-neolithic pottery. *J. Archaeol. Sci.* 11, 147–57.

Meisinger, A. C. 1985a. Diatomite. In *Mineral Facts and Problems,* Preprint Bull. 675. U.S. Department of the Interior, Bureau of a Mines, Washington, D. C. 6 pp.

Meisinger, A. C. 1985b. Diatomite. In *Bureau of Mines Minerals Yearbook,* vol. 1, *Metals and Minerals.* U.S. Bureau of Mines, Washington, DC.

Melton, C. E., and A. A. Giardini. 1984. Petroleum formation and the thermal history of the earth's surface. *J. Pet. Geol.* 7, 303–12.

Nagappa, Y. 1957. Further occurrences of *Botryococcus* in Western Pakistan. *Micropaleontology* 3, 83.

Oehler, J. H., and R. G. Logan. 1977. Microfossils, cherts, and associated mineralization in the Proterozoic McArthur (H.Y.C.) lead-zinc-silver deposit. *Econ. Geol.* 72, 1393–409.

Ozol, M. A. 1984. FGD markets for limestone: real and growing. *Pit and Quarry* 76, 64–8.

Power, T. 1985. Limestone specifications: limiting constraints on the market. *Indust. Minerals* 217, 65–90.

Pressler, J. W., and L. Pelham. 1985. Lime, calcium, and calcium compounds. In *Mineral Facts and Problems.* Bull. 675 Preprint, pp. 1–8. U.S. Bureau of Mines, Washington, DC.

Ramanujam, C. G. K. 1966. Occurrence of *Botryococcus* in the Miocene lignite of South Arcot District, Madras. *Curr. Sci.* 14, 367–8.

Rao, A. R. 1957. Algal remains from the Tertiary lignites of Palana (Eocene), Bikaner. *Curr. Sci.* 6, 177–8.

Schrader, H.-J., and G. Schuette. 1981. Marine diatoms. In *From the Sea,* vol. 7, *The Oceanic Lithosphere,* ed. C. Emiliani, pp. 1179–232. John Wiley and Sons, New York.

Severinghaus, N., Jr. 1983. Fillers, filters, and absorbents. In *Industrial Minerals and Rocks,* 5th ed., vol. 1, ed S. J. Lefond, pp. 243–57. Society of Mining Engineers, New York.

Smith, G. R. S. 1976. How to use rotary, vacuum, precoat filters. *Chem. Eng.* 83, 84–90.

Smith, M. 1984. Calcium carbonate fillers: plastics and paper to grow. *Indust. Minerals* 198, 23–35.

Solti, G. 1980. The oil shale deposit of Várpalota. *Acta Mineralogica-Petrographica* 24, 289–300.

Tappan, H. 1968. Primary production, isotopes, extinctions, and the atmosphere. *Palaeogeog., Palaeoclimat., Palaeoecol.* 4, 187–210.

Tappan, H. 1980. *The Paleobiology of Plant Protists.* W. H. Freeman and Co., San Francisco. 1,028 pp.

Tappan, H., and A. R. Loeblich, III. 1970. Geobiologic implications of fossil phytoplankton evolution and time–space distribution. In *Symposium on Palynology of the Late Cretaceous and Early Tertiary,* ed. R. M. Kosanke and A. T. Cross, pp. 247–340. Geological Society of America, Boulder, CO.

Taylor, H. A., Jr. 1984. Dimension stone. In *Bureau of Mines Minerals Yearbook,* vol. 1, *Metals and Minerals,* pp. 821–5. U.S. Bureau of Mines, Washington, DC.

Tepordei, V. 1985. Crushed stone. In *Mineral Facts and Problems,* Bull. 675 Preprint, pp. 1–12. U.S. Bureau of Mines, Washington, DC.

Tissot, B. P. 1984. Recent advances in petroleum geochemistry applied to hydrocarbon exploration. *Amer. Assoc. Pet. Geol. Bull.* 68, 545–63.

Tissot, B. P., and D. H. Welte. 1984. *Petroleum Formation and Occurrence.* Springer-Verlag, Berlin. 699 pp.

Traverse, A. 1955. Occurrence of the oil-forming alga *Botryococcus* in lignites and other Tertiary sediments. *Micropaleontology* 1, 343–50.

Vimal, K. P. 1953. Occurrence of *Botryococcus* in Eocene lignites of Cutch. *Curr. Sci.* 12, 375–6.

Welte, D. H. 1965. Relation between petroleum and source rock. *Amer. Assoc. Pet. Geol. Bull.* 49, 2246–68.

Williams, L. A. 1984. Subtidal stromatolites in Monterey Formation and other organic-rich rocks as suggested source contributors to petroleum formation. *Amer. Assoc. Pet. Geol. Bull.* 68, 1879–93.

Williams, M. A. J., M. F. Williams, F. Gasse, G. H. Curtis, and D. A. Adamson. 1979. Plio-Pleistocene environments at Gadeb prehistoric site, Ethiopia. *Nature* 282, 29–33.

Wornardt, W. W., Jr. 1969. Diatoms, past, present, future. In *Proceedings of the First International Conference on Planktonic Microfossils, Geneva 1967,* vol. 2, ed. P. Bronnimann and H. H. Renz, pp. 690–713. E. J. Brill, Leiden.

Wornardt, W. W., Jr. 1979. Paleophytogeography of the diatom *Annellus californicus* and its significance in terms of the California Tertiary. In *Cenozoic Paleogeography of the Western U.S. Pacific Coast Paleogeography Symposium 3,* ed. J. M. Armentrout, M. R. Cole, and H. Terbest, pp. 205–18. Society of Economic Paleontologists and Mineralogists, Los Angeles.

Zieglar, D. L., and J. H. Spotts. 1976. Reservoir and source-bed history in Great Valley, California. *Amer. Assoc. Pet. Geol. Bull.* 60, 2192.

Zielinski, R. A. 1982. Uraniferous opal, Virgin Valley, Nevada: conditions of formation and implications for uranium exploration. *J. Geochem. Explor.* 16, 197–216.

14. Algae and agriculture

BLAINE METTING

R & A Plant-Soil, Inc., 24 Pasco-Kahlotus Road, Pasco, WA 99301

WILLIAM R. RAYBURN

The Graduate School, Washington State University, Pullman, WA 99164

PIERRE A. REYNAUD

French Institute for Cooperative Research & Development (ORSTOM), Laboratoire de Microbiologie, Centre de Dakar, B.P. 1386, Dakar, Senegal

CONTENTS

I. Introduction

Utilization of algae in agriculture has a long history, as suggested by early records for the nitrogen-fixing cyanobacteria (N_2-fixing blue-green algae) and seaweeds. Chinese science historians describe documentation of medicinal properties of *Azolla* from the eleventh century, although the exact date when the fern–cyanobacterium symbiosis was first cultivated is not known. Dao and Tran (1979) suggested that *Azolla* was purposefully used to fertilize rice in Vietnam at least from the eleventh century and surely was introduced into China by the early Ming dynasty (fourteenth century). A description of *Azolla* use from 540 B.C. also may refer to intentional cultivation of the plant (Lumpkin and Plucknett 1982).

Seaweed utilization in agriculture is, on the other hand, well documented from ancient times. Stephenson (1974) refers to the fact that the Romans valued seaweed as animal feed supplements and as green manure for vegetable production (see also Harlin and Darley, Chapter 1). However, only in the latter half of our century, specifically the last 15 years, has scientific inquiry been directed at these traditional uses of algae.

Microalgal technologies applicable to agriculture include in situ biological N_2 fixation by free-living cyanobacteria and *Azolla*. To date, these technologies are largely restricted to rice cultivation in the tropics. Eukaryotic, palmelloid (mucilaginous) microalgae have potential as soil conditioners. Of related importance are the ecological phenomenon of the desert algal crust and implications in rangeland ecology and desertification. Plant growth regulators (PGRs) from seaweeds and microalgae are known, and evidence is growing to support the hypothesis that some liquid seaweed products when used in small amounts affect yield or quality of crops.

II. Free-living cyanobacteria and algalization

A. Cyanobacteria and N_2 fixation

The concept of cyanobacterial biofertilizers is inherently attractive. Many cyanobacteria fix nitrogen in aerobic environments but do not compete with crops or the heterotrophic soil microflora for carbon or energy. For decades, the natural fertility of some rice soils has been attributed to cyanobacteria, and interest is great in developing technologies for their mass culture and use with other crops.

Cyanobacteria are morphologically the most diverse and complicated of the prokaryotes (Stanier and Cohen-Baziere 1977). Species with potential as biofertilizers are the heterocystous filaments included by Geitler (1932) and Desikachary (1959) in the botanical orders Nostocales and Stigonematales, and by bacteriologists in Rippka's sections IV and V (Rippka et al. 1979). These cyanobacteria are unique in the microbial world for their ability to simultaneously fix nitrogen in aerobic habitats and carbon by the oxygenic eukaryotic plant mechanism (Haselkorn 1986). This complexity makes them difficult yet important subjects for molecular research. Nonetheless, significant advances have been made including transformation and use of *Anacystis* for photosynthesis and herbicide resistance studies (Golden and Sherman 1984; Golden and Haselkorn 1985) and the successful expression of an *Escherichia coli* gene on a plasmid vector in *Agmanellum* (= *Synechocystis*) (Buzby, Porter, and Stevens 1985).

The biochemistry and molecular genetics of N_2 fixation have been studied intensively in the soil bacteria *Azotobacter, Klebsiella,* and *Rhizobium* (Havelka, Boyle, and Hardy 1982). Relatively less is known about cyanobacteria, although interest is developing rapidly since mutant strains, molecular vectors, and gene libraries have become more common (Haselkorn 1986). Unlike heterotrophic bacteria, reducing equivalents and energy for nitrogen fixation in cyanobacteria are derived from photosynthesis. Added to these costs is the energy required to maintain anaerobic conditions in the heterocyst where nitrogen fixation occurs. Together, these needs are satisfied by expenditure of up to 20 ATP for each molecule of dinitrogen reduced (Bothe 1982). Combined with energy required for growth and reproduction of inocula on soil or in floodwaters, these basic metabolic needs ultimately define the potential of these organisms as biofertilizers. Further research aimed at understanding relationships among heterocyst differentiation, nitrogenase and hydrogenase regulation and function, and overall growth is required. A long-term goal with potentially great rewards is incorporation of maximal energy efficiency of N_2 fixation in fast-growing competitive strains able to do so in the presence of synthetic N fertilizers, which would otherwise inhibit heterocyst development.

B. Distribution and methodology

Cyanobacteria are ubiquitous members of the soil microflora. Species diversity can be large in both temperate and tropical habitats (Metting 1981; Roger and Reynaud 1982). Some studies suggest that species distributions are discontinuous within regions (Granhall 1975; Zimmerman, Metting, and Rayburn 1980). Venkataraman (1975) reported that only 33% of 2,213 Indian soil samples yielded heterocystous cyanobacteria from enrichment culture and that the distribution among and within regions was highly variable. On the other hand, Reynaud (1982a) recorded N_2-fixing cyanobacteria in 86 of 89 rice paddy soils in Senegal. Quite likely, part of these differences arose from use of dissimilar counting and isolation procedures. Indeed, Roger, Grant, and Reddy (1985) reported that more recent unpublished surveys suggest that cyanobacteria are more prevalent in Indian soils than previously thought.

Our meager understanding of the ecology and distribution of free-living edaphic cyanobacteria is due largely to the fact that simple methodologies for accurate, rapid estimation of standing crops or productivities of filamentous species do not exist. Abundance has been determined by direct observation, various plating techniques, pigment extraction, and most-probable-number (MPN) methods (Roger and Reynaud 1979). Areal yields and productivity can be estimated on the basis of multiple measurements of abundance, but these have seldom been performed. The best method for calculating abundance of filamentous cyanobacteria is currently that of Reynaud and Laloe (1985) in which direct calculation of the mean volume of cells, aggregates, and whole or partial filaments of initial soil dilutions is used to enhance the precision of serial dilution methods in liquid or on agar. The major drawback is that it is laborious, but it is the first technique that permits simultaneous quantitative and floristic analysis of mixed cyanobacterial populations from soil.

C. Physiological ecology

Physical, chemical, and biological factors influence growth and N_2 fixation by cyanobacteria on soil and in flooded paddies. Important physical factors include light, temperature, and cycles of wetting and drying (Roger and Reynaud 1979). Heterocystous species are generally intolerant of high irradiance and luminosity, which can reach 110,000 lux in the tropics. The adverse effects of too much light can be ameliorated to some extent in flooded soils by strains able to migrate vertically in the water column. Availability of light influences succession of eukaryotic microalgae and cyanobacteria in the soil algal community. The nature and extent of crop canopy also influence light quality and availability over the course of a growing season.

Cyanobacteria generally grow and fix nitrogen optimally between 30° and 35°C; thus, temperature is not growth-limiting in the tropics. Although evidence from native grassland communities in Canada suggests that cyanobacteria fix significant amounts of nitrogen during episodes of snowmelt (Coxson and Kershaw 1983a, b), temperature probably will be an important consideration in any attempt to introduce a biofertilizer technology to temperate agriculture.

Among chemical properties, pH is the most important factor influencing species composition, growth, and N_2 fixation. Soil cyanobacteria grow best under neutral to alkaline conditions as illustrated by correlations among water pH in flooded paddies, cyanobacterial diversity, and abundance of cyanobacterial spores during the dry season between rice cycles (Roger and Reynaud 1982). Because N_2 fixation is an anaerobic process, soil and paddy oxidation-reduction (redox) potential may be important. Under reduced conditions less energy would be expended to maintain a suitable environment within the heterocyst. However, this ecological factor has not been studied with respect to its influence on cyanobacteria. Watanabe and Roger (1984) describe four microhabitats in the flooded paddy ecosystem within which cyanobacteria can grow and fix nitrogen: the submerged soil surface, the water column, floating, and epiphytic on rice and aquatic weeds. Certainly the redox potential varies among these habitats and influences N_2 fixation by cyanobacteria throughout the rice cycle.

Nutrients and agrichemicals also influence the activities and growth of cyanobacteria. Laboratory and field studies suggest that P limitation and, to a lesser extent, Fe and Mo availability may thwart the inexpensive use of biofertilizers. Okuda and Yamaguchi (1952) incubated 117 submerged soils in a greenhouse and observed that cyanobacterial growth seemed closely related to available P in the soil. Because P fertilizers are often unavailable or prohibitively expensive in developing countries, this might be the most difficult hurdle to overcome for expansion of algalization. Often, addition of lime ($CaCO_3$) to rice fields stimulates growth of cyanobacteria. The response could be due to a combination of increased alkalinity and increased availability of some nutrients (K, Mo) or, in flooded soils, to added C because vigorous algal growth depletes dissolved CO_2 faster than it can be replaced from the air or submerged soil (Roger and Reynaud 1979).

The effects of wetting and drying cycles on succession of native or added cyanobacteria have not been adequately assessed. In general, cyanobacteria are tolerant of extended periods without water, and variation among strains within species appears to be related to habitats of origin (Roger and Reynaud 1982). In common with other microorganisms, the physiological state of a cyanobacterium, temperature, and rate of desiccation together influence drought tolerance (Metting 1981). In Senegal, where the dry season between rice cultivations can last eight months, Roger and Reynaud (1976) observed that spores of heterocystous cyanobacteria constituted more than 95% of the potential algal flora at the end of the dry period. Species morphology also influences tolerance to desiccation. For example, *Cylindrospermum* species forming globose mucilaginous aggregates tend to dominate paddies at the end of the cultivation cycle after fields are drained (Roger and Reynaud 1979). Recovery of nitrogenase activity generally lags behind photosynthesis and growth of cyanobacteria upon rewetting (Rodgers 1977; Reynaud, unpublished data).

Effects of herbicides and other agrichemicals on growth and N_2 fixation by free-living cyanobacteria are variable and differ widely among strains (Pipe and Shubert 1984; Roger and Kulasooriya 1980). Many compounds are inhibitory at high concentrations but are stimulatory when diluted. In general, herbicides that interfere with photosynthesis are detrimental to cyanobacteria at field application rates in liquid culture but less so in soil. Cyanobacteria are generally more tolerant of such herbicides than vascular plants. Most hormone-analogue herbicides (e.g., 2,4-D and other phenoxys) as well as fungicides and insecticides are rarely harmful to cyanobacteria.

Parasitism, antagonism, competition, and grazing have all been implicated as important biotic factors in algalization studies. Parasites, including cyanophages, may be responsible for population crashes, but this relationship has never been confirmed in the field. Antibiotic activities among cyanobacteria and eukaryotic microalgae are well documented from laboratory studies (Metting and Pyne 1986) but also have not been documented in soil. However, laboratory investigations suggest that the actions of lytic streptomycetes (Sharma et al. 1984) and fungi (Al-Maadhidi and Henriksson 1980) may directly influence soil cyanobacteria. Grazing is known to deter development of a significant standing crop some of the time. Invertebrates from many phyla will consume cyanobacteria in rice paddies (Grant and Seegers 1985).

D. Cyanobacteria in rice fields

Roger and Kulasooriya (1980) reviewed 369 citations on free-living cyanobacteria and rice. Rice soils vary widely in standing crop and diversity of cyanobacteria. MPN estimates vary from 10^3 to 10^7 colony-forming units per gram of dry soil. Standing crops range from a few hundred to 16,000 fresh $kg \cdot ha^{-1}$

Dominant species and successional trends change seasonally according to the interactions of the factors listed above. Reynaud and Roger (1978; Reynaud 1987) studied successional patterns in relation to changes in algal biomass, floristics, and environmental parameters through the rice cycle. In Senegalese rice paddies, the initial community is dominated by diatoms, chlorophytes, and unicellular cyanobacteria. This assemblage coincided with full sunlight and remained until the tillering stage in the rice plant's devlopment. Succession thereafter depended on pH change as the crop canopy closed. If the pH decreased to 5 or below as shading increased, the total algal biomass decreased from 100 to 300 fresh kg·ha^{-1} to between 1 and 2 kg, and dominance shifted entirely to eukaryotic species complexes. With or without fertilization with inorganic nutrients, the relative abundance of cyanobacteria increased if the pH increased. Total algal biomass was greatest between tillering and panicle initiation, then decreased to the end of the cultivation cycle. After a brief appearance of N$_2$-fixing unicellular and colonial species at about tillering, uninoculated fields above neutrality were dominated by filamentous heterocystous and nonheterocystous cyanobacteria. At rice maturation, two situations developed on waterlogged or submerged soils. When the canopy was dense and/or the soil waterlogged, only filamentous species were present. Heterocystous species became closely associated with nonheterocystous ensheathed filaments at this time, leading to their protection during the coming dry period. When the vegetative cover was relatively less and/or the soil submerged, the proportions of filamentous green algae, diatoms, and nonfixing cyanobacteria increased. Nitrogen fixation during these successional trends was performed by different strains according to changing pH (Reynaud 1984a). Singh (1961) also described successional patterns for Indian soils, but ecological parameters were not considered in depth.

A wide range of N$_2$ fixation rates was reported for cyanobacterial populations in uninoculated soils and flooded fields (Roger and Kulasooriya 1980). Most values were extrapolated from indirect acetylene reduction assays, which can be reliable provided sampling is adequate to take into account diurnal fluctuations (Roger and Reynaud 1982). Extrapolated rates from 38 studies range from 10 to 90 kg·ha^{-1}·y^{-1}, with an average of 27 kg N (Roger and Kulasooriya 1980).

E. Algalization technology

Algalization technology was developed in India during the 1960s and 1970s, and the term was introduced to denote biofertilization of rice soils with free-living cyanobacteria (Singh 1961; Venkataraman 1972). Efforts in the early years were spearheaded by the All-India Coordinated Project on Algae (AICPA 1979). Technical methods for production of cyanobacterial biomass were first developed in Japan for *Tolypothrix tenuis* (Watanabe 1959; Watanabe et al. 1959). By the early 1980s, similar village-level appropriate technologies also were under development in China (Huang et al. 1981). The recent publication by Roger and Watanabe (1986) treats most of the pertinent literature.

In 1979 AICPA recommended that Indian farmers prepare their own inocula in outdoor soil culture. AICPA supplied a dried starter consisting of *Aulosira, Anabaena, Nostoc, Plectonema,* and *Tolypothrix* to be grown for two months in 2 m^2 ponds. Dry flakes from these ponds were to be added to rice fields at 8–10 kg·ha^{-1} one week after transplanting rice seedlings. Other management practices, such as P fertilization and predator control, were addressed (Venkataraman 1981). In the 1970s, this technology was being applied to 2 million ha in India, 400,000 ha in Burma, and smaller areas in Sri Lanka and Nepal (Venkataraman 1972). For a variety of reasons, including lack of consistent results and poor grower acceptance in some areas, the extent of algalization has not changed in the past decade (Roger et al. 1985). Algalization also was practiced on about 23,000 ha in the early 1980s in China where *Nostoc sphaeroides* and an *Anabeana* sp. called *A. azotica* were employed (Li Shanghao, personal communication).

Few agronomic studies have been reported in which parameters other than grain yield were estimated in algalization trials. Experiments from more than one cropping cycle are lacking as well. Thus, Roger and Watanabe (1986) consider algalization an unproven, developing technology. Unpublished work suggests that, for many Indian rice soils already possessing large numbers of viable spores of heterocystous strains, algalization, using viable cell counts significantly less (three orders of magnitude) than those present, might never prove worthwhile (P. A. Roger, personal communication).

Most reports of algalization trials are from glasshouse experiments with individual rice plants in pots. Extrapolation from pot trials suggests N$_2$ fixation rates of as much as 70 kg N·ha^{-1} in six weeks and increased grain yields of 33%. Stewart (1980) suggested that increased water retention and, perhaps, other factors favorable to cyanobacterial growth in well-tended pots might enhance N$_2$ fixation. Generally, N$_2$ fixation and grain yields judged against controls without added inorganic N in field trials have been about half the average for glasshouse experiments (Roger and Kulasooriya 1980).

Fieldwork has varied with respect to strain composition and application rates of inocula, the latter ranging from 0.2 to 33 dry kg·ha^{-1}. Using individual trials reported to date as replications, differences in grain yields between untreated crops and crops treated with either cyanobacteria and inorganic N or cyanobacteria alone are probably not significant (Table 14-1). Judging from field trials in which algalization was compared to one or two rates of inorganic N application, an average of 30 kg N·ha^{-1} fixed per rice cycle (50% of crop needs) and a potential for 60–90 kg per year in regions supporting two or three crops were suggested by Roger and Kulasooriya (1980).

F. Potential for cyanobacterial biofertilizers

Applied and basic research in a number of disciplines might result in more widespread adoption of algalization technologies. Repeatedly addressed above is the need for a much more thorough understanding

Table 14-1. *Algalization field trials: yield response of rice*

Treatment[a]	Number of cultivation cycles	Mean grain yield (kg·ha^{-1}	Standard deviation[b]	Control (%)
Trials compiled from Roger and Kulasooriya (1980)				
N° A°	22	3,038	802a	100
N° A$^+$	22	3,642	896b	118
N$^+$ A°	11	4,155	823b	135
N$^+$ A$^+$	11	4,734	1063b,c	153
Trials conducted near Dakar, Senegal (Reynaud, unpub. data) (1980–3)				
N° A°	4	5,765	176a	100
N° A$^+$	4	5,690	154a	99
N$^+$ A°	4	8,520	293b	147
N$^+$ A$^+$	4	9,110	253c	158

[a]N° = no inorganic N; N$^+$ = inorganic N added; A° = no algalization; A$^+$ = algalization.
[b]Values followed by the same letter are not significantly different at $P = 0.1$.

of the microbial ecology of cyanobacteria on soil. Impetus in that area will be greatly aided when a rapid, reproducible technique for estimating cyanobacterial biomass on soil is introduced. This would also permit expansion of directed screening and selection programs for the identification of fast-growing strains demonstrating good competitive ability on soil. Other areas in need of additional efforts include basic studies of cyanobacterial molecular biology and biochemistry, applied research to develop an inoculum production technology, and agronomic management practices permitting full expression of the potential of algalization. In particular, replicated plot experiments under various farming regimes must be undertaken if the potential for algalization is ever to be realized and its ultimate usefulness and market share determined. Quality field research necessitates extensive replication of well-defined treatments under controlled agricultural conditions.

As part of this research effort, the potential for use of cyanobacterial biofertilizers in temperate, irrigated agriculture should be assessed. Documented field studies under temperate conditions are limited. Witty (1974) inoculated plots in England sown to wheat with two species of *Nostoc* and an *Anabaena* sp. and measured increased acetylene reduction activity but could not detect a crop response. Anderson et al. (1982) inoculated plots sown to corn with an *Anabaena* sp. and *Tolypothrix tenuis*. Both cyanobacteria began to colonize the irrigated soil surface, as estimated by the MPN technique, but disappeared abruptly after 10 weeks. Significant acetylene reduction activity was not detected. It should be noted that a handful of questionable biofertilizer products purporting to contain cyanobacteria are marketed on a small scale in the United States. Logically, strain selection coupled to production of quality inocula, the same needs as for algalization and rice, could result in products able to make a significant contribution on moist temperate soils. However, the technology does not presently exist. As long as nitrogen fertilizers remain less expensive than the cost of producing, distributing, and applying cyanobacterial inocula that do not perform consistently, the incentive for product development by the private sector will not be strong.

III. *Azolla*

A. *The symbiosis*

In contrast to free-living cyanobacteria, the *Azolla–Anabaena azollae* symbiosis is more widely employed as an appropriate technology in rice production. *Azolla* includes six extant species in two subgenera on the basis of frond morphology and reproductive features. The cyanobacterium *Anabaena azollae* is harbored in a dorsal lobe leaf cavity. The heterocyst frequency of *A. azollae* is 20–30%, compared to 5–10% in nonsymbiotic species. Regulation of heterocyst development and NIF (nitrogen fixation genes) activity in *Anabaena* are related to glutamine synthetase (GS) synthesis and activity (Haselkorn 1986). It is thought that the fern exerts control over heterocyst development and thus over nitrogen fixation by producing an effector substance that inhibits GS synthesis or activity by the cyanobacterium (Peters and Calvert 1982). Molecular biological research with *A. azollae* is scant with but one purported axenic clone only recently available.

    Azolla has a long tradition of use as a green manure and is currently the subject of coordinated worldwide agronomic trials. The advantage enjoyed by *Azolla* over nonsymbiotic cyanobacteria is that it is readily visible to the naked eye so that sophisticated equipment and methods are not required to monitor its growth and influence on rice. Eight hundred scientific articles have been written on *Azolla* in this century, including aspects of botany, distribution of species and strains, physiology, and genetics. Recent proceedings include those from a 1982 workshop in Puerto Rico (Silver and Schroder 1984) and in China (1985, unpublished proceedings).

B. *Physiological ecology*

Azolla species occur in many regions. As shown in Figure 14-1 for *A. pinnata,* strains of a given species adapt to regional climates and local habitat conditions. Therefore, morphological and physiological prop-

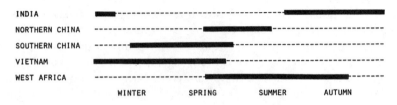

Fig. 14-1. The principal growing season of the ubiquitous *Azolla pinnata* varies widely among the regions in which it is found.

erties, growth, and N_2 fixation vary widely among strains within species (Watanabe and Roger 1984). Important environmental parameters are temperature, light, daylength, nutrients, salinity, pH, atmospheric humidity, desiccation, wind, predators and pests, and density. In many cases, interactions among these parameters are important (Lumpkin and Bartholomew 1986). Nutrient deficiencies result in readily recognizable symptoms in the fern. For example, Ca deficiency is indicated by yellow, erect leaves and reduced root growth whereas purple, bullet-shaped leaves and excessively long roots characterize a shortage of P (Talley, Talley, and Rains 1980). Maximum growth and N_2 fixation by *Azolla* requires greater P fertility than rice, and this is a major barrier to applied use in many instances.

In contrast to other ferns, *Azolla* attracts many pests (Lumpkin and Plucknett 1982). Though many damage the fern, some are also of concern for public health, such as *Limnea natalensis,* a schistosomiasis vector. *Azolla* can be protected from this snail and other pests with pesticides, either synthetic or plant-derived (Reynaud 1986).

Doubling times range from 2 to 10 days for most *Azolla* species and increase with increasing density, presumably because of shading and competition for nutrients (Becking 1979). Maximum standing crops reported in the literature range up to 5.2 dry $mt \cdot ha^{-1}$ and average 2.1 mt. Nitrogen content ranges up to 146 $kg \cdot ha^{-1}$, averaging about 70. Rates of N_2 fixation range from 0.4 to 3.6 kg $N \cdot ha^{-1} \cdot day^{-1}$ (Watanabe and Roger 1984), and extended yields of 11 $g \cdot m^{-2} \cdot day^{-1}$ have been cited (Vincenzini, Margheri, and Sili 1985).

C. Traditional use

China and Vietnam are the only countries with a long tradition of *Azolla* propagation whereas the technology is still under development in India and West Africa. The fern is grown in Vietnam from November through January in fallow, flooded fields for use with the spring rice crop. In China, *Azolla* is usually cultivated from May to June in the North (to 37° N) and from March to April in the South. In India, *Azolla* is grown as

Fig. 14-2. A Mandingo woman collecting *Azolla* (kouban) into a basket prior to transplanting rice from the bundles in the background.

an annual from July to December. These practices largely correspond to ecological requirements of local varieties (Fig. 14-1).

Azolla is used in Vietnam and China both as a green manure and in dual cropping with rice. Field inoculation occurs at rates of from 300 to 500 fresh kg·ha^{-1}. Fronds are often fragmented manually to aid growth and dispersal. If all goes well, the *Azolla* reproduce to about 20 mt·ha^{-1} by 16 to 20 days after inoculation. Sometimes, half the biomass is collected and composted in mounds for later incorporation into the soil with or without fresh biomass one week or so after transplanting the rice (Lumpkin and Plucknett 1982; Watanabe and Roger 1984).

Liu (1979) reported 18% greater rice yield in fields with *Azolla* compared to fields without the fern as being common for China. Various sources estimate that in the 1970s as many as 2 million ha were fertilized by *Azolla* in China and Vietnam (Lumpkin and Plucknett 1980, 1982). Since then it appears that decreased cost and increased availability of fertilizer N has significantly reduced *Azolla* use to perhaps 10% or less of its peak period of utilization (T. A. Lumpkin, personal communication). Details of traditional propagation techniques and agronomic uses of *Azolla* are given by Lumpkin and Plucknett (1982) and Khan (1983).

Use of *Azolla* in West Africa is not as well documented and has been described only within the last few years for traditional (Reynaud 1982b, 1984b; Trinh 1984) and extensive (Diara and van Hove 1984) cultivation. Figure 14-2 shows a Senegalese Mandingo woman collecting *Azolla* biomass before transplanting rice to avoid contributing to lodging (too much *Azolla* will hold seedlings down if they lodge) and for inoculation into other fields. Figure 14-3 shows dual cropping of *Azolla* with seedling clumps prior to transplanting in agronomic trials conducted by

Fig. 14-3. *Azolla pinnata* var. *africana* growing right up to a rice bundle in a nursery area of the paddy.

the French Institute for Cooperative Research and Development (ORSTOM), Dakar.

D. Agronomic research

Substantial agronomic research with *Azolla* began in the mid-1970s. The International Network on Soil Fertility and Fertilizer Evaluation for Rice (INSFFER) has coordinated field trials since 1979 at 19 sites in nine countries. Watanabe (1985) summarized the results of the first four years as follows:

1. Incorporation of one *Azolla* crop (0.2 kg fresh wt. inoculum per ha) into soil before or after transplanting rice is equal to about 30 kg of fertilizer $N \cdot ha^{-1}$.

2. Incorporation of two crops of *Azolla* grown before and after transplanting is equivalent to a split application of 60 kg $N \cdot ha^{-1}$.

3. Yield response of rice to *Azolla* is roughly proportional to fertilizer additions.

4. Spacing of rice plants does not significantly affect *Azolla* growth. However, at some sites it was shown that *Azolla* growth was better after transplanting with 20×20 cm rice spacing than at an earlier stage with rice at 10×40 cm.

Other agronomic research showed that with good water management a residual effect on a second rice crop was obtained by dual-cropping *Azolla* with the first rice crop. With normal water management, N is released from decomposing *Azolla* at two rates. One part is immediately mineralized and assimilated by rice. As the soil becomes anaerobic under waterlogged conditions, mineralization slows, and the remainder of the N in *Azolla* is not available for use until the next aerobic period early in the next cropping cycle. Persistent anaerobic conditions foster production of phytoinhibitory compounds or creation of a C/N balance unfavorable to subsequent N release (Reynaud 1984b). In addition, 20–30% of *Azolla* N is taken up by rice in the first 50 days after transplanting if the biomass is incorporated into the soil, but only 15% is available if *Azolla* is not plowed under (Watanabe, de Guzman, and Cabrera 1981). Therefore, best management practices include sequences of short-term flooding before plowing, inoculation with *Azolla* before transplanting, soil incorporation, draining, then flooding and transplanting.

Talley, Lim, and Rains (1981) summarized their agronomic research with *Azolla* as a green manure crop for temperate rice cultivation in California. Incorporation of 40 and 90 kg *Azolla* N equivalent·ha^{-1}, intended to approximate what would be feasible with aerial seeding, increased rice yield by 1.5–2.6 mt·ha^{-1}, respectively. The response was equal to the addition of 80 and 180 kg·ha^{-1} of NH_4SO_4. *Azolla* treatments required both P and Fe additions for optimal growth. Experiments suggested a potential for companion use with summer rice. The limiting step to use in mechanized rice culture is provision of inocula. Current farm management practices under mechanized conditions do not free enough area for *Azolla* inoculum production during the time of the year when rice is not grown; yet temperatures are adequate for the fern.

E. Future potential

The current *Azolla* technology is very labor-intensive. Methods for propagation have not changed because the fern must be grown in the rice fields themselves. This is the first of a number of problems that must be addressed scientifically before the technology can be improved and used on a large scale, both in traditional centers and in new areas.

Azolla is maintained in standing water. Because varieties are locally adapted and because fresh material always must be on hand, central mass culture of quality inocula is not feasible. It is not even clearly understood whether this is desirable. The barrier to solving this dilemma is controlling sporulation of the fern. Peters and Calvert (1982) state that, if sporulation can be induced and if methods for harvesting and storage are developed, the potential for *Azolla* use will be significant.

Strain selection is another important consideration. Most locations probably already harbor specifically adapted strains; however, Cheng et al. (1981) showed that an introduced *A. filiculoides* outgrew the native *A. pinnata* in field trials and that an introduced *A. caroliniana* was more tolerant to pests.

Improved P use efficiency is an important goal. *Azolla* cannot require more of this nutrient than the crop it is supplying with N. Phosphorus is nonrenewable, and it is not logical or economical to sacrifice this resource for biological production of a renewable nutrient.

Inexpensive pest control is needed. Also required is a better understanding of the symbiosis itself, if it is to be improved with modern molecular techniques. Axenic cultures of the symbiont must become more widely available.

Should these areas receive adequate attention the future for *Azolla* use will hinge on the same issues as those for free-living cyanobacterial biofertilizers. Both technologies are currently labor-intensive, and labor is not always inexpensive relative to commodity output. The popularity and extent of use of these technologies will ultimately be dictated by the price of energy, which directly determines the availability and cost of inorganic N fertilizer. The recent example of decreased use of *Azolla* in China makes this clear.

IV. Microalgal soil conditioners

The size distribution, conformation, and stability of soil aggregates influence soil porosity, aeration, water movement, compaction, ease of tillage, fertilizer use, root development, and erosion potential by wind or water. Modification of soil structure to optimize aggregate stability to slaking by water and to mechanical disruption by wind and other forces is an important goal. Soil conditioners are materials that improve soil structure or alter the surface tension of water to ease infiltration. Among the former, microbial polysaccharides are generally regarded as the most important natural factors in formation and stabilization of aggregates (Cheshire 1979; Burns and Davies 1986). More attention has been paid to bacteria and fungi (Lynch and Bragg 1985); however, the influence of microalgae on soil aggregate stability is also well documented (Metting 1981). Communities of microalgae develop as surface veneers on soil where there is adequate light and moisture. Cyanobacteria dominate on tropical and alkaline soils whereas eukaryotic microalgae tend to displace them on temperate and acidic soils. The influence of cyanobacteria on aggregation of rice and semiarid soils resulting from natural blooms or algalization trials has been reported (Metting

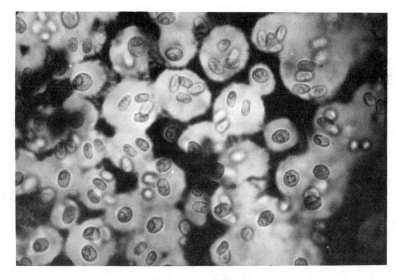

Fig. 14-4. A palmelloid species of *Chlamydomonas* showing polysaccharide sheaths investing individual and groups of cells.

1981, 1988). A comparative study of the aggregating effects of *Azolla pinnata* and the AICPA algalization mixture of five asymbiotic cyanobacteria revealed that only the free-living forms had a measurable influence on soil structure (Roychoudhury et al. 1979). This supports the idea that polysaccharides and readily oxidizable substrates (indirectly via heterotrophic metabolism) more markedly influence the dynamics of aggregation than do complex or highly oxidized organic compounds (Lynch and Bragg 1985).

A potential exists for conditioning irrigated farmland in temperate regions using mass-cultured palmelloid (polysaccharide-producing) green microalgae (Metting 1985). Palmelloid green microalgae belong to 12 genera, distinguished in part by the morphology of their extracellular sheaths (Metting 1981). All of the applied research with palmelloid soil-conditioning microalgae has been with species in the genus *Chlamydomonas*, although a U.S. patent (Porter and Nelson 1975) claimed aggregating effects in soil with nonmucilaginous mass-cultured *Chlorella*. Lewin (1977) and Metting and Rayburn (1983) documented empirical evidence to suggest the commercial potential for conditioning irrigated soils with palmelloid chlamydomonads.

Lewin (1956) first estimated the extent and makeup of extracellular polysaccharides manufactured by *Chlamydomonas mexicana* (Fig. 14-4) and other palmelloid genera and species of green microalgae. Barclay and Lewin (1985) estimated polysaccharide production by *C. mexicana*

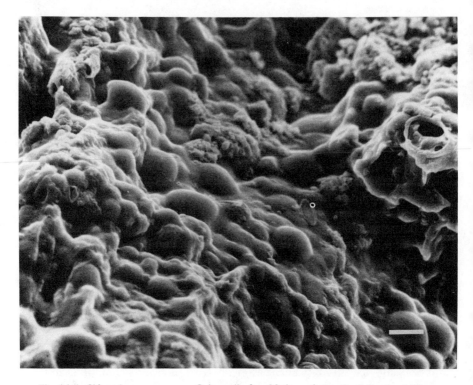

Fig. 14-5. *Chlamydomonas* sp. on a Galva soil after 28 days of growth. Cells are visible as "bumps" embedded in mucilage. Scale = 10 μm. (F. Williams, Iowa State University, Ames.)

and *C. sajao* to amount to 75% of dry weight yields in stationary-phase liquid culture. Kroen and Rayburn (1984) demonstrated that polysaccharide production was related to the growth phase. The exopolymer of *C. mexicana* was composed largely of glucose, galactose, fucose, and arabinose whereas galactose comprised 75% of the neutral sugars of the exopolymer of *C. sajao*. Uronic acids made up about 12% of the extracellular polysaccharide made by each species (Barclay and Lewin 1985). These acidic residues probably effect aggregate stabilization by means of cation bridges between mineral particles that also possess a net negative charge. Figure 14-5 shows polysaccharide materials and cells of *Chlamydomonas* in soil.

Shimmel and Darley (1985) used in situ $^{14}CO_2$ experiments to study productivity of native algal communities on a temperate soil cropped to sorghum. The algal contribution was 78 dry $g \cdot m^{-2} \cdot y^{-1}$, with peak yields of about 3.6 $g \cdot m^{-2} \cdot day^{-1}$. In the laboratory, Kroen (1984) reported growth rates of *C. mexicana* on soil of 1.8 divisions $\cdot day^{-1}$, slightly less than the $2.0 \cdot \text{day}^{-1}$ by the same microalga in liquid culture (Kroen and Rayburn 1984). Monitoring reproduction of *C. sajao* on replicated plots in the field by both chlorophyll extraction and MPN estimates, Metting (1986a) showed peak growth rates of $0.7–0.8$ division $\cdot \text{day}^{-1}$ corresponding to a maximum weekly average of $1.5 \text{ g} \cdot \text{m}^{-2} \cdot \text{day}^{-1}$. Visual estimates of the extent of coverage varied from 50% to 75% on plots inoculated with application rates commercially feasible through center pivot sprinklers. By comparison, Barclay and Lewin (1985) estimated 95% coverage by *C. sajao* on three of six replicates of two soil types and 85% on another when kept near 100% of field capacity in a glasshouse. Both species of *Chlamydomonas* reproduce on soil between 0.0 and -0.5 bars total water potential in the laboratory. On one soil, *C. sajao* produced a visible growth at -1.0 bars (Metting 1986b). To be useful, soil-conditioning microalgae must consistently reproduce a sufficient biomass on the surface. Synthetic conditioners are applied at about 0.1% of the soil (root zone) by weight. Foster (1981) demonstrated that relatively small volumes of polysaccharide were responsible for aggregate stability on an agronomic scale by virtue of their large length-to-cross-section ratio and apparent inaccessibility to microbial attack after periods of soil wetting and drying.

In field studies Metting and Rayburn (1983) used wet and dry aggregate stability tests to estimate the influence of a commercial microalgal conditioner on dry-land and irrigated temperate soils in eastern Washington State. Dry-land soils were not influenced because low availability of water limited reproduction and polysaccharide synthesis by the inoculated algae. Wet stability of the irrigated soil was increased by about 20% in two light soils and by 4% in a silt loam. Dry stability was increased by like amounts in all three. Results of a three-year replicated plot experiment were similar (Metting 1987) and suggest that applied research in other regions should be initiated.

V. Rangeland ecology, reclamation, and algal crusts

Soil surfaces in hot and cold deserts and semiarid steppes are commonly consolidated by cryptogamic communities, of which algal crusts are important constituents. Excluding basic and applied research with cyanobacteria and rice, the literature on soil crusts is the most extensive dealing with terrestrial algae. Especially voluminous are descriptions of crusts in the American West (e.g., Johansen, Rushforth, and Brotherson 1981), the Middle East (Friedmann and Galun 1974), and the Soviet Union (Gollerbach and Shtina 1969). Important contributions were made by the Jet Propulsion Laboratory (United States), which sponsored expeditions to the most remote hot and cold deserts on earth in order to build a foundation of understanding of extreme environments preliminary to the Viking landings on Mars (see Wharton, Smernoff, and Averner, Chapter 19). Algal crusts were found to be ubiquitous desert components (Cameron and Blank 1966).

Functions of algal crusts include consolidation of the crust, mediation of infiltration and retention of water, and nitrogen input. Desert and steppe surfaces that support algal crusts are consolidated by the combined aggregating effects of mucilaginous sheaths and the filamentous nature of dominant cyanobacteria (Bond and Harris 1964; Bailey, Mazurak, and Rosowski 1973). The investments of ensheathed cyanobacteria retain water, which buffers against rapid desiccation. Algal crusts can also fix appreciable N (Rychert and Skujins 1974).

Faust (1970) demonstrated that an important property of algal crusts is promotion of water infiltration. Cornet (1981) showed that a *Scytonema* crust in the Senegalese desert reduced loss of soil moisture to the air nearly 18-fold and that the relative humidity under the first cm was 8–9% with a crust and 1.3% without.

The influence of algal crusts on erosion potential, water relations, and soil fertility points to their importance in rangeland ecology. Comparative studies within regions have shown that excessive grazing or fire can damage crusts, resulting in severe erosion (Johansen and St. Clair 1984). Therefore, crusts are probably of long-term importance in the stability of fragile desert and steppe habitats that are used as rangelands. Destruction of crusts may hasten desertification in regions where short-duration intensive grazing is practiced, thus sacrificing renewable productivity for short-term gains. Intact crusts in Senegal may inhibit seedling establishment and limit growth of cattle grass (Dulieu, Gaston, and Darley 1977). However, Cornet (1981) found that dry grass yields of 280 kg·ha^{-1} resulted from slight disruption of algal crusts compared to only 80 kg in the absence of a crust. Cryptogamic crusts in the American Great Basin Desert are associated with enhanced seedling establishment (St. Clair et al. 1984).

In addition to their importance as established crusts, microalgae and cyanobacteria are early colonizers of primary and secondary substrates. They quickly colonize volcanic materials (Metting 1981; Rayburn, Mack, and Metting 1982), and their importance in secondary plant community succession has also been described (Booth 1941; Willson and Forest 1957). Starks and Shubert (1982) studied successional patterns in soil algal communities on disturbed lands and related these to chemical properties by means of a conceptual model.

Although it is not now practical to enhance the positive influences of microalgae in promoting development of mature soils from disturbed and spoiled lands, prospects should not be abandoned without experimental inquiry such as that reported by St. Clair, Johansen, and Webb (1986) in which stabilization of fire-disturbed sites was effected by inoculation with soil/crust slurries. Factors most likely to be limiting are ecological and economical. Inoculum size, delivery systems, and soil moisture are important criteria. Rangelands are less productive than farmlands and, therefore, less valuable. The return (meat production) on investment (microalgal technology) on large areas of low-production rangeland probably is not justified economically even if efficacious products were available. Nevertheless, there is precedent from use of natural blooms of cyanobacteria in India for reclamation.

Singh (1961) described a method for reclaiming saline-alkaline soils, called "Usar" soils, by construction of artificial impoundments within which cyanobacterial blooms were grown during the monsoon. The biomass was subsequently incorporated into the soil to improve physical properties that influence infiltration and drainage (Kaushik, Krishna Murti, and Venkataraman 1981). When the method was employed and gypsum ($CaSO_4$) was also incorporated into a sodic soil, exchangeable Na and electrical conductivity were lowered and hydraulic conductivity was increased. The ability to wash away accumulated salts and exchange divalent cations for Na are keys to reclamation of saline and sodic soils. In situations where aggregation improves penetration and drainage of water and promotes distribution of gypsum within the profile, the technology is feasible. Unfortunately, it appears that adoption of the technique has not become widespread in the nearly 30 years since its description (Roger et al. 1985).

VI. Microalgal plant growth regulators

A. Cyanobacterial PGR production

Plant growth regulators (PGRs) are synthetic or naturally occurring chemicals that exhibit positive or negative effects on plant growth, development, or metabolism that are exclusive of nutritional effects. PGRs include herbicides, some antibiotics, and other compounds with hormone-like activities. Ever since the discovery of endogenous phytohormones and their control of plant metabolism, scientists have searched for novel compounds that exhibit useful herbicidal, growth-promoting, or growth-altering properties. Real and implied PGR production by cyanobacteria, microalgae, and seaweeds is the topic of this and the following section.

The antibiotic and toxic properties of microalgae were reviewed by Metting and Pyne (1986). Antibacterial, algicidal, antifungal, and anti-protozoan compounds from cyanobacteria have been identified. Included were bromophenols, debromoaplysiatoxins, malyngolides, fatty acids, other organic acids, carbohydrates, and unidentified compounds (see Gorham and Carmichael, Chapter 16, for additional information on cyanobacterial toxins). Of potential agricultural significance were studies showing that extracts of a strain of *Nostoc muscorum* inhibited mycelial development of the soil-borne phytopathogen *Cunninghamella blakesleana* (Gaire et al. 1976; Mulé et al. 1977). Few studies have been directed at substances produced by cyanobacteria that exert a positive influence on other microorganisms. One of these was the demonstration by Hosmani (1976) that aqueous extracts of *Microcystis aeruginosa* enhanced growth of *Chlorella vulgaris*.

Most of the circumstantial evidence for the production of natural PGRs by cyanobacteria has been obtained from algalization studies. Pot culture experiments and field trials with cyanobacteria often generate results showing greater rice grain and/or straw yield increases than can be explained by N_2 fixation alone. Venkataraman (1972) and Roger and Kulasooriya (1980) list more than a dozen such studies. Additional research suggests that cyanobacteria have an additive effect on yield of algalization in the presence of full complements of inorganic N fertilizer (Roger and Kulasooriya 1980). However, few well-designed experiments exist, and other mechanisms may be responsible for the PGR effects. For example, cyanobacteria have been shown to release P from minerals (Bose et al. 1971), and oxygen production by cyanobacteria might reduce sulfide damage to submerged parts by indirectly inhibiting metabolism by anaerobic S-reducing bacteria (Jacq and Roger 1977). Pedurand and Reynaud (in press) conducted an in-depth survey of 133 unialgal cyanobacterial cultures to screen for PGR activity on rice seed germination and early development. Of the 133 cyanobacteria, 90 had a significantly deleterious effect on germination, whereas growth of rice was stimulated by 21% of the isolates and inhibited by 12%. Further investigation revealed that only an *Anabaena* strain had a significant positive PGR effect when extracts were administered or when added as an axenic culture.

Glasshouse experiments with crops other than rice have yielded plant responses attributed to cyanobacterial PGR production. These include increased yield and vitamin C content of tomatoes (Aiyer, Venkataraman, and Sundara Rao 1964; Kaushik and Venkataraman 1979), increased yield and N content of chili peppers and lettuce (Dadhich, Varma, and Venkataraman 1969), accelerated germination and growth of wheat (Kushwaha and Gupta 1972; Bharati, Bongale, and Hosmani 1978), and increased growth of tomatoes, radishes, and potatoes (Rodgers and Addiscott 1977; Rodgers et al. 1979).

Cyanobacterial PGR effects have been attributed to production of amino acids, vitamins, and hormones. Hosmani (1976) hypothesized that stimulation of *Chlorella* growth was due to DL-3,4-dihydroxyphenyl alanine extracted from *Microcystis*. Venkataraman and Neelakantan (1967) demonstrated production of $1 \cdot 5 \ \mu g \cdot g^{-1}$ dry wt of a material with vitamin B_{12}-like activity by a *Cylindrospermum* species and presented circumstantial evidence that this vitamin amendment plus ascorbic acid synthesis accounted for PGR influence on rice seedlings. Roger and Kulasooriya (1980) cite Russian references to PGR effects of *Calothrix* spp., *Anabaena* sp., and *Nostoc* sp. that suggest varied auxin-like responses. Gibberellin-like enhanced root development was reported for extracts of *Phormidium* (Gupta and Shukla 1964) and *Aulosira fertilissima* (Singh and Trehan 1973). Cytokinins were implied to be the active ingredients for effects of cyanobacterial extracts on growth of vegetables because heating did not destroy the response (Rodgers et al. 1979). Also, Huang and Chow (1984) demonstrated ethylene production by an axenic *Hapalosiphon* sp. Clearly, much research remains to be done; however, it appears that certain cyanobacteria might well manufacture PGRs.

B. PGRs from eukaryotic microalgae

Much less attention has been directed at PGR production by eukaryotic microalgae. As with cyanobacteria, studies suggest the potential for antibiotic or toxin production (Metting and Pyne 1986). A nonaxenic *Chlorella* culture had general cytokinin-like effects on germination and development of barley (Davis and Bigler 1973). Growth of *Rhizobium japonicum* in culture and *Agaricus bisporus* in a composted horse manure/straw mixture was enhanced by different species of *Scenedesmus* (Curto and Favelli 1972; Fingerhut, Webb, and Soeder 1984). At least three U.S. patents have been awarded for use of various *Chlorella* extracts as PGRs (Bigler, Camp, and Nelson 1974; Maeda and Tanaka 1984; Tenzer 1985).

VII. Seaweed use in agriculture and horticulture

A. Seaweed meals as feed supplements and soil additives

Where local supplies are plentiful, historical records show the use of seaweeds as animal fodder and soil amendments (Stephenson 1974). Brown algae (Phaeophyta) are most commonly employed in modern feed supplement preparations although some traditional foraging of green algae and brown algae is known. Nutritional experiments have been conducted with cattle, sheep, swine, and poultry. Seaweeds are not suitable materials for caloric content because algal polysaccharides are difficult to digest; however, there is evidence that some ruminants will develop a compatible microflora and derive energy (Black 1955). Chapman and Chapman (1980) concluded that supplementing many feeding regimes with up to 10% of washed, dried, and ground seaweed meal can be beneficial. The principal value of seaweeds as supplement is as a source of trace minerals. Vitamins and vitamin precursors, including carotenoids and xanthophylls, also were shown to be available in seaweed meals.

Seaweed use in agricultural and horticultural crop production is an old practice. Traditionally, fresh seaweeds have been used as green manures where readily available, such as France's Brittany coast where annual application provides as much as 50 kg $N \cdot ha^{-1}$. In some districts of Ireland, farmers place rocks into sandy substrata for attachment of seaweeds, which are later harvested and transported as far inland as 10 km (Chapman and Chapman 1980).

Processed seaweed products for crop use are of three kinds: (1) meals for supplementing soil in large volumes or for blending into defined rooting media for glasshouse crops, and (2) powdered and (3) liquid extracts and concentrates employed both as root dips (or soil drenches) and as foliar sprays. Soil addition of meals has slow-release fertilizer value equal to the nutrient content of the seaweed. This use is uncommon, however, because of the greater cost efficiency of synthetic fertilizer elements (N-P-K). Generally, seaweed meals provide an approximately equivalent amount of N, less P, but more K, totals salts, and readily available microelements compared to most animal manures. Addition of sodium alginate to soil at 0.1% by weight improved aeration and aggregate stability, most probably by the same mechanisms as other polysaccharides (Simpson and Hayes 1958). However, this is not a common practice because of the great cost of applying seaweed meals on an agricultural scale.

Seaweed meal and liquid seaweed manures were evaluated alone and together for their influence on yield, quality, and fertilizer use efficiency of many ornamental, vegetable, fruit, and commodity crops at Clemson University (South Carolina) beginning in 1959. Numerous experiments with treatments varying from incorporation of meal into soil for field plot research (100–500 $kg \cdot ha^{-1}$), to use in preparation of rooting media for glasshouse crops, to application of foliar sprays with and without addition to soil were undertaken over the subsequent 15 years. Norwegian *Ascophyllum nodosum* was used. Most treatments resulted in insignificant or nonreproducible crop responses, but a number of beneficial influences were observed. These included enhanced fertilizer use efficiency for treatments that included soil and foliar seaweed application, an area of research that deserves further inquiry. Other significant results included accelerated seed respiratory activities, increased yields of some crops, increased soluble solids of tomatoes and lambrusca (juice) grapes, and improved shelf life of peaches (Senn and Kingman 1978). Evidence that composted seaweeds added to soil increased the uptake of major nutrients in much the same proportions as meals suggested a chelating mechanism in the soil. Increased availability of P, N, and Fe in relation to their concentrations in meals was reported by Castillo (1966) and Caiozzi, Peirano, and Zunino (1968). Since the 1960s, however, commercial use of seaweed products as foliar sprays has increased more rapidly than has use as soil additives.

B. PGR activity of seaweed products

Use of foliar-applied seaweed extracts predated widespread acceptance of foliar application of inorganic fertilizers and PGRs other than herbicides (Stephenson 1974). The majority of studies were on products made from *Ascophyllum nodosum* or *Ecklonia maxima*.

Many seaweeds have been shown to contain PGRs, including auxins (Augier 1976a; Sumera and Cajipe 1981), gibberellins (Augier 1976b; Taylor and Wilkinson 1977; Wildgoose, Blunden, and Jewers 1978), abscisic acid (Hussain and Boney 1973), and quaternary ammonium compounds that might exhibit antigibberellin action (Blunden et al. 1981). However, most crop responses to seaweed use are thought to be due primarily to cytokinins, a diverse group of PGRs that influence cell division. Brian et al. (1973) were the first to use classical bioassays to implicate seaweeds as sources of cytokinin-like PGRs. Combining different solvent techniques with gas–liquid chromatography, Kingman and Moore (1982) demonstrated that an *Ascophyllum* product contained several PGRs and hormone precursors such as purines and their ribosides, free and bound cytokinins, indoleacetic acid, and abscisic acid. Zeatin-like substances were found in an *Ecklonia* product (Finnie and van Staden 1985).

A number of references in the literature are to experiments with seaweed products in which responses were consistent, at least partly, with a cytokinin effect (Blunden 1972; Blunden, Jones, and Passam 1978; Blunden, Wildgoose, and Nicholson 1979; Wilczek and Ng 1982; Abetz and Young 1983; Featonby-Smith and van Staden 1983a; Nelson and van Staden 1984a; Finnie and van Staden 1985).

Senn and Kingman (1978) also reported a number of trials by other researchers in which the use of seaweed sprays or root drenches were thought responsible for disease prevention (Tarjan 1977; Featonby-Smith and van Staden 1983b). It is known that some seaweeds produce antifungal compounds (Hoppe 1979); however, direct examination of products for antibiotic activity has not been reported although Booth (1965) observed a reduced incidence of fungal attack on crops with seaweed use.

There are also crop responses to foliar seaweed sprays with commercial potential that are not readily explained by direct cytokinin-like action. These include the enhanced efficiency of some herbicides and increased culm development and subsequent yield of wheat. Brain et al. (1973) first described a beneficial interaction, in terms of weed control, between a seaweed extract and the auxinlike herbicide mecoprop (2-[2-methyl-4 chlorophenoxy] propionic acid). Used as an adjuvant the seaweed enhanced the herbicidal effect on two broadleaf weeds by a factor of two but did not increase the detrimental effect of the herbicide on barley. Cytokinins were suspected because substitution of kinetin for seaweed gave similar results. Following application of ^{14}C-labeled MCPA (2-methyl-4-chlorophenoxyacetic acid) with and without an *Ecklonia* extract or a detergent, Erasmus, Nelson, and van Staden (1982) concluded that seaweed at some but not all concentrations promoted MCPA uptake into bean but not wheat leaves.

Nelson and van Staden (1984b) demonstrated in pot experiments that *Ecklonia* extract applied to wheat as either a root drench or a foliar spray significantly increased culm diameter, total number of spikelets per ear, and grain yield per ear and plant. Increased grain yield was associated with a beneficial effect on cell division within the developing endosperm (Nelson and van Staden 1984a). The concentration and timing of application were critical, a finding that is consistent with a PGR hypothesis for the mechanism(s) of action.

The plant hormone ethylene is known to influence stem development, and ethylene-releasing chemicals are used to prevent lodging of cereal crops (Bruinsma 1982). Using thin layer and gas chromatographic techniques, Nelson and van Staden (1985) detected nanomole quantities of the ethylene-releasing compound 1-amino-cyclopropane-1-carboxylic acid in an *Ecklonia* product, suggesting that some of the beneficial effects of seaweed use with agricultural and horticultural crops are not totally due to cytokinins.

C. Potential for seaweed use in agriculture

Discovery and development of synthetic PGR products by industry have yielded a multitude of useful herbicides and some chemicals with growth-promoting properties. Natural sources have not received as much attention, largely because of proprietary considerations and partly because corporations are mostly equipped and staffed to produce and test thousands of novel organic chemicals annually. Sales of seaweed products have grown from an interest in minimizing potential environmental damage and because natural products with traditional use histories are not subject to stringent regulatory control. Presently, seaweed products are not used on an agriculturally significant scale; their total world market share (largely Europe and America) is not large. The larger market segment for seaweed products consists of sale for use in glasshouse and intensive outdoor horticultural operations, such as orchards. Use on an agricultural scale is possible should seaweed products be shown to perform consistently with commodity grains by increasing herbicide use efficiency or promoting yields.

VIII. Summary

The real and potential uses for algal products in agriculture and horticulture are many and varied. Current and projected uses for microalgae and cyanobacteria are with living material, though nonliving cyanobacterial biomass might find a market niche as an organic houseplant and garden fertilizer. Use of living material presents a galaxy of technical and ecological considerations not common to processed seaweed products, which can be manufactured, distributed, and used in the same manner as agrichemicals. For the eukaryotic and prokaryotic algae, technologies have been reviewed that are applicable to a wide range of agronomic problems and situations: input of nitrogen, maintenance of soil structure, consolidation of soil against erosion, and PGR production.

The extent to which these technologies will grow or fade in the marketplace is dependent on economic forces that vary among countries according to their degree of mechanization, available wealth for purchase of agrichemicals, and traditional farming practices. All are somewhat dependent on the price of oil in that fertilizer production and the costs of manufacture and use of machinery are energy-intensive.

Also important to the future of algal products in agriculture is scientific research, particularly on living microalgal products. Too little is understood about the microbial ecology of soil algae and cyanobacteria. Indeed, it is not possible even to predict with a favorable degree of accuracy the market size for microalgal products. The extent to which strain selection is important is not known. Will one or two cyanobacterial strains dominate the marketplace because they have wider ranges of environmental tolerance? What is the minimum amount of continuously available water needed for effective colonization of erodable soils by palmelloid microalgae? The answers to these questions will establish the consistency of N input from algae for rice and the feasibility of applying soil conditioners to arable land variously irrigated with overhead sprinklers of different configurations as well as to rill-irrigated (gravity) fields. And lastly, how widely useful are seaweed PGR products? Concentrated today in the horticultural sector, efforts should be initiated to define their active mechanisms and determine whether and how they might benefit large-scale agriculture.

IX. Acknowledgments

This work was supported, in part, by the French Institute for Cooperative Research and Development (ORSTOM), by the Northwest Association of Colleges and Universities for Science (NORCUS), by U.S. Department of Agriculture small business grant 85-SBIR-8-0067 to

R&A Plant-Soil, Inc., and by R&A Plant-Soil, Inc. Thanks are due to Jane Cooper, librarian at the Tri-City University Center, Richland, Washington, without whose assistance we could not have written this chapter, and to P. A. Roger, T. A. Lumpkin, and W. R. Nelson for sharing unpublished information.

X. References

Abetz, P., and C. L. Young. 1983. The effect of seaweed sprays derived from *Ascophyllum nodosum* on lettuce and cauliflower crops. *Bot. Mar.* 26, 487–92.

AICPA [All-India Coordinated Project on Algae]. 1979. *Algal Biofertilizers for Rice.* Indian Council for Agricultural Research, New Delhi.

Aiyer, R. S., G. S. Venkataraman, and W. V. B. Sundara Rao. 1964. Effect of nitrogen-fixing blue-green algae and *Azotobacter chroococcum* on vitamin C content of tomato fruits. *Sci. and Cult.* 30, 557.

Al-Maadhidi, J., and E. Henriksson. 1980. Effects of the fungi *Trichoderma harzianum* and *Aspergillus flavus* on the nitrogen fixation and growth of the alga *Anabaena variabilis. Oikos* 35, 115–19.

Anderson, D. B., K. R. Hanson, B. Metting, and P. M. Molton. 1982. *Assessment of Blue-Green Algae as Living Sources of Fertilizer Nitrogen.* Pacific Northwest Laboratory Technical Rep. no. 4187, Richland, WA. 41 pp.

Augier, H. 1976a. Les hormones des algues: état actuel des connaisance. I. Recherches et tentatives d'identification des auxines. *Bot. Mar.* 19, 127–43.

Augier, H. 1976b. Les hormones des algues: état actuel des connaisances. II. Recherches et tentatives d'identification de gibbérellines, des cytokinines, et de diverses autres substances de nature hormonales. *Bot. Mar.* 19, 245–54.

Bailey, D., A. P. Mazurak, and J. R. Rosowski. 1973. Aggregation of soil particles by algae. *J. Phycol.* 9, 99–101.

Barclay, W. R., and R. A. Lewin. 1985. Microalgal polysaccharide production for the conditioning of soils. *Plant and Soil* 88, 159–69.

Becking, J. H. 1979. Environmental requirements of *Azolla* for use in tropical rice production. In *Nitrogen and Rice,* pp. 345–75. International Rice Research Institute, Los Baños, Philippines.

Bharati, S. G., U. D. Bongale, and S. P. Hosmani. 1978. The combined effect of presoaking seed and spraying with extract of *Hapalosiphon welwitschii* W. & G. S. West on wheat. *J. Karnatak Univ.* 23, 55–7.

Bigler, E. R., T. L. Camp, and P. I. Nelson. 1974. *Pre-germination Seed Treatment* U.S. Patent no. 3,820,281.

Black, W. A. P. 1955. Seaweeds and their constituents in foods for man and animal. *Chem. Ind.* 51, 1640–5.

Blunden, G. 1972. The effects of aqueous seaweed extract as a fertilizer additive. *Proc. Int. Seaweed Symp.* 7, 584–9.

Blunden, G., M. M. El Barouni, S. M. Gordon, W. F. McClean, and D. J. Rogers. 1981. Extraction, purification, and characterization of Dragendorf-positive compounds from British marine algae. *Bot. Mar.* 24, 451–6.

Blunden, G., E. M. Jones, and H. C. Passam. 1978. Effects of post-harvest treatment of fruit and vegetables with cytokinin-active seaweed extracts in kinetin solutions. *Bot. Mar.* 21, 237–40.

Blunden, G., P. B. Wildgoose, and F. E. Nicholson. 1979. The effects of aqueous seaweed extract on sugar beet. *Bot. Mar.* 22, 539–41.

Bond, R. D., and J. R. Harris. 1964. The influence of the microflora on physical properties of soils. I. The effects of filamentous algae and fungi. *Aust. J. Soil Res.* 2, 111–22.

Booth, C. O. 1965. The manurial value of seaweed. *Bot. Mar.* 8, 138–43.

Booth, W. E. 1941. Algae as pioneers in plant community succession and their role in erosion control. *Ecology* 22, 38–46.

Bose, P., U. S. Nagpal, G. S. Venkataraman, and S. K. Goyal. 1971. Solubilization of tri-calcium phosphate by blue-green algae. *Curr. Sci.* 7, 165–6.

Bothe, H. 1982. Nitrogen fixation. In *The Biology of Cyanobacteria*, Botanical Monograph, vol. 19, ed. N. G. Carr and B. A. Whitton, pp. 87–104. University of California Press, Berkeley.

Brain, K. R., M. C. Chalopin, T. D. Turner, G. Blunden, and P. B. Wildgoose. 1973. Cytokinin activity of commercial aqueous seaweed extract. *Plant Sci. Lett.* 1, 241–5.

Bruinsma, J. 1982. Plant growth regulators in field crops. In *Chemical Manipulation of Crop Growth and Development*, ed. J. S. McClaren, pp. 3–11. Butterworths, London.

Burns, R. G., and J. A. Davies. 1986. The microbiology of soil structure. *Biol. Agric. Hort.* 3, 95–113.

Buzby, J. S., R. D. Porter, and S. E. Stevens, Jr. 1985. Expression of the *Escherichia coli lacZ* gene on a plasmid vector in a cyanobacterium. *Science* 230, 805–7.

Caiozzi, M., P. Peirano, and H. Zunio. 1968. Effect of seaweed on the levels of available phosphorus and nitrogen in a calcareous soil. *Agron. J.* 60, 324–6.

Cameron, R. E., and G. B. Blank. 1966. *Desert Algae: Soil Crusts and Diaphanous Substrata as Algal Habitats.* Jet Propulsion Laboratory Tech. Rep. no. 32-971. Pasadena, CA. 41 pp.

Castillo, N. O. 1966. Effect of the brown alga, *Macrocystis integrefolia*, in increasing iron availability of a calcareous soil. *An. Fac. Quim. Far., Univ. Chile*, 18, 120–6.

Chapman, V. J., and D. J. Chapman. 1980. *Seaweeds and Their Uses.* 3rd ed. Chapman and Hall, London.

Cheng, J.-J., B.-A. Ke, Y.-C. Cheng, J.-H. Chen, and W.-Z. Cheng. 1981. Observations on reproduction and multi-resistance of *Azolla caroliniana* during summer cultivation. *Fujian Nongye Keji* 6, 30–2.

Cheshire, M. V. 1979. *Nature and Origin of Carbohydrates in Soils.* Academic Press, London.

Cornet, A. 1981. *Le bilan hydrique et son role dans la production de la strate herbacée de quelques Phytocénoses Sahéliennes au Sénégal.* Languedoc University of Science and Technology, Languedoc, France.

Coxson, D. S., and K. A. Kershaw. 1983a. Rehydration response of nitrogenase activity and carbon fixation in terrestrial *Nostoc commune* from *Stipa-Bouteloa* grassland. *Can. J. Bot.* 61, 2658–68.

Coxson, D. S., and K. A. Kershaw. 1983b. The pattern of in situ nitrogenase activity in terrestrial *Nostoc commune* from *Stipa-Bouteloa* grassland. *Can. J. Bot.* 61, 2686–93.

Curto, S., and F. Favelli. 1972. Stimulative effect of certain micro-organisms (bacteria, yeasts, micro-algae) upon fruit-body formation of *Agaricus bisporus* (Lange) Sing. *Mushroom Sci.* 8, 67–74.

Dadhich, K. S., A. K. Varma, and G. S. Venkataraman. 1969. The effect of *Calothrix* inoculation on vegetable crops. *Plant and Soil* 46, 499–510.

Dao, T.-T., and Q.-T. Tran. 1979. Use of *Azolla* in rice production in Vietnam. In *Nitrogen and Rice*, pp. 395–405. International Rice Research Institute, Los Baños, Philippines.

Davis, C. H., and E. R. Bigler. 1973. Effect of a phytoplankton culture on the germination and seedling development of barley, sugar beets, and lettuce. *Agron. J.* 65, 462–4.

Desikachary, T. V. 1959. *Cyanophyta*. Indian Council of Agricultural Research, New Delhi.

Diara, H. F., and C. van Hove. 1984. *Azolla*: a water saver in irrigated rice fields? In *Practical Application of Azolla for Rice Production*, ed. W. S. Silver and E. C. Schroeder, pp. 115–18. Martinus Nijoff, Dordrecht.

Dulieu, D., A. Gaston, and J. Darley. 1977. La dégradation des paturages de la région de N'Djaména (Republique du Tchad) en relation avec la présence de cyanophycées psammophiles: étude preliminare. *Rev. Elev. Med. Vet. Pays Trop.* 12, 1270–83.

Erasmus, D. J., W. R. Nelson, and J. van Staden. 1982. Combined use of a selective herbicide and seaweed concentrate. *S. Afr. J. Sci.* 78, 423–4.

Faust, W. F. 1970. *The Effect of Algal-Mold Crusts on the Hydrologic Processes of Infiltration, Runoff, and Soil Erosion under Simulated Rainfall Conditons*. M. S. thesis, University of Arizona, Tucson. 60 pp.

Featonby-Smith, B. C., and J. van Staden. 1983a. The effect of seaweed concentrate and fertilizer on the growth of *Beta vulgaris*. *Z. Pflanzenphysiol.* 112, 155–62.

Featonby-Smith, B. C., and J. van Staden. 1983b. The effect of seaweed concentrate on the growth of tomatoes in nematode-infested soil. *Sci. Hort.* 20, 137–46.

Fingerhut, U., L. E. Webb, and C. J. Soeder. 1984. Increased yields of *Rhizobium japonicum* by an extract of the green alga, *Scenedesmus obliquus* (276-3a). *Appl. Microbiol. Biotechnol.* 19, 358–60.

Finnie, J. F., and J. van Staden. 1985. Effect of seaweed concentrate and applied hormones on *in vitro* cultured tomato roots. *J. Plant Physiol.* 120, 215–22.

Foster, R. C. 1981. Polysaccharides in soil fabrics. *Science* 214, 665–7.

Friedmann, E. I., and M. Galun. 1974. Desert algae, lichens, and fungi. In *Desert Biology II*, ed. G. W. Brown, Jr., pp. 165–212. Academic Press, London.

Gaire, G. Z. de, M. C. Z. de Mulé, S. Daollo, D. R. de Halperin, and L. de Halperin. 1976. Acción de extractos acuosos y etéreos de *Nostoc muscorum* Ag (No. 79a). I. Efecto sobre plántulas de mijo (*Panicum milaceum* L.) mediante tratamiento de sus semillas. *Bull. Bot. Soc. Argentina* 17, 289–300.

Geitler, L. 1932. Cyanophyceae. In *Kryptogammenflora von Deutschland, Oster-*

reich, under de Sweitz, ed. L. Rabenhorst, vol. 14, pp. 673–1056. Akademische Verlagsgesellschaft, Leipzig.

Golden, S., aṅd R. Haselkorn. 1985. Mutation to herbicide resistance maps within the psbA gene of *Anacystitis nidulans* R2. *Science* 229, 1104–7.

Golden, S., and L. A. Sherman. 1984. Optimal conditions for genetic transformation of the cyanobacterium *Anacystis nidulans* R2. *J. Bacteriol.* 158, 36–42.

Gollerbach, M. M., and E. Shtina. 1969. *Soil Algae.* Akad. Sci. USSR, Moscow. [In Russian.]

Granhall, U. 1975. Nitrogen fixation by blue-green algae in temperate soils. In *Nitrogen Fixation by Free-Living Micro-organisms,* ed. W. D. P. Stewart, pp. 189–97. Cambridge University Press, Cambridge.

Grant, I. F., and R. Seegers. 1985. Movement of straw and algae facilitated by tubificids (Oligochaeta) in lowland rice soil. *Soil Biol. Biochem.* 5, 729–30.

Gupta, A. B., and A. C. Shukla. 1964. The effect of algal hormones on the growth and development of rice seedlings. *J. Sci. Technol.* 2, 204.

Haselkorn, R. 1986. Organization of the genes for nitrogen fixation in photosynthetic bacteria and cyanobacteria. *Ann. Rev. Microbiol.* 40, 525–47.

Havelka, U. D., M. G. Boyle, and R. W. F. Hardy. 1982. Biological nitrogen fixation. In *Nitrogen in Agricultural Soils,* ed. F. J. Stevenson, pp. 365–422. American Society of Agronomists, Madison, WI.

Hoppe, H. A. 1979. Marine algae and their products and constituents in pharmacy. In *Marine Algae in Pharmaceutical Science,* ed. H. A. Hoppe, T. Levring, and Y. Tanaka, pp. 25–119. de Gruyter, Berlin.

Hosmani, S. P. 1976. Effect of extract of *Microcystis aeruginosa* on the growth of *Chlorella vulgaris. Beitr. Biol. Pflanzenphysiol.* 51, 321–4.

Huang, T.-C., and T.-J. Chow. 1984. Ethylene production by blue-green algae. *Bot. Bull. Acad. Sin.* 25, 81–6.

Huang, X.-X., G.-R. Fang, Y.-T. Yan, T. Wang, and G.-F. Su. 1981. Studies on the preparation of dried inoculum of blue-green algae and its application in paddy fields. In *Proceedings of Symposium On Paddy Soil,* ed. Institute of Soil Science of the Academy Sinica, pp. 356–62. Springer-Verlag, Berlin.

Hussain, A., and A. D. Boney. 1973. Hydrophobic growth inhibitors from *Laminaria* and *Ascophyllum. New Phytol.* 72, 403–10.

Jacq, V., and P. A. Roger. 1977. Decrease of losses due to sulfate-reducing processes in the spermosphere of rice by presoaking seeds in a culture of blue-green algae. *Cah. ORSTOM Ser. Biol.* 12, 101–8.

Johansen, J. R., S. R. Rushforth, and J. D. Brotherson. 1981. Subaerial algae of Navajo National Monument, Arizona. *Great Basin Naturalist* 41, 433–9.

Johansen, J. R., and L. L. St. Clair. 1984. Recovery patterns of cryptogamic soil crusts in desert rangelands following fire disturbance. *Bryologist* 87, 238–43.

Kaushik, B. D., G. S. Krishna Murti, and G. S. Venkataraman. 1981. Influence of blue-green algae on saline-alkaline soils. *Sci. and Cult.* 47, 169–70.

Kaushik, B. D., and G. S. Venkataraman. 1979. Effect of algal inoculation on the yield and vitamin C content of two varieties of tomato. *Plant and Soil* 52, 135–7.

Khan, M. M. 1983. *A Primer on Azolla Production and Utilization in Agriculture,* International Rice Research Institute, Los Baños, Philippines.

Kingman, A. R., and J. Moore. 1982. Isolation, purification, and quantitation of several growth-regulating substances in *Ascophyllum nodosum* (Phaeophyta). *Bot. Mar.* 25, 149–53.

Kroen, W. K. 1984. Growth and synthesis of extracellular polysaccharide by the green alga *Chlamydomonas mexicana* (Chlorophyceae) on soil. *J. Phycol.* 20, 615–17.

Kroen, W. K., and W. R. Rayburn. 1984. Influence of growth status and nutrients on extracellular polysaccharide synthesis by the soil alga *Chlamydomonas mexicana* (Chlorophyceae). *J. Phycol.* 20, 253–7.

Kushwaha, A. S., and A. B. Gupta. 1972. Effect of algal growth-promoting substances of *Phormidium foveolum* on seedlings of some varieties of wheat. *Hydrobiologia* 35, 324–32.

Lewin, R. A. 1956. Extracellular polysaccharides of green algae. *Can. J. Microbiol.* 2, 665–72.

Lewin, R. A. 1977. The use of algae as soil conditioners. *Centros Invest. Baja Calif. Scripps Inst. Oceanogr.* 3, 33–5.

Liu, Z.-Z. 1979. Use of *Azolla* in rice production in China. In *Nitrogen and Rice,* pp. 375–94. International Rice Research Institute, Los Baños, Philippines.

Lumpkin, T. A., and D. P. Bartholomew. 1986. Predictive models for the growth response of eight *Azolla* accessions to climatic variables. *Crop. Sci.* 26, 107–11.

Lumpkin, T. A., and D. L. Plucknett. 1980. *Azolla:* botany, physiology, and use as a green manure. *Econ. Bot.* 34, 111–53.

Lumpkin, T. A., and D. L. Plucknett. 1982. *Azolla as a Green Manure: Use and Management in Crop Production.* Westview Tropical Agricultural Press, Boulder, CO.

Lynch, J. M., and E. Bragg. 1985. Microorganisms and soil aggregate stability. *Adv. Soil Sci.* 2, 133–71.

Maeda, T., and K. Tanaka. 1984. *Method of Plant Tissue and Cell Culture.* U.S. patent no. 4,431,738.

Metting, B. 1981. The systematics and ecology of soil algae. *Bot. Rev.* 47, 195–312.

Metting, B. 1985. Soil microbiology and biotechnology. In *Biotechnology: Applications and Research,* ed. P. N. Cheremisinoff and R. P. Ouellette, pp. 196–214. Technomics, Lancaster, PA.

Metting, B. 1986a. Population dynamics of *Chlamydomonas sajao* and its influence on soil aggregate stabilization in the field. *Appl. Envir. Microbiol.* 51, 1161–4.

Metting, B. 1986b. *Palmelloid Microalgae as Soil-conditioning Agents.* U.S. Department of Agriculture Technical Rep. no. 85-SBIR-8-0067. 76 pp.

Metting, B. 1987. Dynamics of wet and dry aggregate stability from a three-year microalgal soil-conditioning experiment in the field. *Soil Sci.* 143, 139–43.

Metting, B. 1988. Micro-algae and agriculture. In *Micro-algal Biotechnology,* ed. M. A. Borowitzka and L. A. Borowitzka, pp. 288–304, Cambridge University Press, Cambridge.

Metting, B., and J. W. Pyne. 1986. Biologically active compounds from microalgae. *Enzyme Microbial Technol.* 8, 386–94.

Metting, B., and W. R. Rayburn. 1983. The influence of a microalgal conditioner on selected Washington soils: an empirical study. *Soil Sci. Soc. Amer. J.* 47, 682–5.

Mulé, M. C. de, G. Z. de Caire, S. Doallo, D. R. de Halperin, and L. de Halperin. 1977. Acción de extractos algales acuosos y etéreos de *Nostoc muscorum* Ag. (No. 79a). II. Effecto sobre el desarollo del hongo *Cunninghamella blakesleana* (-) en el medio Mehlich. *Bull. Bot. Soc. Argentina* 28, 121–8.

Nelson, W. R., and J. van Staden. 1984a. The effect of seaweed concentrate on growth of nutrient-stressed greenhouse cucumbers. *HortScience* 19, 81–2.

Nelson, W. R., and J. van Staden. 1984b. The effect of seaweed concentrate on wheat culms. *J. Plant Physiol.* 115, 433–7.

Nelson, W. R., and J. van Staden. 1985. 1-aminocyclopropane-1-carboxylic acid in seaweed concentrate. *Bot. Mar.* 28, 415–17.

Okuda, A., and M. Yamaguchi. 1952. Algae and atmospheric nitrogen fixation in paddy soils. II. Relation between the growth of blue-green algae and physical or chemical properties of soil and effect of soil treatments and inoculation on the nitrogen fixation. *Mem. Res. Inst. Food Sci.* 4, 1–11.

Pedurand, P., and P. A. Reynaud. In press. Do cyanobacteria enhance germination and growth of rice? *Plant and Soil.*

Peters, G. A., and H. E. Calvert. 1982. The *Azolla–Anabaena* symbiosis. In *Advances in Agricultural Microbiology,* ed. N. S. Subba Rao, pp. 191–218. Butterworths, London.

Pipe, A. E., and L. E. Shubert. 1984. The use of algae as indicators of soil fertility. In *Algae as Ecological Indicators,* ed. L. E. Shubert, pp. 213–33. Academic Press, New York.

Porter, K. E., and P. L. Nelson. 1975. *High Density Treatment Product.* U.S. patent no. 3,889,418.

Rayburn, W. R., R. N. Mack, and B. Metting. 1982. Conspicuous algal colonization of the ash from Mount St. Helens. *J. Phycol.* 18, 537–43.

Reynaud, P. A. 1982a. Fixation d'azote chez les cyanobactéries libres ou en symbiose *Azolla:* possibilités d'utilisation agronomique en Afrique tropicale. *Bull. Pedol. F.A.O.* 47, 64–79.

Reynaud, P. A. 1982b. The use of *Azolla* in West Africa. In *Biological Nitrogen Fixation Technology for Tropical Agriculture.* ed. P. H. Graham and S. C. Harris, pp. 565–6. Centro Internationale de Agric. Tropical, Cali, Colombia.

Reynaud, P. A. 1984a. Influence of pH on nitrogen fixation of cyanobacteria from acid paddy fields. In *Advances in Nitrogen Fixation Research,* ed. C. Veeger and W. E. Newton, p. 356. Martinus Nijhoff, Dordrecht.

Reynaud, P. A. 1984b. *Azolla pinnata* var. *africana:* background, ecophysiology, plot assays. In *Practical Applications of Azolla for Rice Production,* ed. W. S. Silver and E. C. Schroder, pp. 73–91. Martinus Nijhoff, Dordrecht.

Reynaud, P. A. 1986. Control of *Azolla* pest *Limnea natalensis* with molluscicides of plant origin. *Int. Rice Res. Newsletter* 11, 27–8.

Reynaud, P. A. 1987. Ecology of N_2-fixing cyanobacteria under dry tropical conditions. *Plant and Soil* 98, 203–20.

Reynaud, P. A., and F. Laloe. 1985. La méthode des suspensions-dilutions adaptée à l'estimation des populations algales dans une rizière. *Rev. Ecol. Biol. Sol* 22, 161–92.

Reynaud, P. A., and P. A. Roger. 1978. N₂-fixing algal biomass in Senegal rice fields. *Ecol. Bull. Stockholm* 26, 148–57.

Rippka, R., J. Deruelles, J. B. Waterbury, M. Herdman, and R.Y Stainer. 1979. Generic assignments, strain histories, and properties of pure cultures of cyanobacteria. *J. Gen. Microbiol.* 111, 1–61.

Rodgers, G. A. 1977. Nitrogenase activity in *Nostoc muscorum:* recovery from desiccation. *Plant and Soil* 46, 671–4.

Rodgers, G. A., and W. Addiscott. 1977. Effects of blue-green algae on growth of potato plants. *Rothamsted Rep. for 1977* 1, 290–1.

Rodgers, G. A., B. Bergman, E. Henriksson, and M. Urdis. 1979. Utilisation of blue-green algae as biofertilizers. *Plant and Soil* 52, 99–107.

Roger, P. A., I. F. Grant, and P. M. Reddy. 1985. *Blue-Green Algae in India: A Trip Report.* International Rice Research Institute, Los Baños, Philippines. 93 pp. (Unpublished.)

Roger, P. A., and S. A. Kulasooriya. 1980. *Blue-green Algae and Rice.* International Rice Research Institute, Los Baños, Philippines.

Roger, P. A., and P. A. Reynaud. 1976. Dynamique de la population algale au cours d'un cycle de culture dans une rizière Sahélienne. *Rev. Ecol. Biol. Sol* 13, 545–60.

Roger, P. A., and P. A. Reynaud. 1979. Ecology of blue-green algae in rice fields. In *Nitrogen and Rice,* pp. 289–309. International Rice Research Institute, Los Baños, Philippines.

Roger, P. A., and P. A. Reynaud. 1982. Free-living blue-green algae in tropical soils. In *Microbiology of Tropical Soils and Plant Productivity,* ed. Y. R. Dommergues and H. G. Diem, pp. 147–68. Martinus Nijhoff, Dordrecht.

Roger, P. A., and I. Watanabe. 1986. Technologies for utilizing biological nitrogen fixation in wetland rice: potentialities, current usage, and limiting factors. *Fertilizer Res.* 9, 39–77.

Roychoudhury, P., B. D. Kaushik, G. S. R. Krishna Murti, and G. S. Venkataraman. 1979. Effect of blue-green algae and *Azolla* application on the aggregation status of the soil. *Curr. Sci.* 48, 454–5.

Rychert, R. C., and J. Skujins. 1974. Nitrogen fixation by blue-green algae-lichen crusts in the Great Basin Desert. *Soil Sci. Soc. Amer. Proc.* 38, 768–71.

Senn, T. L., and A. R. Kingman. 1978. *Seaweed Research in Crop Production.* National Technical Information Service Publication no. PB290101, Washington, DC. 161 pp.

Sharma, P., E. Henriksson, T. Rosswal, and D. V. Vadhra. 1984. Soil colloidal particles as carriers of inhibitory agents against the cyanobacterium *Anabaena* in an Indian soil. *Oikos* 43, 235–40.

Shimmel, S. M., and W. M. Darley. 1985. Productivity and density of soil algae in an agricultural system. *Ecology* 66, 1439–47.

Silver, W. S., and E. C. Schroder. 1984. *Practical Application of Azolla for Rice Production.* Martinus Nijhoff, Dordrecht.

Simpson, K., and S. F. Hayes. 1958. The effect of soil conditioners on plant growth and soil structure. *J. Sci. Food Agric.* 9, 163–70.

Singh, R. N. 1961. *Role of Blue-green Algae in Nitrogen Economy of India.* Indian Council of Agricultural Research, New Delhi.

Singh, V. P., and T. Trehan. 1973. Effects of extracellular products of *Aulosira fertilissima* on the growth of rice seedlings. *Plant and Soil* 38, 457–64.

Stanier, R. Y., and G. Cohen-Baziere. 1977. Phototrophic prokaryotes: the cyanobacteria. *Ann. Rev. Microbiol.* 31, 225–74.

Starks, T. L., and L. E. Shubert. 1982. Colonization and succession of algae and soil-algal interactions associated with disturbed areas. *J. Phycol.* 18, 99–107.

St. Clair, L. L., J. R. Johansen, and B. L. Webb. 1986. Rapid stabilization of fire-disturbed sites using a soil crust slurry: inoculation studies. *Reclamation Revegetation Res.* 4, 261–9.

St. Clair, L. L., B. L. Webb, J. R. Johansen, and G. B. Nebeker. 1984. Cryptogamic soil crusts: enhancement of seedling establishment in disturbed and undisturbed areas. *Reclamation Revegetation Res.* 3, 129–36.

Stephenson, W. A. 1974. *Seaweed in Agriculture and Horticulture.* 3rd ed. Bargyla and Glyver Rateaver Conservation Gardening and Farming Ser. C Reprints, Pauma Valley, CA. 241 pp.

Stewart, W. D. P. 1980. Systems involving blue-green algae. In *Methods for Evaluating Biological Nitrogen Fixation,* ed. F. J. Bergersen, pp. 583–635. Wiley Interscience, New York.

Sumera, F. C., and G. J. B. Cajipe. 1981. Extraction and partial characterization of auxin-like substances from *Sargassum polycystum* C. Ag. *Bot. Mar.* 24, 157–63.

Talley, S. N., E. Lim, and D. W. Rains. 1981. Application of *Azolla* in crop production. In *Genetic Engineering of Symbiotic Nitrogen Fixation,* ed. J. M. Lyons, R. C. Valentine, D. A. Phillips, D. W. Rains, and R. C. Huffaker, pp. 363–84. Plenum Press, New York.

Talley, S. N., B. J. Talley, and D. W. Rains. 1980. Nitrogen fixation by *Azolla* in rice fields. In *Genetic Engineering for Nitrogen Fixation,* ed. A. Hollander, pp. 259–81. Plenum Press, New York.

Tarjan, A. C. 1977. Kelp derivatives for nematode-infested citrus trees. *J. Nematology* 9, 287.

Taylor, I. E. P., and A. J. Wilkinson. 1977. The occurrence of gibberellin-like substances in algae. *Phycologia* 16, 37–42.

Tenzer, A. I. 1985. *Microbial Plant Growth Regulator.* U.S. Patent no. 4,551,164.

Trinh, T.-T. 1984. *Azolla* for rice production in tropical Africa. In *Practical Applications of Azolla for Rice Production,* ed. W. S. Silver and E. C. Schroder, pp. 215–16. Martinus Nijhoff, Dordrecht.

Venkataraman, G. S. 1972. *Algal Biofertilizers and Rice Cultivation.* Today and Tomorrow's Printers and Publishers, New Delhi.

Venkataraman, G. S. 1975. The role of blue-green algae in tropical rice production. In *Nitrogen Fixation by Free-Living Micro-organisms,* ed. W. D. P. Stewart, pp. 207–18. Cambridge University Press, Cambridge.

Venkataraman, G. S. 1981. Blue-green algae as a possible remedy to nitrogen scarcity. *Curr. Sci.* 50, 253–6.

Venkataraman, G. S., and S. Neelakantan. 1967. Effect of the constituents of the nitrogen-fixing blue-green alga *Cylindrospermum musicola* on growth of rice seedlings. *J. Gen. Microbiol.* 13, 53–8.

Vincenzini, M., M. C. Margheri, and C. Sili. 1985. Outdoor mass culture of

Azolla spp.: yields and efficiencies of nitrogen fixation. *Plant and Soil* 86, 57–67.

Watanabe, A. 1959. On the mass culturing of a nitrogen-fixing alga, *Tolypothrix tenuis. J. Gen. Microbiol.* 5, 85–91.

Watanabe, A., T. Hattori, T. Fujita, and T. Kiyohara. 1959. Large-scale culture of a blue-green alga, *Tolypothrix tenuis*, utilizing hot spring and natural gas as heat and carbon dioxide sources. *J. Gen Microbiol.* 5, 51–7.

Watanabe, I. 1985. *Summarized Report on INSFFER Azolla Program.* International Rice Research Institute, Los Baños, Philippines.

Watanabe, I., M. R. de Guzman, and D. A. Cabrera. 1981. The effect of nitrogen fertilizer on N_2-fixation in paddy field measured by *in situ* acetylene reduction assay. *Plant and Soil* 49, 135–9.

Watanabe, I., and P. A. Roger. 1984. Nitrogen fixation in wetland rice fields. In *Current Developments in Biological Nitrogen Fixation*, ed. N. S. Subba Rao, pp. 237–76. Oxford and IBH, New Delhi.

Wilczek, C. A., and T. J. Ng. 1982. Promotion of seed germination in table beet seed by an aqueous seaweed extract. *HortScience* 17, 629–30.

Wildgoose, P. B., G. Blunden, and K. Jewers. 1978. Seasonal variation in gibberellin activity of some species of Fucaceae and Laminariaceae. *Bot. Mar.* 21, 63–5.

Willson, D. L., and H. S. Forest. 1957. An exploratory study on soil algae. *J. Ecol.* 38, 309–13.

Witty, J. F. 1974. *Algal Nitrogen Fixation on Solid Surfaces and Temperate Agricultural Soils.* Ph.D. dissertation, University College, London. 188 pp.

Zimmerman, W. L., B. Metting, and W. R. Rayburn. 1980. The occurrence of blue-green algae in Whitman County, Washington, soils. *Soil Sci.* 130, 11–18.

ADVERSE IMPACTS OF ALGAE

15. Marine dinoflagellate blooms: dynamics and impacts

KAREN A. STEIDINGER

Florida Department of Natural Resources, Bureau of Marine Research, St. Petersburg, FL 33701

GABRIEL A. VARGO

Department of Marine Science, University of South Florida, St. Petersburg, FL 33701

CONTENTS

I. Introduction

Dinoflagellates are eukaryotic microalgae that constitute the division Pyrrhophyta. They are either freshwater or marine with representatives in both pelagic and benthic environments. Dinoflagellate habitats extend from boreal to tropical waters, and many species are cosmopolitan.

Marine dinoflagellates are best known for causing red tides, massive fish kills, and shellfish poisonings such as paralytic shellfish poisoning (PSP), neurotoxic shellfish poisoning (NSP), and diarrhetic shellfish poisoning (DSP). The species associated with such events are all "bloom" species, meaning that populations can increase from background levels (approx. 10^3 cells$\cdot$L^{-1}) to concentrations of 10^5–10^7 cells$\cdot$L^{-1}. Sunlight, when scattered or absorbed by these high concentrations of dinoflagellate cells can cause visible surface discolorations that are red, orange, yellow, brown, or a variety of hues in between.

The presence of near-monospecific dinoflagellate blooms over thousands of km^2 of coastal and oceanic waters lasting weeks to months is indicative of the successful competition strategies and adaptations of this group. No one cell characteristic, however, allows a bloom-forming species to outcompete other phytoplankton. Rather, a combination of behavioral and physiological adaptations interacting with environmental conditions gives dinoflagellates a competitive advantage over nonmotile phytoplankton.

This chapter will provide an overview of dinoflagellate blooms with emphasis on the distinctive cellular characteristics of dinoflagellates, the dynamics of bloom initiation and maintenance, and the ecological and economic impacts of such blooms. The epithets used in this chapter are those used by the respective authors. When a particular epithet is used in a general context, the most recent revised name has been used. A partial list of toxic dinoflagellate synonymies for species referred to in this chapter is given in Table 15-1.

II. Cell characteristics

Dinoflagellates range in size from <10 μm to about 2 mm, but most are <200 μm and exist as single cells or colonies. They are free-living or

Table 15-1. *A partial list of dinoflagellate synonymy*

1. *Protogonyaulax tamarensis* (Lebour emend. Taylor) Taylor, 1979
   Synonyms: *Gonyaulax tamarensis* Lebour 1925
   Gonyaulax tamarensis var. *tamarensis* Lebour, 1925
   Gessnerium tamarensis (Lebour) Loeblich & Loeblich, 1979
   Alexandrium tamarense (Lebour) Balech, 1985
2. *Protogonyaulax excavata* (Braarud) Taylor, 1979
   Synonyms: *Gonyaulax tamarensis* var. *excavata* Braarud, 1945
   Gonyaulax excavata (Braarud) Balech, 1971
   Gessnerium tamarensis (Lebour) Loeblich & Loeblich, 1979
   Alexandrium excavatum (Balech) Balech & Tangen, 1985
3. *Protogonyaulax catenella* (Whedon & Kofoid) Taylor, 1979
   Synonyms: *Gonyaulax catenella* Whedon & Kofoid, 1936
   Gessnerium catenellum (Whedon & Kofoid) Loeblich & Loeblich, 1979
   Alexandrium catenella (Whedon & Kofoid) Balech, 1985
4. *Gonyaulax monilata* Howell, 1953
   Synonyms: *Gessnerium mochimaensis* Halim, 1969
   Gessnerium monilatum (Howell) Loeblich, 1970
   Pyrodinium monilatum (Howell) Taylor, 1976
   Alexandrium monilatum (Howell) Taylor, 1979
5. *Pyrodinium bahamense* var. *compressa* (Bohm) Steidinger, Tester, & Taylor
   Synonyms: *Pyrodinium bahamense* Plate 1906, in part
   Pyrodinium bahamense f. *compressa* Bohm, 1931
   Gonyaulax schilleri Matzenauer, 1933
   Pyrodinium schilleri (Matzenauer) Schiller, 1937
6. *Ptychodiscus brevis* (Davis) Steidinger, 1979
   Synonym: *Gymnodinium breve* Davis, 1948
7. *Prorocentrum lima* (Ehrenberg) Dodge, 1975
   Synonyms: *Cryptomonas lima* Ehrenberg, 1859
   Exuviaella lima (Ehrenberg) Butschli, 1885
   Dinopyxis laevis Stein, 1883
   Exuviaella laevis (Stein) Schroeder, 1900
   Exuviaella marina Tsenkovskiy, 1881
   Exuviaella marina var. *lima* (Ehrenberg) Schiller, 1933

symbiotic. Many free-living species are photosynthetic and contain chlorophylls *a* and c_2 (e.g., *Gonyaulax, Amphidinium, Cachonina*). Some genera contain only a few photosynthetic species (e.g., *Protoperidinium* and *Polykrikos*), and other genera contain strictly heterotrophic species (e.g., *Oxyrrhis, Warnowia,* and *Nematodinium*). Still other genera include an as yet undetermined number of heterotrophic species (e.g., *Gymnodinium* and *Gyrodinium*). The nonphotosynthetic free-living species have both dinoflagellate and "plant" characteristics (dinosterol, polysaccharide cell coverings including cellulose, cyst walls of sporopollenin-like material, and characteristic dinoflagellate morphology and life cycles).

Cytologically, the most unique feature is the nucleus with continually condensed chromosomes (up to 500) attached to a permanent nuclear membrane; mitosis is typically closed. Chloroplasts have a 2–3 membrane envelope and contain chlorophylls, xanthophylls, and, in some species, phycobiliproteins (Wilcox and Wedemayer 1985). Not all photosynthetic species have pyrenoids, but if they occur they are associated with the chloroplast (see Steidinger and Cox 1980). At least one stage in the life cycle is motile, and the motile cells typically have two dissimilar flagella.

Both photosynthetic and nonphotosynthetic free-living species are haplontic, and where sexual cycles have been documented meiosis is usually postzygotic. Some neritic and estuarine species have an alternation of dimorphic generations with a haploid, motile vegetative phase and a diploid benthic resting stage. The benthic resting stages, or hypnozygotes, play a key role as "seed" populations for the initiation of blooms (Steidinger and Haddad 1981; Anderson, Chisholm, and Watras 1983).

III. Bloom dynamics

The life cycles, behavior, physiology, and growth characteristics of dinoflagellates that affect humans ultimately determine the seasonality, geographic extent, and duration of the impact. These biological attributes and species adaptations contribute to the development and maintenance of dinoflagellate blooms that cause mass marine mortalities and shellfish poisonings. The mere occurrence of toxic species does not lead to impacts. Rather, it is the population abundance above background levels and the distribution of these "blooms" along coasts and in estuaries that cause problems. If all these blooms were to develop at sea and dissipate, there would be no impact to coastal regions and minimal impact to man.

A. *Initiation*

Steidinger and Haddad (1981) speculated that all dinoflagellates that produce recurrent blooms in estuarine and coastal waters have a benthic resting stage, probably a hypnozygote. The hypnozygote is typically a thick-walled cell with reduced organelles and is presumed to have reduced metabolic activity. It contains metabolic reserves that may be utilized during dormancy (Anderson 1980; Steidinger 1983). A hypnozygote is produced by a sexual process that starts with the fusion of isogametes or anisogametes to form a motile zygote or planozygote. The planozygote can produce a hypnozygote (historically, the dinocyst). The process has been called encystment; however, it is not really the encystment of a haploid cell but rather the transformation of a diploid cell into a resistant structure. Hypnozygotes play a key role as "seed" populations for the initiation of blooms (Steidinger and Haddad 1981; Anderson, Coats, and Tyler 1985). This phenomenon was demonstrated for *Protogonyaulax tamarensis* in New England waters (Anderson et al. 1983). As noted by Steidinger and Baden (1984), the involvement of benthic resting stages in dinoflagellate life cycles "are examples of recognized survival strategies in that hypnozygotes withstand suboptimal water column conditions, provide genetic diversity, provide dispersal mechanisms (cyst transport), and constitute a permanent source stock."

Anderson and Keafer (1985) report thousands of cysts·cm^{-3} and 6 $\times$ 10^7 cysts·m^{-2} for certain areas. Higher concentrations of cysts are found in silt and clay sediments. They also noted that cysts can be buried up to 12 cm or more, anoxic conditions inhibit germination, imminent germination in the field can be monitored by detection of autofluorescence associated with developing chloroplasts, and some species can exhibit protracted excystment over months or restricted excystment during a specific season. Excystment of some species is influenced by temperature changes and exposure to light (Stosch 1973; Anderson and Keafer 1985). Not all cysts are viable, nor is excystment 100% successful in the production of meiocytes. However, very low numbers of cysts are actually required to initiate a bloom. Steidinger and Haddad (1981) calculated that it would take 4 cysts·cm^{-2} to seed a 2 km^2 area to a water column concentration of 24,576 cells·L^{-1} if the vegetative cells divided at 0.5 day^{-1} for 14 days and excysted cells had 50% motility success. Similarly, Lewis, Tett, and Dodge (1985) demonstrated that approximately 12.5 cysts·cm^{-2} of sediment could have seeded the 1983 *Gonyaulax polyedra* bloom in a Scottish sea loch.

Anderson et al. (1982) documented the distribution of *Protogonyaulax tamarensis* cysts along the northeastern United States and found cysts in areas not known to have PSP outbreaks, for example, Connecticut and Long Island. Although the potential for PSP outbreaks occurs in these areas, documented differences in toxicity among strains of *Protogonyaulax* (Maranda, Anderson, and Shimizu 1985) could account for lack of shellfish poisoning outbreaks. Balech (1985) recently revised *Protogonyaulax* and identified consistent morphological differences among strains of *P. tamarensis* sufficient to warrant the naming of new species. He suggested that some of the differences in toxicity and growth parameters may be due to differences in species (Balech, personal communication, 1985).

B. Growth

After initiation, the asexual growth characteristics of these organisms determine if a bloom will occur. The study of cell adaptations for the utilization of light and nutrients, for the production of allelopathic growth factors, and for migratory behavior provides us with clues as to how dinoflagellates may take advantage of environmental conditions to outcompete other species to form near-monospecific blooms with high population densities.

1. Nutrients. After excystment, vegetative dinoflagellate cells must compete for available nutrients with all other forms of phytoplankton. Dinoflagellates are generally accepted as having lower growth (asexual division) rates than diatoms (Sournia 1974; Brand and Guillard 1981), a condition most often associated with higher dark respiration rates (Falkowski and Owens 1978; Chan 1980; Raven and Beardall 1981). Therefore, they should be less successful competitors for available nutrients. However, inspection of the available growth-rate literature indicates considerable overlap in diatom and dinoflagellate growth rates. Different dinoflagellate species, and even the same species exposed to different environmental variables, have documented division rates of 0.1–2.67 d^{-1} (Chan 1978; Weiler and Eppley 1979; Rivkin and Swift 1985; and others). Also, Smayda and Karentz (1982) found that the growth rate of 14 species (16 clones) grown outdoors all had μ_{max} values higher than any reported from laboratory cultures and ranged from 0.4 to 2.48 d^{-1} (0.58–3.54 doublings·d^{-1}). Maximum growth rates exceeded 1.0 d^{-1} in 13 of 16 clones and 3 d^{-1} in 6 clones. However, it is generally accepted that species that form massive blooms (e.g., *Ptychodiscus brevis, Gyrodinium aureolum, Protogonyaulax tamarensis,* or *Gonyaulax monilata*) have normal dinoflagellate division rates (≤ 1.0 d^{-1}). For example, light-saturated division rates for *Ptychodiscus brevis* are in the range of 0.1 to 0.5 d^{-1} (Shanley 1985) whereas *Protogonyaulax tamarensis* generally has a growth rate of one doubling in three days (Yentsch, Cole, and Salvaggio 1975). Roberts (1979) estimated that the measured division rates of *Ptychodiscus brevis* were sufficient to yield the observed in situ population increase during a bloom.

Dinoflagellates do not display any unusual nutritional requirements. Their use of inorganic or organic nitrogen and phosphorus or trace elements is not unlike other phytoplankton although most dinoflagellates are auxotrophs and require vitamins (e.g., Lewin 1962; Provasoli and MacLaughlin 1963; Loeblich 1967). The larger cell volume and subsequently lower A/V may, however, influence their nutritional requirements and uptake rates (Shuter 1978). Generally, dinoflagellates are considered to have a lower potential for uptake of inorganic nitrogen than diatoms during the light period (Eppley, Rogers, and McCarthy 1969; Sherr et al. 1982; Paasche, Bryceson, and Tangen 1984). Some dinoflagellates, including bloom species, overcome this deficit with the ability to assimilate inorganic nitrogen in the dark (Harrison 1976; MacIsaac 1978; Dortch and Maske 1982). However, MacIsaac et al. (1979) argued against the hypothesis that dinoflagellates had a special ability to take up nitrate and ammonia in the dark. Their work demonstrated that *Gonyaulax excavata* had low dark uptake rates and that nitrate uptake was more light-dependent than ammonium uptake. *Pyrocystis notiluca* has the ability to take up inorganic phosphorus (P) at low, *in situ* concentrations (Rivkin and Swift 1982, 1985), but, generally, bloom-forming dinoflagellates do not appear to be well adapted for nutrient uptake at low *in situ* levels. However, ambient water-column concentrations do not necessarily support observed growth rates if nutrient fluxes or remineralization rates are high or cells are growing on internal nutrient stores.

Organic nitrogen and phosphorus sources may also significantly contribute to dinoflagellate growth. The cellular yield and growth rates of *Protogonyaulax catenella* (Norris and Chew 1975) and *Ptychodiscus brevis* (Vargo, unpublished data) on urea are similar to those obtained on other inorganic nitrogen sources. Uptake rates for urea were somewhat lower than for ammonia and nitrate during a *Peridinium cinctum* bloom in Lake Kinneret, Israel, but urea was second only to ammonia as the preferred nitrogen source (McCarthy, Wynne, and Berman 1982).

The presence of alkaline and acid phosphatases in many species of phytoplankton (Kuenzler and Perras 1965), including dinoflagellates, suggests that a variety of organic phosphorus compounds can be used. Cell-surface-bound alkaline phosphatase can play a significant role in meeting cellular P requirements for photosynthesis and growth in oceanic dinoflagellates (Rivkin and Swift 1980) and during near-shore blooms (Vargo and Shanley 1985) when *in situ* inorganic sources are depleted or flux cannot meet demand.

2. Growth factors. Dinoflagellate blooms, particularly of toxic species, tend to become monospecific or nearly monospecific at their peaks (Sweeney 1979; Steidinger and Baden 1984; Holligan 1985). Since enhanced growth alone cannot account for the superior competitive ability of dinoflagellates, other explanations have been sought. The production of growth-inhibiting (allelopathic or allelochemical) substances has been proposed as one explanation for the high degree of monospecificity in blooms. Inhibition of phytoplankton by dinoflagellates or their extracts has been demonstrated by Pincemin (1969, 1971), Uchida (1977), Freeberg, Marshall, and Heyl (1979), and others. However, both Elbrächter (1976, 1977) and Kayser (1979) point out the problems encountered with attempting to demonstrate allelopathy from experimental data such as the use of high cell densities and the growth stage from which extracts are derived.

Potential self-stimulation or preconditioning of water masses to allow growth is yet another aspect of the unialgal nature of dinoflagellate blooms. Freeberg et al. (1979) and Steidinger (1983) noted that use of

Ptychodiscus brevis preconditioned media for growth of this species reduces the lag phase. Provasoli (1979) elegantly summarized the potential requirements for preconditioning water masses for the inception and growth of dinoflagellate blooms. The role of chelators and trace metals such as iron (Martin and Martin 1973; Glover 1978) and copper (Anderson and Morel 1978; Graneli, Persson, and Edler 1986) cannot be ignored and may be of special importance in population growth stimulation or inhibition.

3. Light. The motility and vertical swimming patterns (taxes) of dinoflagellates mean that they are exposed to a wide range of light quality and irradiance levels. As a compensating mechanism, they possess pigment systems that absorb over a wide range of wavelengths (Jeffrey, Sielicki, and Haxo 1975; Jeffrey 1981; Prézelin 1981). Dinoflagellates respond to variation in irradiance by changing intracellular concentrations of light-harvesting pigment complexes (Prézelin and Sweeney 1979; Falkowski and Owens 1980; Perry, Talbot, and Alberte 1981) and the ratio of accessory pigments to chlorophyll to maintain photosynthetic efficiency in a process termed *photoadaptation* (Prézelin 1981; Richardson, Beardall, and Raven 1983). Photoadaptation is an attempt by the cell to maintain maximal photosynthetic output and, therefore, growth rates.

Photoadaptation in response to light quality has been observed in *Prorocentrum mariae-lebouriae*. In this species the cellular chl. *a*, chl. *c*, and peridinin content, and the peridinin:chl. *a* ratio increased with decreasing intensity of green wavelengths (500–600 nm). However, growth rates were maximal in blue light (400–500 nm) at low irradiances (Faust, Senger, and Meeson 1982). Although *Prorocentrum mariae-lebouriae* responds to spectral quality, the principal photoadaptive response is to photon flux density (Meeson and Faust 1985). Cells grown in blue-green light require less total photosynthetic energy to produce a new cell than in white or orange light.

Exposure to low irradiance of white light also results in an increase in the peridinin:chl. *a* and chl. *a*:chl. *c* ratios, which increases the use efficiency of blue wavelengths that dominate marine waters (Prézelin 1976; Faust et al. 1982; see Richardson et al. 1983 for additional references). Richardson et al. (1983) rank dinoflagellates as having a lower light preference, with lower irradiance levels for the onset of photoinhibition, than diatoms and chlorophytes. Field evidence supports this conclusion. Blasco (1978) noted that five of six dinoflagellate species off Baja California exhibited subsurface maximums and surface avoidance by *Ceratium* and other species has been documented (Hasle 1950; Dodge and Hart-Jones 1977; Harris, Heaney, and Talling 1979). Increased xanthophyll:chlorophyll ratios also protect chlorophyll from photooxidation (Krinsky 1979; Song et al. 1980). This may offer a distinct advantage to bloom-forming species that migrate to or accumulate in near-surface waters where high irradiance and/or ultraviolet radiation may significantly reduce photosynthesis rates, oxidize pigment systems, or result in cell death.

The response time for photoadaptation is critical if cells are to maintain maximum photosynthetic and growth rates (Richardson et al. 1983). Some species are able to adjust in hours (Prézelin and Matlick 1980) whereas one to two generation times are required by others, for example, *Protogonyaulax tamarensis* (Maranda 1985) and *Ptychodiscus brevis* (Shanley 1985). With vertical mixing rates in the euphotic zone on time scales of minutes to hours (Rivkin, Voytek, and Seliger 1982; Denman and Gargett 1983), species such as *P. brevis* may adapt to some integrated light field that allows maintenance of maximum photosynthetic and growth rates.

4. Migratory behavior. During bloom inception and growth, the diel light–dark cycle may trigger vertical migration. After excystment, migration toward the surface is characteristic of most bloom-forming species. Interactions between vertical migration and the physical structure of water masses have been postulated as concentrating mechanisms for blooms and physical transport (see Tyler and Seliger 1978, 1981; Holligan 1985).

Gran (1912, 1929) and Tschirn (1920) initially described dinoflagellate migratory responses as phototactic. Forward (1976) defined phototaxis as a directional movement in relation to the direction of the light stimulus. Most dinoflagellates exhibit a positive phototactic response (e.g., Hasle 1950, 1954; Eppley, Holm-Hansen, and Strickland 1968; Forward 1976), although a negative response has also been described (Taylor et al. 1966; Wheeler 1966; Blasco 1978). Offsets in the timing of vertical movement from the onset or end of the light–dark cycle in species such as *Gymnodinium sanguineum* (Cullen and Horrigan 1981), *Ceratium furca* (Weiler and Karl 1979), *C. hirundinella* (Heaney and Furnass 1980), *Gonyaulax polyedra* (Eppley et al. 1968), and *Ptychodiscus brevis* (Heil 1986) suggest the involvement of a geotactic response possibly coupled with an endogenous rhythm (e.g., Heaney and Eppley 1981; Cullen and Horrigan 1981). Persistence of migration patterns in complete darkness in some of the above species implies a control mechanism not directly coupled to the light–dark cycle.

Rapid changes in temperature and salinity may modify vertical migratory behavior through alterations of swimming speeds. A $5°/\infty$ reduction in salinity depressed the swimming rate and elicted circular swimming motion in *Prorocentrum mariae-lebouriae* (Tyler and Seliger 1981). Thermal stratification reduced the swimming speed of *Cachonina niei* during dawn migrations (Epply et al. 1968; Kamykowski and Zentara 1977). Some species such as *Ptychodiscus brevis* can migrate upward through a weak thermal gradient (Heil 1986) or a halocline (e.g., *Ceratium furca* and *Prorocentrum micans:* Edler and Olsson 1985) but, once entrained above the gradient, will not migrate down through it.

Motility has long been associated with high P/R ratios in dinoflagellates. However, Raven and Beardall (1981) estimated that flagellar movement requires only a small fraction (approximately 1%) of the total maintenance respiration in *Gonyaulax polyedra*. Therefore, migratory behavior may not exact an onerous energetic cost on dinoflagellates.

C. Maintenance

Possession of a positive tactic response (i.e., upward movement), whether governed by light, gravity, or an internal clock, brings motile species into surface waters where growth must exceed losses from advection, predation, and natural mortality for blooms to persist. Retention of populations within convection cells (Evans and Taylor 1980) or along boundaries associated with temperature, salinity, or shear gradients contributes to bloom maintenance (Holligan 1985).

Temperature discontinuities act to modify migratory behavior by altering swimming speeds. A sharp reduction in temperature will reduce swimming speed (Hand, Collard, and Davenport 1965), but species-specific differences in this response occur (Kamykowski and McCollum 1986). Kamykowski (1981) also found that light interacting with temperature and the absolute temperatures rather than the thermal gradient itself yielded variations in the migratory behavior of *Prorocentrum micans.*

Salinity barriers, whether a low-salinity surface lens (e.g., Le Fevre and Grall 1970; Tangen 1979) or deeper haloclines in estuaries and bays (Seliger et al. 1970; Tyler and Seliger 1978, 1981) may restrict, concentrate, maintain, or transport dinoflagellate populations. On a small scale, populations can be maintained in a restricted area for long periods as a result of migratory behavior interacting with internal waves or seiches (Kamykowski 1974, 1978) and fronts set up by the interaction of tidal excursions with river flow, salinity, and bottom topography (Incze and Yentsch 1981). Populations of *Protogonyaulax tamarensis* and *Heterocapsa triquetra* were retained in a Massachusetts salt pond by avoidance of the surface layer that was exchanged at a rate of 0.5 d^{-1}, the restriction of exchange in the lower layer by the inlet sill, and the maintenance of a salinity gradient by freshwater from springs (Anderson and Stolzenbach 1985).

On a larger scale, restriction to a particular water mass has been identified as the transport mechanism for movement of *Ptychodiscus brevis* patches from the west to the east coast of Florida (Murphy et al. 1975;

Roberts 1979). Formation and maintenance of *Gyrodinium aureolum* blooms in the English Channel are associated with fronts induced by sharp thermal boundaries (see Holligan 1979), whereas Tangen (1979) suggests that blooms of this species were associated with surface stratification caused by runoff following nutrient enrichment by either sewage or upwelling.

Turbulence and shear forces at the boundaries of stable water columns may act as negative selection factors for dinoflagellates. Kemp and Mitsch (1979) considered shear forces and their effect on cellular integrity and metabolism as important factors in maintaining species diversity. White (1976) determined that high-speed rotary shaking (125 rpm) caused death and disintegration of *Protogonyaulax excavata* and that slower shaking caused cell mortality or decreased growth rates (Pollingher and Zemel 1981). The association of dinoflagellates with stable, stratified water columns (Margalef 1968; Wyatt and Horwood 1973; Margalef, Estrada, and Blasco 1979) may therefore be a consequence of the effects of turbulence on growth and cellular integrity.

The lack of vertical mixing in a stable water column restricts nutrient flux. High nutrient removal rates by elevated population densities within a patch may yield nutrient-depleted conditions if nutrient flux is restricted (Seliger, Tyler, and McKinley 1979). Eppley et al. (1968), Eppley and Harrison (1975), and Cullen and Horrigan (1981) proposed that downward migration to the nitrocline at night where dark uptake of nitrogen can occur followed by upward migration during the daylight hours when the stored nitrogen can be used to maintain photosynthesis could explain how some species of dinoflagellates are able to maintain nutrient sufficiency in highly concentrated blooms.

Utilization of organic phosphorus compounds (Seliger, Loftus, and Subba Rao 1975; Rivkin and Swift 1979, 1980; Vargo and Shanley 1985) or bacterial-derived organic substrates as in *Ptychodiscus brevis* (Baden and Mende 1978) can contribute to bloom maintenance, but whether sufficient organic substrates are available for continued cell division remains to be determined. Being K-strategists, dinoflagellates are characterized by high storage capacities (e.g., Wyatt 1974; Kilham and Kilham 1980). With a high storage capacity, a cell should have enhanced ability for growth or maintenance under nutrient-depleted conditions. This hypothesis is not supported by Dortch et al. (1984), who found that *Amphidinium carterae* and several diatoms of varying size exhibited little difference in the number of divisions after nutrient depletion. However, a variety of dinoflagellates studied by Hitchcock (1982) did contain higher levels of carbohydrates, protein, and lipid than diatoms of equivalent cell volume. Storage of cellular phosphorus allowed *Pyrocystis noctiluca* to maintain growth at a reduced rate for 25 days after orthophosphate in the media declined below 0.1 μM (Rivkin and Swift 1985), and *Gonyaulax polyedra* continued to divide for up to 2 generation times using stored carbon when exposed to suboptimal light intensities (Rivkin et al. 1982). Similarly, *Prorocentrum mariae-lebouriae* uses carbon fixed during exposure to low light to replenish carbon stores during subsurface transport from the mouth to the upper Chesapeake Bay (see Tyler and Seliger 1981), allowing the cell to resume maximum growth rates quickly upon reintroduction into optimum light fields (Rivkin et al. 1982).

D. Termination

Grazing, cell death, advection, nutrient limitation, parasitism, and shifts in the life cycle have been offered as possible explanations for the demise of blooms in general (e.g., Smayda 1973, 1980; Anderson 1984; Watras et al. 1985). Given the apparently intimate relationships between hydrographic factors and bloom initiation/maintenance, degradation of these physical conditions would probably dissipate the bloom. This might be due to the direct effect of increased turbulence on cell growth, as seen in *Peridinium cinctum* in Lake Kinneret (Pollingher and Zemel 1981) or *Gonyaulax excavata* in Argentina (Carreto et al. 1986), or to destabilization of the frontal systems that concentrate these organisms (Haddad and Carder 1979) with subsequent dispersal and dilution (Steidinger and Haddad 1981).

Macronutrient limitation has been offered as an explanation for the termination of a *Pyrodinium bahamense* bloom in Port Moresby Harbor, New Guinea (Maclean 1977). Cyst formation in batch cultures of *Gonyaulax tamarensis* was triggered by nitrogen and phosphorus limitation but was negligible in nutrient-replete media even at nonoptimal light and temperature conditions (Anderson, Kulis, and Binder 1984).

Herbivorous zooplankton may be excluded from toxic dinoflagellate blooms. Grazing by planktonic larvae of the polychaete *Polydora ligni* and the tintinnid *Favella* sp. was offered as an alternate explanation for the demise of *Protogonyaulax excavata* blooms in two Cape Cod, Massachusetts, embayments and for the absence of blooms in some years (Watras et al. 1985). Similarly, the phagotrophic dinoflagellate *Polykrikos schwartzii* appeared to replace the normal zooplankton herbivores in a *Gonyaulax excavata* bloom in Argentina (Carreto et al. 1986).

Other evidence, however, reveals the complex nature of dinoflagellate–zooplankton interactions in regulating blooms. Huntley (1982) found a significant reduction in the grazing rate of *Calanus pacificus* in a bloom of *Gymnodinium flavum*. In a later study (Huntley, Barthel, and Star 1983), reduced grazing on *Protogonyaulax tamarensis*, *Scrippsiella trochoidea*, *Gonyaulax grindleyi*, and *Ptychodiscus brevis* could not be explained on the basis of particle size or food concentration. The chemosensory ability of zooplankton has been amply demonstrated (Poulet and Marsot 1978, 1980; Donaghay and Small 1979), which suggests that at least some dinoflagellates may produce substances that suppress ingestion.

Some toxic dinoflagellates are also bioluminescent. Suppression of grazing has been offered as one explanation for the adaptive significance of bioluminescence in dinoflagellates (Tett and Kelly 1973; Porter and Porter 1979). Laboratory studies by Esaias and Curl (1972) and White (1979) demonstrated reduced ingestion rates of bioluminescent dinoflagellates. They suggest that the flash acts as a visual display that startles or confuses the grazing copepod sufficiently to allow the dinoflagellate to escape. Buskey, Mills, and Swift (1983) noted a decrease in the slow, circular swimming patterns characteristic of normal grazing behavior in the estuarine copepod, *Acartia hudsonica*, with a concomitant increase in swimming speed and in high-speed swimming bursts (escape behavior) in the presence of bioluminescent clones of *Protogonyaulax tamarensis*. They suggest that these behavioral modifications would lead to reduced grazing efficiency.

For blooms that occur in shallow waters, such as estuaries and ponds, predation by benthic or sessile filter feeders may also contribute to bloom regulation. Bivalves and tunicates are capable of removing particles down to 1 μm in size at filtering rates on the order of liters·hr^{-1} (Jorgensen 1966; Winter 1978). Recent studies emphasize the importance of the benthos in regulating water-column phytoplankton biomass (Elmgren et al. 1980; Cloern 1981; Officer, Smayda, and Mann 1982). The fact that bivalves and other filter feeders are the intermediaries in human PSP, NSP, and DSP (Steidinger and Baden 1984) suggests that their role in bloom regulation must be evaluated in estuaries and shallow coastal waters.

IV. Ecological and economic impacts

A. Negative impacts

Dinoflagellate blooms have been associated with marine fish and invertebrate mortalities worldwide. Most of these events are caused by dinoflagellates that produce neurotoxins capable of killing animals directly or via toxic prey items (Mills and Klein-MacPhee 1979; White 1984). In other cases, oxygen depletion caused by community respiration may cause mortalities. An example of the latter occurred in the New York Bight during the summer of 1976 when a bloom of nontoxic *Ceratium tripos* covered 13,000 km^2 and caused a 69% mortality of offshore fisheries stocks (Mahoney and Steimle 1979). Toxic dinoflagellate blooms can be a major threat to concentrated mariculture populations of salmon, trout, and bivalves with potential losses in the millions of dollars. For example, a 1984 *Gonyaulax excavata* red tide in the Faroe Islands impacted production of sea-reared rainbow trout and salmon; 27–35 metric tons were lost through mortality (Mortensen 1985). This red tide seeded benthic areas of fish cage culture with *G. excavata* cysts (Mortensen 1985). Since Faroe Island trout and salmon production is expected to be in the thousands of metric tons per year, this could be a considerable economic loss. Such blooms can also cause losses to multispecies fisheries in coastal waters, particularly if they involve demersal and pelagic fishes or bivalves such as oysters, clams, and mussels. The economic impact is always negative, but the ecological impact may be a necessary part in the adaptation of some systems to stress and perturbations.

Toxic marine dinoflagellates cause paralytic shellfish poisoning, neurotoxic shellfish poisoning, and diarrhetic shellfish poisoning. Human symptoms of poisoning can include nausea, vomiting, diarrhea, tingling of the lips, fingers, and toes, numbness, disorientation, hot/cold sensation reversals, pupil dilation, incoordination, and paralysis. Normally, symptoms occur within several hours of ingestion and last only two to three days; however, over 350 human mortalities from PSP have been documented (See Lo Cicero 1975; Taylor and Seliger 1979; Anderson, White, and Baden 1985). Paralytic shellfish poisoning caused by *Protogonyaulax* spp., *Pyrodinium bahamense* var. *compressa*, *Gymnodinium catenatum*, and others has been documented from the United States, Mexico, Canada, South America, South Africa, Europe, Scandinavia, and Southeast Asia and is literally a worldwide public health problem.

Canada initiated public health monitoring programs for toxins in shellfish (mouse bioassay) in the 1940s. Between the 1940s and 1975, 236 illnesses and 10 deaths were recorded in Canada; however, no deaths and only 5% of the illnesses were from commercially produced shellfish. Recreational shellfish harvesters, either local residents or visitors, create the major public health problem (Bond 1975). Established monitoring programs reduce human intoxications by providing the capability to ban the harvest of bivalves and post warning signs. In areas with no history of shellfish poisoning, knowledge of the possible causative agents and the ability to implement immediate protective measures can save lives, for example, the 1972 Massachusetts PSP outbreak (Bicknell and Walsh 1975). In Canada and Alaska, permanent bans have been placed on the harvest of particular bivalve species, and specific areas have been closed because of year-round toxicity (Lutz and Incze 1979; Neve and Reichardt 1984). Such closures, whether temporary or permanent, have significant economic impacts. In Alaska, millions of pounds of shellfish meats are not harvested and marketed because of toxic shellfish closures (Neve and Reichardt 1984). Canada's loss represents about 2 million Canadian dollars annually (Lutz and Incze

1979). In the Philippines, an eight-month period of closure represented about a $500,000 loss (White, Anraku, and Hudi 1984). In addition, bivalves can suffer stress and even death (Gilfillan and Hanson 1975; Shumway et al. 1985), thus adding another potential economic loss factor.

Fish kills and NSP in Florida can cause economic stress to local communities and a number of industries. The 1971 red tide caused an estimated economic loss of $20 million to the tourist industry alone, and a 1973–4 red tide was estimated to cause a $15 million loss to that industry. In addition, sports fishing, wholesale and retail seafood sales, and real estate sales are affected. As Jensen (1975) pointed out, an "economic halo" effect occurs with red tides because the public notoriety causes buyer resistance to all seafood products, even if they are not affected or were harvested in safe areas. This halo effect can extend far beyond the country or state involved, but the economic impacts of this overreaction have not yet been calculated.

Ptychodiscus brevis red tides in the Gulf of Mexico cause negative economic impact through losses in tourism, fishing expenditures, real estate sales, seafood sales, and closure of shellfish beds to commercial and recreational harvest. These losses are due to brevetoxins causing NSP and fish kills. Additionally, an irritating aerosol is associated with *P. brevis* red tides. The combination of NSP, fish kills, and the toxic aerosol makes *P. brevis* red tides unique. The irritating aerosol is not a gas. It has a particulate component that causes tearing eyes, coughing, nasal irritation, and tingling lips in humans exposed to sea spray, generated in surf zones, where *P. brevis* and/or brevetoxins exist. Brevetoxins have been documented from aerolized *P. brevis* seawater (Richard Pierce, Mote Marine Laboratory, personal communication, 1987) and are probably associated with either cell debris or salt crystals.

Interstate and international trade presents a potential health problem if shellfish harvest or production areas are not routinely monitored. An example is the case of epidemic PSP in western Europe in 1976. PSP was reported from Spain, France, Italy, Switzerland, and Germany in October and November 1976. The source was mussels shipped from northwestern Spain (Lüthy 1979). Spain produces 200,000 metric tons of shellfish yearly using raft culture. This practice was started in 1940, but a toxic dinoflagellate problem was not recognized until 1976. The Spanish government now has an extensive public health program that monitors PSP and DSP in shellfish. The government also initiated a comprehensive research program to determine the toxic species involved, their distribution, and physical, chemical, and biological factors leading to blooms (Campos et al. 1982).

Diarrhetic shellfish poisoning is an only recently recognized public health problem. It is documented from the Netherlands, Spain, Japan,

United States, Norway, Chile, and other areas. Shellfish can become toxic from *Dinophysis* spp., which contain bioactive fatty acid derivatives that cause gastrointestinal distress. Counts as low as 20,000 cells·L^{-1} of seawater in the Netherlands caused toxicity (Kat 1985). The same is true for PSP; counts of 20,000 toxic *Protogonyaulax* cells·L^{-1} can cause toxicity. Since bivalves are filter feeders, toxins can be bioaccumulated rapidly. In France and Japan, shellfish beds are closed when certain *Dinophysis* species reach 200 cells·L^{-1} of seawater (Lassus et al. 1985). In addition to *Dinophysis* spp., planktonic *Prorocentrum* spp. are suspect. Benthic *Prorocentrum* spp. also have been implicated in ciguatera, a tropical fish poisoning. *Prorocentrum lima* produces maitotoxin (associated with tropical fish poisonings) and okadaic acid, a fatty acid associated with DSP (Yasumoto 1985).

Total economic and living resource losses are difficult to determine. Dead fish sink until their body cavities fill with gases and they float to the surface. Bottom surveys are rarely done to determine losses. Some shellfish exposed to toxins stop filtering and survive as long as they can keep their valves tightly closed and metabolize anaerobically. Economically, cost data on tourism, real estate sales, export markets, hospitalization for victims, shellfish monitoring programs, fish cleanup activities, unharvested resources due to bans, seafood rejection by the consumer, and other factors would have to be determined for each episode to project total economic loss. Economic losses are high, but probably lower on an event basis compared to floods, droughts, hurricanes, earthquakes, and other natural environmental impacts.

B. Positive impacts

Most dinoflagellates that form blooms are photosynthetic, and all dinoflagellates that cause toxic red tides are photosynthetic. As phytoplankton, they are primary producers at the base of many food webs. They provide a high-quality food for herbivores (Paffenhofer 1970; Morey-Gaines 1979; Hitchcock 1982). However, the herbivore must have enzymes capable of digesting and assimilating complex carbohydrates such as cellulose and other polysaccharides (Hitchcock 1982). In addition to providing particulate organic carbon, dinoflagellates are "leaky" and release 2–30% of their daily carbon as amino acids, carbohydrates, polypeptides, enzymes, and other biochemicals (Provasoli 1979). Upon death, they create detrital particles and release more dissolved organics. In nearly monospecific blooms such as those of *Gyrodinium aureolum* in the western English Channel, this species constitutes 90% of the total microalgal biomass, and chlorophyll biomass values have exceeded 500 mg·m^2 (Holligan 1985). Consequently, dinoflagellate blooms, toxic or not, provide large carbon inputs to marine systems, particularly if the blooms last several months. Unfortunately, little information is available

on the potential carbon input from red tide blooms except for speculations of increased secondary production due to *Protogonyaulax* in eastern Canada (White et al. 1984) and dinoflagellate blooms in Chesapeake
Bay (Sellner and Olson 1985). A preliminary study by Vargo et al.
(1987) suggests that *Ptychodiscus brevis* can contribute a significant portion (20%) of the annual phytoplankton carbon production in the central West Florida Shelf water column. However, the lack of long-term
seasonal data during nonbloom periods did not allow an assessment of
what production would be if *P. brevis* were not present.

Periodic perturbations, such as red tides caused by *P. brevis* off the
west coast of Florida, may keep patch reefs and other bottom communities in continual successional phases and create less mature systems
with simplified interactions. Fish and invertebrate kills at various areas
along the shelf could, theoretically, decrease species diversity, reduce
structural complexity, create new niches, shorten food chains or reduce
the complexity of food webs, reduce interspecific and intraspecific competition, recycle nutrients, eliminate weak and diseased animals, and
increase turnover and production. Overall, this could increase system
efficiency (Margalef 1968; Vogel 1980; Rapport, Regles, and Hutchinson 1985). Mature systems, typically high in diversity, have complicated
species interactions and narrow tolerances; they can be more susceptible to collapse following a perturbation event (e.g., flood, temperature
extreme, or red tide). Vogel (1980) and others have pointed out that
certain systems are dependent on pulsed or periodic stresses that sim-
plify the system and reduce its vulnerability to a catastrophic event.
They are called stress- or perturbation-dependent ecosystems. Multiple
stresses may even be involved in some systems. This is not to say, however, that all successful ecosystems are stress-dependent.

Reef and intertidal defaunation associated with *P. brevis* blooms do
provide space and habitat for new colonizers or recolonizers (Smith
1975; Dauer and Simon 1976). Recolonization can take two to five years
and appears to affect corals more than fishes.

V. Conclusions

To interpret and understand dinoflagellate blooms requires an interdisciplinary approach to identify physical-chemical driving forces and biological responses. Fronts and density gradients along water masses alter
the migratory pattern, tactic behavior, and motility of dinoflagellates
and can, at times, concentrate and restrict large populations to particular areas of the water column. These organisms themselves can modify
the quantity and quality of their pigments and so maximize their photosynthetic rates under varying light and environmental conditions to
maintain growth rates. Migratory behavior combined with high

nutrient-storage capacity could play a key role in survival and growth. Production of possible allelopathic substances and, in some species, toxins and even bioluminescence also may affect growth by minimizing losses from competition and grazing. Despite lower growth rates than diatoms and other phytoplankton, dinoflagellates maintain a competitive advantage through a combination of factors.

Reports of near-monospecific blooms of toxic dinoflagellates have increased. Whether this is the result of increased public awareness of the phenomenon with subsequent reports to local agencies, an increase in the number of scientists working in this field, or an actual increase in the incidence of blooms remains to be demonstrated. The number of "new" areas experiencing toxic red tides in the last 15 years has been substantial (Massachusetts, Papau New Guinea, Brunei, Thailand, Spain, eastern Indonesia, Uruguay, Chile, Argentina, England, and others).

In Argentina, the first case of PSP was documented in 1980 off Valdés Peninsula, and by 1983 the entire Argentine coast was within the toxicity zone. Carreto, Negri, and Benavides (1985) demonstrated that *Protogonyaulax* blooms originated along a frontal system 50 nmi offshore. They postulated that the front provided the mechanism for initiation and transport. Areas with no documented history of shellfish poisoning can have an immediate public health problem because no monitoring and regulatory programs are in place inshore and offshore (e.g., Argentina). The first Argentine outbreak resulted in seven human deaths, 12 serious illnesses, and a number of intoxications. The Argentina incident is a typical example of (1) physical-biological interactions in red tides, (2) the need for biological monitoring, and (3) the need for public health monitoring of shellfish areas. Based on the worldwide distribution of toxic *Protogonyaulax* species causing PSP and *Dinophysis* species (and probably *Prorocentrum* species) causing DSP, all governmental public health or natural resource agencies responsible for shellfish should monitor their coastal sediments for cysts of potentially toxic dinoflagellates. Steidinger (1975) suggested that recurrent blooms of toxic dinoflagellates were due to resident cysts in sediments and that these seedbeds could and should be monitored.

Toxic red tides can restructure marine communities through selective mortality. This perturbation alone, or in combination with other stressors, can adversely affect human economic systems. Losses in excess of US $15 million for a single event have been estimated. The amount of economic loss depends on when and where the bloom occurs and its duration. Whether this stress has a positive or negative ecological impact on total system production and survival requires long-term analyses. Theoretically, stressors can be important in maintaining particular ecosystems. Often humans think and react to nature as observers and relate natural events to their economic well-being; however, people are participants. As participants, they need to understand and appreciate natural systems, their adaptability, interactions, and value.

VI. References

Anderson, D. M. 1980. Effects of temperature conditioning on development and germination of *Gonyaulax tamarensis* (Dinophyceae) hypnozygotes. *J. Phycol.* 16, 166–72.

Anderson, D. M. 1984. Shellfish toxicity and dormant cysts in toxic dinoflagellate blooms. In *Seafood Toxins*, ed. E. P. Ragelis, pp. 125–38. American Chemical Society, Symposium Ser., no. 262. Washington, DC.

Anderson, D. M., S. W. Chisholm, and C. J. Watras 1983. The importance of life cycle events in the population dynamics of *Gonyaulax tamarensis*. *Mar. Biol.* 76, 179–90.

Anderson, D. M., D. W. Coats, and M. A. Tyler. 1985. Encystment of the dinoflagellate *Gyrodinium uncatenum:* temperature and nutrient effects. *J. Phycol.* 21, 200–6.

Anderson, D. M., and B. A. Keafer. 1985. Dinoflagellate cyst dynamics in coastal and estuarine waters. In *Toxic Dinoflagellates*, ed. D. M. Anderson, A. W. White, and D. G. Baden, pp. 219–24. Elsevier, New York.

Anderson, D. M., D. M. Kulis, and B. J. Binder. 1984. Sexuality and cyst formation in the dinoflagellate, *Gonyaulax tamarensis:* cyst yield in batch cultures. *J. Phycol.* 20, 418–25.

Anderson, D. M., D. M. Kulis, J. A. Orphanos, and A. R. Ceurvels. 1982. Distribution of the toxic dinoflagellate *Gonyaulax tamarensis* in the southern New England region. *Estuar. Coast. Shelf Sci.* 14, 447–58.

Anderson, D. M., and F. F. Morel. 1978. Copper sensitivity of *Gonyaulax tamarensis*. *Limnol. Oceanogr.* 23, 283–95.

Anderson, D. M., and K. D. Stolzenbach. 1985. Selective retention of two dinoflagellates in a well-mixed estuarine embayment: the importance of diel vertical migration and surface avoidance. *Mar. Ecol. Prog. Ser.* 25, 39–50.

Anderson, D. M., A. W. White, and D. G. Baden (eds.). 1985. *Toxic Dinoflagellates*. Elsevier, New York. 561 pp.

Baden, D. G., and T. J. Mende. 1978. Glucose transport and metabolism in *Gymnodinium breve*. *Phytochemistry* 17, 1553–8.

Balech, E. 1985. The genus *Alexandrium* or *Gonyaulax* of the Tamarensis group. In *Toxic Dinoflagellates*, ed. D. W. Anderson, A. W. White and D. G. Baden, pp. 33–8. Elsevier, New York.

Bicknell, W. J., and D. C. Walsh. 1975. The first "red-tide" in recorded Massachusetts history. In *Proceedings of the First International Conference on Toxic Dinoflagellate Blooms*, ed. V. R. Lo Cicero, pp. 447–58. Massachusetts Science and Technology Foundation, Wakefield.

Blasco, D. 1978. Observations on the diel migration of marine dinoflagellates off the Baja California Coast. *Mar. Biol.* 46, 41–7.

Bond, R. M. 1975. Management of PSP in Canada. In *Proceedings of the First International Conference on Toxic Dinoflagellate Blooms*, ed. V. R. Lo Cicero, pp. 473–82. Massachusetts Science and Technology Foundation, Wakefield.

Brand, L. E., and R. R. L. Guillard. 1981. The effects of continuous light and light intensity on the reproduction rates of twenty-two species of marine phytoplankton. *J. Exp. Mar. Biol. Ecol.* 50, 119–32.

Buskey, E., L. Mills, and E. Swift. 1983. The effects of dinoflagellate bioluminescence on the swimming behavior of a marine copepod. *Limnol. Oceanogr.* 28, 575–9.

Campos, M. J., S. Fraga, J. Marino, and F. J. Sanchez. 1982. *Red Tide Monitoring Programme in NW Spain*. Report of 1977–81. International Council for the Exploration of the Sea, Biological Oceanographic Committee, CM 1982;/L:27. 7 pp.

Carreto, J. I., H. R. Benavides, R. M. Negri, and P. D. Glorioso. 1986. Toxic red tide in the Argentine Sea: phytoplankton distribution and survival of the toxic dinoflagellate *Gonyaulax excavata* in a frontal area. *J. Plank. Res.* 8, 15–28.

Carreto, J. I., R. M. Negri, and H. R. Benavides. 1985. Toxic dinoflagellate blooms in the Argentine Sea. In *Toxic Dinoflagellates*, ed. D. M. Anderson, A. W. White, and D. G. Baden, pp. 147–57. Elsevier, New York.

Chan, A. T. 1978. Comparative physiological studies of marine diatoms and dinoflagellates in relation to irradiance and cell size. I. Growth under continuous light. *J. Phycol.* 14, 396–402.

Chan, A. T. 1980. Comparative physiological study of marine diatoms and dinoflagellates in relation to irradiance and cell size. II. Relationship between photosynthesis, growth, and carbon/chlorophyll *a* ratio. *J. Phycol.* 16, 428–32.

Cloern, J. E. 1981. Does the benthos control phytoplankton biomass in South San Francisco Bay? *Mar. Ecol. Prog. Ser.* 9, 191–202.

Cullen, J. J., and S. G. Horrigan. 1981. Effects of nitrate on the diurnal vertical migration, carbon to nitrogen ratio, and the photosynthetic capacity of the dinoflagellate *Gymnodinium splendens*. *Mar. Biol.* 62, 81–9.

Dauer, D. M., and J. L. Simon. 1976. Repopulation of a polychaete fauna of an intertidal habitat following natural defaunation: species equilibrium. *Oecologia* 22, 99–117.

Denman, K. L., and A. E. Gargett. 1983. Time and space scales of vertical mixing and advection of phytoplankton in the upper ocean. *Limnol. Oceanogr.* 28, 801–15.

Dodge, J. D., and B. Hart-Jones. 1977. The vertical and seasonal distribution of dinoflagellates in the North Sea. *Bot. Mar.* 20, 307–11.

Donaghay, P., and L. Small. 1979. Food selection capabilities of the estuarine copepod *Acartia clausi*. *Mar. Biol.* 52, 137–46.

Dortch, Q., J. R. Clayton, S. S. Thoresen, and S. I. Ahmed. 1984. Species differences in accumulation of nitrogen pools in phytoplankton. *Mar. Biol.* 81, 237–50.

Dortch, Q., and H. Maske. 1982. Dark uptake of nitrate and nitrate reductase activity of a red-tide population off Peru. *Mar. Ecol. Prog. Ser.* 9, 299–303.

Edler, L., and P. Olsson. 1985. Observations on diel migration of *Ceratium furca* and *Prorocentrum micans* in a stratified bay on the Swedish west coast. In *Toxic Dinoflagellates,* ed. D. M. Anderson, A. W. White, and D. G. Baden, pp. 195–200. Elsevier, New York.

Elbrächter, M. 1976. Population dynamic studies on phytoplankton cultures. *Mar. Biol.* 35, 201–9.

Elbrächter, M. 1977. On population dynamics in multi-species cultures of diatoms and dinoflagellates. *Helgoländ. Wiss. Meeresunters.* 30, 192–200.

Elmgren, R., G. A. Vargo, J. F. Grassle, J. P. Grassle, D. R. Heinle, G. Langlois, and S. L. Vargo. 1980. Trophic interactions in experimental marine ecosystems perturbed by oil. In *Microcosms in Ecological Research,* ed. J. P. Giesy, pp. 779–800. U.S. Department of Energy Symposium Series, Augusta, GA, Nov. 8–10, 1978, CONF781101, NTIS.

Eppley, R. W., and W. G. Harrison. 1975. Physiological ecology of *Gonyaulax polyedra,* a red water dinoflagellate of Southern California. In *Proceedings of the First International Conference on Toxic Dinoflagellate Blooms,* ed. V. R. Lo Cicero, pp. 11–22. Massachusetts Science and Technology Foundation, Wakefield.

Eppley, R. W., O. Holm-Hansen, and J. D. H. Strickland. 1968. Some observations on the vertical migration of dinoflagellates. *J. Phycol.* 4, 333–40.

Eppley, R. W., J. N. Rogers, and J. J. McCarthy. 1969. Half-saturation constants for growth and nitrate uptake of marine phytoplankton. *Limnol. Oceanogr.* 14, 912–20.

Esaias, W. E., and H. C. Curl, Jr. 1972. Effect of dinoflagellate bioluminescence on copepod ingestion rates. *Limnol. Oceanogr.* 17, 901–6.

Evans, G. T., and F. J. R. Taylor. 1980. Phytoplankton accumulation in Langmuir cells. *Limnol. Oceanogr.* 25, 840–5.

Falkowski, P. G., and T. G. Owens. 1978. Effects of light intensity on photosynthesis and dark respiration in six species of marine phytoplankton. *Mar. Biol.* 45, 289–95.

Falkowski, P. G., and T. G. Owens. 1980. Light-shade adaptation: two strategies in marine phytoplankton. *Plant Physiol.* 66, 592–5.

Faust, M. A., J. C. Senger, and B. W. Meeson. 1982. Response of *Prorocentrum mariae-lebouriae* (Dinophyceae) to light of different spectral qualities and irradiances: growth and pigmentation. *J. Phycol.* 18, 349–56.

Forward, R. B. 1976. Light and diurnal vertical migration: photobehavior and photophysiology of plankton. *Photobiol. Rev.* 1, 157–209.

Freeberg, D. L., A. Marshall, and M. Heyl. 1979. Interrelationships of *Gymnodinium breve* (Florida red tide) within the phytoplankton community. In *Toxic Dinoflagellate Blooms: Developments in Marine Biology,* ed. D. L. Taylor and H. H. Seliger, pp. 139–44. Elsevier, New York.

Gilfillan, E. S., and S. A. Hanson. 1975. Effects of paralytic shellfish poisoning toxin on the behavior and physiology of marine invertebrates. In *Proceedings of the First International Conference on Toxic Dinoflagellate Blooms,* ed. V. R. Lo Cicero, pp. 367–75. Massachusetts Science and Technology Foundation, Wakefield.

Glover, H. E. 1978. Iron in Maine coastal waters: seasonal variation and its apparent correlation with a dinoflagellate bloom. *Limnol. Oceanogr.* 23, 534–7.

Gran, H. H. 1912. Plankton. *Handworter Buch der Naturwissenschaffen*. Siebenter Band.

Gran, H. H. 1929. Investigation of the production of plankton outside the Romsdals-fjiord, 1926–1927. In *Rapp. et proces.-verbaux. LVI. Cons. Intern. pour l'explor. de la mer.* Copenhagen.

Graneli, E., H. Persson, and L. Edler. 1986. Connection between trace metals, chelators, and red tide blooms in the Laholm Bay, S.E. Kattegat – an experimental approach. *Mar. Envir. Res.* 18, 61–78.

Haddad, K. D., and K. L. Carder. 1979. Oceanic intrusion: one possible initiation mechanism of red tide blooms on the West Florida coast. In *Toxic Dinoflagellate Blooms: Developments in Marine Biology,* ed. D. L. Taylor and H. H. Seliger, pp. 269–74. Elsevier, New York.

Hand, W., P. A. Collard, and D. Davenport. 1965. The effects of temperature and salinity changes on swimming rates in the dinoflagellates *Gonyaulax* and *Gyrodinium. Biol. Bull.* 128, 90–101.

Harris, G. P., S. I. Heaney, and J. F. Talling. 1979. Physiological and environmental constraints in the ecology of the planktonic dinoflagellate, *Ceratium hirundinella. Freshwater Biol.* 9, 413–28.

Harrison, W. G. 1976. Nitrate metabolism of the red tide dinoflagellate *Gonyaulax polyedra* Stein. *J. Exp. Mar. Biol. Ecol.* 21, 199–209.

Hasle, G. R. 1950. Phototactic migration in marine dinoflagellates. *Oikos,* 2, 162–75.

Hasle, G. R. 1954. More on phototactic diurnal migration in marine dinoflagellate. *Nytt. Mag. Bot.* 2, 139–47.

Heaney, S. I., and R. W. Eppley. 1981. Light, temperature, and nitrogen as interacting factors affecting diel vertical migrations in dinoflagellates in culture. *J. Plank. Res.* 3, 331–44.

Heaney, S. I., and T. I. Furnass. 1980. Laboratory models of diel migration in the dinoflagellate *Ceratium hirundinella. Freshwater Biol.* 10, 163–70.

Heil, C. A. 1986. Vertical migration of *Ptychodiscus brevis* (Davis) Steidinger. M.S. thesis, University of South Florida. 118 pp.

Hitchcock, Gary L. 1982. A comparative study of the size dependent organic composition of marine diatoms and dinoflagellates. *J. Plank. Res.* 4, 363–77.

Holligan, P. M. 1979. Dinoflagellate blooms associated with tidal fronts around the British Isles. In *Toxic Dinoflagellate Blooms: Developments in Marine Biology,* ed. D. L. Taylor and H. H. Seliger, pp. 249–56. Elsevier, New York.

Holligan, P. M. 1985. Marine dinoflagellate blooms: growth strategies and environmental exploitation. In *Toxic Dinoflagellates,* ed. D. M. Anderson, A. W. White, and D. G. Baden, pp. 133–9. Elsevier, New York.

Huntley, M. E. 1982. Yellow water in La Jolla Bay, California, July 1980. II. Suppression of zooplankton grazing. *J. Exp. Mar. Biol. Ecol.* 63, 81–91.

Huntley, M. E., K.-G. Barthel, and J. L. Star. 1983. Particle rejection by *Calanus pacificus:* discrimination between similarly sized particles. *Mar. Biol.* 74, 151–60.

Incze, L. S., and C. M. Yentsch. 1981. Stable density fronts and dinoflagellate patches in a tidal estuary. *Estuar. Coast. Mar. Sci.* 13, 547–56.

Jeffrey, S. W. 1981. An improved thin-layer chromatographic technique for marine phytoplankton pigments. *Limnol. Oceanogr.* 26, 191–7.

Jeffrey, S. W., M. Sielicki, and F. T. Haxo. 1975. Chloroplast pigment patterns in dinoflagellates. *J. Phycol.* 11, 374–85.

Jensen, A. C. 1975. The economic halo of a red-tide. In *Proceedings of the First International Conference on Toxic Dinoflagellate Blooms,* ed. V. R. Lo Cicero, pp. 507–16. Massachusetts Science and Technology Foundation, Wakefield.

Jorgensen, C. B. 1966. Biology of suspension feeding. In *International Series Monographs in Pure and Applied Biology,* 27. Pergamon Press, New York.

Kamykowski, D. 1974. Possible interactions between phytoplankton and semi-diurnal tides. *J. Mar. Res.* 32, 67–89.

Kamykowski, D. 1978. Organism patchiness in lakes resulting from the interaction between internal seiche and plankton diurnal migration. *Ecol. Mod.* 4, 197–210.

Kamykowski, D. 1981. Laboratory experiments on the diurnal vertical migration of marine dinoflagellates through temperature gradients. *Mar. Biol.* 62, 57–64.

Kamykowski, D., and S. A. McCollum. 1986. The temperature acclimatized swimming speed of selected marine dinoflagellates. *J. Plank. Res.* 8, 275–87.

Kamykowski, D., and S. J. Zentara. 1977. The diurnal vertical migration of phytoplankton through temperature gradients. *Limnol. Oceanogr.* 22, 148–51.

Kat, M. 1985. *Dinophysis acuminata* blooms, the distinct cause of Dutch mussel poisoning. In *Toxic Dinoflagellates,* ed. D. M. Anderson, A. W. White, and D. G. Baden, pp. 73–7. Elsevier, New York.

Kayser, H. 1979. Growth interactions between marine dinoflagellates in multi-species culture experiments. *Mar. Biol.* 52, 357–69.

Kemp, W. M., and W. J. Mitsch. 1979. Turbulence and phytoplankton diversity: a general model of the "paradox of the plankton." *Ecol. Mod.* 7, 201–22.

Kilham, P., and S. S. Kilham. 1980. The evolutionary ecology of phytoplankton. In *The Physiological Ecology of Phytoplankton,* ed. I. Morris, pp. 571–97. University of California Press, Berkeley.

Krinsky, N. I. 1979. Carotenoid pigments: multiple mechanisms for coping with the stress of photosensitized oxidations. In *Strategies of Microbial Life in Extreme Environment,* vol. 13, ed. M. Shilo, pp.163–87. Gustav Fisher, New York.

Kuenzler, E. J., and J. P. Perras. 1965. Phosphatases of marine algae. *Biol. Bull.* 128, 271–84.

Lassus, P., M. Bardouil, I. Truquet, P. Truquet, C. Le Baut, and M. J. Pierre. 1985. *Dinophysis acuminata* distribution and toxicity along the southern Brittany Coast (France): correlation with hydrological factors. In *Toxic Dinoflagellates,* ed. D. M. Anderson, A. W. White, and D. G. Baden, pp. 159–64. Elsevier, New York.

Le Fevre, J., and J. R. Grall. 1970. On the relationship of *Noctiluca* swarming off the western coast of Brittany with hydrological features and plankton characteristics of the environment. *J. Exp. Mar. Biol. Ecol.* 4, 287–306.

Lewin, R. A. 1962. *Physiology and Biochemistry of Algae.* Academic Press, New York. 929 pp.

Lewis, J., P. Tett, and J. D. Dodge. 1985. The cyst-theca cycle of *Gonyaulax polyedra (Lingulodinium machaerophorum)* in Creran, a Scottish West Coast sea-loch. In *Toxic Dinoflagellates*, ed. D. M. Anderson, A. W. White, and D. G. Baden, pp. 85–90. Elsevier, New York.

Lo Cicero, V. R. (ed.). 1975. *Proceedings of the First International Conference on Toxic Dinoflagellate Blooms*. Massachusetts Science and Technology Foundation, Wakefield. 527 pp.

Loeblich, A. R., III. 1967. Aspects of the physiology and biochemistry of the Pyrrhophyta. *Phykos* 5, 216–55.

Lüthy, J. 1979. Epidemic paralytic shellfish poisoning in Western Europe, 1976. In *Toxic Dinoflagellate Blooms: Developments in Marine Biology*, ed. D. L. Taylor and H. H. Seliger, pp. 15–22. Elsevier, New York.

Lutz, R. A., and L. S. Incze. 1979. Impact of toxic dinoflagellate blooms on the North American shellfish industry. In *Toxic Dinoflagellate Blooms: Developments in Marine Biology*, ed. D. L. Taylor and H. H. Seliger, pp. 476–83. Elsevier, New York.

McCarthy, J. J., D. Wynne, and T. Berman. 1982. The uptake of dissolved nitrogenous nutrients by Lake Kinneret (Israel) microplankton. *Limnol. Oceanogr.* 27, 673–80.

MacIsaac, J. J. 1978. Diel cycles of inorganic nitrogen uptake in a natural phytoplankton population dominated by *Gonyaulax polyedra*. *Limnol. Oceanogr.* 23, 1–9.

MacIsaac, J. J., G. S. Grunseich, H. E. Glover, and C. M. Yentsch. 1979. Light and nutrient limitation in *Gonyaulax excavata*: nitrogen and carbon trace results. In *Toxic Dinoflagellate Blooms: Developments in Marine Biology*, ed. D. L. Taylor and H. H. Seliger, pp. 107–10. Elsevier, New York.

Maclean, J. L. 1977. Observation on *Pyrodinium bahamense* Plate, a toxic dinoflagellate in Papua, New Guinea. *Limnol. Oceanogr.* 22, 234–54.

Mahoney, J. B., and F. W. Steimle, Jr. 1979. A mass mortality of marine animals associated with a bloom of *Ceratium tripos* in the New York Bight. In *Toxic Dinoflagellate Blooms: Developments in Marine Biology*, ed. D. L. Taylor and H. H. Seliger, pp. 225–30. Elsevier, New York.

Maranda, L. 1985. Response time of *Gonyaulax tamarensis* to changes in irradiance. In *Toxic Dinoflagellates*, ed. D. M. Anderson, A. W. White, and D. G. Baden, pp. 91–6. Elsevier, New York.

Maranda, L., D. M. Anderson, and Y. Shimizu. 1985. Comparison of toxicity between populations of *Gonyaulax tamarensis* of eastern North American waters. *Estuar. Coast Shelf Sci.* 21, 401–10.

Margalef, R. 1968. *Perspectives in Ecological Theory*. University of Chicago Press, Chicago. 111 pp.

Margalef, R., M. Estrada, and D. Blasco. 1979. Functional morphology of organisms involved in red tides as adapted to decaying turbulence. In *Toxic Dinoflagellate Blooms: Developments in Marine Biology*, ed. D. L. Taylor and H. H. Seliger, pp. 89–94. Elsevier, New York.

Martin, D. F., and B. B. Martin. 1973. Implications of metal-organic interactions in red tide outbreaks. In *Trace Metals and Metal-organic Interactions in Natural Waters*, ed. P. Singer, pp. 339–61. Ann Arbor Science Publ., Ann Arbor MI.

Meeson, B. W., and M. A. Faust. 1985. Response of *Protocentrum minimum*

(Dinophyceae) to different spectral qualities and irradiances: growth and photosynthesis. In *Marine Biology of Polar Regions and Effects of Stress on Marine Organisms*, ed. J. S. Gray and M. E. Christiansen, pp. 445–61. John Wiley and Sons, New York.

Mills, L. J., and G. Klein-MacPhee. 1979. Toxicity of the New England red tide dinoflagellate to winter flounder larvae. In *Toxic Dinoflagellate Blooms: Developments in Marine Biology*, ed. D. L. Taylor and H. H. Seliger, pp. 389–94. Elsevier, New York.

Morey-Gaines, G. 1979. The ecological role of red tides in the Los Angeles–Long Beach Harbor food web. In *Toxic Dinoflagellate Blooms: Developments in Marine Biology*, ed. D. L. Taylor and H. H. Seliger, pp. 315–19. Elsevier, New York.

Mortensen, A. M. 1985. Massive fish mortalities in the Faroe Islands caused by *Gonyaulax excavata* red-tide. In *Toxic Dinoflagellates*, ed. D. M. Anderson, A. W. White, and D. G. Baden, pp. 165–70. Elsevier, New York.

Murphy, E., K. Steidinger, B. Roberts, J. Williams, and J. Jolley, Jr. 1975. An explanation for the Florida east coast *Gymnodinium breve* red tide in November, 1972. *Limnol. Oceanogr.* 20, 481–4.

Neve, R. A., and P. B. Reichardt. 1984. Alaska's shellfish industry. In *Seafood Toxins*, ed. E. P. Ragelis, pp. 53–8. American Chemical Society Symposium Series, no. 262. Washington, DC.

Norris, L., and I. K. Chew. 1975. Effect of environmental factors on growth of *Gonyaulax catenella*. In *Proceedings of the First International Conference on Toxic Dinoflagellate Blooms*, ed. V. R. Lo Cicero, pp. 143–52. Massachusetts Science and Technology Foundation, Wakefield.

Officer, C. B., T. J. Smayda, and R. Mann. 1982. Benthic filter feeding: a natural eutrophication control. *Mar. Ecol. Prog. Ser.* 9, 203–10.

Paasche, E., I. Bryceson, and K. Tangen. 1984. Interspecific variation in dark nitrogen uptake by dinoflagellates. *J. Phycol.* 20, 394–401.

Paffenhofer, G. A. 1970. Cultivation of *Calanus helgolandicus* under controlled conditions. *Helgoländ. Wiss. Meeresunters.* 20, 346–59.

Perry, M. J., M. C. Talbot, and R. S. Alberte. 1981. Photoadaption in marine phytoplankton: response of the photosynthetic unit. *Mar. Biol.* 62, 91–101.

Pincemin, J. M. 1969. Apparition d'une eau rouge a *Cochlodinium* sp. devant. Juan les-Pins. *Rev. Int. Oceanogr. Med.* 13/14, 205–16.

Pincemin, J. M. 1971. Telemediateurs chimiques et equilibre biologique oceanique. Troisieme partie. Étude in vitro de relations entre populations phytoplanctoniques. *Rev. Int. Oceanogr. Med.* 22/23, 165–96.

Pollingher, U., and E. Zemel. 1981. *In situ* and experimental evidence of the influence of turbulence on cell division processes of *Peridinium cinctum* forma *westii* (Lemm.) Lefevre. *Brit. Phycol. J.* 16, 281–7.

Porter, K. G., and J. W. Porter. 1979. Bioluminescence in marine plankton: a coevolved antipredation system. *Amer. Nat.* 114, 458–61.

Poulet, S. A., and P. Marsot. 1978. Chemosensory grazing by marine calanoid copepods (Arthropoda: Crustacea). *Science* 200, 1403–5.

Poulet, S. A., and P. Marsot. 1980. Chemosensory feeding and food-gathering by omnivorous marine copepods. In *Evolution and Ecology of Zooplankton Com-*

munities, ed. C. Kerfoot, pp. 198–218. American Society of Limnology and Oceanography, Special Symposium, no. 3. University Press of New England, Hanover, NH.

Prézelin, B. B. 1976. The role of peridinin-chlorophyll *a*-proteins in the photosynthetic light adaptation of the marine dinoflagellate, *Glenodinium* sp. *Planta* 130, 225–33.

Prézelin, B. B. 1981. Light reactions in photosynthesis. In *Physiological Bases of Phytoplankton Ecology,* ed. T. Platt, pp. 1–43. Canadian Bulletin of Fisheries and Aquatic Sciences, no. 210, Ottawa.

Prézelin, B. B., and H. A. Matlick. 1980. Time course of photoadaptation in the photosynthesis-irradiance relationship of a dinoflagellate exhibiting photosynthetic periodicity. *Mar. Biol.* 58, 85–96.

Prézelin, B. B., and B. M. Sweeney. 1979. Photoadaptation of photosynthesis in two bloom-forming dinoflagellates. In *Toxic Dinoflagellate Blooms: Developments in Marine Biology,* ed. D. L. Taylor and H. H. Seliger, pp. 101–6. Elsevier, New York.

Provasoli, L. 1979. Recent progress, an overview. In *Toxic Dinoflagellate Blooms: Developments in Marine Biology,* ed. D. L. Taylor and H. H. Seliger, pp. 1–14. Elsevier, New York.

Provasoli, L., and J. J. A. McLaughlin. 1963. Limited heterotrophy of some photosynthetic dinoflagellates. In *Symposium on Marine Microbiology,* ed. C. H. Oppenheimer, pp. 105–13. C. C. Thomas, New York.

Rapport, D. J., H. A. Regles, and T. C. Hutchinson. 1985. Ecosystem behavior under stress. *Amer. Nat.* 125, 617–40.

Raven, J. A., and J. Beardall. 1981. Respiration and photorespiration. In *Physiological Bases of Phytoplankton Ecology,* ed. T. Platt, pp. 55–82. Canadian Bulletin of Fisheries and Aquatic Sciences, no. 210, Ottawa.

Richardson, K., J. Beardall, and J. A. Raven. 1983. Adaptation of unicellular algae to irradiance: an analysis of strategies. *New Phytol.* 93, 157–91.

Rivkin, R. B., and E. Swift. 1979. Diel and vertical patterns of alkaline phosphatase activity in the oceanic dinoflagellate *Pyrocystis noctiluca. Limnol. Oceanogr.* 24, 107–16.

Rivkin, R. B., and E. Swift. 1980. Characterization of alkaline phosphatase and organic phosphorus utilization in the oceanic dinoflagellate *Pyrocystis noctiluca. Mar. Biol.* 61, 1–8.

Rivkin, R. B., and E. Swift. 1982. Phosphate uptake by the oceanic dinoflagellate *Pyrocystis noctiluca. J. Phycol.* 18, 113–20.

Rivkin, R. B., and E. Swift. 1985. Phosphorus metabolism of oceanic dinoflagellates: phosphate uptake, chemical composition, and growth of *Pyrocystis noctiluca. Mar. Biol.* 88, 189–98.

Rivkin, R. B., M.Voytek, and H. H. Seliger. 1982. Phytoplankton division rates in light-limited environments: two adaptations. *Science* 215, 1123–5.

Roberts, B. S. 1979. Occurrence of *Gymnodinium breve* red tides along the west and east coasts of Florida during 1976 and 1977. In *Toxic Dinoflagellate Blooms: Developments in Marine Biology,* ed. D. L. Taylor and H. H. Seliger, pp. 199–202. Elsevier, New York.

Seliger, H. H., J. H. Carpenter, M. E. Loftus, and W. D. McElroy. 1970. Mechanisms for the accumulations of high concentrations of dinoflagellates in a bioluminescent bay. *Limnol. Oceanogr.* 15, 234–45.

Seliger, H. H., M. E. Loftus, and D. V. Subba Rao. 1975. Dinoflagellate accumulations in Chesapeake Bay. In *Proceedings of the First International Conference on Toxic Dinoflagellate Blooms,* ed. V. R. Lo Cicero, pp. 181–205. Massachusetts Science and Technology Foundation, Wakefield.

Seliger, H. H., M. A. Tyler, and K. R. McKinley. 1979. Phytoplankton blooms and red tides resulting from frontal circulation patterns. In *Toxic Dinoflagellate Blooms: Developments in Marine Biology,* ed. D. L. Taylor and H. H. Seliger, pp. 239–48. Elsevier, New York.

Sellner, K. G., and M. M. Olson. 1985. Copepod grazing in red tides of Chesapeake Bay. In *Toxic Dinoflagellates,* ed. D. M. Anderson, A. W. White, and D. G. Baden, pp. 245–50. Elsevier, New York.

Shanley, E. 1985. Photoadaptation in the red tide dinoflagellate *Ptychodiscus brevis.* M.S. thesis, University of South Florida. 122 pp.

Sherr, E. B., B. F. Sherr, T. Berman, and J. J. McCarthy. 1982. Differences in nitrate and ammonia uptake among components of a phytoplankton population. *J. Plank. Res.* 4, 961–5.

Shumway, S. E., T. L. Cucci, L. Gainey, and C. M. Yentsch. 1985. A preliminary study of the behavioral and physiological effects of *Gonyaulax tamarensis* on bivalve molluscs. In *Toxic Dinoflagellates,* ed. D. M. Anderson, A. W. White, and D. G. Baden, pp. 389–94. Elsevier, New York.

Shuter, B. J. 1978. Size dependence of phosphorus and nitrogen subsistence quotas in unicellular organisms. *Limnol. Oceanogr.* 23, 1248–55.

Smayda, T. J. 1973. The growth of *Skeletonema costatum* during a winter–spring bloom in Narragansett Bay, Rhode Island. *Nor. J. Bot.* 20, 219–47.

Smayda, T. J. 1980. Phytoplankton species succession. In *The Physiological Ecology of Phytoplankton,* ed. I. Morris, pp. 493–570. University of California Press, Berkeley.

Smayda, T. J., and D. Karentz. 1982. Rapid growth of some marine dinoflagellates. *EOS* 63, 945.

Smith, G. B. 1975. The 1971 Red tide and its impact on certain reef communities in the mid-eastern Gulf of Mexico. *Envir. Lett.* 9, 141–52.

Song, P. S., E. B. Walker, J. Jung, R. A. Auerbach, G. W. Robinson, and B. B. Prezelin. 1980. Primary processes of phytobiological receptors. In *New Horizons in Biological Chemistry,* pp. 79–94. Acad. Publ. Cent., Tokyo.

Sournia, A. 1974. Circadian periodicities in natural populations of marine phytoplankton. *Adv. Mar. Biol.* 12, 325–89.

Steidinger, K. A. 1975. Implications of dinoflagellate life cycles on initiation of *Gymnodinium breve* red tides. *Envir. Lett.* 9, 129–39.

Steidinger, K. A. 1983. A re-evaluation of toxic dinoflagellate biology and ecology. In *Progress in Phycological Research,* vol. 2., ed. F. E. Round and D. J. Chapman, pp. 147–88. Elsevier, New York.

Steidinger, K. A., and D. G. Baden. 1984. Toxic marine dinoflagellates. In *Dinoflagellates,* ed. L. Spector, pp. 201–61. Academic Press, New York.

Steidinger, K. A., and E. R. Cox. 1980. Free-living dinoflagellates. In *Phytoflagellates,* ed. E. R. Cox, pp. 407–32. Elsevier, New York.

Steidinger, K. A., and K. Haddad. 1981. Biologic and hydrographic aspects of red tides. *BioScience* 31, 814–19.

Stosch, H. A. von. 1973. Observations on vegetative reproduction and sexual life cycles of two freshwater dinoflagellates, *Gymnodinium pseudopalustre* Schiller and *Woloszynskia apiculata* sp. nov. *Brit. Phycol. J.* 8, 105–34.

Sweeney, B. M. 1979. The organisms, opening remarks. In *Toxic Dinoflagellate Blooms: Developments in Marine Biology,* ed. D. L. Taylor and H. H. Seliger, pp. 37–40. Elsevier, New York.

Tangen, K. 1979. Dinoflagellate blooms in Norwegian waters. In *Toxic Dinoflagellate Blooms: Developments in Marine Biology,* ed. D. L. Taylor and H. H. Seliger, pp. 179–82. Elsevier, New York.

Taylor, D. L., and H. H. Seliger (eds.). 1979. *Toxic Dinoflagellate Blooms: Developments in Marine Biology,* vol. 1. Elsevier, New York. 505 pp.

Taylor, W. R., H. H. Seliger, W. G. Fastie, and W. D. McElroy. 1966. Biological and physical observations on a phosphorescent bay in Falmouth Harbor, Jamaica, W.I. *J. Mar. Res.* 24, 28–43.

Tett, P. B., and M. G. Kelly. 1973. Marine bioluminescence. *Oceanogr. Mar. Biol. Ann. Rev.* 11, 89–173.

Tschirn, E. 1920. Biologische Studien an Ceratien der Kieler Fohrde. Dissertation. Kiel, Auszug.

Tyler, M. A., and H. H. Seliger. 1978. Annual subsurface transport of a red tide dinoflagellate to its bloom area: water circulation patterns and the organism distribution in Chesapeake Bay. *Limnol. Oceanogr.* 23, 227–46.

Tyler, M. A., and H. H. Seliger. 1981. Selection for a red tide organism: physiological responses to a physical environment. *Limnol. Oceanogr.* 26, 310–24.

Uchida, T. 1977. Excretion of a diatom-inhibitory substance by *Prorocentrum micans.* Ehrenberg. *Jap. J. Ecol.* 27, 1–4.

Vargo, G. A., K. L. Carder, W. Gregg, E. Shanley, C. Heil, K. A. Steidinger, and K. D. Haddad. 1987. The potential contribution of primary production by red-tides to the west Florida shelf ecosystem. *Limnol. Oceanogr.* 32, 762–7.

Vargo, G. A., and E. Shanley. 1985. Alkaline phosphatase activity in the red tide dinoflagellate, *Ptychodiscus brevis. Mar. Ecol. Publ. Stn. Zool. Napoli.* 6, 251–64.

Vogel, R. J. 1980. The ecological factors that produce perturbation-dependent ecosystems. In *The Recovery Process in Damaged Ecosystems,* ed. J. Cairns, Jr., pp. 63–94. Ann Arbor Science Publ., Ann Arbor, MI.

Watras, C. J., V. C. Garcon, R. J. Olson, S. W. Chisholm, and D. M. Anderson. 1985. The effect of zooplankton grazing on estuarine blooms of the toxic dinoflagellate *Gonyaulax tamarensis. J. Plank. Res.* 7, 891–908.

Weiler, C. S., and R. W. Eppley. 1979. Temporal pattern of division in the dinoflagellate genus *Ceratium* and its application to the determination of growth rate. *J. Exp. Mar. Biol. Ecol.* 39, 1–24.

Weiler, C. S., and D. M. Karl. 1979. Diel changes in phase-division cultures of *Ceratium furca* (Dinophyceae): nucleotide triphosphate, adenylate energy charge, cell carbon, and patterns of vertical migration. *J. Phycol.* 15, 384–91.

Wheeler, B. 1966. Phototactic vertical migration in *Exuviella baltica. Bot. Mar.* 9, 15–7.

White, A. W. 1976. Growth inhibition caused by turbulence in the toxic marine dinoflagellate *Gonyaulax excavata. J. Fish. Res. Bd. Can.* 33, 2598–602.

White, A. W. 1984. Paralytic shellfish toxins and finfish. In *Seafood Toxins*, ed. E. P. Ragelis, pp. 171–80. American Chemical Society Symposium Series, no. 262. Washington, DC.

White, A. W., M. Anraku, and K. Hudi (eds.). 1984. *Toxic Red Tides and Shellfish Toxicity in Southeast Asia*. Southeast Asia Fisheries Development Center and the International Development Research Center, Singapore and Ottawa. 133 pp.

White, H. H. 1979. Effects of dinoflagellate bioluminescence on the ingestion rates of herbivorous zooplankton. *J. Exp. Mar. Biol. Ecol.* 36, 217–24.

Wilcox, L. W., and G. J. Wedemayer. 1985. *Gymnodinium acidotum* Nygaard (Pyrrophyta), a dinoflagellate with an endosymbiotic cryptomonad. *J. Phycol.* 20, 236–42.

Winter, J. E. 1978. A review of the knowledge of suspension feeding in lamellibranchiate bivalves, with special reference to artificial aquaculture systems. *Aquaculture* 13, 1–33.

Wyatt, T. 1974. Red tides and algal strategies. In *Ecological Stability*, ed. M. B. Usher and M. H. Williamson, pp. 35–9. Chapman and Hall, London.

Wyatt, T., and J. Horwood. 1973. A model which generates red tides. *Nature* [London] 244, 238–40.

Yasumoto, T. 1985. Recent progress in the chemistry of dinoflagellate toxins. In *Toxic Dinoflagellates*, ed. D. M. Anderson, A. W. White, and D. G. Baden, pp. 259–70. Elsevier, New York.

Yentsch, C. M., E. J. Cole, and M. G. Salvaggio. 1975. Some of the growth characteristics of *Gonyaulax tamarensis* isolated from the Gulf of Maine. In *Proceedings of the First International Conference on Toxic Dinoflagellate Blooms*, ed. V. R. Lo Cicero, pp. 163–80. Massachusetts Science and Technology Foundation, Wakefield.

16. Hazards of freshwater blue-green algae (cyanobacteria)

PAUL R. GORHAM

Department of Botany, University of Alberta, Edmonton, Alberta T6G 2E9, Canada

WAYNE W. CARMICHAEL

Department of Biological Sciences, Wright State University, Dayton, Ohio 45435

CONTENTS

I. Introduction

Since the first scientific account by Francis (1878) of a blue-green algal bloom (*Nodularia spumigena* Mertens) in Australia that caused the death of sheep, horses, pigs, and dogs, many similar occurrences have been reported from all parts of the world (Table 16-1). These accounts describe sickness and death of livestock, pets, and wildlife following ingestion of algal blooms dominated not only by *Nodularia* (Main et al. 1977; Edler et al. 1985) but also by other species of freshwater blue-green algae. Some reports contain indications of human toxicity (Carmichael 1981a; Carmichael et al. 1985).

Blue-green algae are prokaryotes, have cell walls composed of peptidoglycan and lipopolysaccharide layers instead of cellulose, and possess a genome size range like the gram-negative bacteria (Herdman et al. 1979a). They are frequently referred to as cyanobacteria, and many scientists classify them as eubacteria. It has been proposed that all cyanobacteria be assigned to four genera (*Aphanocapsa, Nostoc, Fischerella* and *Synechococcus*) on the basis of genome affinities (Rippka et al. 1979; Herdman et al. 1979b; Fox et al. 1980). In this review, however, the nomenclature used is based on the morphological descriptions of Geitler (1932), Komarek (1958), Desikachary (1959), and Prescott (1962).

Episodes of poisoning by blue-greens are unpredictable both in occurrence and in duration. Early investigations of toxic blooms were conducted with varying degrees of thoroughness; consequently, progress in understanding poisonings was slow. Attempts to isolate and grow toxic strains of bloom-forming species in the laboratory were either unsuccessful or unsatisfactory because the minimum lethal dose for mice injected intraperitoneally (LD_{100} i.p. mouse) was too high and variable (650–3,400 mg·kg^{-1} body weight) (Olson 1951), and signs of intoxication or survival times were not always recorded.

Dependence on toxic blooms for experimental material was altered by the successful isolation and mass culture of a toxic colonial isolate termed NRC-1, of *Microcystis aeruginosa* Kütz. *emend.* Elenkin, from a water bloom in Little Rideau Canal, Ontario, Canada (Hughes, Gorham, and Zehnder 1958). From this culture, Bishop, Anet, and Gorham

[404]

Table 16-1. *Known occurrences of toxic cyanobacteria in fresh or marine waters*

Argentina	India
Australia	Israel
Bangladesh	Japan
Bermuda	New Zealand
Brazil	Okinawa (marine)
Canada	People's Republic of China
Alberta	South Africa
Manitoba	USA
Ontario	Colorado
Saskatchewan	Florida
	Hawaii (marine)
Europe	Illinois
Czechoslovakia	Iowa
Denmark	Michigan
East Germany	Minnesota
Finland	Montana
Great Britain	Nevada
Hungary	New Hampshire
Netherlands	New Mexico
Norway	New York
Poland	North Dakota
Portugal	Oregon
Sweden	Pennsylvania
West Germany	South Dakota
	Texas
	Washington
	Wisconsin
	USSR
	Ukraine

Source: From Carmichael et al. 1985.

(1959) isolated a fast-death factor, later called microcystin (Konst et al. 1965). This toxin was not an alkaloid as suggested by Louw (1950) but a dialyzable, stable, acidic, probably cyclic peptide of 1,300 to 2,600 daltons composed of seven amino acids and having an LD_{100} i.p. mouse of 0.47 mg·kg^{-1}.

Gorham (1964) concluded, on the basis of knowledge available at the time, that toxic blue-green blooms in public water supplies were nuisances, frequently with negative economic consequences, rather than public health hazards. This view is no longer tenable. Eutrophication of water supplies and recreational water bodies has increased, and with increases in productivity have come increases in the number of known poisoning episodes. Recognition and reporting of toxic blue-green episodes have broadened, and commercial production and marketing of blue-greens as food for animals and humans are expanding (see Jassby, Chapter 7). Since 1979 phycotoxins from blue-greens have been considered sufficiently hazardous to warrant their inclusion in the triennial international symposium "Mycotoxins and Phycotoxins," sponsored by the International Union of Pure and Applied Chemistry. Also, a recent report of the NATO Committee on Challenges of Modern Society titled *Drinking Water Microbiology* (Cliver et al. 1984) includes a section on "Toxic Cyanobacteria in Raw Water Supplies."

The review that follows considers neurotoxins, hepatotoxins, and other toxins produced by freshwater blue-greens. This is followed by a discussion of the hazards to humans from accidental ingestion of or contact with these organisms. It concludes with an assessment of the hazards of using various species and strains of freshwater blue-greens as single-cell protein in human diets.

II. Neurotoxins

Toxins from freshwater blue-greens isolated and identified to date are alkaloids or low-molecular-weight peptides. No system for naming them has been adopted. Most have been given names based upon the genus that produces them. The alkaloid toxins act more rapidly than the peptide toxins because they target the neuromuscular system. They paralyze skeletal and respiratory muscles and cause death from respiratory arrest within minutes or, at most, one or two hours after ingestion. They are produced by strains of *Anabaena (An.) flos-aquae* (Lyngb.) De Bréb. and *Aphanizomenon (Aph.) flos-aquae* (L.) Ralfs.

A. Anatoxins

Toxins produced by *An. flos-aquae* are called anatoxins. The six isolated toxins were given letter designations on the basis of strains that produced different signs of toxicity when injected intraperitoneally into mice, rats, and chicks (Carmichael and Gorham 1978). Anatoxin-*a* (antx-*a*) was the first chemically defined toxin from a freshwater blue-green. It is a secondary amine, 2-acetyl-9-azabicyclo (4-2-1) non-2-ene (Huber 1972; Devlin et al. 1977) with a molecular weight of 166 daltons (Fig. 16-1). Antx-*a* has been synthesized by ring expansion of cocaine (Campbell, Edwards, and Kolt 1977) from cyclooctadiene (Campbell et al. 1979), iminium salts (Bates and Rapoport 1979; Petersen, Toteberg-Kaulen, and Rapoport 1984; Koskinen and Rapoport 1985), nitrone (Tufariello, Meckler, and Senaratne 1984, 1985), and 4-cycloheptenone or tetrabromotricyclooctane (Danheiser, Morin, and Salaski 1985).

Antx-*a* is a potent, postsynaptic, depolarizing, neuromuscular blocking agent that mimics the effects of acetylcholine and acts at both nico-

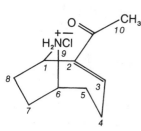

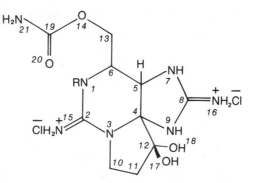

anatoxin-a hydrochloride

R = H; saxitoxin dihydrochloride
R = OH; neosaxitoxin dihydrochloride

Fig. 16-1. Structures of anatoxin-*a*, saxitoxin, and neosaxitoxin.

tinic and muscarinic receptors (Carmichael, Biggs, and Gorham 1975; Carmichael, Biggs, and Peterson 1979; Spivak, Witkop, and Albuquerque 1980; Aronstam and Witkop 1981). Poisoning of pets, livestock, wild animals, and birds occurs after ingestion of an algal bloom or scum dominated by toxic strains of the *An. flos-aquae/lemmermanii/circinalis* complex (Olson 1951; Gorham et al. 1964; May 1981). Field reports of poisoning symptoms include staggering, muscle fasciculations (small involuntary contractions), gasping, convulsions, and opisthotonus (spasms causing head and limbs to bend backward) (in birds), with death occurring in minutes to hours depending on species, dosage, and prior food consumption. The LD_{100} i.p. mouse for pure antx-*a* is 250 $\mu g \cdot kg^{-1}$ (Devlin et al. 1977), but different species have different susceptibilities. The oral minimal lethal dose for ducks and calves is about one-quarter that for mice and rats (Carmichael and Gorham 1977). It has been estimated that a 60 kg calf would die from ingestion of 1.2 L of a heavy bloom (20 g dry wt$\cdot L^{-1}$) composed entirely of *An. flos-aquae* with a toxicity equivalent to that of strain NRC-44-1 (Carmichael, Gorham, and Biggs 1977; Carmichael and Gorham 1981). Juday et al. (1981) described such a toxic bloom in the Grayling arm of the Hegben Reservoir, Montana, that killed 30 cows and 8 dogs in June 1977.

The neuromuscular blockade produced by antx-*a* is very long-lasting, making artificial respiration impractical and development of an antidote unlikely. The bioassay in general use is not very sensitive or definitive but does detect mammalian neurotoxicity in a water bloom. It consists of coarse-filtering a bloom sample to remove debris, freezing and thawing to destroy cell membranes, injecting 0.5–1.0 ml intraperitoneally into two or more mice with appropriate controls, and observing signs and survival times for one to two hours. Astrachan and Archer (1981) developed a reverse-phase high-performance liquid chromatography (HPLC) method that, when combined with the extraction procedure, achieved an overall sensitivity of 0.1 ppm. They used it to show that chronic subacute dosage of rates with antx-*a* did not cause any apparent toxic signs.

Anatoxin-*a*(s) (antx-*a*(s)) from *An. flos-aquae* strain 525-17 is different from antx-*a* produced by strain NRC-44-1 (Carmichael and Gorham 1978). It produces opisthotonus in chicks like antx-*a* but also induces pronounced viscous salivation(s) and lachrymation (discharge of tears) in mice, chromodacryorrhea (shedding of bloody tears) in rats, and urinary incontinence and defecation in mice and rats prior to respiratory arrest and dose-dependent fasciculation of limbs for one to two minutes after death. Antx-*a*(s) was purified by column chromatography and HPLC methods developed for antx-*a* (Mahmood and Carmichael 1986b), but its structure is still unknown. The LD_{50} i.p. mouse is 40–50 $\mu g \cdot kg^{-1}$, which is five times more potent than antx-*a*. Survival time is 30–60 min. Mahmood and Carmichael (1986b) concluded that the toxinological and pharmacological signs produced by antx-*a*(s) indicate excessive cholinergic stimulation. More recent work (Mahmood and Carmichael, 1987) shows that it acts as an irreversible anticholinesterase. No antidote or treatments are currently available.

Mahmood, Carmichael, and Pfahler (1988) identified antx-*a*(s) as the cause of death for five dogs, eight pups, and two calves that ingested *An. flos-aquae* from a bloom in Richmond Lake, South Dakota, in late summer 1985. One dog exhibited no lesions, but another, which died one hour after vomiting and diarrhea, exhibited severe intestinal lesions when autopsied.

B. Aphantoxins

Aphantoxins (aphtxs) were discovered as products of blooms of nonfasciculate (single-filament) *Aphanizomenon (Aph.) flos-aquae* that occurred in lakes and ponds in New Hampshire (and in three strains isolated from these sources) from 1966 to 1980. Chromatographic and pharmacologic evidence showed that aphtxs consist mainly of two known neurotoxic alkaloids: neosaxitoxin (neostx), 80%, and saxitoxin (stx), 20% (see Fig. 16-1; Jackim and Gentile 1968; Sawyer, Gentile, and Sasner 1968; Sasner, Ikawa, and Foxall 1984; Mahmood and Carmichael 1986a). The original toxic strains have been lost, and current research is centered on toxic strains NH-1 and NH-5, isolated from a small pond near Durham, New Hampshire, in 1980 (Carmichael and Mahmood 1984).

Saxitoxins were first reported as products of two marine dinoflagellates, *Protogonyaulax catenella* and *P. tamarensis*, associated with "red tides." They are concentrated in the food chain by shellfish and are responsible for paralytic shellfish poisoning (PSP) of humans (see Steidinger and Vargo, Chapter 15). The structures of stx and neostx were first established by Schantz et al. (1975) and Shimizu et al. (1978), respectively. The LD_{50} i.p. mouse for both neostx and stx is 10 $\mu g \cdot kg^{-1}$ (Hall 1982; Carmichael et al. 1985). Potency varies widely with animal species and route of uptake. Substances that increase intestinal absorption or peripheral circulation tend to intensify the response (Carmichael et al. 1985).

Aphtxs and stxs have a fast-acting neuromuscular action that inhibits nerve conduction by blocking sodium channels without affecting permeability to potassium, the transmembrane resting potential, or membrane resistance (Gentile 1971; Sasner et al. 1981; Adelman et al. 1982). The chemical structures have not been determined directly for aphtxs isolated from *Aph. flos-aquae*. Mahmood and Carmichael (1986a) showed, using thin layer chromatography and HPLC, that aphtxs I and II from strain NH-5 correspond to authentic neostx and stx standards, respectively. Shimizu et al. (1984) studied the biosynthesis of stx analogues with strain NH-1 using ^{13}C- and ^{15}N-labeled precursors.

Since stxs and aphtxs leave the system quickly, resuscitation is an effective treatment. The mouse assay has been used mainly for aphtxs, but Barton, Foster, and Johnson (1980) suggested a more sensitive bioassay based on loss of viability of human leukocytes exposed to an extract from a toxic *Aphanizomenon* bloom from Eagle Nest Lake, New Mexico. Ikawa et al. (1981) developed a sensitive fluorescence assay for aphtxs based on the peroxide oxidation to fluorescing derivatives used for stxs (Sullivan and Wekell 1984). Carmichael (1986) suggested that HPLC methods used to purify neurotoxins of both *Anabaena* and *Aphanizomenon* might be modified to detect low levels of toxins in reservoirs or recreational waters.

III. Hepatotoxins

Low-molecular-weight peptide toxins act more slowly than alkaloid toxins because they target the liver. They cause extensive necrosis of this organ, resulting in death by hemorrhagic shock and/or liver failure after a few hours to a few days. They are currently known or suspected to be produced by strains of *Microcystis aeruginosa* (= *Anacystis cyanea*), *M. wesenbergii*, *Nodularia spumigena*, *Oscillatoria agardhii/rubescens* group, *Anabaena flos-aquae*, *An. spiroides*, *Aphanizomenon flos-aquae*, and *Gomphosphaeria lacustris*. Six chemically related cyclic peptides have been identified to date from two of these species, *M. aeruginosa* and *An. flos-aquae* (Carmichael 1988). These toxins are responsible for the great majority of poisonings of pets, livestock, and wildlife that have been documented from all parts of the world (Schwimmer and Schwimmer 1968; Gentile 1971; Kirpenko and Kirpenko 1980; Scott, Barlow, and Hauman 1981; Runnegar and Falconer 1981; Watanabe, Oishi, and Nakao 1981; Skulberg, Codd, and Carmichael 1984; Carmichael et al. 1985; Carmichael 1986).

The signs exhibited after animal dosing or ingestion of toxic quantities of these peptides were summarized by Falconer et al (1981), Carmichael and Schwartz (1984), and Jackson et al. (1984). A lethal dose by the oral or intraperitoneal route is followed by a latent period of 10–30 min. Pallor, prostration, and sometimes convulsions and twitching then develop. Respiration becomes labored and irregular. Death usually occurs within one to three hours. Upon necropsy there may be mild jaundice, dark, slowly clotting blood, widespread pinpoint hemorrhages, and excess yellow clotted fluid in the body cavities. An enlarged, pale, mottled or hemorrhagic liver that shows centrilobular to massive hepatocyte necrosis is a persistent diagnostic characteristic.

Two sets of terminology are currently in use to name peptide hepatotoxins, creating confusion among people trying to describe and understand these new natural product toxins. It is hoped that one of these basic terms (either microcystin or cyanoginosin) plus a letter prefix will shortly be adopted once the full range of structures is understood.

A. Microcystin

A peptide toxin produced by strain NRC-1(SS-17) of *Microcystis aeruginosa*, called fast-death factor, was first identified by Bishop et al. (1959) and later named microcystin (mcyst) (Konst et al. 1965). Subsequent isolations of mcyst were made from the same strain (Murthy and Capindale 1970; Rabin and Darbre 1975) and from *M. aeruginosa* blooms in South Africa (Toerien, Scott, and Pitout 1976) and Australia (Elleman et al. 1978; Runnegar and Falconer 1981). Although toxic properties were similar, hydrolysates of partially pure mcyst had substantially different amino acid compositions (Carmichael 1982), indicating important differences in purity or chemical structure of the peptide toxin(s) from one strain or among different strains. Amann and Eloff (1980) reported strain UV-001 (= NRC-1(SS-17)) to be toxic after 26 years in culture. It and several other *M. aeruginosa* strains produced a single toxic peptide having the same molecular weight (approx. 1,300 daltons). One strain, UV-006 (= WR88), produced two toxins of different molecular weight. Eloff, Siegelman, and Kycia (1982a) grew 10 toxic and three nontoxic strains of *M. aeruginosa* under standard conditions. They isolated toxin(s) by solvent extraction, gel filtration, and HPLC and identified seven that were chemically related. Some strains produced a single toxin (e.g., UV-001 (= NRC-1(SS-17)), others yielded four (e.g., UV-010 (= WR70)), and some as many as six. However, one or two toxins generally accounted for 90% of total toxicity. All hydrolysates consisted of β-methyl aspartate, glutamate, alanine, methylamine, and one of seven different pairs of other amino acids that accounted for the seven different toxin types. The two toxins produced by strain UV-006 (= WR88) had the common amino acids in the D form whereas those of the variable pairs were in the L form (Eloff, Siegelman, and Kycia 1982b).

B. Cyanoginosin

Botes and his colleagues (Botes, Kruger, and Viljoen 1982; Botes et al. 1982a, b) and Santikarn et al. (1983) were the first to provide the complete structure of one of four toxins (designated as *Microcystis* toxin BE-4) produced by strain WR70 (= UV-010). They concluded that it was monocyclic and had, in addition to the common D-amino acids and one of the pairs of L-amino acids, two other unusual amino acids. These were N-methyldehydroalanine (Mdha) (which gives methylamine on hydrolysis) and a nonpolar side chain blocking group of 20 carbon atoms with a mass of 313 daltons (a novel β-amino acid, 3-amino-9-methoxy-2,6,8-trimethyl-10-phenyldeca-4,6-dienoic acid (adda)). Based on fast atom bombardment mass spectrometry (FABMS) and nuclear magnetic resonance (NMR) studies, Botes et al. (1984) concluded that the BE-4 toxin was a cyclic heptapeptide (m.w. 909 daltons) having the structure given in Figure 16-2.

The LD_{50} i.p. mouse of cyanoginosin-LA is 50 $\mu g \cdot kg^{-1}$ (Botes, Kruger, and Viljoen 1982; Botes et al. 1982a). Instead of calling it a microcystin or a *Microcystis* toxin with either alphabetical or numerical suffixes to indicate chromatographic elution order, Botes et al. (1984) proposed the generically derived designation "cyanoginosin" (cygsn) followed by a two-letter suffix to indicate the two variant amino acid residues in the otherwise invariant cyclic heptapeptide. Botes et al. (1985) also reported the structure of four other cyanoginosins termed cyanoginosin-LR (leucine-arginine); -YR (tyrosine-arginine); -YA (tyrosine-alanine); and -YM (tyrosine-methionine).

Krishnamurthy et al. (1986) used FABMS and ^{13}C-NMR methods to show that the toxin called Akerstox, from a bloom of *M. aeruginosa* collected from Akersvatn (Akers Lake), Norway, has a basic structure similar to cygsn-LA but possesses the variants leucine and arginine. Its designation, according to Botes et al. (1984), would be cygsn-LR.

Krishnamurthy, Carmichael, and Sarver (1986) found that *M. aeruginosa* 7820 from Scotland and *Anabaena flos-aquae* S-23-g-1 from Canada produced cyclic heptapeptide toxins with the same molecular weight (994 daltons) and toxicity (LD_{50} i.p. mouse = 50 $\mu g \cdot kg^{-1}$) as Akerstox from Norway. Mcyst produced by *M. aeruginosa* NRC-1-SS-17

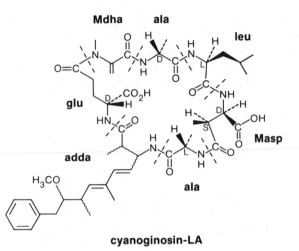

cyanoginosin-LA

Fig. 16-2. Structure of cyanoginosin-LA (toxin BE-4) (LD_{50} i.p. mouse = 50 $\mu g \cdot kg^{-1}$), an example of one of six currently known, related, monocyclic heptapeptide toxins (including microcystin) produced by strains of *Microcystis aeruginosa* and *Anabaena flos-aquae:* adda = 3-amino-9-methoxy-2,6,8-trimethyl-10-phenyldeca-4,6-dienoic acid; ala = alanine; glu = glutamate; leu = leucine; Masp = methyl iso-aspartate; Mdha = methyldehydroalanine; D = dextrorotatory, L = laevorotatory; S = sinister rotation; broken lines across the ring = peptide bonds.

probably has the cygsn-LR structure exhibited by Akerstox (Siegelman, Kycia, and Eloff, pers. comm.). A summary of these cyclic peptide toxins is given in Table 16-2.

The molecular basis of toxin production in these cyanobacteria is not known. Some suggest possible control by extrachromosomal DNA or plasmids (Hauman 1981a, b) whereas others have found no evidence to support a plasmid control mechanism (Vakeria et al. 1985). This important aspect of toxin production will have to be elucidated as more becomes known about the ecological and toxicological role of these secondary chemicals.

C. Mode of action of cyanoginosins and microcystins

The mode of action of cygsns or mcysts is not yet clear. The N-methyl-dehydroalanine residue may be important in target cell specificity, and the apolar side chain may facilitate toxin penetration of cell membranes (Botes et al. 1982b).

In laboratory studies of histopathological changes in affected animals rapid and extensive centrilobular necrosis of the liver occurs with loss of characteristic architecture of the hepatic cords. Both hepatocytes and endothelial cells are destroyed with extensive fragmentation and vesiculation of cell membranes (Østensvik, Skulberg, and Søli, 1981; Run-

Table 16-2. *Mono-cyclic heptapeptide hepatotoxins of Anabaena flos-aquae and Microcystis aeruginosa (LD$_{50}$ i.p. mouse = 50 μg·kg^{-1})*

Species/strain/source	Toxins
A. *flos-aquae* Strain S-23-g-1 (Canada: Saskatchewan)	microcystin (mcyst)
M. *aeruginosa* Strain WR-70 (South Africa: Transvaal)	cyanoginosins[a] (cygsns)
M. *aeruginosa* (Waterbloom, Australia: New South Wales)	cyanoginosin[b]
M. *aeruginosa* (Waterbloom, U.S.: Wisconsin)	microcystin
M. *aeruginosa* Strain NRC-1(SS-17) (Canada: Ontario)	microcystin
M. *aeruginosa* Strain 7820 (Scotland: Loch Balgaves)	microcystin
M. *aeruginosa* (Waterbloom, Norway: Lake Akersvatn)	microcystin

[a]Cygsn-LA; cygsn-LR; cygsn-YR; cygsn-YA.
[b]Cygsn-YM.

negar and Falconer 1981; Foxall and Sasner 1981). The liver is the main target organ for both accumulation and excretion of toxin (Falconer, Buckley, and Runnegar, 1986). Death appears to come from hemorrhagic shock caused by massive pooling of blood in the liver following destruction of the sinusoids (Falconer et al. 1981; Runnegar et al. 1981; Theiss and Carmichael 1986).

It has been suggested that blood pooling and congestion in the liver could arise secondarily from right heart failure caused by pulmonary congestion induced by the thrombi detected in lungs of affected animals (Slatkin et al. 1983). There is considerable evidence against this hypothesis, however. The pulmonary thrombi are not extensive enough to be life-threatening and appear long after signs of liver damage can be detected (Falconer et al. 1981; Theiss and Carmichael 1986). Blood pressure measurements made by cannulation in anaesthetized rats dosed with mcyst failed to show venous congestion but showed, instead, a fall in systemic arterial pressure to values indicative of hypovolemic shock (Theiss and Carmichael 1986). Mcysts and cygsns are therefore considered to act primarily as rapid-acting hepatotoxins. Hepatocyte necrosis was also observed in cases of chronic sublethal exposure (Falconer et al. 1981; Runnegar and Falconer 1981). Although the hepatic hemorrhagic necrosis caused by blue-green peptides is similar to that of phalloidin (a bicyclic peptide toxin from the poisonous mushroom *Amanita phalloides*), the histological and cellular effects are different (Runnegar and Falconer 1981).

Pure mcyst has no hemagglutination activity or effect on synthesis of proteins or nucleic acids. It causes freshly isolated hepatocytes to deform and develop characteristic blebs (swellings) on their surfaces (Runnegar and Falconer 1982). Because lysis does not follow blebbing, mcyst probably affects a cytoskeletal component instead of destroying the cell membrane directly (Runnegar and Falconer 1986). Mcyst uptake, like that of phalloidin, probably takes place at transport sites for bile acids on hepatocyte membranes because cell deformation *in vitro* is blocked by the presence of sodium deoxycholate in the medium (Runnegar, Falconer, and Silver 1981).

There is no known antidote for poisoning by mcysts. Gastric lavage, if applied early, has been used with some success as a treatment, and sublethal chloroform or hydrocortisone prevented or delayed deaths of mice (Adams et al. 1985).

Rapid and sensitive assay procedures for detection of these toxins in water supplies are now receiving some attention. A method involving HPLC fractionation and spectral absorbance at 240 nm was used by Eloff (1982) and Eloff, Siegelman, and Kycia (1982c). Siegelman et al. (1984) and Berg et al. (1987) developed a similar micromethod for assaying toxin from small amounts of cell material. A microtiter cell culture assay more sensitive than the mouse bioassay was developed and suggested for use in water-quality screening by Grabow et al. (1982). Kfir, Johannsen, and Botes (1986) described a method for conjugating cygsn-LA with polylysine and muramyldipeptide to form a complex of a hapten, a carrier, and a vaccine. They used this complex to prepare an anti-cygsn-LA monoclonal antibody. With the use of these antibodies, it should be possible to identify target cells, elucidate mode of action, and detect minute toxin quantities in water supplies and distribution systems.

IV. Other toxins

A. Neurotoxins, hepatotoxins, and unknowns

A number of other freshwater blue-green toxins have been recognized by signs and survival times produced when ingested by or injected into various animal species. Many of these have not yet been isolated, and none has been chemically identified. Some may prove to be chemically related to known toxins and some to be altogether new. Four other anatoxin types, in addition to antx-*a* and antx-*a*(s), were recognized by their different physiological effects (Carmichael and Gorham 1978, 1980). These were called antx-*b*, antx-*b*(s), antx-*c*, and antx-*d*. They are produced by *Anabaena flos-aquae* strains isolated in 1972–5 from lake blooms in western Canada. Types that produce salivation and lachrymation in mice, rats, and chicks prior to respiratory arrest are designated by the suffix (s). Antx-*d* produces salivation and lachrymation, and also chromodacryorrhea in rats. Survival times of mice injected with a minimal lethal dose are 4–12 min for all types except antx-*c*, which has a 1–2 h survival time. This suggests neurotoxins and a hepatotoxin, respectively.

Carmichael (1982, 1986) cites reports by Grigor'eva et al. (1977) and Kirpenko and Kirpenko (1980) of a peptide toxin (called TAB-22) produced by *M. aeruginosa* from the Ukraine which has neurotoxic effects on different systems (e.g., cardiac muscle and blood hemostasis). Unlike mcysts (Bishop et al. 1959; Foxall and Sasner 1981; Grabow et al. 1982), this toxin has a wide spectrum of antimicrobial activity. *Microcystis* toxin from a 1977 bloom in the Ukraine was reported by Kirpenko, Sirenko, and Kirpenko (1981) to have embryolethal, teratogenic, gonadotoxic, and mutagenic effects. As measured by the Ames *Salmonella* test, however, mcysts are not mutagenic (Grabow et al. 1982; Runnegar and Falconer 1982).

Strain *M. aeruginosa* M228 from Japan produces a single toxic peptide that has a molecular weight in the range of 770–2,950 daltons and is composed mainly of glutamate, aspartate, alanine, glycine, serine, arginine, and leucine plus six other amino acids in smaller amounts (Watanabe, Oishi, and Nakao 1981; Watanabe and Oishi 1982, 1985). Unlike mcysts, this toxin, purified by HPLC, has an LD_{100} i.p. mouse of 2,600 $\mu g \cdot kg^{-1}$, is ninhydrin-positive, and is inactivated by proteases. Like mcysts or antx-*c*, it causes lethargy, hair erection, pallor, and enlarged livers in mice but is nontoxic by the oral route, even at very high doses. Survival times of one hour were reported for mice injected interperitoneally. Mice are much more sensitive to crude extracts administered intraperitoneally than are rats. Phenoxybenzamine, an alpha-adrenergic blocker, coadministered intraperitoneally prolongs survival times significantly (Oishi and Watanabe 1986), indicating that the toxin affects the sympathetic nervous system.

Lincoln and Carmichael (1981) described growths of a toxic *Synechocystis* sp. in a large outdoor facility for recycling swine wastes in Florida. Poultry fed a feed containing 10–20% of these coccoid blue-greens died, and those on 5% became paralyzed after three days. However, swine fed 16.5% by weight for 30 days showed no signs of poisoning. The LD_{100} i.p. mouse was 500–750 $mg \cdot kg^{-1}$ with a survival time of 5 to 7 h and liver damage of the antx-*c* kind.

In 1981, 10 of 65 sows that ingested a bloom of *Anabaena spiroides* var. *contracta* Kleb. in a farm pond in Illinois died (Beasley et al. 1983). One sow that died was observed closely. After vomiting approximately four liters of dark green liquid, visible trembling progressed to violent shaking, and the animal died within 30 minutes. Necropsy revealed generalized cyanosis, flaccid myocardium, mild emphysema, and ecchymotic hemorrhages of the stomach and intestine. Definitive histological lesions were not detected.

Codd and Bell (1985) found toxic blooms in a number of lakes, ponds, and reservoirs in Scotland during surveys in 1982 and 1983. Several of the lethal blooms injected intraperitoneally into mice were dominated by *Gloeotrichia echinulata* (J. E. Smith) Richter. The LD_{50} i.p. mouse ranged from 113 to 1,500 mg·kg^{-1}. Survival time was about 24 h but was shortened to 1 h by a massive dosage of 7,500 mg·kg^{-1}. Signs of poisoning were lethargy, labored breathing, erection of the hair, and occasional muscular spasms. Emphysema in the gut, an enlarged heart, and blood in the pericardium were found. Unlike mcyst poisoning, there was no pallor or swollen liver. Characterization of the *Gloeotrichia* toxin(s) is in progress.

Toxic strains of *Pseudanabaena catenata* Lauterb. were isolated as contaminants of fasciculate colonies of *Aphanizomenon flos-aquae* collected from a lake in Alberta, Canada (Gorham, McNicholas, and Allen 1982; McNicholas 1983). Toxins from two strains (subsequently lost) caused violent convulsions but not death of mice injected intraperitoneally with lyophilized cells. Seizures began 15–20 s after injection and lasted 3–10 min. When sacrificed and autopsied after 3.5 h, the livers were normal in size but unusually pale around the margins of the lobes; the other organs appeared normal.

B. Endotoxins

Bacterial endotoxins are lipopolysaccharides (LPS) found in the outer layer of the cell wall of gram-negative bacteria. They are responsible for various medical problems such as toxemias in debilitated patients or infants (Keleti et al. 1981). High concentrations of endotoxin were noted in a drinking water reservoir involved in a water-borne outbreak of gastroenteritis in 1975 that affected approximately 5,000 persons in Sewickley, Pennsylvania. An accumulation of the blue-green *Schizothrix calcicola* (Ag.) Gom. in the uncovered, treated water supply caused investigators to suspect it as the likely source of LPS and cause of the gastroenteritis (Lippy and Erb 1976). Sykora and Keleti (1981) measured total, bound, and free LPS in six drinking water systems and concluded that blue-greens in uncovered finished water reservoirs were a significant source of LPS. They recommended such reservoirs be covered. Raziuddin, Siegelman, and Tornabene (1983) showed that the hepatotoxin (cygsn-(LR)) from *M. aeruginosa* UV-006 (= WR88) was distinct from LPS. Keleti et al. (1981) found that LPS from *Schizothrix calcicola* UPB 1 was nontoxic (i.p. mouse), even at a very high dosage of 5 mg per mouse. LPS from *Oscillatoria brevis* UTEX B1567 was toxic only at high dosage (LD_{50} i.p. mouse 3.83 mg per mouse). Keleti et al. (1981) examined the effects of LPS from blue-greens on a range of biological reactions and found it was much weaker than LPS from bacteria. As a result, public health concerns about the hazard of LPS from blue-greens in drinking water systems diminished.

V. Medical aspects

A. Human poisoning

Acute poisoning of humans by freshwater blue-greens, as occurs with PSP, has almost never been reported. Humans are probably just as susceptible as pets, livestock, or wildlife, but people are repelled by the idea of eating or drinking from an algal bloom. Furthermore, there are no vectors, like shellfish, to concentrate toxins of freshwater blue-greens in the normal food chain (Carmichael 1981b). The susceptibility of humans to such toxins is supported by indirect and direct evidence. Tustin, von Rensburg, and Eloff (1973) demonstrated that primates are susceptible to mcysts. Suspensions of toxic *Microcystis* from a bloom were administered by intubation at $100–500$ mg·kg^{-1}·wk^{-1} to vervet monkeys. Depending upon dosage, death occurred after four or more weeks, accompanied by panlobular hepatic necrosis with hemorrhage.

Dillenberg and Dehnel (1960) described cases of human poisoning from accidental ingestion of blooms from lakes in Saskatchewan, Canada. In one case a physician accidentally fell into a lake with a heavy algal bloom and swallowed about 250 ml. Three to five hours later he experienced stomach pains, vomiting, and painful diarrhea. Later, he developed fever, severe headache, pain in muscles and joints, and weakness. His stool contained many *Microcystis* cells and some *Anabaena circinalis* chains but no bacterial pathogens. Other cases of gastroenteritis and one case of coma with labored breathing associated with ingestion of blooms of *Aphanizomenon, Anabaena,* and *Microcystis* were observed and described by Dillenberg according to Schwimmer and Schwimmer (1964, 1968).

Sargunar and Sargunar (1979) published a short abstract of a study in India that involved 48 subjects. Thirty-seven of these ingested a quantity of *Microcystis aeruginosa* bloom known to be toxic to fish but not to affect dogs, cats, or fowls. Two subjects suffered nausea, vomiting, giddiness, stomachache, diarrhea, cramps, vague body pains, and weakness.

Aziz (1974) reported a nondialyzable factor obtained from blooms and a laboratory strain of *Microcystis aeruginosa* collected from a city pond in Dacca, Bangladesh, that caused diarrhea. This factor caused fluid accumulation in ligated loops of guinea pig intestine (a test used for cholera). The dialyzable fraction caused no fluid accumulation but killed rats given 0.5 ml intraperitoneally. Two toxins, distinctly different in molecular size and physiological effect, were produced by these *M. aeruginosa* cells. Aziz (1980, personal communication to W. W. Carmichael) found *Microcystis* cells in some stool samples of patients with diarrhea but had no direct evidence this alga was the cause of the ailment.

In 1979, an outbreak of hepatoenteritis occurred on Palm Island, Queensland, Australia. It involved 148 people (mostly children), many of whom required hospitalization. The illnesses began three days after copper sulfate was used to control a blue-green bloom in Solomon Dam, the reservoir for the domestic water supply. The epidemiological investigation that followed led to the discovery of a toxic strain of a new species, *Cylindrospermopsis raciborskii* (Wolosz,) Seenaya and Subba Raju (Hawkins et al. 1985). Blooms in Solomon Dam were regularly composed of *C. raciborskii* and *An. circinalis,* but the latter species was nontoxic when tested.

Lyophilized cells of toxic *C. raciborskii* had an LD_{50} i.p. mouse of 64 ± 5 mg·kg^{-1} with a survival time of 19 h. Upon necropsy, livers of mice were pale, sometimes with white foci (caused by fibrin thrombi) that were frequently associated with focal hemorrhages in the lungs. There was centrilobular to massive hepatocyte necrosis with variable degrees of injury in the kidneys, adrenals, and intestines as well as in the lungs.

Circumstantial evidence implicates toxic *C. raciborskii* as the likely producer of the toxin that, when released into the water supply by the copper sulfate treatment, was responsible for the outbreak of human gastroenteritis among users. The longer survival time with mice and different effects on the liver from those produced by mcysts suggest that similarities, if any, may be with slower-acting antx-*c* toxin.

Wheeler, Lackey, and Schott (1942) found that toxins produced by *Microcystis* blooms were not completely destroyed or removed by the laboratory equivalent of regular water purification processes such as alum coagulation, filtration, chlorination, and treatment with activated carbon. Massive amounts of activated carbon were needed to render the effluent nontoxic. Hoffman (1976) confirmed that mcysts are not removed by normal water treatment practices.

Østensvik, Skulberg, and Søli (1981) reported that rats given lethal doses of mcysts showed significant increases in blood plasma levels of aspartate amino transferase (but not alanine amino transferase or bilirubin), which correlated with increased liver damage. Falconer, Beresford, and Runnegar (1983) obtained evidence of chronic liver damage among residents of Armidale, N.S.W., Australia, whose drinking water was supplied from the Malpas Dam Reservoir. They compared the plasma levels of hepatic enzymes of this group before, during, and after a heavy bloom of toxic *M. aeruginosa* in the dam (Feburary–March 1981) with the levels of people whose water came from other sources. The Malpas Dam group showed a statistically significant rise in a sensitive indicator of liver damage, gammaglutamyl transferase, that was not observed with the other group. This case and the Solomon Dam case emphasize the need for better methods to monitor and remove toxins produced by blue-greens in eutrophic reservoirs or water bodies used for drinking water.

In a pilot plant study Falconer, Runnegar, and Huynh (1983) found that a 7–8 cm layer of granular activated carbon on top of a sand filter removed mcysts. They suggested that a 10–15 cm layer would be sufficient for five to seven days of peak load of blue-greens and that this should be used prior to treating a reservoir with copper sulfate.

B. Contact dermatitis

Billings (1981) documented four episodes during the summer of 1979 that involved significant numbers of children and adults who developed illnesses while swimming or within 2–12 h after contact with *Anabaena* blooms in two lakes in Pennsylvania. Symptoms ranged from gastroenteritis to hay-fever-like eye irritation, sore throat, earache, sneezing, running nose, and swollen lips. Tests for bacterial or viral pathogens were negative, and giardiasis (from the protozoan *Giardia*) was eliminated as a possible agent.

Many years before, Cohen and Reif (1953) described repeated episodes of erythrematous papulovescicular dermatitis among swimmers who had come in contact with *Anabaena* in Lake Carey, Pennsylvania. They demonstrated cutaneous sensitization to a phycocyanin-containing fraction from *Anabaena* and concluded that blue-greens were responsible for itching, swelling, and conjunctivitis experienced by many swimmers. Heise (1949, 1951) also documented repeated episodes of hay-fever-like symptoms involving itchy eyes, nasal discharge and blockage, asthma, and generalized urticaria in two patients who came in contact with species of *Oscillatoria* or *Microcystis* while swimming in Wisconsin lakes.

Episodes of dermatitis and/or irritation from contact with freshwater blue-greens have occurred with increasing frequency as more eutrophic waters are used for recreational purposes. Investigations in the United States, Canada, Scotland, and Norway during the last decade have shown that dermatotoxic blooms may be dominated by *Anabaena* spp., *Aphanizomenon flos-aquae*, the *Oscillatoria agardhii/rubescens* group, and *Gloeotrichia echinulata* (Skulberg, Codd, and Carmichael 1984; Carmichael et al. 1985; Codd and Bell 1985; Codd and Carmichael, unpublished data; Gorham, unpublished data). In some bioassays, blooms were hazardous to animals if ingested, though in other cases they were not. Signs and survival times indicated high to low titers of hepatotoxins like mcysts or antx-*c* toxin or of neurotoxins like antx-*a* and antx-*a*(s). Such variability suggests that other agents, perhaps located in the outer layers of the cell walls, are responsible for the contact dermatitis experienced by susceptible or sensitized individuals.

Filamentous marine blue-greens (*Lyngbya majuscula, Schizothrix calcicola,* and *Oscillatoria nigroviridis*) produce inflammatory agents responsible for a severe contact dermatitis known as "swimmer's itch" (Moore 1981). Two of these agents, lyngbyatoxin A (an indole alkaloid) and debromoaplysiatoxin (a phenol), are also potent tumor-producing compounds. What relation, if any, exists between the dermatotoxins of freshwater and marine species is still unknown.

VI. Food use

As the foregoing has indicated, strains of many species of freshwater blue-greens produce potent neurotoxins, hepatotoxins, and other toxins whose properties are not yet established. The chemical structures of a few toxins are known, others are partly known, and still others are unknown. As time goes on it is likely that more new toxins will be discovered and that more will be learned about the structures, toxicological properties, and modes of action of currently recognized toxins.

Blue-green algal toxins pose a hazard to humans when sufficient amounts are ingested (accidentally or by way of drinking water systems), when there is contact and contact sensitivity, or when blue-greens are grown and eaten as food without adequate monitoring and observation of sanitary precautions during culture, harvest, distribution, and storage of the product (see also Jassby, Chapter 8).

Most blue-green blooms are nontoxic, but the environmental conditions that select for toxicity are not understood (Carmichael and Gorman 1981) nor are the factors regulating the stability and persistence of toxicity in laboratory cultures and natural water bodies. Transmission of toxigenicity may occur from one strain or species to another, but this has not been proven. Until more is known about toxic blue-greens and until adequate health standards for production and marketing of blue-greens for single-cell protein (SCP) are adopted and observed, there will be a degree of hazard in using them as food for human or animal consumption.

A. Spirulina

Spirulina platensis (Nordst.) Gom. and *S. maxima* (Setch. et Gardner) Geitl. are two closely related species that have been used as food by the Kanembu in Chad and the Aztecs in Mexico (Ortega 1972; Furst 1978;

Switzer 1982; Jassby, Chapter 7). These algae are consumed either alone or in conjunction with cereals, particularly under famine conditions. Interest in *Spirulina* as an innovative food source first occurred in 1965. Since then many strains of *Spirulina* have been collected from different parts of the world. They contain 60–70% protein and 5% nucleic acid on a dry weight basis, and the nutritional quality is high (Switzer 1982; Jassby, Chapter 7).

No toxicity has been recorded for *Spirulina* (Switzer 1982). Tests on rats have been negative (Bourges et al. 1971; Pirie 1975; Contreras et al. 1979), but more than 5% *Spirulina* in supplementary diets retarded growth of chickens (Clément 1975; Contreras et al. 1979). A group of Mexican athletes took 20–40 g *Spirulina* daily for two periods of 30–45 successive days with no detectable negative effects (Pirie 1975). Clinical human feeding trials have also taken place in the United States, Japan, Germany, and France (Switzer 1982). The sanitary standards of the industry are stated to be within the range applied for whole milk, with the product kept below 7% moisture to prevent spoilage (Switzer 1982).

The Protein Advisory Group of the United Nations recommends using SCP from sources such as *Spirulina* as a dietary supplement but not as a sole source, because yeasts, fungi, bacteria, or algae with unusually high protein and/or nucleic acid contents cause nausea, vomiting, diarrhea, and undesirably high uric acid levels (Scrimshaw 1975).

Starting in 1979, widespread publicity by producers and distributors of *Spirulina* throughout large parts of the United States and Canada triggered a great amount of interest in it as a nutritious food for humans as well as a safe diet pill. A few grams of *Spirulina* taken one-half to one hour before meals causes a marked reduction in appetite (Switzer 1982).

In 1981–2 two separate but related cases of human illness involving dietary consumption of a brand of *Spirulina* pills occurred in British Columbia, Canada. Within 20–90 min after taking two to five pills (6 g each) on an empty stomach or with additional protein supplement, both adults repeatedly experienced nausea and vomiting or severe diarrhea. Official investigation of the first lot of pills showed the product to be sterile. The second lot of pills, obtained from the retailer and from the other consumer who became ill, was negative for adulteration with other species of blue-greens, for bacterial toxins, and for blue-green toxins. However, the patient's pills, but not the retailer's, had high counts of bacteria. This pointed to spoilage after purchase.

B. Anabaena and other species

The production and marketing of *Spirulina* as a health food has prompted other enterprises of a similar nature utilizing species of blue-greens that are potentially toxic. One of these enterprises states in its promotional literature the "[Brand name] is a U.S.-grown, fresh-water, blue-green micro-algae *(anabaena)* [*sic*] freeze-dried and specially formulated with extra glutamic acid for increased fuel to the brain." This is potentially hazardous, because no indication is given about the species or strain(s) of *Anabaena* being used and whether or not an adequate monitoring program is being carried out to guarantee that the product is nontoxic.

Commerical harvest, drying, and sale of natural blooms for use as animal feed have also occurred. Because of variable and uncontrolled composition and the inability to distinguish toxic from nontoxic strains, this type of product poses a potentially serious hazard to animals and humans. Harvest and sale of natural blooms should be restricted or banned, at least for the present, until more research has been done to identify and assess the hazard of inadequately known and possibly dangerous toxins.

Rapid, sensitive, and reliable assay methods need to be developed or adapted to monitor the levels of toxins produced by freshwater bluegreens. These should be applied to the production and marketing of *Spirulina* or other species as SCP. Only regularly tested nontoxic strains and carefully controlled culture conditions should be used for SCP production. A combination of all these safeguards is needed to make bluegreens safe and reliable as SCP for human or animal consumption.

VII. References

Adams, W. H., R. D. Stoner, D. G. Adams, D. W. Slatkin, and H. W. Siegelman. 1985. Pathophysiologic effects of a toxic peptide from *Microcystis aeruginosa*. *Toxicon* 23, 441–7.

Adelman, W. J., Jr., J. F. Fohlmeister, J. J. Sasner, Jr., and M. Ikawa. 1982. Sodium channels blocked by aphantoxin obtained from the blue-green alga, *Aphanizomenon flos-aquae*. *Toxicon* 28, 513–16.

Amann, M. J., and J. N. Eloff. 1980. A preliminary study on the toxins of different *Microcystis* strains. *S. Afr. J. Sci.* 70, 419–20.

Aronstam, R. S., and B. Witkop. 1981. Anatoxin-*a* interactions with cholinergic synaptic molecules. *Proc. Natl. Acad. Sci. USA* 78, 4639–43.

Astrachan, N. B., and B. G. Archer. 1981. Simplified monitoring of anatoxin-*a* by reverse-phase high performance liquid chromatography and the sub-acute effects of anatoxin-*a* in rats. In *The Water Environment: Algal Toxins and Health.* ed. W. W. Carmichael, pp. 437–46. Plenum Press, New York.

Aziz, K. M. S. 1974. Diarrhea toxin obtained from a waterbloom-producing species, *Microcystis aeruginosa* Kütz. *Science* 183, 1206–7.

Barton, L. L., E. W. Foster, and G. V. Johnson. 1980. Viability changes in human neutrophils and monocytes following exposure to toxin extracted from *Aphanizomenon flos-aquae*. *Can. J. Microbiol.* 26, 272–4.

Bates, H. A., and H. Rapoport. 1979. Synthesis of anatoxin-*a* via intramolecular cyclization of iminium salts. *J. Amer. Chem. Soc.* 101, 1259–67.

Beasley, V. R., R. W. Coppock, J. Simon, R. Ely, W. B. Buck, R. A. Corley, D. M. Carlson, and P. R. Gorham. 1983. Apparent blue-green algae poisoning in swine subsequent to ingestion of a bloom dominated by *Anabaena spiroides*. *J. Amer. Vet. Med. Assoc.* 1982, 423–4.

Berg, K., W. W. Carmichael, O. M. Skulberg, C. Benestad, and B. Underdal. 1987. Investigation of a toxic water-bloom of *Microcystis aeruginosa* (Cyanophyceae) in Lake Akersvatn, Norway. *Hydrobiologia* 144, 97–103.

Billings, W. H. 1981. Water-associated human illness in Northeast Pennsylvania and its suspected association with blue-green algae blooms. In *The Water Environment: Algal Toxins and Health*, ed. W. W. Carmichael, pp. 243–55. Plenum Press, New York.

Bishop, C. T., E. F. L. J. Anet, and P. R. Gorham. 1959. Isolation and identification of the fast-death factor in *Microcystis aeruginosa* NRC-1. *Can. J. Biochem. Physiol.* 37, 453–71.

Botes, D. P., H. Kruger, and C. C. Viljoen. 1982. Isolation and characterization of four toxins from the blue-green alga, *Microcystis aeruginosa*. *Toxicon* 20, 945–54.

Botes, D. P., A. A. Tuinman, P. L. Wessels, C. C. Viljoen, H. Kruger, D. H. Williams, S. Santikarn, R. J. Smith, and S. J. Hammond. 1984. The structure of cyanoginosin-LA, a cyclic heptapeptide toxin from the cyanobacterium *Microcystis aeruginosa*. *J. Chem. Soc. Perkin Trans.* 1, 2311–18.

Botes, D. P., C. C. Viljoen, H. Kruger, P. L. Wessels, and D. H. Williams. 1982a. Structure of toxins of the blue-green alga *Microcystis aeruginosa*. *S. Afr. J. Sci.* 78, 378–9. (Abstract.)

Botes, D. P., C. C. Viljoen, H. Kruger, P. L. Wessels, and D. H. Williams. 1982b. Configuration assignments of the amino acid residues and the presence of N-methyldehydroalanine in toxins from the blue-green alga, *Microcystis aeruginosa*. *Toxicon* 20, 1037–42.

Botes, D. P., P. L. Wessels, H. Kruger, M. T. C. Runnegar, S. Santikarn, R. J. Smith, J. C. J. Barna, and D. H. Williams. 1985. Structural studies on cyanoginosins-LR,-YR,-YA and -YM, peptide toxins from *Microcystis aeruginosa*. *J. Chem. Soc. Perkin Trans.* 1, 2747–8.

Bourges, H., A. Sotomayor, E. Mendoza, and A. Chávez. 1971. Utilization of the alga *Spirulina* as a protein source. *Nutr. Rep. Int.* 4, 31–43.

Campbell, H. F., O. E. Edwards, J. W. Elder, and R. J. Kolt. 1979. Total synthesis of DL-anatoxin-*a* and DL-isoanatoxin-*a*. *Pol. J. Chem.* 53, 27–37.

Campbell, H. F., O. E. Edwards, and R. Kolt. 1977. Synthesis of nor-anatoxin-*a* and anatoxin-*a*. *Can. J. Chem.* 55, 1372–9.

Carmichael, W. W. (ed.). 1981a. *The Water Environment: Algal Toxins and Health*. Environmental Science Research, vol. 20. Plenum Press, New York. 491 pp.

Carmichael, W. W. 1981b. Freshwater blue-green algae (cyanobacteria) toxins: a review. In *The Water Environment: Algal Toxins and Health*, ed. W. W. Carmichael, pp. 1–13. Plenum Press, New York.

Carmichael, W. W. 1982. Chemical and toxicological studies of the toxic freshwater cyanobacteria *Microcystis aeruginosa*, *Anabaena flos-aquae*, and *Aphanizomenon flos-aquae*. *S. Afr. J. Sci.* 78, 367–72.

Carmichael, W. W. 1986. Algal toxins. *Adv. Bot. Res.* 12, 47–101.

Carmichael, W. W. 1988. Toxins of freshwater algae. In *Handbook of Natural*

Toxins, vol. 3, *Marine Toxins and Venoms,* ed. A. T. Tu, pp. 121–47. Marcel Dekker, New York.

Carmichael, W. W., D. F. Biggs, and P. R. Gorham. 1975. Toxicology and pharmacological action of *Anabaena flos-aquae* toxin. *Science* 187, 542–4.

Carmichael, W. W., D. F. Biggs, and M. A. Peterson. 1979. Pharmacology of anatoxin-*a,* produced by the freshwater *Anabaena flos-aquae* NRC-44-1. *Toxicon* 17, 229–36.

Carmichael, W. W., and P. R. Gorham. 1977. Factors influencing the toxicity and animal susceptibility of *Anabaena flos-aquae* (Cyanophyta) blooms. *J. Phycol.* 13, 97–101.

Carmichael, W. W., and P. R. Gorham. 1978. Anatoxins from clones of *Anabaena flos-aquae* isolated from lakes of western Canada. *Mitt. Int. Verein. Limnol.* 21, 285–95.

Carmichael, W. W., and P. R. Gorham. 1980. Freshwater cyanophyte toxins: types and their effects on the use of micro algae biomass. In *Algae Biomass: Production and Use,* ed. G. Shelef and C. J. Soeder, pp. 437–48. Elsevier/North Holland Biomedical Press, Amsterdam.

Carmichael, W. W., and P. R. Gorham. 1981. The mosaic nature of toxic blooms of cyanobacteria. In *The Water Environment: Algal Toxins and Health,* ed. W. W. Carmichael, pp. 161–72. Plenum Press, New York.

Carmichael, W. W., P. R. Gorham, and D. F. Biggs. 1977. Two laboratory case studies on the oral toxicity to calves of the freshwater cyanophyte (blue-green alga) *Anabaena flos-aquae* NRC-44-1. *Can. Vet. J.* 18, 71–5.

Carmichael, W. W., C. L. A. Jones, N. A. Mahmood, and W. C. Theiss. 1985. Algal toxins and water-based diseases. *CRC Crit. Rev. in Envir. Contr.* 15, 275–313.

Carmichael, W. W., and N. A. Mahmood. 1984. Toxins from freshwater cyanobacteria. In *Seafood Toxins,* ed. E. Ragelis, pp. 377–89. American Chemical Society Symposium Series, no. 262. Washington, D.C.

Carmichael, W. W., and L. D. Schwartz. 1984. *Preventing Livestock Deaths from Blue-green Algae Poisoning.* USDA Farmers' Bulletin no. 2275. 12 pp.

Clément, G. 1975. Producing *Spirulina* with CO_2. In *Single Cell Protein II,* ed. S. R. Tannenbaum and D. I. C. Wang, pp. 467–74. MIT Press, Cambridge, MA.

Cliver, D. O., R. A. Newman, R. D. Pickford, and P. S. Berger (eds.). 1984. *Drinking Water Microbiology.* EPA 570/9-94-006; CCMS 128, U.S. Environmental Protection Agency, Washington, D.C. 545 pp.

Codd, G. A., and S. G. Bell. 1985. Eutrophication and toxic cyanobacteria in freshwaters. *Water Poll. Contr.* 84, 225–32.

Cohen, S. G., and C. B. Reif. 1953. Cutaneous sensitization to blue-green algae. *J. Allergy* 24, 452–7.

Contreras, A., D. C. Herbert, B. G. Grubbs, and I. L. Cameron. 1979. Blue-green alga *Spirulina* as the sole dietary source of protein in sexually maturing rats. *Nutr. Rep. Int.* 19, 749–63.

Danheiser, R. L., J. M. Morin, Jr., and E. J. Salaski. 1985. Efficient total synthesis of (±)-anatoxin-*a. J. Amer. Chem. Soc.* 107, 8066–73.

Desikachary, T. V. 1959. *Cyanophyta.* Indian Council of Agricultural Research, New Delhi. 686 pp.

Devlin, J. P., O. E. Edwards, P. R. Gorham, N. R. Hunter, R. K. Pike, and B. Stavric. 1977. Anatoxin-*a*, a toxic alkaloid from *Anabaena flos-aquae* NRC-44h. *Can. J. Chem.* 55, 1367–71.

Dillenberg, H. O., and M. K. Dehnel. 1960. Toxic waterbloom in Saskatchewan, 1959. *Can. Med. Assoc. J.* 83, 1151–4.

Edler, L., S. Ferno, M. G. Lund, R. Lundberg, and Per Olle Nilsson. 1985. Mortality of dogs associated with a bloom of the cyanobacterium *Nodularia spumigena* in the Baltic Sea. *Ophelia* 24, 103–9.

Elleman, T. C., I. R. Falconer, A. R. B. Jackson, and M. T. Runnegar. 1978. Isolation, characterization, and pathology of the toxin from a *Microcystis aeruginosa* = *(Anacystis cyanea)* bloom. *Aust. J. Biol. Sci.* 31, 209–18.

Eloff, J. N. 1982. A specific ultraviolet absorbance of *Microcystis aeruginosa* toxins. *J. Limnol. Soc. S. Afr.* 8, 5–7.

Eloff, J. N., H. W. Siegelman, and H. Kycia. 1982a. Comparative study on the toxins from several *Microcystis aeruginosa* isolates. *S. Afr. J. Sci.* 78, 377. (Abstract.)

Eloff, J. N., H. W. Siegelman, and H. Kycia. 1982b. Amino acid sequence of two toxic peptides from *Microcystis aeruginosa* UV-006. *S. Afr. J. Sci.* 78, 377. (Abstract.)

Eloff, J. N., H. W. Siegelman, and H. Kycia. 1982c. Extraction, isolation, and stability of toxins from *Microcystis aeruginosa*. *S. Afr. J. Sci.* 78, 377–8. (Abstract.)

Falconer, I. R., A. M. Beresford, and M. T. C. Runnegar. 1983. Evidence of liver damage by toxin from a bloom of the blue-green alga, *Microcystis aeruginosa*. *Med. J. Aust.* 1, 511–14.

Falconer, I. R., T. Buckley, and M. T. C. Runnegar. 1986. Biological half-life, organ distribution, and excretion of ^{125}I-labelled toxic peptide from the blue-green alga *Microcystis aeruginosa*. *Aust. J. Biol. Sci.* 39, 17–21.

Falconer, I. R., A. R. B. Jackson, J. Langley, and M. T. Runnegar. 1981. Liver pathology in mice in poisoning by blue-green alga *Microcystis aeruginosa*. *Aust. J. Biol. Sci.* 34, 179–87.

Falconer, I. R., M. T. C. Runnegar, and V. L. Huynh. 1983. Effectiveness of activated carbon in the removal of algal toxin from potable water supplies: a pilot plant investigation. In *Australian Water and Wastewater Association, Tenth Federal Convention*, Technical Papers, pp. 26-1–26-8. Sydney, April.

Fox, G. E., E. Stackebrandt, R. B. Hespell, J. Gibson, J. Maniloff, T. A. Dyer, R. S. Wolfe, W. E. Balch, R. S. Tanner, L. J. Magrum, L. B. Zablen, R. Blakemore, R. Gupta, L. Bonen, B. J. Lewis, D. A. Stahl, K. R. Luehrsen, K. N. Chen, and C. R. Woese. 1980. The phylogeny of prokaryotes. *Science* 209, 457–63.

Foxall, T. L., and J. J. Sasner, Jr. 1981. Effects of hepatic toxin from the cyanophyte *Microcystis aeruginosa*. In *The Water Environment: Algal Toxins and Health*, ed. W. W. Carmichael, pp. 365–87. Plenum Press, New York.

Francis, G. 1878. Poisonous Australian lakes. *Nature* 18, 11–12.

Furst, P. T. 1978. *Spirulina*: a nutritious alga, once a staple of Aztec diets, could feed many of the world's hungry people. *Human Nature* 1, 60–5.

Geitler, L. 1932. *Cyanophyceae*. Akad. Verlag m.b.H., Leipzig; Johnson Reprint Corp., New York. 1st reprinting (1971).

Gentile, J. H. 1971. Blue-green and green algal toxins. In *Microbial Toxins VII: Algal and Fungal Toxins,* ed. S. Kadis, A. Ciegler, and S. J. Ajl, pp. 27–66. Academic Press, New York.

Gorham, P. R. 1964. Toxic algae as a public health hazard. *J. Amer. Water Works Assoc.* 56, 1481–8.

Gorham, P. R., J. McLachlan, U. T. Hammer, and W. K. Kim. 1964. Isolation and culture of toxic strains of *Anabaena flos-aquae* (Lyngb.) de Bréb. *Verh. Int. Verein. Limnol.* 15, 796–804.

Gorham, P. R., S. McNicholas, and E. A. D. Allen. 1982. Problems encountered in searching for new strains of toxic planktonic cyanobacteria. *S. Afr. J. Sci.* 28, 357–62.

Grabow, W. O. K., W. C. DuRandt, O. W. Prozesky, and W. E. Scott. 1982. *Microcystis aeruginosa* toxin: cell culture toxicity, hemolysis, and mutagenicity assays. *Appl. Envir. Microbiol.* 43, 1425–33.

Grigor'eva, L. A., Yu. A. Kirpenko, V. M. Orlovskiy, and V. V. Stankevitch. 1977. On the antimicrobic effect of toxic metabolites of some blue-green algae. *Gidrobiologicheskii Zh.* 13, 57–62. (In Ukrainian.)

Hall, S. 1982. *Toxins and Toxicity of Protogonyaulax from the Northeast Pacific.* Ph.D. thesis, University of Alaska, Fairbanks. 196 pp.

Hauman, J. H. 1981a. Is a plasmid(s) involved in the toxicity of *Microcystis aeruginosa?* In *The Water Environment: Algal Toxins and Health,* ed. W. W. Carmichael, pp. 97–102. Plenum Press, New York.

Hauman, J. H. 1981b. *A Molecular Biological Characterization of Microcystis aeruginosa Kütz. emend Elenkin.* M. S. thesis. University of Pretoria. 101 pp.

Hawkins, P. R., M. T. C. Runnegar, A. R. B. Jackson, and I. R. Falconer. 1985. Severe hepatotoxicity caused by the tropical cyanobacterium (blue-green alga) *Cylindrospermopsis raciborskii* (Wolosz) Seenaya and Subba Raju isolated from a domestic water supply reservoir. *Appl. Envir. Microbiol.* 50, 1292–5.

Heise, H. A. 1949. Symptoms of hay fever caused by algae. *J. Allergy* 20, 383–5.

Heise, H. A. 1951. Symptoms of hay fever caused by algae. II. *Microcystis:* Another form of algae producing allergenic reactions. *Ann. Allergy* 9, 100–1.

Herdman, M., M. Janvier, R. Rippka, and R. Y. Stanier. 1979a. Genome size of cyanobacteria. *J. Gen. Microbiol.* 111, 73–85.

Herdman, M., M. Janvier, J. B. Waterbury, R. Rippka, and R. Y. Stanier. 1979b. Deoxyribonucleic acid base composition of cyanobacteria. *J. Gen. Microbiol.* 111, 63–71.

Hoffman, J. R. H. 1976. Removal of *Microcystis* toxins in water purification processes. *Water S. Afr.* 2, 58–60.

Huber, C. S. 1972. The crystal structure and absolute configuration of 2, 9-diacetyl-9-azabicyclo (4,2,1) non-2,3-ene. *Acta Crystallogr.* B78, 2577–82.

Hughes, E. O., P. R. Gorham, and A. Zehnder. 1958. Toxicity of a unialgal culture of *Microcystis aeruginosa. Can. J. Microbiol.* 4, 225–36.

Ikawa, M., K. Wegner, T. L. Foxall, J. J. Sasner, Jr., P. W. Carter, and N. H. Shoptaugh. 1981. Studies on the fluorometric determination of the toxins of the blue-green alga *Aphanizomenon flos-aquae.* In *The Water Environment: Algal Toxins and Health,* ed. W. W. Carmichael, pp. 415–25. Plenum Press, New York.

Jackim, E., and J. Gentile. 1968. Toxins of a blue-green alga: similarity to saxitoxin. *Science* 162, 915–16.

Jackson, A. R. B., A. McInnes, I. R. Falconer, and M. T. C. Runnegar. 1984. Clinical and pathological changes in sheep experimentally poisoned by the blue-green alga *Microcystis aeruginosa*. *Vet. Pathol.* 21, 102–13.

Juday, R. E., E. J. Keller, A. Horpestad, L. L. Bahls, and S. Glasser. 1981. A toxic bloom of *Anabaena flos-aquae* in Hegben Reservoir, Montana, in 1977. In *The Water Environment: Algal Toxins and Health*, ed. W. W. Carmichael, pp. 103–12. Plenum Press, New York.

Keleti, G., J. L. Sykora, L. A. Maiolie, D. L. Doerfler, and I. M. Campbell. 1981. Isolation and characterization of endotoxin from cyanobacteria (blue-green algae). In *The Water Environment: Algal Toxins and Health,* ed. W. W. Carmichael, pp. 447–64. Plenum Press, New York.

Kfir, R., E. Johannsen, and D. P. Botes. 1986. Monoclonal antibody specific for cyanoginosin-LA: preparation and characterization. *Toxicon* 24, 543–52.

Kirpenko, Yu. A., and N. I. Kirpenko. 1980. Biological activity of the algotoxin of blue-green algae-pathogens of water "blooming." *Gidrobiologicheskii Zh.* 16, 53–7. (In Ukrainian.)

Kirpenko, Yu. A., L. A. Sirenko, and N. I. Kirpenko. 1981. Some aspects concerning remote after-effects of blue-green algae toxin impact on warm-blooded animals. In *The Water Environment: Algal Toxins and Health.* ed. W. W. Carmichael, pp. 257–69. Plenum Press, New York.

Komarek, J. 1958. Die taxomomische revision der planktischen blaualgen der Tschechoslowakei. In *Algologische Studien*, ed. J. Komarek and H. Ettl. Czechoslovakia Academy of Science, Prague.

Konst, H., P. D. McKercher, P. R. Gorham, A. Robertson, and J. Howell. 1965. Symptoms and pathology produced by toxic *Microcystis aeruginosa* NRC-1 in laboratory and domestic animals. *Can. J. Comp. Med. Vet. Sci.* 29, 221–8.

Koskinen, M. P., and H. Rapoport. 1985. Synthetic and conformational studies on anatoxin-*a:* a potent acetylcholine agonist. *J. Med. Chem.* 28, 1301–9.

Krishnamurthy, T., W. W. Carmichael, and E. W. Sarver. 1986. Investigations of freshwater cyanobacteria (blue-green algae) toxic peptides. I. Isolation, purification, and characterization of peptides from *Microcystis aeruginosa* and *Anabaena flos-aquae*. *Toxicon* 24, 865–73.

Krishnamurthy, T., L. Szafraniec, E. W. Sarver, D. F. Hunt, S. Shabanowitz, W. W. Carmichael, S. Missler, O. Skulberg, and G. Codd. 1986. Amino acid sequences of freshwater blue-green algal toxic peptides by fast atom bombardment tandem mass spectrometer ^{13}C technique. In *Proceeding of the 34th Annual Conference on Mass Spectrography and Allied Topics,* Cincinnati, OH, pp. 93–4.

Lincoln, E. P., and W. W. Carmichael. 1981. Preliminary tests of toxicity of *Synechocystis* sp. grown on wastewater medium. In *The Water Environment: Algal Toxins and Health,* ed. W. W. Carmichael, pp. 223–30. Plenum Press, New York.

Lippy, E. C., and J. Erb. 1976. Gastrointestinal illness at Sewickley, Pa. *J. Am. Water Works Assoc.* 68, 606–10.

Louw, P. G. J. 1950. The active constituent of the poisonous algae *Microcystis toxica* Stephens. *S. Afr. Ind. Chem.* 4, 62–6.

McNicholas, S. 1983. *The Isolation of Functionally Axenic Filaments of Pseudanabaena catenata and Other Planktonic Cyanobacteria.* M.S. thesis, University of Alberta, Edmonton. 98 pp.

Mahmood, N. A., and W. W. Carmichael. 1986a. Paralytic shellfish poisons produced by the freshwater cyanobacterium *Aphanizomenon flos-aquae* NH-5. *Toxicon* 24, 175–86.

Mahmood, N. A., and W. W. Carmichael. 1986b. The pharmacology of anatoxin-*a*(s), a neurotoxin produced by the freshwater cyanobacterium *Anabaena flos-aquae* NRC 525-17. *Toxicon* 24, 425–34.

Mahmood, N. A., and W. W. Carmichael. 1987. Anatoxin-*a*(s), an anticholinesterase from the cyanobacterium *Anabaena flos-aquae* NRC 525-17. *Toxicon* 25, 1221–7.

Mahmood, N. A., W. W. Carmichael, and D. Pfahler. 1988. Anticholinesterase poisonings in dogs from a cyanobacteria (blue-green algae) bloom dominated by *Anabaena flos-aquae*. *Amer. J. Vet. Res.* 49:500–3.

Main, D. C., P. H. Berry, R. L. Peet, and J. P. Robertson. 1977. Sheep mortalities associated with the blue-green alga *Nodularia spumigena*. *Aust. Vet. J.* 53, 578–81.

May, V. 1981. The occurrence of toxic cyanophyte blooms in Australia. In *The Water Environment: Algal Toxins and Health*, ed. W. W. Carmichael, pp. 127–42. Plenum Press, New York.

Moore, R. 1981. Toxins from marine blue-green algae. In *The Water Environment: Algal Toxins and Health*, ed. W. W. Carmichael, pp. 15–23. Plenum Press, New York.

Murthy, J. R., and J. B. Capindale. 1970. A new isolation and structure for the endotoxin from *Microcystis aeruginosa* NRC-1. *Can. J. Biochem.* 48, 508–10.

Oishi, S., and M. F. Watanabe. 1986. Acute toxicity of *Microcystis aeruginosa* and its cardiovascular effects. *Environ. Res.* 40, 518–24.

Olson, T. A. 1951. Toxic plankton. In *Proceedings of Inservice Training Course in Water Works Problems*, Feb. 15 and 16, pp. 86–95. University of Michigan School of Public Health, Ann Arbor.

Ortega, M. M. 1972. Study of the edible algae of the Valley of Mexico. *Bot. Mar.* 15, 162–6.

Østensvik, O., O. M. Skulberg, and N. E. Søli. 1981. Toxicity studies with blue-green algae from Norwegian inland waters. In *The Water Environment: Algal Toxins and Health*, ed. W. W. Carmichael, pp. 315–24. Plenum Press, New York.

Petersen, J. S., S. Toteberg-Kaulen, and H. Rapoport. 1984. Synthesis of (±)-w-Aza[x.y.l] bicycloalkanes by an intramolecular Mannich reaction. *J. Org. Chem.* 49, 2948–53.

Pirie, N. W. 1975. The *Spirulina* algae. In *IBP 4: Food Protein Sources*, ed. N. W. Pirie, pp. 33–9. Cambridge University Press, Cambridge.

Prescott, G. W. 1962. *Algae of the Western Great Lakes Area.* Wm. C. Brown, Dubuque, IA.

Rabin, P., and A. Darbre. 1975. An improved extraction procedure for the endotoxin from *Microcystis aeruginosa* NRC-1. *Biochem. Soc. Trans.* 3, 428–30.

Raziuddin, S., H. W. Siegelman, and T. G. Tornabene. 1983. Lipopolysaccharides of the cyanobacterium *Microcystis aeruginosa. Eur. J. Biochem.* 137, 333–6.

Rippka, R., J. Deruelles, J. B. Waterbury, M. Herdman, and R. Y. Stanier. 1979. Generic assignments, strain histories, and properties of pure cultures of cyanobacteria. *J. Gen. Microbiol.* 111, 1–61.

Runnegar, M. T. C., and I. R. Falconer. 1981. Isolation, characterization, and pathology of the toxin from the blue-green alga *Microcystis aeruginosa.* In *The Water Environment: Algal Toxins and Health,* ed. W. W. Carmichael, pp. 325–42. Plenum Press, New York.

Runnegar, M. T. C., and I. R. Falconer. 1982. The *in vivo* and *in vitro* biological effects of the peptide hepatotoxin from the blue-green alga *Microcystis aeruginosa. S. Afr. J. Sci.* 78, 363–6.

Runnegar, M. T. C., and I. R. Falconer. 1986. Effect of toxin from the cyanobacterium *Microcystis aeruginosa* on ultrastructural morphology and actin polymerization in isolated hepatocytes. *Toxicon* 24, 109–15.

Runnegar, M. T. C., I. R. Falconer, and J. Silver. 1981. Deformation of isolated rat hepatocytes by a peptide hepatotoxin from the blue-green alga *Microcystis aeruginosa. Naunyn-Schieberg's Arch. Pharmacol.* 317, 268–72.

Santikarn, S., D. H. Williams, R. J. Smith, S. J. Hammond, D. P. Botes, A. Tuinman, P. L. Wessels, C. C. Viljoen, and H. Kruger. 1983. A partial structure for the toxin BE-4 from the blue-green algae, *Microcystis aeruginosa. J. Chem. Soc., Chem. Commun.* 12, 652–4.

Sargunar, H. T. P., and A. A. A. Sargunar. 1979. Phycotoxins from *Microcystis aeruginosa:* their implication in human and animal health. Chemische Rundschau. Abstract 610. 4th Int. IUPAC Symp. on Mycotoxins and Phycotoxins, Lausanne, Aug. 29–31.

Sasner, J. J., Jr., M. Ikawa, and T. L. Foxall. 1984. Studies on *Aphanizomenon* and *Microcystis* toxins. In *Seafood Toxins,* ed. E. P. Ragelis, pp. 391–406. American Chem. Society Symposium Ser. no. 262. Washington, D.C.

Sasner, J. J., Jr., M. Ikawa, T. L. Foxall, and W. H. Watson. 1981. Studies on aphantoxin from *Aphanizomenon flos-aquae* in New Hampshire. In *The Water Environment: Algal Toxins and Health,* ed. W. W. Carmichael, pp. 389–403. Plenum Press, New York.

Sawyer, P. J., J. H. Gentile, and J. J. Sasner, Jr. 1968. Demonstration of a toxin from *Aphanizomenon flos-aquae* (L). Ralfs. *Can. J. Microbiol.* 14, 1191–204.

Schantz, E. J., V. E. Ghazarossian, H. K. Schnoes, F. M. Strong, J. P. Springer, J. O. Pezzanite, and J. Clardy. 1975. The structure of saxitoxin. *J. Amer. Chem. Soc.* 97, 1238–9.

Schwimmer, D., and M. Schwimmer. 1964. Algae and medicine. In *Algae and Man,* ed. D. F. Jackson, pp. 368–412. Plenum Press, New York.

Schwimmer, M., and D. Schwimmer. 1968. Medical aspects of phycology. In *Algae, Man, and the Environment,* ed. D. F. Jackson, pp. 278–358. Syracuse University Press, Syracuse, NY.

Scott, W. E, D. J. Barlow, and J. H. Hauman. 1981. Studies on the ecology, growth, and physiology of toxic *Microcystis aeruginosa* in South Africa. In *The Water Environment: Algal Toxins and Health,* ed. W. W. Carmichael, pp. 49–69. Plenum Press, New York.

Scrimshaw, N. S. 1975. Single-cell protein for human consumption: an over-view. In *Single Cell Protein II,* ed. S. R. Tannenbaum and D. I. C. Wang, pp. 24–45. MIT Press, Cambridge, MA.

Shimizu, Y., C. P. Hsu, W. E. Fallon, Y. Oshima, I. Miura, and K. Nakanishi. 1978. Structure of neosaxitoxin. *J. Amer. Chem. Soc.* 100, 6791–3.

Shimizu, Y., M. Norte, A. Hori, A. Genenah, and M. Kobayashi. 1984. Biosynthesis of saxitoxin analogues: the unexpected pathway. *J. Amer. Chem. Soc.* 106, 6433–4.

Siegelman, H. W., N. H. Adams, R. D. Stoner, and D. W. Slatkin. 1984. Toxins of *Microcystis aeruginosa* and their hematological and histopathological effects. In *Seafood Toxins,* ed. E. P. Ragelis, pp. 407–13. American Chemical Society Symposium Series, no. 262. Washington, DC.

Skulberg, O. M., G. A. Codd, and W. W. Carmichael. 1984. Toxic blue-green algal blooms in Europe: a growing problem. *Ambio* 13, 244–7.

Slatkin, D. N., R. D. Stoner, W. H. Adams, J. H. Kycia, and H. W. Siegelman. 1983. Atypical pulmonary thrombosis caused by a toxic cyanobacterial peptide. *Science* 220, 1383–5.

Spivak, C. E., B. Witkop, and E. X. Albuquerque. 1980. Anatoxin-*a:* a novel, potent agonist at the nicotinic receptor. *Mol. Pharmacol.* 18, 384–94.

Sullivan, J. J., and M. M. Wekell. 1984. Determination of paralytic shellfish poisoning toxins by high pressure liquid chromatography. In *Seafood Toxins,* ed. E. P. Ragelis, pp. 197–205. American Chemical Society Symposium Series, no. 262. Washington, DC.

Switzer, L. 1982. *Spirulina: The Whole Food Revolution.* Bantam Books, Toronto. 134 pp.

Sykora, J. L., and G. Keleti. 1981. Cyanobacteria and endotoxins in drinking water supplies. In *The Water Environment: Algal Toxins and Health,* ed. W. W. Carmichael, pp. 285–301. Plenum Press, New York.

Theiss, W. W., and W. W. Carmichael. 1986. Physiological effect of a peptide toxin produced by the freshwater cyanobacteria (blue-green algae) *Microcystis aeruginosa* strain 7820. In *Mycotoxins and Phycotoxins,* ed. P. S. Steyn and R. Vleggaar, pp. 353–63. Elsevier/North Holland Biomedical Press, Amsterdam.

Toerien, D. F., W. E. Scott, and M. J. Pitout. 1976. *Microcystis* toxins: isolation, identification, implications. *Water S. Afr.* 2, 160–2.

Tufariello, J. J., H. Meckler, and K. P. A. Senaratne. 1984. Synthesis of anatoxin-*a:* very fast death factor. *J. Amer. Chem. Soc.* 106, 7979–80.

Tufariello, J. J., H. Meckler, and K. P. A. Senaratne. 1985. The use of nitrones in the synthesis of anatoxin-*a,* very fast death factor. *Tetrahedron* 41, 3447–53.

Tustin, R. C., S. J. von Rensburg, and J. N. Eloff. 1973. Hepatic damage in the primate following ingestion of toxic algae. In *Liver,* ed. J. Saunders and J. Terblanche, pp. 383–5. Proceedings of International Liver Congress. Pitman Medical, London.

Vakeria, D., G. A. Codd, S. G. Bell, K. A. Beattie, and I. M. Priestly. 1985. Toxicity and extrachromosomal DNA in strains of the cyanobacterium *Microcystis aeruginosa. FEMS Microbiol. Lett.* 29, 69–72.

Watanabe, M. F., and S. Oishi. 1982. Toxic substance from a natural bloom of *Microcystis aeruginosa. Appl. Envir. Microbiol.* 43, 819–22.

Watanabe, M. F., and S. Oishi. 1985. Effects of environmental factors on toxicity of a cyanobacterium *(Microcystis aeruginosa)* under culture conditions. *Appl. Envir. Microbiol.* 49, 1342–4.

Watanabe, M. F., S. Oishi, and T. Nakao. 1981. Toxic characteristics of *Microcystis aeruginosa. Verh. Int. Verein. Limnol.* 21, 1441–3.

Wheeler, R. E., J. B. Lackey, and S. A. Schott. 1942. Contribution on the toxicity of algae. *Pub. Health Rep.* 57, 1695–701.

17. Marine biofouling

L. V. EVANS

Department of Pure and Applied Biology, University of Leeds, Leeds LS2 9JT, UK

CONTENTS

I. Introduction

Development of biofouling on surfaces immersed in water is a familiar universal phenomenon. It represents a type of ecological succession starting with an organic film of bacteria, diatoms, and their cellular products (microfouling) and, in the absence of suitable control measures, filamentous algae, animal larvae, and larger algae and animals (macrofouling). Rapid colonization occurs not only on inshore rock surfaces, pilings, piers, and aquatic vegetation but also on offshore man-made structures. With reference to the latter, micro- and macrofouling cause problems on ships' hulls, submarines, offshore oil and gas installations, ocean thermal energy conversion (OTEC) platforms, navigational buoys, moored oceanographic instruments, fishnets and cages, and underwater sound equipment. In addition, biofouling causes problems in all kinds of aquaculture systems and on pipes and hulls carrying seawater on ships and in marine aquariums (for an overview, see Harderlie 1984).

Fouling on ships' hulls and propellors increases frictional drag and causes significant speed and power losses, increased fuel consumption, self-noise, and vibration. On the legs and other supports of static offshore platforms, which are not normally treated to reduce and/or prevent fouling, biological growths substantially increase the loading forces caused by waves and currents, necessitating regular and costly cleaning operations for safety reasons. Fouling also seriously impedes inspection and maintenance. In addition, fouling influences corrosion directly by local modification of the steel environment and enhances corrosion indirectly by providing conditions suitable for sulfate-reducing bacterial activities (see, e.g., Freeman 1977; Terry and Edyvean 1986). Macrofouling on the hulls and mooring systems of OTEC platforms is likely to increase surface roughness and contribute substantially to drag, and the presence of microfouling on the heat exchanger tubes clogs inflow and outflow portals, reducing efficiency of operation (Thorhaug and Marcus 1981). The increased drag caused by heavy macrofouling of navigational buoys and their anchor chains causes the buoys to dip under so that they become almost invisible on the surface. Biofilms in tubes and pipes greatly increase fluid frictional resistance to flow by (1) reducing the cross-sectional area, (2) increasing surface roughness, and (3) increasing drag by their viscoelastic properties (Characklis and Cooksey 1983). Fouling of seawater intakes of ships and power stations therefore restricts water flow to the condensers preventing development of full power. Fouling of other pipe systems, such as those leading to fire mains, can result in reduced pressure and clogging (see, e.g., Harderlie 1984). Periodic cleaning of seawater, plumbing systems (e.g., aquaculture systems) increases costs, and fouling of aquaculture structures themselves by diatoms, *Enteromorpha,* or kelps such as *Laminaria* and *Nereocystis* causes increased drag and competition within the system, necessitating removal.

When fouling affects economic efficiency or safety, as is usually the case, suitable protective systems must be developed. Since publication of *Marine Fouling and Its Prevention* (Woods Hole Oceanographic Institute 1952), a great deal of research aimed at achieving a better understanding of the fundamental biology of fouling organisms as a basis for the development of improved methods of control has been carried out. This chapter discusses algal fouling and the methods used to control it, with particular reference to ships' hulls (container ships, very large crude carriers, passenger liners, military and cargo vessels), an area where the author has extensive experience. In 1976 the cost of fouling to the U.S. military and commercial shipping industry was reported to be in excess of US $1.3 billion per annum (Ochiltree 1985). More recently, it was reported that an antifouling program for the entire U.S. Navy fleet resulting in a 15% reduction in fuel consumption would save some $150 million per annum (Spalding 1986). However, the same principles and preventive technologies developed for ships' hulls may to a large extent be applied to other submerged surfaces subject to biofouling.

II. Sequence of biofouling

A. *Bacterial biofouling*

A prerequisite to development of marine fouling communities is generally believed to be the formation of bacterial biofilms on the immersed surface. Biofilm formation involves the following sequence of events: (1) initial surface conditioning by adsorption of a monolayer of polymeric material from the sea (this is considered to occur more or less instantaneously and changes the properties of the wetted surface); (2) chemical attraction of motile bacteria to this film; (3) reversible adsorption of both motile and nonmotile bacteria, holding them weakly at the surface; (4) irreversible attachment, growth, and development following production of bacterial extracellular polymer substances (EPS), for example, polysaccharide or glycoprotein polymers, which act as strong adhesives; (5) a further stage of development whereby a second microbial population develops, characterized by the presence of large numbers of stalked and filamentous bacteria (see, e.g., Characklis and Cooksey 1983; Baier 1984; Mitchell and Kirchman 1984). The first four stages are believed to occur very rapidly, being complete within hours. Development of the final stage may take days or weeks.

B. Microalgal biofouling

It is difficult to establish conclusively whether a bacterial film is an essential prerequisite for diatom colonization. However, this is unlikely and in any case is irrelevant in practical terms because such a film will always be present on any surface submerged in the sea. Diatoms and cyanobacteria (blue-green algae) become enmeshed in the pioneer bacterial polymeric biofilm when it is 1–2 μm thick and grow intermixed with it (Cooksey et al. 1984; Cooksey and Cooksey 1986). This results in increased film thickness, and diatom films 500 μm thick have been recorded on supertanker hulls (Characklis and Cooksey 1983).

C. Macroalgal biofouling

Surfaces covered with a semirigid bacterial/microalgal slime provide a natural substrate for settlement of macroalgal spores (e.g., *Enteromorpha, Ectocarpus, Ulothrix*), which may germinate and grow intermixed with it. Prodigious numbers of spores are available for colonization and, in addition to the motility of green and brown algal spores, dispersal is aided by water currents. In the case of ships' hulls, settlement generally occurs while the vessels are stationary, although seaweed spores can settle in water flows of up to 10 knots. Spores of smaller macroalgae such as *Ectocarpus* and *Enteromorpha* may be brought to the vicinity of offshore installations by reproductive plants growing on supply or container ships. Ships also may be the source of fouling kelps (e.g., *Laminaria, Alaria*) on platforms such as those in the North Sea, since it is unlikely that motile spores could reach these platforms from coastal populations 150 or more km distant (Moss, Tovey, and Court 1981). However, it is possible that gametophytes (or even young sporophytes) may be carried over long distances by water currents, thus functioning as perennating/dispersal stages.

III. Control treatments

Effective methods of keeping ships' hulls free of fouling (and preventing penetration by wood-boring shipworms) have been sought since the fifth century B.C. Preventive methods tried include the use of animal grease, animal hair tar mixtures, arsenic/sulfur mixtures in oil, brimstone, charring of the surface to a depth of several inches, pitch, sulfur, tar, tallow, and thin copper or lead sheets (Major 1970). Following the first successful use of copper sheathing on British naval ships in the mid-eighteenth century, copper became widely used as an antifoulant and in shipworm prevention. However, with the introduction of iron-hulled ships galvanic action between the two metals prevented use of copper. Although shipworm penetration was no longer a problem, the need to develop a system to prevent or reduce fouling (and corrosion) was still paramount.

Some 300 patents for antifouling compositions were issued in the United Kingdom between 1835 and 1865, but the first (1860) notably successful product was "McInnes," a metallic soap containing copper sulfate as toxin and applied over a primer of iron oxide pigment and resin varnish (Major 1970). By 1900 several antifouling compositions were being made, but research aimed at improved control of biofouling on ships' hulls and other submerged structures continued. Research efforts over the last 30 years have been particularly intensive, with emphasis on the need to find longer-term, more efficient, environmentally acceptable methods.

Because voyage patterns and mode of operation of modern supertankers favor settlement and growth of macroalgae (notably *Enteromorpha* and *Ectocarpus*), microalgae (notably diatoms such as *Amphora* and *Achnanthes*), and bacteria rather than animal fouling forms (Christie and Evans 1975), much of the research has centered on these organisms. Various solutions for the control of biofouling emerged from this research including (1) the use of antifouling coatings and elastomeric materials, (2) electrochemical chlorination, and (3) mechanical removal techniques (for a review, see, e.g., Fischer et al. 1984). In addition, biological control methods have been employed. Milkfish *(Chanos chanos)*, grass shrimps *(Penaeus monodon)*, and crabs *(Scylla serrata)* are used to graze fouling epiphytes from pond-grown agar-producing *Gracilaria* in Taiwan, thus increasing economic efficiency (Chiang 1981). Limpets and abalone have also been used in outdoor culture tanks to help keep the system free of fouling algae (Waaland, personal communication). However, the present chapter will be concerned primarily with the state of the art with respect to the first of these three methods, namely, the use of antifouling coatings and elastomeric materials.

A. Antifouling coatings

The three types of conventional antifouling coatings (AFs) are the soluble matrix, contact leaching, and diffusion systems (see, e.g., Evans 1981). These are all based on the presence of toxic metallic or organometallic compounds (biocides) physically incorporated into the paint matrix, which leach out at a relatively well-controlled rate and form a layer of high toxin concentration (the boundary layer), which kills cells that settle at the surface. Following a high initial release, the rate of toxin release decreases exponentially in all three systems with time (i.e., occurs at a diminishing rate and therefore with reduced effectiveness). In soluble matrix and contact leaching systems, copper compounds (mainly cuprous oxide, Cu_2O) are the main toxic component, although metallic and organometallic compounds of mercury, lead, and arsenic, now excluded on environmental grounds, also were once used (see Fischer et al. 1984). In the diffusion system, the toxic moiety is an organotin compound. Toxin leaching rate varies with factors such as coating age (the rate of release is excessive at first), water temperature, salinity, and velocity.

Naturally occurring copper levels range from 0.3 (open ocean) to 3.0 $\mu g \cdot dm^{-3}$ (coastal waters). Copper toxicity depends on factors such as age of the organism, ambient physical and chemical seawater conditions, and presence of other biocides (Swain, Farrar, and Hutton 1982). A copper leaching rate of 20 $\mu g \cdot cm^{-2} \cdot day^{-1}$ was reported to be required to prevent settlement of diatom and bacterial slimes (Banfield 1980), but since such a rate would not be economical over a long period, the critical leaching rate for AF compositions based on copper in practice is approximately 10 $\mu g \cdot cm^{-2} \cdot day^{-1}$ (van Londen 1969). In vitro toxicity of copper in seawater of normal pH is due to the cupric ion (Cu^{2+} or CuII) since, starting from a water soluble copper compound, this is the only stable form. However, copper metal and cuprous oxide (Cu_2O) are extremely effective in the prevention of fouling. The first stage in the dissolution of copper metal is the formation of cuprous oxide; the first stage in the dissolution of cuprous oxide in seawater is the formation of CuCl, $CuCl_2^-$, and $CuCl_3^{2-}$. Therefore, in the dynamic situation cuprous ion (CuI) and copper in anionic form are possible (short-lived) toxic species (A. Milne, personal communication). However, the comparative toxicity of different copper species in seawater requires reexamination. (For a review of copper toxicity, see Kuwabara 1986.) A primary toxic effect of copper in the diatom *Amphora coffeae-formis* is alteration of membrane permeability resulting in loss of cell potassium. In addition, copper affects rates of light-induced O_2 evolution, dark O_2 uptake, and light $NaH^{14}CO_3$ uptake in this diatom, as well as inhibiting whole chain electron transport (French 1985).

In spite of the efficacy of copper as a biocide, conventional copper-oxide-based AFs used extensively in the 1950s have relatively short working lives (12–14 months) and are unable to control the growth of the ubiquitous copper-tolerant green fouling macroalga *Enteromorpha*. Swain et al. (1982) attribute the failure of copper as an antifoulant to insufficient dissolution rates (possibly as a result of electrochemical constraints), formation of insoluble corrosion products, or precipitation of cathodic chalks at the paint surface. However, use (in the diffusion system) of organotin compounds instead of copper permits effective control of *Enteromorpha*. Triorganotin compounds have the general formula R_3SnX, where R is propyl, butyl, or phenyl and X the inorganic group. Triphenyltin chloride (TPTC) or tributyltin oxide (TBTO), for example, can effectively prevent the growth of *Enteromorpha* through their action as energy transfer inhibitors in respiration and photosynthesis (Millner and Evans 1980, 1981). (For a review of the use of organotins as antifoulants, see Good and Monaghan 1984.) However, other algae, notably the filamentous brown alga *Ectocarpus*, are considerably more resistant to organotins (Mearns 1973). In the absence of *Enteromorpha*, *Ectocarpus* can become the major fouling organism on organotin AFs (Clitheroe and Evans 1975). The filamentous green alga *Ulothrix* also occurs on these AFs due to its high degree of resistance; in equivalent TPTC concentrations inhibition of photosynthesis was 100% greater in *Enteromorpha* than in *Ulothrix* (Taylor 1976; Millner and Evans 1980).

B. Recent improvements in chemical control methodology

Greatly improved control of weed fouling of all kinds was possible with the introduction in the early 1970s of the copolymer AF system (see, e.g., Evans 1981; de la Court 1984). In this system triorganotins are polymerized with unsaturated monomers to produce film-forming polymers with a high molar proportion of chemically incorporated toxic organotin moieties (Fig. 17-1). At the paint/seawater interface the copolymer hydrolyzes in a linear manner and at a controlled rate (after an initial high rate of release) to release the principal biocide, a triorganotin such as TBTO. Release rate is related to factors such as copolymer composition, vessel speed, water temperature, and pH. As well as being linear, such release is slower and therefore gives a considerably longer effective lifetime than when organotin is used in conventional systems. Other biocides ("pigments"), such as cuprous oxide (or cuprous thiocyanate), zinc oxide, triphenyltin fluoride, or triphenyltin hydroxide, may also be physically incorporated in the polymer matrix and are also released linearly as the polymer hydrolyzes (see, e.g., Sghibartz 1985). Addition of cuprous oxide considerably improves the antifouling performance compared with that of paint containing only organotin, and the additional presence of zinc oxide increases the leaching rate of copper relative to that from paint containing only cuprous oxide (French, Evans, and Dalley 1985; French and Evans 1986). Since the surface is also self-polishing (i.e., surface roughness and therefore frictional resistance decrease as the polymer is eroded (Fig. 17-2)), the composition is known as self-polishing copolymer (SPC). However, as in all AF paints, gross surface defects of any kind should be avoided as

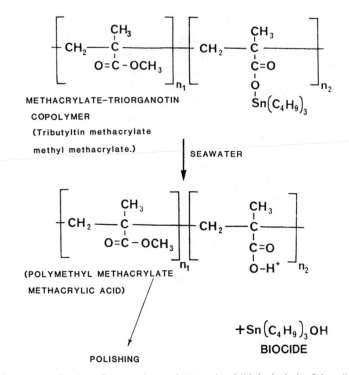

METHACRYLATE-TRIORGANOTIN
COPOLYMER
(Tributyltin methacrylate
methyl methacrylate.)

SEAWATER

(POLYMETHYL METHACRYLATE
METHACRYLIC ACID)

POLISHING

$+Sn(C_4H_9)_3OH$
BIOCIDE

Fig. 17-1. Mode of action of organotin copolymer paint. With hydrolysis of the tributyltin methacrylate/methyl methacrylate copolymer in seawater, the chemically bound tributyltin radical is released in a controlled manner, forming tributyltin hydroxide (or chloride), which is toxic to fouling organisms. The polymer backbone is gradually lost as it becomes water-soluble due to formation of Na^+ and K^+ salts.

these will foul rapidly, irrespective of the composition of the surrounding paint (Callow, Wood, and Evans 1982). As well as being widely used on tankers and commercial and naval oceangoing vessels, organotin-based AF paints are used extensively on smaller pleasure craft as well as on fishnets used in salmon mariculture installations (Wood 1986).

Studies on the speciation of the leachates from SPC AFs show that they appear to be the trialkyltin species, probably as the cation or associated with chloride or hydroxide in the aqueous solution (Fig. 17-1). The organotin moiety also appears to be relatively stable; this, however, is inconsistent with earlier reports that rapid decomposition to inorganic tin oxides occurs (see Good and Monaghan 1984). In spite of a relatively low organotin (TBTO) leaching rate from SPC AFs (ca. 5 μg $Sn \cdot cm^{-2} \cdot day^{-1}$; de la Court and de Vries 1976), some buildup of TBTO levels is likely to occur in coastal/estuarine areas where large numbers of yachts and pleasure craft or salmon cages are moored. Concentra-

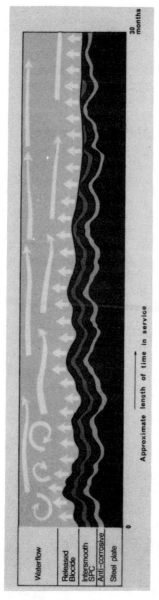

Fig. 17-2. Diagram of self-polishing copolymer (SPC) antifouling paint to show increasing smoothness during movement through water with time, as a result of the increased polishing rate (due to greater turbulence) on peaks compared with depressions. Biocides chemically and physically combined with the copolymer paint medium are linearly released during hydrolysis of the polymer and protect the surface from fouling. The use of alternate coats of SPC of different colors helps to identify areas of differential polishing.

tions equaling or exceeding those known to have lethal or sublethal effects on a range of marine organisms have been reported, and in such areas there is evidence from the United Kingdom, France, and the United States of inhibitory or disruptive effects on the development of coastal marine invertebrates cultivated for human consumption (see, e.g., Stebbing 1985; Pollution Paper No. 25 1986; Wood 1986). Consequently, the use of tributyltin by the U.S. Navy has been blocked by the Senate until the Environmental Protection Agency certifies that it does not pose an unacceptable hazard to marine organisms, in particular shellfish (Spalding 1986). The use of antifouling paints containing organotin compounds on boats less than 25 m long is also forbidden in France (Alzieu and Heral 1984). Sale of copolymer paints is now prohibited in the United Kingdom if the dried paint contains more than 7.5% tin by weight. Sale of other types of antifouling paint is prohibited if the dried film contains more than 2.5% tin by weight (Pollution Paper No. 25 1986).

Movement of a ship through water is essential for optimal self-polishing and fouling control, and in the absence of this, when the copolymer is not self-polishing optimally, microscopic biological slimes occur (Christie, Evans, and Callow 1976). Although often visually insignificant, these contribute substantially to the frictional resistance of a moving ship; a slime layer 1 mm thick was shown to cause a 15% loss in ship speed and an 80% increase in skin friction compared with a clean hull (Lewthwaite, Molland, and Thomas 1985). In addition, slime may retard the rate of copolymer hydrolysis and therefore biocide release from the paint.

As stated earlier, slime films are composed of an aggregation of organisms, chiefly bacteria and diatoms, held together in a semirigid mucilage (Fig. 17-3). However, selection of fouling forms occurs, and effective AF paints reduce species diversity (Robinson, Hall, and Voltolina 1985). In a recent worldwide survey using static test panels (Callow 1986a), diatoms of the genus *Achnanthes* preferentially colonized surfaces where organotin was the only biocide present, in particular *A. subsessilis* (now known as *A. brevipes* var. *intermedia*) (Fig. 17-4), which is highly resistant to organotins (Callow and Evans 1981). On the other hand, species of the diatom genera *Amphora* (Fig. 17-5) (in particular *A. veneta, A. coffeaeformis* and its var. *perpusilla,* and *A. bigibba*), *Amphiprora,* and to a lesser extent *Stauroneis* were the most frequently encountered diatoms on AFs containing only copper or organotin and copper (see also Robinson et al. 1985; French and Evans 1986). *Amphora* spp. also predominate in the slimes on oceangoing ships with copper and tributyltin-containing SPC AFs (Callow 1986b). In addition, *Amphora* has been found to be highly resistant to SPC leachates (Thomas and Robinson 1986). *Amphora* and *Amphiprora* species are highly copper-resis-

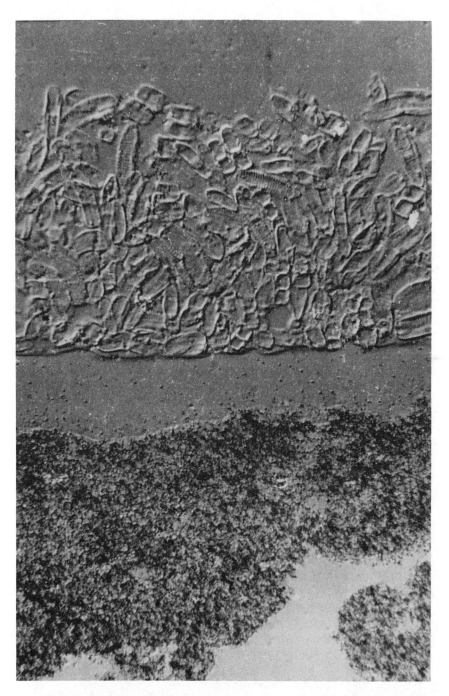

Fig. 17-3. Section to show a substantial diatom slime accumulation (upper layer: ca. 100 μm) on an organotin-containing copolymer varnish (middle layer: ca. 25 μm) applied over an anticorrosive coating (lower layer), 21 months after immersion. Interference contrast light microscopy: ×750; 1 μm section.

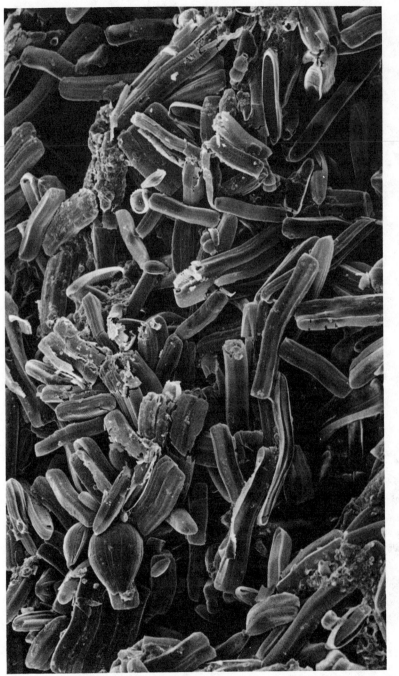

Fig. 17-4. Scanning EM of an SPC antifouling paint with cuprous thiocyanate after 12 months exposure on in-service tanker trading on the Malay Peninsula. Slime film composed mainly of the diatom *Achnanthes* (probably *A. subsessilis*, currently known as *A. brevipes* var. *intermedia*). × 850. (From Callow 1986b.)

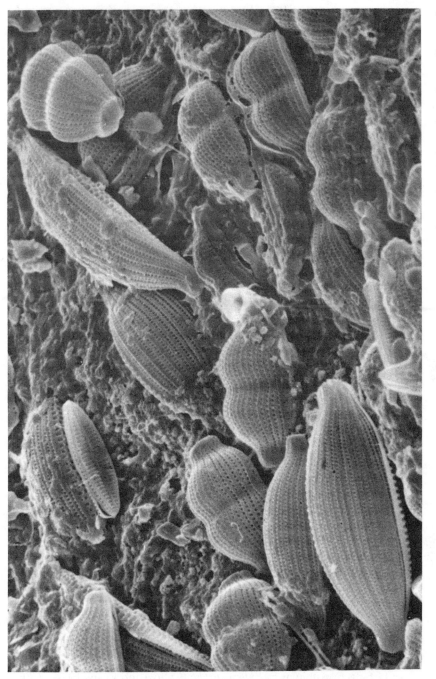

Fig. 17-5. Scanning EM of an SPC antifouling paint with cuprous thiocyanate after 12 months exposure on in-service tanker trading on the Malay Peninsula. Slime film composed mainly of the diatoms *Amphora coffeaeformis* var. *perpusilla* (e.g., bottom left) and *Amphora biggiba* (e.g., top right). ×4250. (From Callow 1986b.)

tant (see, e.g., French 1985). In *Amphora veneta,* such resistance resulted from immobilization of the metal within membrane-bound intracellular bodies, thus keeping cytoplasmic levels low (Daniel and Chamberlain 1981). Resistance may also result from the ability of an organism to exclude toxic metals and production of detoxifying substances or by sequestering compound(s).

Under average ship-operating conditions slime films will develop on SPC AFs if the polishing rate (biocide leaching) is suboptimal. Research therefore is currently aimed in the longer term at finding new biocides which are less toxic to nontarget organisms and degrade more readily than tributyltin and which may be incorporated into SPC paints in place of the latter to give improved slime control. In addition, development of other ways to prevent fouling that do not involve the use of toxic biocides is under active study. Foremost in this context is the use of biological control or of nontoxic surfaces (see Sections C and D, below). This is particularly important in view of the possible organotin environmental hazard and increasing emphasis on the need for environmentally acceptable control methods. In the shorter term, research is aimed at the development of SPC paint systems (without loss of efficiency) with reduced organotin content; one possibility is the use of a biocide that acts synergistically in combination with triorganotin, thus permitting the use of reduced organotin levels without loss of algicidal activity (see, e.g., Evans, Callow, and Wood 1986).

C. Biological control

As indicated earlier, one of the first steps in the establishment of biofouling is the coating of the surface with film-forming polysaccharides secreted by adhering marine bacteria (e.g., *Vibrio* spp.). The influence exerted by this bacterial film on the attachment and development of algal cells requires further investigation. Metabolic products of attached bacteria attract algae (and invertebrate larvae) to surfaces. The green alga *Dunaliella* is strongly attracted by specific amino acids at concentrations as low as 10^{-6} M (Sjoblad, Chet, and Mitchell 1978). Bacterial metabolites (such as benzoic and tannic acids) also repel algal cells and spores and therefore prevent adhesion to biofilms (Mitchell 1984). The presence of specific bacterial metabolites on the surface may therefore control attachment. Adhesion may also be influenced by bacterial polymers. Colonization of surfaces by *Enteromorpha* zoospores is stimulated by attached marine bacteria while other bacteria reduce settlement (Thomas and Allsopp 1983). Surfaces coated with bacterial films also cause *Dunaliella tertiolecta* cells to attach more rapidly and in greater quantities. Marine-fouling algae produce extracellular adhesives composed of glycoprotein as in *Enteromorpha* (Callow and Evans 1974), *Ulva* (Braten 1975), *Ectocarpus* (Baker and Evans 1973; Clitheroe 1977), and

Ceramium (Chamberlain and Evans 1973, 1981) or various substituted acidic polysaccharides as in diatoms (Chamberlain 1976; Blunn and Evans 1981). Protein-binding sites (lectins) produced by algal cells have been suggested to bind specifically with complementary bacterial exopolysaccharides because minimal settlement occurs when the lectin concanavalin A (glucose- and mannose-specific) is added to the bacterial-coated surface (Mitchell 1984; Mitchell and Kirchman 1984). Genetic control by the bacterial film of lectin-mediated attachment may therefore be a possibility. Genetic engineering techniques are also being used to determine how bacteria initiate fouling. There is good evidence that with 15–20 min of encountering a surface, certain marine bacteria completely change their pattern of gene expression. Once this happens they can efficiently colonize the entire surface. It is hoped that a better understanding of the gene expression process will permit the development of specific inhibitors that are more effective and less toxic than currently used antifoulants (M. I. Simon, quoted in Tucker 1985).

Another possibility for nonbiocidal fouling control is the use of nontoxic compounds produced by marine organisms such as coral in the prevention of surface fouling. Such natural antifouling substances recently were isolated from the sea corals *Leptogorgia virgulata* and *Renilla reniformis;* those from the latter are diterpenes. One of the compounds binds readily to polar surfaces and inhibits settlement of larvae of the barnacle *Balanus amphitrite* (but not larvae of the bryozoan *Bugula neritina*); another binds to nonpolar surfaces and inhibits settlement of *B. neritina* (but not *B. amphitrite*) larvae. Inhibitory activity is believed to occur by nontoxic modification of the settlement surface, since even at high concentrations the fouling organisms are not killed. Research is now aimed at isolating several fragments of the settlement-inhibiting compounds and coupling these covalently to a polymer that would provide a coating that would be easy to apply (Rittschoff et al. 1985; Rittschoff, Hooper, and Costlow 1986; Spalding 1986).

D. Nonadhesive surfaces

Another alternative for the long-term (2–5 years) prevention of marine fouling is the use of nontoxic, nonadhesive surfaces. Although surfaces permitting no adhesion would be difficult to achieve, it is envisaged that surfaces to which fouling is only weakly attached would be easy to clean using mechanical systems such as soft underwater brushes or hoses (see, e.g., Griffith 1985). Low-energy surfaces may be based, for example, on silicone elastomers (Milne and Callow 1985; Callow et al. 1986), fluoropolymers (Bultman et al. 1984; Griffith 1985), or "tethering" of drag-reducing molecules such as polyox (polyoxyethylene) (Gucinski et al. 1984). Differences in rhizoid morphology occur when some algae grow on low-energy surfaces. For example, the green macroalga *Enteromorpha* forms weakly adhering, loose filamentous rhizoids on silane-based coatings with low surface energy properties, compared with strongly adhering, compact rhizoid discs on high-energy surfaces (Fletcher and Baier 1984; Fletcher, Baier, and Fornalik 1984). How such surfaces work in relation to weakening adhesion, however, is not a biological problem but one in physical chemistry. The physical chemistry of bioadhesion is touched upon briefly by Milne and Callow (1985) (with particular regard to silicone elastomers) where further references to the topic may also be found.

Other tested materials include polytetrafluorethylene (PTFE) (a refractory insoluble plastic on which adhesion was found to be lower than in the controls) and "no-foul rubber" (elastomeric materials, usually neoprene rubbers in which biocides such as organotin compounds are dispersed) (see Ochiltree 1985; Milne and Callow 1985). However, materials in liquid form are required by the paint maker for economic application to large areas. During tests, the nonbiocide component of elastomers in liquid form was shown to have good antifouling properties. Silicone polymers are based on a backbone of repeating (-Si-O-) units, and the nonbackbone valencies are attached to saturated organic radicals. Further tests of a number of commercially available elastomers and mussel byssus (attaching threads), using the bacterium *Pseudomonas fluorescens* and the diatom *Amphora coffeaeformis* as test organisms for adhesion measurements, showed it was possible to improve the antifouling performance of RTV (room-temperature vulcanized) silicone elastomers by addition of, for example, PMS (phenyl-methyl silicone fluids) (Milne and Callow 1985; Callow, Pitchers, and Milne 1986). However, their physical properties (poor abrasion resistance and tear strength) so far prevent their use as general AFs in the marine environment, although they currently have uses where such properties are of less importance (OTEC, fish farming) and may have uses where their lack of toxicity may be advantageous (e.g., in water intakes of coastal power stations and aquaculture systems from which food for human consumption is harvested).

IV. Current situation

Considerable advances have been made in the control of marine fouling over recent years, especially with the development of controlled-release SPC compositions. However, in spite of the large numbers of compounds shown over the years to have antifouling potential (see, e.g., Fischer et al. 1984), prevention of fouling by marine organisms (plant and animal) at the present time depends almost entirely upon the use of copper and organotin compounds. Copper is a good antifoulant, though of limited usefulness when used on its own. Triorganotins are excellent antifoulants, but their environmental acceptability is already in question, especially in relation to coastal waters, and copper could also become environmentally unacceptable in the future. Since the use of organotins, at least in coastal waters, will almost certainly be further restricted or discontinued in the near future, the need to find suitable alternative "safe" biocides to replace organotins in SPC systems is considerable. However, as observed by Milne and Callow (1985), the costs of screening and gaining environmental clearance for new materials are great, and as fouling is so biologically diverse and the required lifetime of an antifouling coating is from two to five years, a broad-spectrum persistent biocide of very high potency is almost the antithesis of a safe and harmless material! Therefore, in addition to finding other means of selectively poisoning settled spores and cells of target fouling marine algae (and animal larvae), the long-term answer must ultimately lie in preventing settlement of these organisms in the first place without the use of toxic materials. Genetically modified surfaces are one (biological) possibility, and the use of natural nontoxic antifoulants is another. A third possibility is the use of low-surface-energy materials, where quite good progress has been made already. We have clearly come a long way, but equally clearly we have a long way to go.

V. References

Alzieu, C., and M. Heral. 1984. Ecotoxological effects of organotin compounds on oyster culture. In *Ecotoxicological Testing for the Marine Environment,* vol. 2, ed. G. Persoone, E. Jaspers, and C. Claus, pp. 187–96. State University, Gent, and Institute for Marine Scientific Research, Bredene, Belgium.

Baier, R. E. 1984. Initial events in microbial film formation. In *Marine Biodeterioration: An Interdisciplinary Study,* ed. J. D. Costlow and R. C. Tipper, pp. 57–62. U.S. Naval Research Institute, Annapolis, MD.

Baker, J. R. J., and L. V. Evans. 1973. The ship-fouling alga *Ectocarpus.* I. Ultrastructure and cytochemistry of plurilocular reproductive stages. *Protoplasma* 77, 1–13.

Banfield, T. A. 1980. Marine finishes, pt. 2. OCCA Monograph no. 1. *J. Oil Col. Chem. Assoc.* 69, 93–100.

Blunn, G. W., and L. V. Evans. 1981. Microscopical observations on *Achnanthes subsessilis,* with particular reference to stalk formation. *Bot. Mar.* 24, 193–9.

Braten, T. 1975. Observations on mechanisms of attachment in the green alga *Ulva mutabilis* Foyn. *Protoplasma* 84, 161–73.

Bultman, J. D., J. R. Griffith, and D. E. Field. 1984. Fluoropolymer coatings for the marine environment. In *Marine Biodeterioration: An Interdisciplinary Study,* ed. J. D. Costlow and R. C. Tipper, pp. 237–43. U.S. Naval Research Institute, Annapolis, MD.

Callow, M. E. 1968a. A world-wide survey of slime formation on anti-fouling paints. In *Algal Biofouling,* ed. L. V. Evans and K. D. Hoagland, pp. 1–20. Elsevier, Amsterdam.

Callow, M. E. 1986b. Fouling algae from "in-service" ships. *Bot. Mar.* 29, 351–7.

Callow, M. E., and L. V. Evans. 1974. Studies on the ship-fouling alga *Enteromorpha*. III. Cytochemistry and autoradiography of adhesive production. *Protoplasma* 80, 15–27.

Callow, M. E., and L. V. Evans. 1981. Some effects of triphenyltin chloride on *Achnanthes subsessilis. Bot. Mar.* 24, 201–5.

Callow, M. E., R. A. Pitchers, and A. Milne. 1986. The control of fouling by non-biocidal systems. In *Algal Biofouling,* ed. L. V. Evans and K. D. Hoagland, pp. 145–85. Elsevier, Amsterdam.

Callow, M. E., K. R. Wood, and L. V. Evans. 1982. Paint surface defects and fouling. *Shipping World and Shipbuilder* 177, 353–5.

Chamberlain, A. H. L. 1976. Algal settlement and secretion of adhesive materials. In *Proceedings of the Third International Biodegradation Symposium,* ed. J. M. Sharpley and A. M. Kaplan, pp. 417–32. Applied Science, London.

Chamberlain, A. H. L., and L. V. Evans. 1973. Aspects of spore production in the red alga *Ceramium. Protoplasma* 76, 139–59.

Chamberlain, A. H. L., and L. V. Evans. 1981. Chemical and histochemical studies on the spore adhesive of *Ceramium. Proc. Int. Seaweed Symp.* 8, 539–42.

Characklis, W. G., and K. E. Cooksey. 1983. Biofilms and microbial fouling. *Adv. Appl. Microbiol.* 29, 93–138.

Chiang, Y-M. 1981. Cultivation of *Gracilaria* (Rhodophycophyta, Gigartinales) in Taiwan. *Proc. Int. Seaweed Symp.* 10, 556–74.

Christie, A. O., and L. V. Evans. 1975. A new look at marine fouling, pt. 1. *Shipping World and Shipbuilder* 168, 953–5.

Christie, A. O., L. V. Evans, and M. E. Callow. 1976. A new look at marine fouling, pt. 4. *Shipping World and Shipbuilder* 169, 121–4.

Clitheroe, S. B. 1977. *Fine Structural and Cytochemical Studies on the Brown Seaweed Ectocarpus.* Ph.D. thesis, University of Leeds. 142 pp.

Clitheroe, S. B., and L. V. Evans. 1975. A new look at marine fouling, pt. 3. *Shipping World and Shipbuilder* 168, 1123–4.

Cooksey, B., K. E. Cooksey, C. A. Miller, J. H. Paul, R. W. Rubin, and D. Webster. 1984. The attachment of microfouling diatoms. In *Marine Biodeterioration: An Interdisciplinary Study,* ed. J. D. Costlow and R. C. Tipper, pp. 167–71. U.S. Naval Research Institute, Annapolis, MD.

Cooksey, K. E., and B. Cooksey. 1986. Adhesion of fouling diatoms to surfaces: some biochemistry. In *Algal Biofouling,* ed. L. V. Evans and K. D. Hoagland, pp. 41–53. Elsevier, Amsterdam.

Daniel, G. F., and A. H. L. Chamberlain. 1981. Copper immobilization in fouling diatoms. *Bot. Mar.* 24, 229–43.

de la Court, F. H. 1984. Erosion and hydrodynamics of biofouling coatings. In *Marine Biodeterioration: An Interdisciplinary Study,* ed. J. D. Costlow and R. C. Tipper, pp. 230–6. U.S. Naval Research Institute, Annapolis, MD.

de la Court, F. H., and H. J. de Vries. 1976. The leaching mechanisms of some organotin toxicants from antifouling paints. In *Proceedings of the Fourth International Congress on Marine Corrosion and Fouling,* pp. 113–18. Centre de Recherches et d'Etude Oceanographique, Boulogne.

Evans, L. V. 1981. Marine algae and fouling: a review, with particular reference to ship-fouling. *Bot. Mar.* 24, 167–71.

Evans, L. V., M. E. Callow, and K. R. Wood. 1986. Synergism between anti-fouling biocides. In *Algal Biofouling,* ed. L. V. Evans and K. D. Hoagland, pp. 55–64. Elsevier, Amsterdam.

Fischer, E. C., V. J. Castelli, S. D. Rodgers, and H. R. Bleile. 1984. Technology for control of marine biofouling: a review. In *Marine Biodeterioration: An Interdisciplinary Study,* ed. J. D. Costow and R. C. Tipper, pp. 261–99. U.S. Naval Research Institute, Annapolis, MD.

Fletcher, R. L., and R. E. Baier. 1984. Influence of surface energy on the development of the green alga *Enteromorpha. Mar. Biol. Lett.* 5, 251–4.

Fletcher, R. L., R. E. Baier, and M. S. Fornalik. 1984. The influence of surface energy on spore development in some common marine fouling algae. In *Sixth International Congress on Marine Corrosion and Fouling,* pp. 129–44. Athens.

Freeman, J. H. 1977. *The Marine Fouling of Fixed Offshore Installations.* Department of Energy Offshore Technology Paper, ISSN 0309-8184. CIRIA, Storey's Gate, London.

French, M. S. 1985. *Copper and Zinc in Anti-fouling Paint and Their Effects upon the Diatoms Amphora and Amphiprora.* Ph.D. thesis, University of Leeds. 245 pp.

French, M. S., and L. V. Evans. 1986. Fouling on paints containing copper and zinc. In *Algal Biofouling,* ed. L. V. Evans and K. D. Hoagland, pp. 79–100. Elsevier, Amsterdam.

French, M. S., L. V. Evans, and R. Dalley. 1985. Raft trial experiment on leaching from antifouling paints. In *Polymers in a Marine Environment,* ed. R. Smith, pp. 127–30. Institute of Marine Engineers, London.

Good, M. L., and C. P. Monaghan. 1984. The chemical and physical characterisation of marine coatings containing organotin antifoulants. In *Marine Biodeterioration: An Interdisciplinary Study,* ed. J. D. Costlow and R. C. Tipper, pp. 244–9. U.S. Naval Research Institute, Annapolis, MD.

Griffith, J. R. 1985. The fouling-release concept: a viable alternative to toxic antifouling coatings. In *Polymers in a Marine Environment,* ed. R. Smith, pp. 235–6. Institute of Marine Engineers, London.

Gucinski, H., R. E. Baier, A. E. Meyer, M. S. Fornalik, and R. W. King, 1984. Surface microlayer properties affecting drag phenomena in seawater. In *Sixth International Congress on Marine Corrosion and Fouling,* pp. 585–604. Athens.

Harderlie, E. C. 1984. A brief overview of the effects of macrofouling. In *Marine Biodeterioration: An Interdisciplinary Study,* ed. J. D. Costlow and R. C. Tipper, pp. 163–6. U.S. Naval Research Institute, Annapolis, MD.

Kuwabara, J. S. 1986. Physico-chemical processes affecting copper, tin, and zinc toxicity to algae: a review. In *Algal Biofouling,* ed. L. V. Evans and K. D. Hoagland, pp. 129–44. Elsevier, Amsterdam.

Lewthwaite, J. C., A. F. Molland, and K. W. Thomas. 1985. An investigation into the variation of ship skin frictional resistance with fouling. *Trans. Roy. Inst. Naval Architects* 127, 269–84.

Major, A. 1970. Preventing ship fouling. *Meccano Mag.* 55 (12), 639–41.

Mearns, R. D. 1973. Vinyl weed-resistant antifouling: a practical approach to laboratory formulations. *J. Oil Col. Chem. Assoc.* 50, 201–8.

Millner, P. A., and L. V. Evans. 1980. The effects of triphenyltin chloride on respiration and photosynthesis in the green algae *Enteromorpha intestinalis* and *Ulothrix pseudoflacca. Plant Cell Environ.* 3, 339–48.

Millner, P. A., and L. V. Evans. 1981. Uptake of triphenyltin chloride by *Enteromorpha intestinalis* and *Ulothrix pseudoflacca. Plant Cell Environ.* 4, 383–9.

Milne, A., and M. E. Callow. 1985. Non-biocidal antifouling processes. In *Polymers in a Marine Environment,* ed. R. Smith, pp. 229–33. Institute of Marine Engineers, London.

Mitchell, R. 1984. Colonization by higher organisms. In *Microbial Adhesion and Aggregation,* ed. K. C. Marshall, pp. 189–200. Springer-Verlag, New York.

Mitchell, R., and D. Kirchman. 1984. The microbial ecology of marine surfaces. In *Marine Biodeterioration: An Interdisciplinary Study,* ed. J. D. Costlow and R. C. Tipper, pp. 49–56. U.S. Naval Research Institute, Annapolis, MD.

Moss, B. L., D. Tovey, and P. Court. 1981. Kelps as fouling organisms on North Sea platforms. *Bot. Mar.* 24, 207–9.

Ochiltree, B. C. 1985. Rubber compositions for the prevention of marine fouling. In *Polymers in a Marine Environment,* ed. R. Smith, pp. 221–7. Institute of Marine Engineers, London.

Pollution Paper No. 25. 1986. Organotin in antifouling paints: environmental considerations, pp. 1–65. CDEP, Department of the Environment, HMSO, London.

Rittschof, D., I. R. Hooper, E. S. Branscomb, and J. D. Costlow. 1985. Inhibition of barnacle settlement and behaviour by natural products from whip corals, *L. virgulata* (Lamarck, 1815). *J. Chem. Ecol.* 11, 551–63.

Rittschoff, D., I. R. Hooper, and J. D. Costlow. 1986. Barnacle settlement inhibitors from sea pansies *(Renilla reniformis). Bull. Mar. Sci.* 39, 376–82.

Robinson, M. G., B. D. Hall, and D. Voltolina. 1985. Slime films on antifouling paints: short-term indicators of long-term effectiveness. *J. Coatings Technol.* 57, 35–41.

Sghibartz, C. M. 1985. Organotin polymers: the state of the art. In *Polymers in a Marine Environment,* ed. R. Smith, pp. 217–9. Institute of Marine Engineers, London.

Sjoblad, R. D., I. Chet, and R. Mitchell. 1978. Chemoreception in the green alga *Dunaliella tertiolecta. Curr. Microbiol.* 1, 305–7.

Spalding, B. J. 1986. Antifouling compounds from coral. *Chemical Week,* July 9, 62–65.

Stebbing, A. R. D. 1985. Organotin and water quality: some lessons to be learned. *Mar. Poll. Bull.* 16, 383–90.

Swain, G. W. J., R. A. Farrar, and S. P. Hutton. 1982. The use of controlled copper dissolution as an antifouling system. *J. Materials Sci.* 17, 1079–94.

Taylor, G. E. 1976. *Studies on the Ship-fouling Seaweed Ulothrix.* Ph.D. thesis, University of Leeds. 77 pp.

Terry, L. A., and R. G. J. Edyvean. 1986. Recent investigations into the effect of algae on corrosion: a review. In *Algal Biofouling,* ed. L. V. Evans and K. D. Hoagland, pp. 211–29. Elsevier, Amsterdam.

Thomas, R. W. S. P., and D. Allsopp. 1983. The effects of certain periphytic marine bacteria upon the settlement and growth of *Enteromorpha*, a fouling alga. In *Biodeterioration* 5, ed. T. A. Oxley and S. Barry, pp. 348–61. John Wiley, New York.

Thomas, T. E., and M. G. Robinson. 1986. The physiological effects of an organotin antifouling paint on marine diatoms. *Mar. Envir. Res.* 18, 215–29.

Thorhaug, A., and J. H. Marcus. 1981. The effects of temperature and light on attached forms of tropical and semi-tropical macroalgae potentially associated with OTEC (ocean thermal energy conversion) machine operation. *Bot. Mar.* 24, 393–8.

Tucker, J. B. 1985. Biotechnology goes to sea. *High Technology* 5(2), 34–44.

van Londen, A. M. 1969. An evaluation of testing methods for antifouling paints. *ACS Preprints* 29(2), 54–61.

Wood, E. 1986. Organotin anti-fouling paints, an environmental problem. Marine Conservation Society, 4 Gloucester Road, Ross-on-Wye, UK. 19 pp.

Woods Hole Oceanographic Institute. 1952. *Marine Fouling and Its Prevention.* U.S. Naval Research Institute, Annapolis, MD.

18. Algae as weeds: economic impact, ecology, and management alternatives

CAROLE A. LEMBI and STEVEN W. O'NEAL

Department of Botany and Plant Pathology, Purdue University, West Lafayette, IN 47907

DAVID F. SPENCER

USDA/ARS, Aquatic Weed Research Laboratory, University of California, Davis, CA 95616

CONTENTS

[455]

I. Introduction

Although algae play an essential and beneficial role in maintaining the health of aquatic systems, excessive algal growths can have detrimental effects on both indigenous organisms and potential human uses of these systems. Algal populations that cause adverse effects are referred to as nuisance or, in the case of macrophytic forms, "weedy" algae. However, people have varying perceptions as to what constitutes a nuisance algal problem.

Laymen and scientists concerned with the large issues of pollution and eutrophication seldom think of algae as weeds in the same sense as they might a terrestrial weed. Phytoplankton blooms resulting from eutrophication are devastating to a lake because fish and other animal life are adversely affected. The solutions for such problems often are envisioned as requiring regulations and action by community, state, or federal agencies and even by international cooperative agreements (e.g., the Great Lakes Water Quality Agreement between the United States and Canada). Individuals may not perceive themselves as being directly affected by or as participating in the solution (other than in the collective sense).

To the person who steps out on a boat dock to fish and instead pulls up a wad of charoid algae, or whose boat motor becomes choked with filamentous algal mats, algae become weeds in a very direct and personal sense. Although government agencies are actively involved in managing aquatic weed growths, it is frequently the individual (either alone or as a member of a property owners' association) who will attempt to solve the "weed" problem in his/her situation. These situations range from small farm ponds to properties located on large lakes, and they may involve algal weeds in flowing as well as in static waters.

This chapter will cover the impact, ecology, and management of weedy macrophytic algae in freshwater environments. It should be recognized that "weedy" algae, both macrophytic and microphytic forms, also occur in marine systems. Examples include the green alga *Codium* that infests shellfish beds (Carlton and Scanlon 1985), the biofouling organisms that attach to ship hulls (Evans, Chapter 17), and the unicellular algal blooms that retard growth of seaweeds in confined aquaculture systems (Shacklock and Doyle 1983).

II. Major types of weedy algae

The two major growth forms of freshwater macrophytic algae are filamentous mat formers and charoids. Mat-forming algae are typically
found in shallow water where they may be free-floating (e.g., *Pithophora,
Spirogyra, Hydrodictyon*) or attached, primarily as epiphytic or epilithic
forms. Epilithic algae (e.g., *Cladophora, Ulothrix, Stigeoclonium*) commonly attach to rocks or cement linings; epiphytic algae (e.g., *Oedogonium*) attach to other macrophytes, including algae and submersed vascular plants.

Free-floating forms are generally restricted to static waters such as
ponds and sheltered littoral areas of lakes. Attached forms occupy a
much wider range of habitats. They are found in both static and flowing
systems, including the wave-scoured edges of lakes (e.g., *Cladophora* in
the Laurentian Great Lakes), fast-flowing streams, and the extensive
irrigation systems and aqueducts of the western United States. Attached
algae can become detached from their substrates and, because of their
positive buoyancy when oxygen bubbles are trapped in the mat, form
free-floating rafts of algal material. These mats, in various stages of
senescence and decomposition, may then be moved downstream or
washed up on beaches and shorelines. Epiphytic algae are frequently
protected from washout in a flowing system by remaining entangled in
the foliage of rooted macrophyte beds.

The charoid algae *(Chara, Nitella)* are characterized by a central axis,
whorls of short lateral branches, and rhizoids for anchoring. The plant
body can be 40 cm (Prescott 1962) to as much as a meter or more in
length (personal observations). Charoid algae are seldom problems in
deep, open water areas but can obstruct the uses of ponds, canals, and
shallow areas of lakes where the plants extend to the water surface.

All of the algae cited above are members of the Chlorophyta. Several
genera of blue-greens (Cyanophyta or cyanobacteria) include species
that form mats (e.g., *Oscillatoria, Lyngbya, Phormidium*). Black-colored
mats of *Oscillatoria* are typically found as a layer on bottom sediments
but readily float to the surface when gas bubble accumulation dislodges
both mats and associated sediments (Halfen and McCann 1975). Several
species of *Phormidium* form the "black algae" growths that attach to
cement linings of swimming pools (Fitzgerald 1959; Adamson and Sommerfeld 1978).

III. Problems caused by weedy algae

A. *Impact on human resources*

The economic impact of weedy algae is virtually impossible to quantify.
Water often has no monetary value assigned to it, and the costs from

human impacts, although significant, are difficult to assess. Some information is available from the amount of algicide used for algae control, but even these values are inaccurate because the labor, machinery, and other costs of applying the algicide are not taken into consideration. Thus, a description of the problems caused by macrophytic algae tends to be more qualitative than quantitative and may underestimate the actual impact.

Even qualitative information is difficult to obtain because little distinction is made between algae and aquatic flowering plants when problems are described in the aquatic plant literature. In general, we know that excessive growths of macrophytic algae, either alone or in combination with vascular plants, impair recreational activities such as swimming, fishing, and boating. Swimming beaches fouled with algal mats are not only unappealing but hazardous when ladders, rocks, and submerged concrete are coated by slime-producing species such as *Spirogyra* (Bennett 1971). *Cladophora* growths in the Great Lakes were noted as posing a potential danger to young and inexperienced swimmers who might become entangled in the mats and drown (Herbst 1969). The loss of recreational and aesthetic values can have a significant economic impact on waterfront properties. Ormerod (1970) reported that real estate on Lake Erie fronted with *Cladophora* mats averaged 80–85% of the value of a clean frontage.

Macrophytic algae restrict and greatly reduce the efficiency of culture and harvest activities in fish culture ponds (Tucker, Busch, and Lloyd 1983). They may compete with phytoplankton, thus reducing the base of the aquatic food chain (Boyd 1982). A 50% reduction in fish production in farm ponds was attributed to heavy *Pithophora* growths and concomitant loss of phytoplankton (Lawrence 1954).

Although to our knowledge citations implicating macrophytic algae as direct causes of fish kills are few (see, e.g., Robinson and Hawkes 1986), excessive algal growth presumably adds to the oxygen deficits that result from respiration of submersed plant growth and/or phytoplankton at night, during periods of cloudy weather, or under snow-covered ice. These oxygen deficits can be stressful to fish, causing declines in food consumption and growth and making them more prone to bacterial infections (Boyd 1982). Overpopulation, stunted or reduced growth, and decline in capture efficiency of forage species by predatory fish have been attributed to dense vascular plant growth (Colle and Shireman 1980; Savino and Stein 1982; Shireman et al. 1983), and algal growth is assumed to contribute to this problem. However, the need for some macrophyte growth to provide protection for young fish and habitat for fish food organisms is also recognized (Barnett and Schneider 1974; Wiley et al. 1984). The proper balance of macrophyte coverage and density for optimal fisheries is at this point somewhat controversial, and both it and the role of macrophytic algae as fish habitat require further study (see review in Hinkle 1986).

Cladophora, Stigeoclonium, Oedogonium, and *Ulothrix* have been cited as presenting serious problems in irrigation canals by attaching to concrete canal linings, thus reducing both flow rate and capacity. Mats that break away from the linings float downstream where they foul pump inlets, irrigation siphons, trashracks, and sprinkler heads (Hansen, Oliver, and Otto 1984). *Cladophora glomerata* in particular is a major problem in water delivery systems in the western states. For example, in the 620 km length of the Salt River Project Canal, which supplies water and electricity for the cities of Phoenix and Tempe, Arizona, the control of aquatic weeds is a major operation/maintenance task (Corbus 1982). The current annual budget for aquatic macrophyte control (primarily *Cladophora*) in this system is approximately US $1.5 million.

The clogging of rivers, canals, and drainage ditches by aquatic weeds and algae can prevent adequate drainage so that water backs up, even to the point of causing flooding. In California alone, the U.S. Bureau of Reclamation uses 3,000 mt of copper sulfate per annum for algae control in canals (C. Tennis, personal communication). *Vaucheria,* a filamentous member of the Chrysophyta, has been cited as a major problem for land drainage engineers in the United Kingdom (Dowidar and Robson 1972; Robson, Fowler, and Barrett 1976).

Associated with water conveyance or storage systems, particularly in arid parts of the world, is the potential to lose water through evaporation from floating or emergent plant surfaces. Although considerable data are available showing that substantial loss does occur with coverage of vascular plants such as water hyacinth (Brezny, Mehta, and Sharma 1973), nothing is known of the potential of surface-floating algae to add to the problem.

Algal mats are often viewed as harboring or providing shelter for noxious insects such as mosquito larvae and flies. Filamentous algae have been reported to harbor snails that cause schistosomiasis, a debilitating disease in Egypt (and elsewhere) that affects a large proportion of the population (Van Schayck 1986). Filamentous algae also can be a source of odors, particularly if anaerobic decomposition occurs, and although microscopic diatoms, blue-greens, and chrysophytes are the principal causes of taste and odor problems in potable water, several filamentous greens *(Cladophora, Hydrodictyon, Spirogyra)* and the charoid algae also have been implicated (Palmer 1962).

Macrophytic algae are indicators of nutrient and, in some cases, heavy metal pollution. The increase in *Cladophora* growths near sewage outfalls in the Great Lakes has been correlated with increased phosphorus levels (Neil and Owen 1964; Auer and Canale 1980). *Stigeoclonium* is commonly found in flowing waters contaminated with nitrogen and organic matter (McLean and Benson-Evans 1974; De Vries, Torenbeek, and Hillebrand 1983; Francke and Den Oude 1983), and species of *Stigeoclonium, Hormidium,* and *Mougeotia* have been found to tolerate high concentrations of zinc and copper, particularly in acid-impacted waters (Harding and Whitton 1976; Say, Diaz, and Whitton 1977; Francke and Hillebrand 1980). *Mougeotia* appears to be an indicator organism of aquatic acidification appearing at pH levels from 5 to 5.9 in experimentally acidified lakes in Ontario, Canada (Schindler et al. 1985).

B. Impact on natural resources

Very little is known of the impact, whether positive or negative, of macrophytic algae on natural aquatic systems. Species such as *Spirogyra* and *Cladophora* that appear soon after ice-out and before vascular plant establishment in the spring have been observed to act as temporary cover and habitat for aquatic insects and largemouth bass in Wisconsin lakes (Engel 1985). Seven species of macroinvertebrates, including pygmy backswimmers *(Neoplea),* regularly occupied algal mats until summer water temperatures above 30°C restricted their use (Engel 1985), and at least 10 species were found associated with *Cladophora* growths in a stream (Neel 1968). Although macrophytic algae themselves do not appear to be a major food source for macroinvertebrates, their attached algae can be grazed by invertebrates (Allanson 1973), as can their detrital remains (Jansson 1967; review in Engel 1985). The charoid algae, in particular, are considered excellent habitat for both littoral invertebrates (Rosine 1955; Quade 1969; Allanson 1973) and fish (Fassett 1957; Schardt 1984). Rosine (1955) found *Chara* to harbor the largest number of invertebrates per unit area in a Colorado lake, frequently supporting 1.5 to 5 times the number of animals found on submersed vascular plants.

Some predator fish species such as bass and bluegill consume small percentages of vegetation including filamentous algae and *Chara.* This plant material may serve as nourishment. Bluegills grew best in aquariums when their diet of either oligochaetes or mealworm larvae was supplemented with dried *Chara* (Kitchell and Windell 1970). *Cladophora* has been noted as a minor component in the diet of fathead minnows (Coyle 1930) and wildfowl (Taft 1975).

Some evidence, however, suggests that dense growths of *Cladophora* reduce invertebrate diversity and may have disrupted shoal spawning of walleye, whitefish, and lake trout in the Great Lakes (Neil 1975). Patrick (1978) noted reductions in reproductive rates of stream-inhabiting snails when they were fed *Spirogyra* or *Cladophora.* Charoids, along with submersed vascular plants, produce repellent (allelopathic) materials that exclude certain limnetic (open water) species of invertebrates (Pennak 1966, 1973) and phytoplankton (Gibbs 1973; Anthoni et al. 1980,

Wium-Andersen et al. 1982). The latter finding may provide a partial explanation for the lack of epiphytes and clear water conditions frequently associated with charoid growths (Crawford 1979; Wium-Andersen et al. 1982).

Algae may have adverse effects on the vascular macrophytes they epiphytize. Epiphytes (composed primarily of diatoms, filamentous greens, and filamentous blue-greens) were recorded as composing as much as 50% of the total macrophyte biomass in a freshwater system in early summer (Jones, Gurevitch, and Adams 1979) and 33% in mid to late summer (Filbin and Barko 1985). Considerable evidence now exists to indicate resource competition between submersed vascular plants and their epiphytes (Simpson and Eaton 1986). In nutrient-poor water, rooted plants can obtain nutrients from the sediments through their roots whereas epiphytic algae may be nutrient-limited. As nutrients in the water increase, algae proliferate and depress photosynthetic performance of macrophytes by shading (Phillips, Eminson, and Moss 1978; Sand-Jensen and Sondergaard 1981) or CO_2/bicarbonate limitation (Sand-Jensen 1977; Spence 1982). Examples in which increased nutrient loadings have resulted in vascular plant replacement by epiphytic algae include the replacement of *Najas marina* by *Spirogyra* in the United Kingdom (Phillips et al. 1978), diverse macrophytes by *Cladophora glomerata* in a river (Bolas and Lund 1974), and *Elodea* by *Cladophora* and *Spirogyra* in a Liverpool canal (Simpson and Eaton 1986).

IV. Extent of problem

Data on the extent and distribution of aquatic weeds, including algae, are difficult to obtain because few governmental agencies, either in the United States or elsewhere, have conducted surveys to establish the proportion of waters infested. Those surveys that have been conducted often emphasize vascular plant species, even though filamentous algae may represent a significant portion of the macrophyte stand. Furthermore, a distinction is seldom made between macrophytic and phytoplankton forms in surveys that do include algae. However, in a 1981 survey conducted by the U.S. Department of Agriculture of aquatic *weed* problems in the southeastern states, 40% of lakes, ponds, and reservoirs in Alabama and North Carolina was estimated as being infested by algae (Aurand 1982). In Florida and Louisiana, 40% and 15%, respectively, of the irrigation and drainage canal area were similarly infested. In a previous survey conducted in 1968, 20% of the total available area of surface water in the state of Georgia was overgrown with algae.

The most thorough survey of aquatic weed populations in the United States is conducted in the state of Florida. Surveys cover both surface area containing aquatic plants (Schardt 1983, 1984) and area treated (mostly chemically and mechanically) for excessive aquatic plant growths (Nelson and Dupes 1985). In 1984 approximately 56% of Florida's reported 1 million hectares of fresh water was surveyed (Schardt 1984). Noticeable quantities of filamentous algae were found in 57% of the canals, 50% of the lakes, and 35% of the rivers. Total area infested nearly doubled from the previous year causing filamentous algae to increase in rank from the fifteenth to the sixth most abundant plant in Florida waters. Most of the increase was monitored in urban waters receiving runoff from streets, fertilized lawns, or agricultural fields.

In general, filamentous algae rank higher in terms of actual control measures taken than in incidence. In state-funded programs for permanent bodies of water (mostly man-made canals, ditches, and lakes accessible to the general public for recreational activities), the Florida Department of Natural Resources listed algae second and *Chara* eighth among 21 species ranked in order of treated area (Nelson and Dupes 1985). The "algal" component (exclusive of *Chara*) represented 20% of the total area treated by the state (15,400 hectares). A 20% figure may also be the case for regions outside of Florida and the Southeast. Algicides accounted for 23% of aquatic herbicides used by state agencies in Illinois in 1981 (Trudeau 1982).

The data provided by governmental agencies do not include waters treated either by individuals or by commercial applicators hired by individuals, property-owner groups, or industry. The area infested and/or treated for macrophytic algae in privately owned waters appears to be substantial. In a 1984 survey, county extension agents in Florida perceived filamentous algae to be the number one problem whereas members of the Florida Aquatic Plant Management Society (FAPMS) ranked algae eighth (Joyce and Summerhill 1985). This discrepancy was attributed to the fact that extension agents have primary contact with private individuals or agricultural interests whereas FAPMS members work more in the public agency and research sectors.

Macrophytic algae appear to invade sites in which submersed vascular plants have been chemically or mechanically removed (Hodgson and Carter 1982; Sheldon 1986), a situation that is easily achieved in man-made bodies of water, most of which tend to be small and shallow. In addition, many privately owned waters (typically surrounded by golf courses, condominiums, and homes where aesthetics is of premium value) may be treated with an algicide several times a year in order to maintain year-long algae control. For example, copper sulfate, the most commonly used algicide, has a relatively short persistence time in water (e.g., 10 days in Manitoba farm ponds: Whitaker et al. 1978); consequently, algal cells not affected by the treatment or germinating from spores provide the source for new infestations. It is not unusual in Florida for copper sulfate treatments to be made 4 to 20 times a year, with 7 to 10 being an average (C. E. Timmer, personal communication). As one of the largest commercial applicators of aquatic herbicides in the southeastern United States, Timmer estimates that 88% of the total area treated by his company is for algae control. This represents 8,100 to 12,150 hectares per annum in Florida treated with copper sulfate by his firm alone.

V. Ecology and growth characteristics

Research on the environmental conditions regulating macrophytic algal growth has emphasized those species with the most noticeable impacts on human utilization of water. *Cladophora,* especially *C. glomerata* (L.) Kütz., has been the most extensively studied because of its wide distribution and infestations in the Great Lakes and irrigation systems in the western United States. Its colonization of the rocky littoral zones of Lakes Michigan, Erie, Huron, and Ontario has been accompanied by weedy growths of two other filamentous algae, the green alga *Ulothrix* and the red alga *Bangia.* However, it is the abundant growth of massive mats of *Cladophora,* averaging 100–400 g dry wt·m^{-2} (Neil and Jackson 1982) with individual filament lengths of 30–50 cm (Blum 1982), that has received the greatest attention.

Pithophora spp. and a form of *Lyngbya,* both free-floating macrophytes, pose major problems in the management of shallow lakes and ponds, including fish culture ponds. The filaments making up the mats are tightly intertwined, producing thick clumps of floating vegetation. Mat thicknesses of up to 0.5 m and biomass values as high as 13,800 g fr wt·m^{-2} (1,440 g dry wt·m^{-2}) have been recorded for *Lyngbya* (Beer, Spencer, and Bowes 1986). Both *Pithophora* and *Lyngbya* have been reported to be extremely difficult to control using algicides such as copper sulfate and simazine (Crance 1974; Beer et al. 1986). Thus, they are regarded as causing more serious management problems than more algicide-susceptible filamentous genera such as *Spirogyra* and *Oedogonium* found in similar habitats.

The characteristics of these two major habitat types (attached versus free-floating) will be examined here in terms of vertical, seasonal, and geographic distribution. In doing so, both the environmental parameters regulating filamentous algal growth and the management problems associated with these two distinct habitats can be compared.

A. Attached macrophytic algae of the Great Lakes

A major environmental feature of the rocky littoral zone of the Great water is the relatively unrestricted water movement at the shoreline. The beneficial effect of a current to break a steep nutrient diffusion gradient and to enhance exchange of materials at the surface of filamentous algae was postulated by Whitford and Schumacher (1961, 1964). The replenishment of nutrients and CO_2 afforded by current and wave motion and the availability of stable attachment substrates at many sites in the Great Lakes provide ideal habitat for epilithic macroalgae.

1. Vertical distribution. The rocky littoral of the Great Lakes shows a distinctive vertical zonation of macrophytic algae (Blum 1982). The distribution pattern consists of *Bangia atropurpurea* (Roth) Ag. dominating the splash zone just above the waterline, *Ulothrix zonata* (Weber and Mohr) Kütz. occurring at the waterline, and *Cladophora glomerata* inhabiting depths to 2 or 3 m.

Bangia is a relatively recent invader of the Great Lakes and was first described in Lake Erie in 1969 (Kishler and Taft 1970). It appears to be conspecific with the marine species *B. fuscopurpurea* (Geesink 1973), and its intrusion into the Great Lakes may have been aided by anthropogenic increases in halide concentrations (Sheath and Cole 1980). Its tolerance to periodic desiccation and a wide range of temperatures (2–26°C) enable it to survive under the extreme environmental conditions characteristic of the upper splash zone (Garwood 1982).

Below the *Bangia* band, light intensity is a major factor in regulating vertical distribution. The higher position of *Ulothrix* in the zonation pattern suggests a greater tolerance for high light levels than *Cladophora*. When measured under laboratory conditions and optimal temperatures for photosynthesis, maximum net photosynthetic rates at $1,100$ $\mu E \cdot m^{-2} \cdot s^{-1}$ for *Ulothrix* (Graham, Kranzfelder, and Auer 1985) and between 300 and 600 $\mu E \cdot m^{-2} \cdot s^{-1}$ for *Cladophora* (Graham et al. 1982) tend to support this hypothesis. In fact, enhancement of respiration at high light levels in *Cladophora* may restrict growth in shallow waters in which light penetration is high (Graham et al. 1982; Canale, Auer, and Graham 1982).

Light is also an important factor in regulating the depth distribution of *Cladophora* in deeper waters. Lester, Adams, and Deltmann (1974) reported a light compensation point ranging from 44 to 104 $\mu E \cdot m^{-2} \cdot s^{-1}$, and Graham et al. (1982) reported that light levels of 35 $\mu E \cdot m^{-2} \cdot s^{-1}$ and lower were limiting to photosynthesis. Field data from Lorenz and Herdendorf (1982) generally substantiate these values. They reported no *Cladophora* growth at depths where irradiances were lower than 50 $\mu E \cdot m^{-2} \cdot s^{-1}$ (approximately the 2 m depth in western Lake Erie). *Cladophora* growth was, however, recorded at >2 m depth in areas where lower turbidity permitted increased light penetration.

2. *Seasonal distribution.* Ulothrix zonata is a cold-water species frequently found in oligotrophic waters (e.g., Lake Superior) throughout the summer (Parker 1979). In the lower Great Lakes it shows a distinct bimodal seasonality. It appears at ice-out in the spring (March in Lakes Huron and Michigan) (Blum 1982; Auer et al. 1982) and dominates the flora until water temperatures rise to 10°C (Taft and Kishler 1973; Lorenz and Herdendorf 1982), which usually occurs in May or June. Above 10°C, *Ulothrix* is replaced by *Cladophora glomerata* and does not reappear until fall.

A temperature optimum of 5°C for net photosynthesis of *Ulothrix* from laboratory studies (Graham, Kranzfelder, and Auer 1985) matches well with field observations. However, *U. zonata* was also observed by these researchers to photosynthesize very well at 15° and 20°C. Graham, Graham, and Kranzfelder (1985) found that a shift from vegetative growth to zoospore production was favored by temperatures near 20°C, irradiances of 520 $\mu E \cdot m^{-2} \cdot ^{-1}$, and increasing daylengths (16:8 h light–dark). Thus, the loss of *Ulothrix* biomass at higher temperatures in the field appears to occur because the filamentous material disintegrates as zoospores are released.

Cladophora also shows a biomodal seasonal distribution in the Great Lakes with growth maximums in the spring and autumn and a period of summer decline (Bellis and McLarty 1967; Herbst 1969; Lorenz and Herdendorf 1982). Maximum photosynthetic rates of *C. glomerata* measured in Green Bay, Lake Michigan, occurred at temperatures between 19° and 24°C. Rates were depressed at temperatures above 25°C (Adams and Stone 1973). The decrease in midsummer standing crop appears to be common in temperate regions and in both static and flowing waters (Bellis and McLarty 1967; Whitton 1970a; Wong et al. 1978). Similar spring–autumn maximums also were reported for *C. fracta* in a shallow lake in Michigan (Cheney and Hough 1983).

Optimal growth rates at 15–20°C were measured in laboratory cultures of *Cladophora* with a drop in growth rate as temperature was increased from 25° to 30°C (Robinson and Hawkes 1986). Graham et al. (1982) found maximum net photosynthesis to occur between 13° and 17°C and 300–600 $\mu E \cdot m^{-2} \cdot s^{-1}$. At temperatures above 25°C net photosynthesis was reduced, but dark respiration was elevated. They postulated that at high temperatures the alga experiences an unfavorable energy balance between photosynthetic oxygen production and respiratory oxygen consumption, a physiological state that results in reduced growth during the summer. Other factors that have been suggested as enhancing the summer decline include increasing cell wall thickness, which inhibits nutrient diffusion and light availability, the lessening of water agitation, and an increase in epiphyte growth (Lorenz and Herdendorf 1982). Evidence that un-ionized ammoniacal nitrogen is toxic to *Cladophora* suggests that under conditions of high pH, temperature, and ammonium concentrations, ammonium toxicity may also account in part for the summer decline (Robinson and Hawkes 1986).

3. Geographic distribution. Although reports of *Cladophora glomerata* flourishing in tropical, desert, and surface waters of ponds with mid-summer temperatures of 27°C or greater (Kachroo 1956; Mason 1965) exist, Fritsch's (1907) contention that *C. glomerata* is primarily a temperate species is supported by the growth studies mentioned above. Within a temperate area, such as the Great Lakes region, geographic distribution appears to be associated with suitable attachment sites and nutrient concentrations, particularly phosphorus (Neil and Owen 1964; Herbst 1969; Auer and Canale 1980). In the phosphorus-rich waters of Lakes Erie and Ontario, *Cladophora* is generally widespread with its spatial distribution primarily regulated by adequate substrate, light, and temperature (Lorenz and Herdendorf 1982; Auer et al. 1982). Cellular nitrogen and phosphorus levels in samples collected from Lake Erie, for example, were higher than the 1.1% N and 0.06% P critical concentrations established as limiting for *C. glomerata*, even during the summer biomass decline period (Mantai, Garwood, and Peglowski 1982).

In Lakes Superior, Huron, and Michigan, where overall phosphorus levels are insufficient to support extensive algal growth, the geographic distribution of both *Cladophora* and *Ulothrix* is restricted to local sites with an abundant supply of phosphorus (Herbst 1969; Parker and Drown 1982; Auer et al. 1982). High levels of internal phosphorus and low phosphorus uptake rates were measured in *Cladophora* collected near the outfall of a Lake Huron wastewater treatment plant, whereas *Cladophora* distant from the plant had low internal phosphorus levels and rapid phosphorus uptake rates (Auer and Canale 1980). The relationships among phosphorus uptake kinetics, external phosphate concentrations, and growth and biomass have been well established (Canale and Auer 1982a). In Lake Superior, *Cladophora* growth is additionally restricted by low water temperatures (midsummer maximum approximately 10°C).

Approaches to managing *Cladophora* in the Great Lakes have understandably centered on phosphorus regulation with the long-term objective of eliminating the cause of the problem. The potential to use algicides to alleviate immediate problems is negated by the water dilution factor, possible environmental hazards, and public pressures. Mathematical models that integrate phosphorus-dependent growth relationships with light and temperature have proven useful in calculating the spatial distribution of biomass (Canale and Auer 1982b). The use of these models to evaluate the utility of various phosphorus management

Fig. 18-1. Shallow lake infested with *Pithophora.*

strategies in reducing nuisance growth has been demonstrated for several sites in Lake Huron (Canale and Auer 1982b; Jackson and Hamdy 1982) and provides the data necessary to set environmental standards for similar bodies of water with a high potential for human impact.

B. Free-floating macrophytic algae

Environmental conditions for free-floating algae are considerably different from those for attached forms. Water flow is generally minimal so that replenishment of essential resources may not be continuous. The free-floating alga is subject to movement that can range from buoyancy at the surface, to sinking below the photic depth, to complete washout from the system. Water level fluctuations in static bodies of water which might, for example, lead to stranding of the alga above the water line in summer are not uncommon. Thus, the alga must be adapted to survive an inconstancy of environmental conditions if it is to persist.

1. Vertical distribution. Most free-floating macrophytic algae begin growth in the spring on the bottom sediments from overwintering spores/zygotes in the case of annual forms (e.g., *Spirogyra*) or from overwintering filaments in perennial forms (e.g., *Pithophora,* which also germinates from akinetes) (O'Neal, Spencer, and Lembi 1983). As oxygen bubbles become trapped in the filaments, the mats rise to the surface to form streamers from the bottom, some portions of which break off to form free-floating rafts (Hillebrand 1983). (See Fig. 18-1.) A study of top and bottom distribution of the blue-green alga *Lyngbya* in shallow Florida lakes and ponds showed that as much as 87% of the midsummer biomass was attached to the bottom sediments. However, it must be pointed out that the 1,800 g fr wt·m^{-2} (140 g dry wt·m^{-2}) in the floating component alone is a substantial standing crop and on a par with those measured for *Cladophora* (Whitton 1970a; Neil and Jackson 1982) or *Pithophora* (O'Neal, Lembi, and Spencer 1985). The large biomass found on the bottom for both *Lyngbya* and *Pithophora* (O'Neal and Lembi 1983) presents a difficult management problem because algicide treatments directed only toward the visible floating biomass disregard the potential for reinfestation from bottom growths.

Although *Pithophora* also can be found below the surface, there is evidence that mats settling into light-limited zones will not regain buoyancy and will eventually die due to low photosynthetic production (O'Neal et al. 1985). However, photosynthesis in both *Lyngbya* (Beer et al. 1986) and *Pithophora* (O'Neal et al. 1985) appears to be well adapted to low light conditions with light compensation points of approximately 1% and 0.3–1% of full sunlight (2,000 μE·m^{-2}·s^{-1}), respectively. These compensation points are 45–81% less than those measured for *Cladophora glomerata* of 1.8% (Graham et al. 1982), 2.2–5.2% (Lester et al. 1974), and 2.5% (Lorenz and Herdendorf 1982) of full sunlight. Low light compensation points would be advantageous to mat formers that are temporarily submerged in light-limited zones and that undergo significant self-shading within the mat itself. In addition, *Pithophora* is able to survive up to 60 days in complete darkness without significant loss of chlorophyll (O'Neal and Lembi 1983).

Floating mats are often bleached on the upper surface, suggesting photodestruction of pigments by high light intensities. In *Pithophora* and *Lyngbya,* where the mats are thick, the middle and bottom portions of the mats are green, appear healthy, and are photosynthetically competent (O'Neal et al. 1985). Other algae that produce thinner mats may be much more severely affected by high light intensities than those that produce thick mats. The filmy mats of *Spirogyra,* for example, often bleach completely when they break away from the bottom sediments (Dobbins and Boyd 1976), a phenomenon that is frequently accompanied by conjugation, zygote formation, and senescence (Prescott 1962).

2. Seasonal distribution. Although little studied, a distinct seasonality of filamentous algal genera is common in static waters of temperate climates. Members of the Zygnemataceae (e.g., *Spirogyra*) are often dominant in the spring flora (Prescott 1962), being supplanted by *Cladophora* (e.g., *C. fracta*) in the late spring–early summer. The summer–early fall flora may be dominated by *Oedogonium* (McCracken, Gustafson, and Adams 1974). Both *Lyngbya* (Beer et al. 1986) and *Pithophora* biomass (O'Neal et al. 1985) peak in the summer–early fall period.

In a midwestern lake, germination of overwintering akinetes of *Pith-*

ophora (which can compose as much as 20% of the overwintering biomass) occurred in response to an increase in water temperatures from 10° to 20°C (Spencer, Volpp, and Lembi 1980). Germination is initially dependent on the respiration of stored carbohydrates and lipids, which allow germination and growth in light-limited microenvironments such as the hydrosoil and within the *Pithophora* mats (O'Neal and Lembi 1983). Photosynthesis rates increased from a low of 0.91 to 2.33 mg $O_2 \cdot$g dry wt$^{-1} \cdot$h^{-1} in midwinter to maximums of 29–39 mg $O_2 \cdot$g dry wt$^{-1} \cdot$h^{-1} in late spring and summer and were positively correlated with temperature and external nitrogen and phosphorus levels (O'Neal et al. 1985). Biomass development lagged behind photosynthesis and peaked in autumn (163–206 g dry wt$\cdot$m^{-2}). Akinete formation occurred during this period and was at least partially a response to low external nutrient conditions (Lembi and Spencer 1981). Approximately 10–25% of the biomass may overwinter; however, the magnitude of biomass loss due to decomposition versus washout from the system is unknown. In Florida, *Lyngbya* can overwinter with biomass values from 120 to 440 g dry wt$\cdot$m^{-2} (Beer et al. 1986). Its behavior in more temperate regions in midwinter is unknown.

Laboratory studies of both *Pithophora* and *Lyngbya* support field measurements. Optimal temperatures for photosynthesis are high: 25° to 35°C for *Pithophora* (Spencer, Lembi, and Graham 1985; O'Neal et al. 1985) and 40°C for *Lyngbya* (Beer et al. 1986). Growth rates of *Pithophora* were maximal at 30°C (highest temperature studied) (Spencer 1983).

3. Geographic distribution. Fritsch's (1907) contention that *Pithophora* is primarily a tropical genus is supported by the high temperature optima for photosynthesis and growth. The "protective" layer of surface filaments allows the organism to survive not only high light intensities but stranding along the shoreline during periods of high water evaporative losses, a condition frequent in the tropics (Fritsch 1907). This layer provides a barrier to loss of moisture from the bottom portions of the mat, thus protecting the bulk of the vegetation from rapid desiccation (Lembi, Pearlmutter, and Spencer 1980). The organism, however, is also adapted to temperate conditions, being able to overwinter successfully in deep waters or in ice (O'Neal et al. 1983) as either filaments or akinetes.

Within its range, *Pithophora* distribution is regulated by nutrient availability and other water-quality parameters. The half-saturation constant (K_s) at 20°C relating filament growth to external concentrations of nitrate-N is 1.23 mg$\cdot$L^{-1} and for phosphate-P, 0.1 mg$\cdot$L^{-1} (Spencer and Lembi 1981). These K_s values are considerably higher than those calculated at 23°C for *Cladophora glomerata* (0.25 and 0.013 mg$\cdot$L^{-1} for nitrate-N and phosphate-P, respectively) (Gerloff and Fitzgerald 1976) and may contribute to the generally narrower distribution of *Pithophora* in the United States in contrast to *Cladophora.*

Both *Pithophora* and *Lyngbya* are typically found in waters of high pH (Spencer and Lembi 1981; Tubea, Hawxby, and Metha 1981) where dissolved inorganic carbon (DIC) diffusion rates tend to be low and bicarbonate (HCO_3^-) is the predominant DIC source. Beer et al. (1986) found that *Lyngbya* utilizes HCO_3^- very efficiently with low DIC values required to saturate photosynthesis and growth. Concentration mechanisms or efficient utilization of HCO_3^- may suppress photorespiration, which in this organism appears to be minimal. Conditions that enhance photorespiration such as high midday temperatures and oxygen levels are characteristic of the floating mat environment; thus, physiological adaptations to these pressures permit *Lyngbya* and possibly other free-floating mat formers (Cheney and Hough 1983) to colonize stagnant bodies of water successfully in midsummer. These adaptations may give filamentous algae a competitive advantage over vascular plants. For example, Beer et al. (1986) noted that the HCO_3^- level required to saturate photosynthesis in submersed vascular plants such as *Hydrilla* and *Myriophyllum* (Van, Haller, and Bowes 1976) is about tenfold higher than it is in *Lyngbya,* and photorespiration was found to be more detrimental to productivity of the submersed vascular plant *Elodea* than either *Cladophora glomerata* or *Spirogyra* (Simpson and Eaton 1986). These observations suggest not only a survival mechanism for the filamentous alga itself under extremely low HCO_3^- conditions but a way in which it can modify its environment and place potential competitors at a disadvantage. It is quite possible that the observed replacement of *Elodea* by *C. glomerata* or *Spirogyra* was due to continued algal photosynthesis that resulted in changes in water quality (e.g., high pH and dissolved oxygen, low CO_2 concentrations) detrimental to the higher plant (Simpson and Eaton 1986).

The importance of the thickness of the mats and the dense, intertwined growth of the filaments of *Pithophora, Lyngbya,* and *Cladophora* cannot be overemphasized from either an ecological or a management standpoint. The interior of the mat must present a unique environment in terms of light, temperature, nutrients, and pH, which could have an impact on associated epiphytes and invertebrates. It is known that the light compensation point for photosynthesis in *Pithophora* is reached between 8 and 10 mm beneath the mat surface (O'Neal et al. 1985). Net negative energy balance plus decomposition of filaments within the mats might release nutrients that would affect not only associated organisms but also the growth of outer mat edges, a situation that would be especially critical in late summer under nutrient-limiting conditions.

The impact of mat morphology on management practices, such as algicide treatments, may be significant. For example, the lack of light penetration into the mat can alter the activity of light-activated algicides such as simazine (O'Neal and Lembi 1983). In *Pithophora* mats, approximately 95% of the incident light is absorbed within 5 mm of the mat surface and represents a light level that is below the critical light-intensity range required to cause triazine toxicity.

Whitton (1970b) suggested that the lack of penetration of mats by copper sulfate was a major factor in the observed copper tolerance of *Cladophora* in the field, even though the cells themselves were highly susceptible. A similar protective function was attributed to the dense mats produced by the blue-green alga *Phormidium inundatum* in swimming pools treated with algicides (Fitzgerald 1959) and by the yellow-green alga *Vaucheria dichotoma* in drainage channels treated with diquat (Dowidar and Robson 1972). Preliminary quantitative studies suggest that these observations are correct. Copper concentration was reduced by 80% within 2 cm of the surface of a *Pithophora* mat and by 90% within 7 cm (unpublished data). At a normal treatment level of 8 μM Cu^{2+}, the concentration at 2 cm would have been below that required to cause a 50% loss in chlorophyll content (1.8 μM Cu^{2+}).

VI. Management practices

Alteration of environmental conditions (e.g., nutrient reduction) regulating growth remains the most effective management tool for macrophytic algae. Mathematical growth models developed for both *Cladophora* (Canale and Auer 1982b) and *Pithophora* (Spencer, O'Neal, and Lembi 1987) suggest that significant biomass reductions result from nutrient reduction practices. However, sites in which nutrient reduction is not currently a feasible option or is a long-term project (e.g., irrigation and drainage systems in which water is drawn from a large watershed, confined aquaculture systems, and shallow lakes and ponds impinged by nonpoint sources of nutrients) may require more direct approaches for the removal of algae to insure use of the water. However, the availability of direct control methods (such as the use of algicides) does not negate the need for research and/or implementation of nutrient-reduction approaches.

A. Algicides

Anderson and Dechoretz (1984) list six active ingredients currently used in the United States for algae control: copper (in several formulations), simazine, endothall, acrolein, diquat, and chlorine. Of these, copper (primarily in the form of copper sulfate pentahydrate) is the most commonly used in "natural" bodies of water. Only copper and diquat have federally established tolerances for potable water (1 ppmw and 0.01 ppmw, respectively) and, with the exception of trout habitat, copper can be used in waters intended for fishing, aquaculture, livestock watering, irrigation, and swimming. In contrast, simazine, endothall, and diquat have time restrictions (ranging from 24 h to 1 year) on various uses of water after treatment. Both acrolein and chlorine are highly toxic to fish and are limited to use in irrigation water conveyance systems and in swimming pools and municipal water treatment facilities, respectively.

Of the herbicides listed above, copper is the only one that can be used selectively for algae control since at normal use rates it has little effect on most submersed vascular plant species. As already noted, differences in copper tolerance among filamentous algal species exist with *Pithophora, Cladophora, Vaucheria, Hydrodictyon,* and *Lyngbya* being most frequently noted as tolerant species and *Spirogyra* and *Oedogonium* as sensitive species. At least part of the tolerance is due to lack of penetration through dense mat growth. Other factors found to affect copper uptake include pH and presence or absence of oxygen, divalent cations, and organic chelators (Stokes 1979; McKnight 1981). In addition, concerns over potential harmful effects on aquatic fauna to continuous exposure to copper have been raised (see review in Andrews 1986) and preclude its use for large-scale treatments in natural lake and wetland environments.

B. Harvesting

Mechanical harvesters have been used successfully in place of or in conjunction with aquatic herbicides for submersed weed control. Epiphytic algae are commonly harvested along with the vascular plants, although Jones, Gurevitch, and Adams (1979) note that both the degree of attachment and the spatial configuration of the epiphyte affect harvest efficiency. Harvesters are seldom used alone for free-floating algae although they may be used successfully in enclosed areas where the algae cannot float away. The major disadvantages of mechanical harvesters for floating-mat algae control include their inability to move in waters less than 1 m depth and a depth of cut of approximately 2–3 m.

C. Biological controls

A number of organisms have been used successfully for macrophytic algae control. These include waterfowl (ducks, geese, and swans), snails (*Marisa* and *Pomacea* sp.), crayfish (*Orconectes* spp.), and herbivorous fish such as tilapia (*Tilapia* and *Sarotherodon* sp.) and grass carp (*Ctenopharyngodon idella* Val.) (for review, see National Academy of Sciences 1976). Almost all of these organisms have limitations for widespread use including the potential for increased nutrient pollution (waterfowl), narrow thermal range for survival (snails, tilapia), and the large numbers of animals required to reduce algae growth adequately (crayfish). The most widely used biocontrol agent at the present time is the sterile triploid grass carp (Cassini and Caton 1986), which feeds on both vascular submersed plants and filamentous algae (Valin 1985). Concern in the United States over the spread of the grass carp into areas of native vegetation and fish-spawning and wildfowl habitat areas has limited its introduction at the present time to 20 states, although it is widely used in Europe (Van Zon 1977).

VII. Conclusions

Filamentous algae have a direct and significant impact on the utilization of water. Their increase, as a result of urbanization, water pollution, and removal of natural vegetation, will require additional funding for research and implementation of management programs. With the current emphasis on chemical control, an important question that remains unanswered is the potential for the increase of algicide-tolerant species. Most reports that this occurs are anecdotal. Furthermore, it is not known whether algicide-tolerant algae are developing resistance from prolonged exposure to algicides or are simply moving in when potential competitors are eliminated. The development of accurate delivery methods to insure that the algicide comes in contact with the target plant for the required exposure period is needed, particularly in flowing irrigation systems, which now must be dosed with large "slugs" of algicide (copper sulfate) or treated with extremely toxic materials (acrolein). Some new methodologies currently being used for aquatic vascular weed control that might be adapted for algae control include the use of adjuvants such as stickers, invert (water in oil) emulsion systems to reduce in-water movement, or controlled-release formulations.

Virtually nothing is known of the ecology and nutrient requirements of some of the most common weedy species (e.g., *Spirogyra, Oedogonium,* and *Hydrodictyon*). This is surprising in view of the fact that these genera are among the most familiar and recognizable members of the green algae. Indeed, even the number and identity of the species within these genera that should be classified as "weedy" are unknown. Until more is known, adequate environmental manipulation approaches cannot be developed. Considerable information, on the other hand, is available on the ecology of the charoid algae (for review, see Round 1981); however, very little has been done to integrate management approaches with life cycle or nutritional requirements.

The potential for the beneficial use of freshwater macrophytic algae appears minor, although some research has shown applications for waste treatment (San Marzano, Naveau, and Nyns 1980; Stom et al. 1980; Sladeckova, Marvan, and Vymazal 1983), farm pond restoration

(Denike and Geiger 1974; Crawford 1977, 1979), and feed supplements for livestock (Boyd 1973) and fish (Gunn, Qadri, and Mortimer 1977; Appler 1985). Management recommendations for plant coverage within fishery lakes range from 10% to 40% (Hinkle 1986), but these estimates are based solely on rooted vascular plant stands rather than filamentous algae. Thus, neither the man-made nor the natural benefits of filamentous algae have been adequately explored, and until they are, management approaches will continue to be one-way, emphasizing control rather than utilization.

VIII. Acknowledgments

We are grateful to Dr. James M. Graham for his review of this manuscript.

IX. References

Adams, M. S., and W. Stone. 1973. Field studies on photosynthesis of *Cladophora glomerata* (Chlorophyta) in Green Bay, Lake Michigan. *Ecology* 54, 853–62.

Adamson, R. P., and M. R. Sommerfeld. 1978. Survey of swimming pool algae of the Phoenix, Arizona, metropolitan area. *J. Phycol.* 14, 519–21.

Allanson, B. R. 1973. The fine structure of the periphyton of *Chara* sp. and *Potamogeton natans* from Wytham Pond, Oxford, and its significance to the macrophyte-periphyton metabolic model of R. G. Wetzel and H. L. Allen. *Freshwater Biol.* 3, 535–42.

Anderson, L. W. J., and N. Dechoretz. 1984. Laboratory and field investigations of a potential selective algicide, PH 4062. *J. Aquat. Plant Manage.* 22, 67–75.

Andrews, J. 1986. *Nuisance Vegetation in the Madison Lakes: Current Status and Options for Control.* Committee Report, University of Wisconsin, Madison. 196 pp.

Anthoni, U., C. Christophersen, J. O. Madsen, S. Wium-Andersen, and N. Jacobsen. 1980. Biological active sulfur compounds from the green alga *Chara globularis. Phytochemistry* 19, 1228–9.

Appler, H. N. 1985. Evaluation of *Hydrodictyon reticulatum* as protein source in feeds for *Oreochromis (Tilapia) niloticus* and *Tilapia zillii. J. Fish. Biol.* 27, 327–34.

Auer, M. T., and R. P. Canale. 1980. Phosphorus uptake dynamics as related to mathematical modeling of *Cladophora* at a site on Lake Huron. *J. Great Lakes Res.* 6, 1–7.

Auer, M. T., R. P. Canale, H. C. Grundler, and Y. Matsuoka. 1982. Ecological studies and mathematical modeling of *Cladophora* in Lake Huron. I. Program description and field monitoring of growth dynamics. *J. Great Lakes Res.* 8, 73–83.

Aurand, D. 1982. *Nuisance Aquatic Plants and Aquatic Plant Management Pro-*

grams in the United States. Vol. 2, *Southeastern Region.* Publ. no. MTR-82W47-02, U.S. Environmental Protection Agency and the Mitre Corp., McLean, VA. 359 pp.

Barnett, B. S., and R. W. Schneider. 1974. Fish populations in dense submerged aquatic plant communities. *Hyac. Contr. J.* 12, 12–14.

Beer, S., W. Spencer, and G. Bowes. 1986. Photosynthesis and growth of the filamentous blue-green alga *Lyngbya birgei* in relation to its environment. *J. Aquat. Plant Manage.* 24, 61–5.

Bellis, V. J., and D. A. McLarty. 1967. Ecology of *Cladophora glomerata* (L.) Kütz. in southern Ontario. *J. Phycol.* 3, 57–63.

Bennett, G. W. 1971. *Management of Lakes and Ponds.* Van Nostrand Reinhold, New York. 375 pp.

Blum, J. L. 1982. Colonization and growth of attached algae at the Lake Michigan water line. *J. Great Lakes Res.* 8, 10–15.

Bolas, P. M., and J. W. G. Lund. 1974. Some factors affecting the growth of *Cladophora glomerata* in the Kentish Stour. *Water Treat. Exam.* 23, 25–51.

Boyd, C. E. 1973. Amino acid composition of freshwater algae. *Arch. Hydrobiol.* 72, 1–9.

Boyd, C. E. 1982. *Water Quality Management for Pond Fish Culture.* Elsevier, Amsterdam. 318 pp.

Brezny, D., I. Mehta, and R. K. Sharma. 1973. Studies of evapotranspiration of some aquatic weeds. *Weed Sci.* 21, 197–204.

Canale, R. P., and M. T. Auer. 1982a. Ecological studies and mathematical modeling of *Cladophora* in Lake Huron. V. Model development and calibration. *J. Great Lakes Res.* 8, 112–25.

Canale, R. P., and M. T. Auer. 1982b. Ecological studies and mathematical modeling of *Cladophora* in Lake Huron. VII. Model verification and system response. *J. Great Lakes Res.* 8, 134–43.

Canale, R. P., M. T. Auer, and J. M. Graham. 1982. Ecological studies and mathematical modeling of *Cladophora* in Lake Huron. VI. Seasonal and spatial variation in growth kinetics. *J. Great Lakes Res.* 8, 126–33.

Carlton, J. T., and J. A. Scanlon. 1985. Progression and dispersal of an introduced alga: *Codium fragile* ssp. *tomentosoides* (Chlorophyta) on the Atlantic coast of North America. *Bot. Mar.* 28, 155–65.

Cassini, J. R., and W. E. Caton. 1986. Efficient production of triploid grass carp *(Ctenopharyngodon idella)* utilizing hydrostatic pressure. *Aquaculture* 55, 43–50.

Cheney, C., and R. A. Hough. 1983. Factors controlling photosynthetic productivity in a population of *Cladophora fracta* (Chlorophyta). *Ecology* 64, 68–77.

Colle, D. E., and J. V. Shireman. 1980. Coefficients of condition for largemouth bass, bluegill, and redear sunfish in *Hydrilla*-infested lakes. *Trans. Amer. Fish. Soc.* 109, 521–31.

Corbus, F. G. 1982. Aquatic weed control with endothall in a Salt River Project Canal. *J. Aquat. Plant Manage.* 20, 1–3.

Coyle, E. E. 1930. The algal food of *Pimephales promelas* (fathead minnow). *Ohio J. Sci.* 30, 23–35.

Crance, J. H. 1974. Observations on the effects of Israeli carp on *Pithophora* and other aquatic plants in Alabama fish ponds. *J. Alabama Acad. Sci.* 45, 41–3.

Crawford, S. A. 1977. Chemical, physical, and biological changes associated with *Chara* succession in farm ponds. *Hydrobiologia* 55, 209–17.

Crawford, S. A. 1979. Farm pond restoration using *Chara vulgaris* vegetation. *Hydrobiologia* 62, 17–31.

Denike, T. J., and R. W. Geiger. 1974. The utilization of *Chara* in water management. *Hyac. Contr. J.* 12, 18–20.

De Vries, P. J. R., M. Torenbeek, and H. Hillebrand. 1983. Bioassays with *Stigeoclonium* Kütz. (Chlorophyceae) to identify nitrogen and phosphorus limitations. *Aquat. Bot.* 17, 95–106.

Dobbins, D. A., and C. E. Boyd. 1976. Phosphorus and potassium fertilization of sunfish ponds. *Trans. Amer. Fish. Soc.* 105, 536–40.

Dowidar, A. R., and T. O. Robson. 1972. Studies on the biology and control of *Vaucheria dichotoma* found in freshwaters in Britain. *Weed Res.* 12, 221–8.

Engel, S. 1985. *Aquatic Community Interactions of Submerged Macrophytes.* Technical Bulletin 156, Department of Natural Resources, Madison, WI. 79 pp.

Fassett, N. C. 1957. *A Manual of Aquatic Plants.* University of Wisconsin Press, Madison. 405 pp.

Filbin, G. J., and J. W. Barko. 1985. Growth and nutrition of submersed macrophytes in a eutrophic Wisconsin impoundment. *J. Freshwater Ecol.* 3, 275–85.

Fitzgerald, G. P. 1959. Bacterial and algicidal properties of some algicides for swimming pools. *Appl. Microbiol.* 7, 205–11.

Francke, J. A., and P. J. Den Oude. 1983. Growth of *Stigeoclonium* and *Oedogonium* species in artificial ammonium-N and phosphate-P gradients. *Aquat. Bot.* 15, 375–80.

Francke, J. A., and H. Hillebrand. 1980. Effects of copper on some filamentous Chlorophyta. *Aquat. Bot.* 8, 285–9.

Fritsch, F. E. 1907. The subaerial and freshwater algal flora of the tropics. *Ann. Bot.* 21, 235–75.

Garwood, P. E. 1982. Ecological interactions among *Bangia, Cladophora,* and *Ulothrix* along the Lake Erie shoreline. *J. Great Lakes Res.* 8, 54–60.

Geesink, R. 1973. Experimental investigations on marine and freshwater *Bangia* (Rhodophyta) from the Netherlands. *J. Exp. Mar. Biol. Ecol.* 11, 239–47.

Gerloff, G. C., and G. P. Fitzgerald. 1976. *The Nutrition of Great Lakes Cladophora.* EPA-600/3-76-044, U.S. Environmental Protection Agency. Duluth, MN. 111 pp.

Gibbs, G. W. 1973. Cycles of macrophytes and phytoplankton in Pukepuke lagoon following a severe drought. *Proc. N.Z. Ecol. Soc.* 20, 13–20.

Graham, J. M., M. T. Auer, R. P. Canale, and J. P. Hoffmann. 1982. Ecological studies and mathematical modeling of *Cladophora* in Lake Huron. IV. Photosynthesis and respiration as functions of light and temperature. *J. Great Lakes Res.* 8, 100–11.

Graham, J. M., L. E. Graham, and J. A. Kranzfelder. 1985. Light, temperature, and photoperiod as factors controlling reproduction in *Ulothrix zonata* (Ulvophyceae). *J. Phycol.* 21, 235–9.

Graham, J. M., J. A. Kranzfelder, and M. T. Auer. 1985. Light and temperature as factors regulating seasonal growth and distribution of *Ulothrix zonata* (Ulvophyceae). *J. Phycol.* 21, 228–34.

Gunn, J. M., S. U. Qadri, and D. C. Mortimer. 1977. Filamentous algae as a food source for the brown bullhead *(Ictalurus nebulosus)*. *J. Fish. Res. Bd. Can.* 34, 396–401.

Halfen, L. N., and M. T. McCann. 1975. Behavioral aspects of benthic communities of filamentous blue-green algae in lentic habitats. *Mich. Bot.* 14, 49–56.

Hansen, G. W., F. E. Oliver, and N. E. Otto. 1984. *Herbicide Manual.* U.S. Department of the Interior, Bureau of Reclamation, Denver, CO. 346 pp.

Harding, J. P. C., and B. A. Whitton. 1976. Resistance to zinc of *Stigeoclonium tenue* in the field and the laboratory. *Brit. Phycol. J.* 13, 65–8.

Herbst, R. P. 1969. Ecological factors and the distribution of *Cladophora glomerata* in the Great Lakes. *Amer. Midl. Nat.* 82, 90–8.

Hillebrand, H. 1983. Development and dynamics of floating clusters of filamentous algae. In *Periphyton of Freshwater Ecosystems,* ed. R. G. Wetzel, pp. 31–9. Dr. W. Junk, The Hague.

Hinkle, J. 1986. A preliminary literature review on vegetation and fisheries with emphasis on the largemouth bass, bluegill, and hydrilla. *Aquatics* 8(4), 9–14.

Hodgson, L. M., and C. C. Carter. 1982. Effect of hydrilla management by herbicides on a periphyton community. *J. Aquat. Plant Manage.* 20, 17–19.

Jackson, M. B., and Y. S. Hamdy. 1982. Projected *Cladophora* growth in southern Georgian Bay in response to proposed municipal sewage treatment plant discharges to the Mary Ward Shoals. *J. Great Lakes Res.* 8, 153–63.

Jansson, A.-M. 1967. The food-web of the *Cladophora* belt fauna. *Helgoländ. Wiss. Meeresunters.* 15, 574–88.

Jones, R. C., A. Gurevitch, and M. S. Adams. 1979. Significance of the epiphyte component of the littoral to biomass and phosphorus removal by harvesting. In *Aquatic Plants, Lake Management, and Ecosystem Consequences of Lake Harvesting,* ed. J. E. Breck, R. T. Prentki, and O. L. Loucks, pp. 51–61. Center for Biotic Systems, University of Wisconsin, Madison. 435 pp.

Joyce, J. C., and W. R. Summerhill. 1985. Aquatic plant research and extension surveys. *Aquatics* 7(1), 11–12.

Kachroo, P. 1956. Plant types of the ponds in the lower Damodar Valley. *J. Indian Bot. Soc.* 35, 430–45.

Kishler, J., and C. E. Taft. 1970. *Bangia atropurpurea* (Roth) Ag. in western Lake Erie. *Ohio J. Sci.* 70, 56–7.

Kitchell, J. F., and J. T. Windell. 1970. Nutritional value of algae to bluegill sunfish, *Lepomis macrochirus. Copeia,* 186–90.

Lawrence, J. M. 1954. Control of a branched alga, *Pithophora,* in farm fish ponds. *Prog. Fish Cult.* 16, 83–6.

Lembi, C. A., N. L. Pearlmutter, and D. F. Spencer, 1980. *Life Cycle, Ecology, and Management Considerations of the Green Filamentous Alga, Pithophora.* Tech. Rep. 130. Purdue University Water Resources Research Center, W. Lafayette, IN. 97 pp.

Lembi, C. A., and D. F. Spencer. 1981. Role of the akinete in the survival of *Pithophora* (Chlorophyceae). *J. Phycol.* 17 (Suppl.), 8.

Lester, W. W., M. S. Adams, and E. H. Deltmann. 1974. *Light and Temperature Effects on Photosynthesis of Cladophora glomerata (Chlorophyta) from Green Bay, Lake Michigan: Model Analysis of Seasonal Productivity.* MS thesis, University of Wisconsin, Madison.

Lorenz, R. C., and C. E. Herdendorf. 1982. Growth dynamics of *Cladophora glomerata* in western Lake Erie in relation to some environmental factors. *J. Great Lakes Res.* 8, 42–53.

McCracken, M. D., T. D. Gustafson, and M. S. Adams. 1974. Productivity of *Oedogonium* in Lake Wingra, Wisconsin. *Amer. Midl. Nat.* 92, 247–54.

McKnight, D. 1981. Chemical and biological processes controlling the response of a freshwater ecosystem to copper stress: a field study of the $CuSO_4$ treatment of Mill Pond Reservoir, Burlington, Massachusetts. *Limnol. Oceanogr.* 26, 518–31.

McLean, R. O., and K. Benson-Evans. 1974. The distribution of *Stigeoclonium tenue* Kütz. in South Wales in relation to its use as an indicator of organic pollution. *Brit. Phycol. J.* 9, 83–9.

Mantai, K. E., P. E. Garwood, and L. E. Peglowski. 1982. Environmental factors controlling physiological changes in *Cladophora* in Lake Erie. *J. Great Lakes Res.* 8, 61–5.

Mason, C. P. 1965. Ecology of *Cladophora* in farm ponds. *Ecology* 46, 421–9.

National Academy of Sciences. 1976. *Making Aquatic Weeds Useful: Some Perspectives for Developing Countries.* National Academy of Science–National Research Council, 2101 Constitution Ave., Washington, DC. 174 pp.

Neel, J. K. 1968. Seasonal succession of benthic algae and their macro-invertebrate residents in a headwater limestone stream. *J. Water Poll. Contr. Fed.* 40, R10–R30.

Neil, J. H. 1975. Ecology of the *Cladophora* niche. In *Cladophora in the Great Lakes,* ed. H. Shear and D. E. Konasewich, pp. 125–7. International Joint Commission Regional Office, Windsor, Ont. 179 pp.

Neil, J. H., and M. B. Jackson. 1982. Monitoring *Cladophora* growth conditions and the effect of phosphorus additions at a shoreline site in northeastern Lake Erie. *J. Great Lakes Res.* 8, 30–4.

Neil, J. H., and G. E. Owen. 1964. Distribution, environmental requirements, and significance of *Cladophora* in the Great Lakes. In *Proceedings of the Seventh Conference on Great Lakes Research,* Publ. 11, pp. 113–21. Great Lakes Research Division, University of Michigan, Ann Arbor.

Nelson, B. V., and J. M. Dupes. 1985. Publicly funded aquatic plant management operations in Florida during 1984. *Aquatics* 7(3), 11–12.

O'Neal, S. W., and C. A. Lembi. 1983. Effect of simazine on photosynthesis and growth of filamentous algae. *Weed Sci.* 31, 899–903.

O'Neal, S. W., C. A. Lembi, and D. F. Spencer. 1985. Productivity of the filamentous alga *Pithophora oedogonia* (Chlorophyta) in Surrey Lake, Indiana. *J. Phycol.* 21, 562–9.

O'Neal, S. W., D. F. Spencer, and C. A. Lembi. 1983. *Integration of Management Methods with the Life Cycle and Ecology of the Filamentous Alga, Pithophora.* Tech. Rep. 154, Purdue University Water Resources Research Center, W. Lafayette, IN. 62 pp.

Ormerod, G. K. 1970. The relationship between real estate values, algae, and water levels. July. Report of Lake Erie Task Force, Department of Public Works, Canada.

Palmer, C. M. 1962. *Algae in Water Supplies.* U.S. Public Health Service, Washington, DC. 88 pp.

Parker, R. D. R. 1979. Distribution of filamentous green algae in nearshore periphyton communities along the north shore of Lake Superior. *Amer. Midl. Nat.* 101, 326–32.

Parker, R. D. R., and D. B. Brown. 1982. Effects of phosphorus enrichment and wave simulation on populations of *Ulothrix zonata* from northern Lake Superior. *J. Great Lakes Res.* 8, 16–26.

Patrick, R. 1978. Effects of trace metals in the aquatic ecosystem. *Amer. Sci.* 66, 185–91.

Pennak, R. W. 1966. Structure of zooplankton populations in the littoral macrophyte zone of some Colorado lakes. *Trans. Amer. Microsc. Soc.* 85, 329–49.

Pennak, R. W. 1973. Some evidence for aquatic macrophytes as repellents for a limnetic species of *Daphnia. Int. Rev. ges. Hydrobiol.* 58, 569–76.

Phillips, G. L., D. Eminson, and B. Moss. 1978. A mechanism to account for macrophyte decline in progressively eutrophicated waters. *Aquat. Bot.* 4, 103–26.

Prescott, G. W. 1962. *Algae of the Western Great Lakes Area.* Wm. C. Brown, Dubuque, IA. 977 pp.

Quade, H. W. 1969. Cladoceran fauna associated with aquatic macrophytes in some lakes in northwestern Minnesota. *Ecology* 50, 170–9.

Robinson, P. K., and H. A. Hawkes. 1986. Studies on the growth of *Cladophora glomerata* in laboratory continuous-flow culture. *Brit. Phycol. J.* 21, 437–44.

Robson, T. O., M. C. Fowler, and P. R. F. Barrett. 1976. Effect of some herbicides on freshwater algae. *Pestic. Sci.* 7, 391–402.

Rosine, W. N. 1955. The distribution of invertebrates on submerged aquatic plant surfaces in Muskee Lake, Colorado. *Ecology* 36, 308–14.

Round, F. E. 1981. *The Ecology of Algae.* Cambridge University Press, Cambridge. 653 pp.

Sand-Jensen, K. 1977. Effects of epiphytes on eelgrass photosynthesis. *Aquat. Bot.* 3, 55–63.

Sand-Jensen, K., and M. Sondergaard. 1981. Phytoplankton and epiphyte development and their shading effect on submerged macrophytes in lakes of different nutrient status. *Int. Rev. ges. Hydrobiol.* 66, 529–52.

San Marzano, C. M. A., H. P. Naveau, and E. J. Nyns. 1980. Production of methane from freshwater macro-algae, an anaerobic two step digestion system. In *Energy from Biomass,* ed. W. Palz, D. O. Chartier, and D. O. Hall, pp. 703–8. Applied Science Publishers, London. 982 pp.

Savino, J. E., and R. A. Stein. 1982. Predator–prey interactions between largemouth bass and bluegills as influenced by simulated, submersed vegetation. *Trans. Amer. Fish. Soc.* 111, 255–66.

Say, P. J., B. M. Diaz, and B. A. Whitton. 1977. Influence of zinc on lotic plants. I. Tolerance of *Hormidium* species to zinc. *Freshwater Biol.* 7, 357–76.

Schardt, J. D. 1983. *1983 Florida Aquatic Plant Survey.* Florida Department of Natural Resources, Tallahassee. 143 pp.

Schardt, J. D. 1984. *1984 Florida Aquatic Plant Survey.* Florida Department of Natural Resources, Tallahassee. 149 pp.

Schindler, D. W., K. H. Mills, D. F. Malley, D. L. Findley, J. A. Sheaver, I. J. Davies, M. A. Turner, G. A. Linsey, and D. R. Cruikshank. 1985. Long-term ecosystem stress: the effects of years of experimental acidification on a small lake. *Science* 228, 1395–401.

Shacklock, P. F., and R. W. Doyle. 1983. Control of epiphytes in seaweed cultures using grazers. *Aquaculture* 31, 141–51.

Sheath, R. G., and K. M. Cole. 1980. Distribution and salinity adaptations of *Bangia atropurpurea* (Rhodophyta), a putative migrant into the Laurentian Great Lakes. *J. Phycol.* 16, 412–20.

Sheldon, S. P. 1986. The effects of short-term disturbance on a freshwater macrophyte community. *J. Freshwater Ecol.* 3, 309–17.

Shireman, J. V., W. T. Haller, D. E. Colle, and D. F. DuRant. 1983. Effects of aquatic macrophytes on native sportfish populations in Florida. In *Proceedings of the International Symposium on Aquatic Macrophytes,* pp. 208–14. Nijmegen, Netherlands.

Simpson, P. S., and J. W. Eaton. 1986. Comparative studies of the photosynthesis of the submerged macrophyte *Elodea canadensis* and the filamentous algae *Cladophora glomerata* and *Spirogyra* sp. *Aquat. Bot.* 24, 1–12.

Sladeckova, A., P. Marvan, and J. Vymazal. 1983. The utilization of periphyton in waterworks pre-treatment for nutrient removal from enriched effluents. In *Periphyton of Freshwater Ecosystems,* ed. R. G. Wetzel, pp. 299–303. Dr. W. Junk, The Hague.

Spence, D. H. N. 1982. The zonation of plants in freshwater lakes. In *Advances in Ecological Research,* vol. 12, ed. A. Macfadyen and E. D. Ford, pp. 37–125. Academic Press, New York.

Spencer, D. F. 1983. Temperature and the growth of *Pithophora oedogonia* Wittr. (Chlorophyta). *Phycologia* 22, 202–5.

Spencer, D. F., and C. A. Lembi. 1981. Factors regulating the spatial distribution of the filamentous alga *Pithophora oedogonia* (Chlorophyceae) in an Indiana lake. *J. Phycol.* 17, 168–73.

Spencer, D. F., C. A. Lembi, and J. M. Graham. 1985. Influence of light and temperature on photosynthesis and respiration by *Pithophora oedogonia* (Mont.) Wittr. (Chlorophyceae). *Aquat. Bot.* 23, 109–18.

Spencer, D. F., S. W. O'Neal, and C. A. Lembi. 1987. A model to describe growth of the filamentous alga *Pithophora oedogonia* (Chlorophyta) in an Indiana lake. *J. Aquat. Plant Manage.* 25, 33–40.

Spencer, D. F., T. R. Volpp, and C. A. Lembi. 1980. Environmental control of *Pithophora oedogonia* (Chlorophyceae) akinete germination. *J. Phycol.* 16, 424–7.

Stokes, P. M. 1979. Copper accumulations in freshwater biota. In *Copper in the Environment I. Ecological Cycling,* ed. J. O. Nriagu, pp. 357–81. John Wiley, New York.

Stom, D. J., S. S. Timofeeva, N. F. Kashina, L. J. Bielykh, S. N. Souslov, V. V. Boutorov, and M. S. Apartzin. 1980. Methods of analyzing quinones in water and their application in studying the effects of hydrophytes on phenols. V.

Elimination of carcinogenic amines from solutions under the action of *Nitella*. *Acta Hydrochim. Hydrobiol.* 8, 241–5.

Taft, C. E. 1975. History of *Cladophora* in the Great Lakes. In *Cladophora in the Great Lakes*, ed. H. Shear and D. E. Konasewich, pp. 9–15. International Joint Commission Regional Office, Windsor, Ont. 179 pp.

Taft, C. E., and W. J. Kishler. 1973. *Cladophora as Related to Pollution and Eutrophication in Western Lake Erie.* Proj. Compl. Rep. no. 332X, 339X. Water Resources Center, Ohio State University, Columbus.

Trudeau, P. 1982. *Nuisance Aquatic Plants and Aquatic Plant Management Programs in the United States.* Vol. 3, *Northeastern and North Central Region.* Publ. no. MTR-82W47-03. U.S. Environmental Protection Agency and the Mitre Corp., McLean, VA. 157 pp.

Tubea, B., K. Hawxby, and R. Metha. 1981. The effects of nutrient, pH, and herbicide levels on algal growth. *Hydrobiologia* 79, 221–7.

Tucker, C. S., R. L. Busch, and S. W. Lloyd. 1983. Effects of simazine treatment on channel catfish production and water quality in ponds. *J. Aquat. Plant Manage.* 21, 7–11.

Valin, D. J. 1985. Herbivorous fish permitting update. *Aquatics* 7(4), 13–14, 21.

Van, T. K., W. T. Haller, and G. Bowes. 1976. Comparison of the photosynthetic characteristics of three submersed aquatic plants. *Plant Physiol.* 58, 761–8.

Van Schayck, C. P. 1986. The effect of several methods of aquatic plant control on two Bilharzia-bearing snail species. *Aquat. Bot.* 24, 303–9.

Van Zon, J. C. J. 1977. Grass carp (*Ctenopharyngodon idella* Val.) in Europe. *Aquat. Bot.* 3, 143–55.

Whitaker, J., J. Barica, H. Kling, and M. Buckley. 1978. Efficiency of copper sulfate in the suppression of *Aphanizomenon flos-aquae* blooms in prairie lakes. *Envir. Poll.* 15, 185–94.

Whitford, L. A., and G. J. Schumacher. 1961. Effect of a current on mineral uptake and respiration by a freshwater alga. *Limnol. Oceanogr.* 6, 423–5.

Whitford, L. A., and G. J. Schumacher. 1964. Effect of a current on respiration and mineral uptake in *Spirogyra* and *Oedogonium*. *Ecology* 45, 168–70.

Whitton, B. A. 1970a. Biology of *Cladophora* in freshwaters. *Water Res.* 4, 457–76.

Whitton, B. A. 1970b. Toxicity of zinc, copper, and lead to Chlorophyta from flowing waters. *Arch. Mikrobiol.* 72, 353–60.

Wiley, M. J., R. W. Gordon, S. W. Waite, and T. Powless. 1984. The relationship between aquatic macrophytes and sport fish production in Illinois ponds: a simple model. *N. Amer. J. Fish. Manage.* 4, 111–19.

Wium-Andersen, S., U. Anthoni, C. Christophersen, and G. Houen. 1982. Allelopathic effects on phytoplankton by substances isolated from aquatic macrophytes (Charales). *Oikos* 39, 187–90.

Wong, S. L., B. Clark, M. Kirby, and R. F. Kosciuw. 1978. Water temperature fluctuations and seasonal periodicity of *Cladophora* and *Potamogeton* in shallow rivers. *J. Fish. Res. Bd. Can.* 35, 866–70.

THE FUTURE OF ALGAE IN HUMAN AFFAIRS

19. Algae in space

ROBERT A. WHARTON, JR.*

Life Science Division, Ames Research Center, National Aeronautics and Space Administration, Moffett Field, CA 94035

DAVID T. SMERNOFF†

Complex Systems Research Center, University of New Hampshire, Durham, NH 03825

MAURICE M. AVERNER

Biological Research Branch, National Aeronautics and Space Administration, Washington, DC 20546

CONTENTS

* Present address: Department of Biology and Desert Research Institute, University of Nevada, Reno, NV 89557

† Present address: Life Science Division, MS 239-4, NASA-Ames Research Center Moffett Field, CA 94035

[485]

I. Introduction

The year is 2025 and the research expedition *Thule* has established an outpost on Mars. In the closed atmosphere of their protective underground habitat, the crew utilizes algae to remove carbon dioxide and replace it with oxygen. In the food production module, algae are grown and processed to provide a protein source for the crew. The expedition's biologist has just returned from a reconnaissance of the polar ice cap looking for algae growing in the ice. Another team of scientists will soon depart to search for algal fossils in the 3.5 billion-year-old sediments of the Valles Marineris. This scenario is not science fiction but one of many possible examples of the role of algae in space exploration.

As suggested in the above example, the role of algae in space involves their use in human life support systems during long-duration space flights and/or in support of extraterrestrial human outposts and as subjects of studies of the presence of extant and extinct life forms on planetary bodies (i.e., planetary biology). In life support systems, algae have the potential to provide air revitalization and food. In planetary biology, algae are frequently used as model organisms for experimental and theoretical studies of the potential of extraterrestrial habitats (e.g., Mars and Europa, a moon of Jupiter) to support life forms.

This chapter discusses the role of algae in life support systems and planetary biology. We will consider these topics from both historical and state-of-the-art perspectives. Finally, we conclude with a discussion of future research directions.

II. Algae and life support systems

Upon leaving the relative security of the earth's surface, humans require technologically sophisticated equipment to provide basic necessities of life, namely, air, water, and food. Also important is the capability to remove waste products (e.g., CO_2, feces, and urine). On short-duration space flights, such as *Apollo, Skylab,* and space shuttle missions, life support requirements are met by a combination of physical and chemical

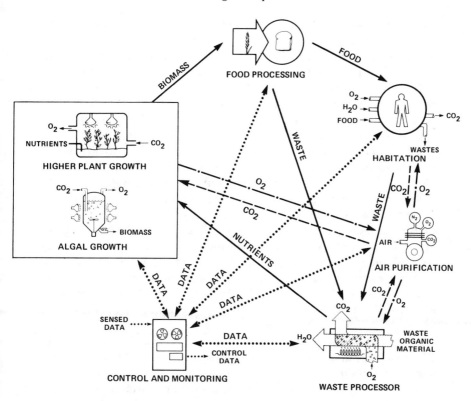

Fig. 19-1. Conceptual diagram of a bioregenerative life support system showing interactions among systems that include higher plants and algae, food processing, human habitation, air purification and waste processing, and control and monitoring.

techniques, and the entire complement of air, water, and food is carried aboard the spacecraft at launch. During these missions waste products are stored for subsequent disposal upon return to earth.

Longer-duration missions, such as an orbiting space station or lunar and Martian bases, would require resupply of air, water, and food and recycling of wastes. As in most human endeavors, economics quickly enters into the picture. For example, it would be expensive to launch and send adequate supplies of air, water, and food for a three-year round-trip mission to Mars. An attractive, less costly alternative to the reliance on resupply from earth is to develop a recycling life support system (Fig. 19-1) based on biological processes to regenerate oxygen, water, and food from waste materials (Gustan and Vinopal 1982; MacElroy 1983). Such a bioregenerative life support system is envisaged to include photosynthetic organisms (i.e., higher plants and algae) for air revitalization and food production and a waste-processing system to regenerate potable water for the crew and CO_2 and nutrient solutions for the higher plants and algae. Both biological and physical/chemical methods for processing waste materials are currently under investigation. These include activated sludge techniques, wet-air oxidation, and supercritical water oxidation (Oleson et al. 1986). Ideally, a bioregenerative life support system conserves mass and requires only an energy source to operate. Figure 19-2 shows an example of a ground-based research facility designed to study algae in a bioregenerative life support system. In this system, designed to support two people, algae remove CO_2, provide O_2, and grow on processed human wastes.

Research on bioregenerative life support systems began in the 1950s when the possibility of prolonged space missions raised the question of whether a biologically based life support system could be developed. Much of the early work conducted by the U.S. Air Force, National Aeronautics and Space Administration (NASA), and aviation companies has been reviewed by Taub (1974). The extensive Soviet program has been summarized by Gitelson et al. (1975; Gitelson 1977). In 1978, NASA established a research program called the Controlled Ecological Life Support System (CELSS) program to integrate efforts toward developing a bioregenerative life support system. The current status of the CELSS program and its research is described in MacElroy, Smernoff, and Klein (1985).

Algae are potentially important elements in research on bioregenerative life support systems (Myers 1954; Krauss 1968; Taub 1974). As early as 1953, Bowman suggested that algal systems for atmosphere control and food would be of use for extended space travel. Several features of algae, particularly blue-green (Cyanophyta) and green (Chlorophyta) algae, make them attractive candidates for inclusion in a bioregenerative system. These algae typically grow rapidly, have metabolisms that can be controlled, produce a high ratio of edible to nonedible biomass, and have gas-exchange characteristics compatible with human requirements. In addition, many strains of blue-green algae convert gaseous nitrogen to ammonium, a biologically useful form of nitrogen.

Unfortunately, a number of potential problems are associated with the use of algae in life support systems (Taub 1974; Averner, Karel, and Radmer 1984). Such problems include adequacy and acceptability of algal-derived food, harvesting and processing algae, and long-term stability of algal cultures. Nonetheless, scientists generally agree that algae will continue to play an important role in the development of bioregenerative life support systems and will have an actual role in systems used in space (Averner, Karel, and Radmer 1984).

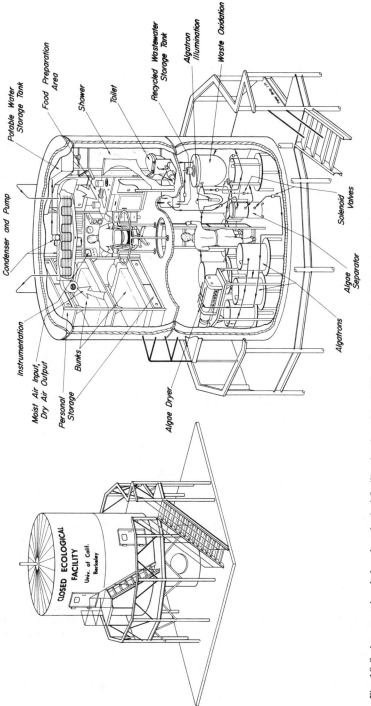

Fig. 19-2. An operational closed ecological facility developed by William Oswald at the University of California, Berkeley. This facility incorporates the essential elements of a bioregenerative life support system, including algae grown in "algatrons." A system such as this might be used as a ground-based demonstrator prior to developing flight hardware. (Used with permission.)

A. Algae and air revitalization

As photosynthetic organisms, algae are potential key components in bioregenerative life support systems for maintaining proper concentrations of O_2 and CO_2 during long-duration space missions. Algae utilize CO_2 from the spacecraft or habitat and replace it with photosynthetically derived O_2. Although algae do use a small amount of O_2 for respiration, there will be a net gain of O_2. In effect, algae act as photosynthetic gas exchangers with the crew. Desired atmospheric gas concentrations to support a crew range from 0% to 0.4% for CO_2 and from 18.4% to 21.8% for O_2. Approximate values for loss of human function are greater than 3% for CO_2, and below 13% and above 37% for O_2 (Miller and Heppner 1985).

Most studies to date have involved experiments on gas exchange between algae and mice, not humans. Early experiments quickly showed that the atmospheres of gas-closed algae/heterotroph systems are unstable. For example, Doney and Myers (1958) conducted one of the first quantitative studies coupling photosynthesis and respiration in a gas-closed system with two mice and *Chlorella ellipsoidea* Gerneck. Their 30-day experiment showed CO_2 levels varying between 0.5% and 1% with O_2 levels rising from 21% to 25%. In gas-exchange experiments coupling *Anacystis nidulans* (Richt.) Drouet and Daily with mice, photosynthesis exceeded respiration, and O_2 concentration in the system increased by more than 30% (Gafford and Craft 1959).

Bowman and Thomae (1960) utilized a gas-closed system in which *Chlorella pyrenoidosa* Chick and mice were grown in separate vessels but shared a common atmosphere. Of the eight experiments attempted, five were terminated after an average duration of seven days due to an unacceptable accumulation of CO_2. Data on the most successful trial of 28 days indicated that the atmospheric concentration of CO_2 increased from 0.03% to a final concentration of 0.6%, and O_2 concentration decreased from 20.9% to 18.0%. Additional experiments by Bowman and Thomae (1961) included a 66-day run in which atmospheric O_2 concentration exhibited a steady increase from an initial concentration of 21% to a final concentration of 63%. Carbon dioxide concentration ranged from an initial concentration of 0.4% to a final concentration of 0.13% with a maximum of 5%. No toxic effects were observed on either the algae or the mice.

In a more complete and sophisticated series of gas-exchange experiments, pure cultures of *Chlorella ellipsoidea* and a dwarf variety of mouse were maintained and monitored in a closed chamber (Eley and Myers 1964). The contribution of bacterial activity to gas exchange due to culture contamination or accidental wetting of accumulated mouse feces by urine was minimized by careful design. In addition, the integrity of the experimental system was examined for leaks. Although attempts were made to match more closely the absolute rates of gas uptake and release by algae and the mouse, as well as gas-exchange ratios, oxygen concentration invariably increased suggesting the potential for deleterious consequences (e.g., toxic effects on humans at O_2 concentrations greater than 37%).

An extensive study of the behavior of a gas-closed system containing mice, algae, and bacteria, carried out by Golueke and Oswald (1963) and Oswald, Golueke, and Horning (1965), initially resulted in rapid increases in CO_2 and decreases in O_2. Subsequent modifications of algal growth conditions allowed the system to perform satisfactorily for up to six weeks. During this time, O_2 concentration fluctuated in the mid-20% range and CO_2 remained below 1%.

Zuraw (1962) reported gas exchange experiments involving *Chlorella pyrenoidosa* and a cebus monkey in a gas-closed system. During a run of 50 hours the O_2 concentration increased from 21% to 25%, and the CO_2 concentration cycled between 0.2% and 1%. The rates of change varied with the activity of the monkey. When the monkey was awake and active, O_2 concentration decreased with a concomitant rise in CO_2. When the monkey was asleep, the concentration of O_2 increased steadily and CO_2 remained relatively constant at a low level.

These examples of gas-exchange experiments utilizing algae illustrate that in the course of a gas-closed run fluctuations in CO_2 and/or O_2 concentrations occur in the system's atmosphere. The causes of these fluctuations can be traced to one or several of the following mechanisms: (1) an imbalance between the rates of photosynthesis and respiration (rate instability); (2) a mismatch between the assimilatory quotient (AQ = moles CO_2 consumed/moles O_2 produced) of the algae and the respiratory quotient (RQ = moles CO_2 produced/moles O_2 consumed) of the heterotroph (ratio instability); (3) anaerobic and/or aerobic respiration by bacteria in the system; (4) leakage in the closure of the system caused either by the lack of balance between inputs (e.g., food) and removals (e.g., algal harvest) or by physical leaks in the experimental apparatus. For example, Gafford and Craft (1959) attributed fluctuations in their system to lack of balance between photosynthesis and respiration. Doney and Myers (1958), Bowman and Thomae (1960, 1961), Zuraw (1962), and Eley and Myers (1964) attributed the lack of stability in their systems to mismatch between AQ and RQ.

Recently, we conducted an experimental and modeling study to investigate atmosphere behavior in a gas-closed algal/mouse system (Averner, Moore, et al. 1984; Smernoff, Wharton, and Averner 1987). These studies focused on the dynamics of gas exchange between *Chlorella pyrenoidosa* and a dwarf mouse under a variety of algal growth conditions including differing optical densities, irradiance levels, and nitrogen

EXPERIMENTAL MOUSE-ALGAL SYSTEM

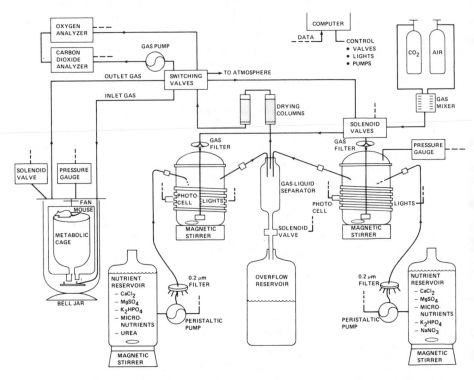

Fig. 19-3. Diagram of experimental apparatus used to study atmospheric behavior and environmental control techniques in a gas-closed algae/mouse system. During closed experiments gases are circulated through the system without further input from cylinders. (From Smernoff et al. 1987.)

sources in the algal medium. The experimental system (Fig. 19-3) was composed of three gas-tight reactors, two that supported continuous algal growth and one that housed a mouse. Results from our experiments showed that the dynamics of gas exchange can be controlled by manipulating the algal cultures. For example, Figure 19-4 indicates the effect of varying the irradiance of algal cultures on overall system gas dynamics. At the beginning of this run, irradiance was $700 \ \mu E \cdot m^{-2} \cdot s^{-1}$. Under these conditions O_2 percentage in the gas stream increased and CO_2 decreased. At 2.1 hours the light was reduced to $170 \ \mu E \cdot m^{-2} \cdot s^{-1}$, resulting in a reduced photosynthesis rate and a rapid drop in O_2 percentage and increase in CO_2 percentage. At 3.4 hours light was increased to $395 \ \mu E \cdot m^{-2} \cdot s^{-1}$, and the rate of change of O_2 and CO_2 percentage in the gas stream was reduced. Results such as these suggest

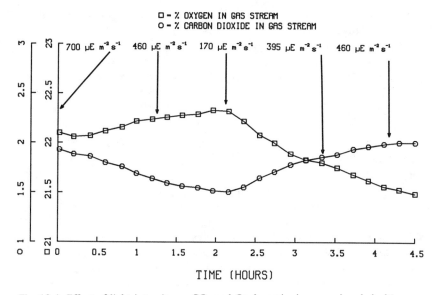

Fig. 19-4. Effect of light intensity on CO_2 and O_2 dynamics in a gas-closed algal/mouse system. (From Smernoff 1985.)

that gas concentrations in a gas-closed algal/mouse system can be controlled by manipulating the algal environment (by varying light levels in this case).

What about using algae to supply the air revitalization requirements of a life support system for humans? Bovee et al. (1962) studied the change in gas concentration in a gas-closed *Chlorella pyrenoidosa*/human system. Both O_2 and CO_2 concentrations fluctuated throughout the 56-hour experiment with O_2 levels dropping toward the end of the experiment. Soviet workers (Kirensky et al. 1967, 1968), reporting on gas exchange between *Chlorella vulgaris* Beij. and a human in a gas-closed system, attempted to control the imbalance between algal AQ and human RQ by modifying the human's diet. Their results showed a dependence of gas concentrations on the degree of human activity. Specifically, CO_2 accumulated and O_2 decreased during the day, and CO_2 decreased and O_2 increased at night. They reported relative stability of the gas concentrations within their system over the 30-day period. Unfortunately, the description of the apparatus and methodology used in these experiments is not adequate to judge the significance of the results.

No toxicological effects from these algae/human experiments were reported. However, the accumulation of trace contaminants from biological and nonbiological sources may occur within a gas-closed system (Garavelli 1985; Schwartz and Oldmark 1986). Some of these trace contaminants are toxic if accumulated over time.

In spite of this rather limited data set, air revitalization performed by algae in a life support system for humans seems probable. A human in space will require somewhere between 700 and 836 g of O_2 per day and produce 840 to 1,000 g of CO_2 per day (Adams 1963; Gustan and Vinopal 1982). A review by Miller and Ward (1966) of numerous algal growth systems showed that the requirements of an algal reactor to provide gas exchange for one human varied from 3.5 to 3,000 liters of algal culture, from 2.5 to 71 m^2 reactor surface area, and from 7 to 100 kw electrical power. In order to optimize reactor design, dense cultures requiring large surface areas but small volumes should be compared to less dense cultures requiring a smaller surface area but greater volumes (Taub 1974). Using an apparatus called a Recyclostat, Krauss (1978) reported that 436 liters of *Chlorella vannielli* S. & K. grown at an optical density of 1.0 would be required to support the daily O_2 requirements of a human. Using *C. pyrenoidsa*, we determined that 280 liters of algal culture growing at an optical density of 1.7 are required to yield 700 g of O_2 per day (Smernoff et al., unpublished data). We expect that culture volume can be further reduced by optimizing growth conditions and reactor design.

Clearly, more experiments are required on algal/human air revitalization systems that incorporate environmental control techniques such as those suggested by Averner, Moore, et al. (1984) and Smernoff et al. (1987). Also, consideration must be given to developing algal reactors that conform to the expected volume and surface area constraints of a proposed space habitat (e.g., space station). Experiments also must be conducted in space to test the effect of the space environment on algal physiology as well as to test the effectiveness of reactor design.

B. Algae as space food

Much of the work on algae as a food source in bioregenerative life support systems, recently summarized by Kamarei, Nakhost, and Karel (1985), demonstrates the challenge of producing safe, nutritious, and acceptable fabricated foods from algae in space. Algal biomass is envisaged as providing single-cell protein, lipid, and/or carbohydrate to supplement the diet of a crew during a long-duration space mission (Karel 1980; Karel and Kamarei 1984).

Research on algae as a food source has centered primarily on species of *Spirulina*, *Chlorella*, and *Scenedesmus*. As pointed out by Kamarei et al. (1985), direct consumption of algae without purification is not feasible because of physiological (e.g., nondigestible components of the cell wall, excess nucleic acids, unknown toxins and allergins) and organoleptic (e.g., unpleasant flavor, color, and texture) problems. These problems must be overcome in order to utilize algal biomass as a food source. Kamarei et al. (1985) developed methodology to eliminate or largely reduce three undesirable components (cell walls, nucleic acids, and pigments) in *Scenedesmus obliquus* (Turp.) Kütz. A protein isolate with a reasonably high yield was obtained. Cell suspensions were subjected to homogenization with glass beads to disrupt the cell walls. After freeze-drying and rehydration the suspension was centrifuged to remove cell walls, organelles, and fragments. The remaining supernatant was treated with RNase and DNase and centrifuged to remove soluble nucleotides, peptides, and free lipids. The final step was removal of pigments and lipid with an ethanol extraction. The yield of algal protein isolate was 36.6% with a protein concentration of 70.5% and a protein recovery of 49.1%. Although the total recovery and protein concentration seem reasonable, Kamarei et al. (1985) suggest that attempts should be made to increase these values.

Maximizing algal productivity and biomass also is a major concern if algae are to be used as a food source. To this end, Radmer et al. (1985) established a series of long-term cultures of *Scenedesmus obliquus* in an annular airlift column operated as a turbidostat. The glass apparatus consisted of three concentric, cylindrical chambers, the innermost chamber housing the light source, the middle chamber containing the algal culture, and the outer chamber acting as a temperature-regulated water jacket. A mathematical model describing the relationship of culture productivity to cell concentration of light-limited cultures was developed. Predicted productivity as a function of cell concentration agreed well with experimental data, suggesting that models can permit accurate predictions of culture productivity given the growth parameters of the culture system.

The issue of maximizing culture productivity and biomass may at times run counter to the use of algal cultures for air revitalization. Indeed, gas-exchange requirements may dictate that productivity be minimized. A single algal culture system could provide both functions; however, a single system would have to be well regulated to balance food and air requirements. Two culture systems may be necessary, one for producing algal biomass for food and another for revitalizing the atmosphere in the space habitat.

Will algae be part of a space explorer's diet? The answer is a cautious yes. Given that algae will probably be utilized in an air revitalization system, it seems reasonable that if an edible product can be processed the excess algal biomass from the air revitalization system can be recycled at least partially in the form of edible biomass. Toward this end, it is important to (1) establish the feasibility of extracting purified edible components from algal biomass, (2) establish the feasibility of converting these components to food, (3) modify relevant processes for space conditions, and (4) analyze the relationship between the degree of biomass utilization and the weight of equipment required for growth and processing (Averner, Karel, and Radmer 1984).

C. Algae in the space environment

Relatively few studies have focused on the effects of the space environment on algae. Effects of space radiation and prolonged weightlessness or microgravity on growth and development of algae are of primary importance in the bioengineering evaluation of algae for human life support systems (Miller and Ward 1966; Ward, Wilks, and Craft 1970). Phillips (1962), Ward and Guerra (1962), and Ward and Phillips (1968) studied the relatively short-term (50–72 hours) effects of a space capsule (*Discoverer* satellites) environment on photosynthetically inactive cultures of *Chlorella pyrenoidosa*. Due to power and volume limitations on these satellites a lighting system was not available. These authors detected no significant genetic or physiological changes in the cultures upon return to earth. Similar results were obtained by Soviet workers Semenenko and Vladimirova (1963) and Meshcheryakova et al. (1967). In these studies, radiation exposures were well below those known to be detrimental to organisms. The problem of harmful ionizing radiation may best be avoided by housing algal culture systems in areas of the spacecraft or habitat protected by special radiation-shielding materials.

At present there are few data concerning the effect of microgravity on algae. Indeed, more studies focusing on the physiology and maintenance of algal culture systems in microgravity are urgently needed. Averner, Karel, and Radmer (1984) proposed a possible scenario for testing algae in microgravity. Following ground-based physiological characterization of a variety of algae, multiple cultures of these algae could be placed in earth orbit. Cultures would be retrieved from orbit at six-month intervals, returned to earth, and subjected to physiological characterization to determine any changes resulting from exposure (6 months, 1 year, 1.5 years, etc.) to microgravity. For longer-term missions that incorporate algae in life support systems, backup cultures of algae will be necessary on board, since it is unlikely that a single continuous culture will operate optimally and without accident for the duration of a mission.

Microgravity may have no effect on algal physiology per se; however, long-term exposure to the space environment may affect culture stability or viability. For example, the space shuttle cabin atmosphere contains volatile organic compounds outgassed from structural materials (Schwartz and Oldmark 1986). Continual exposure of algal cultures to such compounds may allow accumulation to toxic levels. Other potential problems include mutagenic effects of ionizing radiation, microbial contamination of the algal cultures, and inefficient heat and mass transfer. Biological parameters that might be affected include rates of growth, photosynthesis, and respiration; algal chemical composition (proteins, lipids, and carbohydrates); and excretion of organic and inorganic compounds by the algae. To assess the long-term behavior of algal cultures, space flight experiments lasting for several hundred algal generations are required.

An algal growth reactor is of central importance to successful utilization of algae in a bioregenerative life support system in space. Some of the engineering problems in developing a reactor include gas–liquid separation, behavior and transport of culture medium, cell adherence to surfaces, optimal lighting techniques and configuration, and heat transfer minimization. Present ground-based algal growth reactors use processes that are dependent on gravity such as gas-bubble sparging and mixing and overflow harvesting. These processes will have to be replaced with functionally analogous processes that can operate in the space environment. Thus, the effects of microgravity on particle, fluid, and gas-bubble behavior, as they affect culture aeration and mixing, must be determined. Other considerations include reactor size and weight, culture illumination, and reactor monitoring and control. Similar problems, especially the effects of microgravity on both particle and fluid properties and flow behavior, are associated with the harvesting and processing of algae for food.

III. Algae and planetary biology

Planetary biology (or exobiology) is the study of the origin, evolution, and distribution of life and life-related molecules throughout the universe (De Vincenzi 1984). Algae are important in planetary biology from both terrestrial and extraterrestrial perspectives. On earth, fossil blue-green algae (cyanobacteria) in Precambrian stromatolites represent one of the earliest evidences of life (see also Harlin and Darley, Chapter 1) and consequently are important in studying the emergence of life on earth (Schopf 1983). Such algae frequently are used as model organisms for experimental and theoretical studies of the potential of extraterrestrial environments to support life.

Mars is probably the most often cited extraterrestrial environment with the potential for algal growth. Averner and MacElroy (1976) examined the possibility of using Mars as a habitat for terrestrial life, including humans. They suggested that blue-green algae or lichens could be seeded on Mars with the objective of producing an O_2 atmosphere. A strategy was proposed that called for an increase in Martian atmospheric pressure by volatilizing water and CO_2 from the northern polar ice cap (Fig. 19-5). A water vapor cloud would be formed that, together

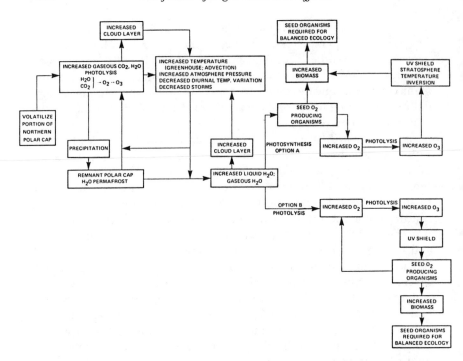

Fig. 19-5. Proposed strategy for alteration of Martian atmosphere to produce an environment suitable for terrestrial life, including humans. (From Averner and MacElroy 1976.)

with the thermal properties of the increased CO_2 concentration, would cause a temperature increase via a "greenhouse" effect. Once pressure and temperature were higher and more water was present in the atmosphere, UV photolysis was expected to cause an increase in the ozone layer density, thus decreasing the UV flux at the Martian surface. At that point, implanted autotrophic organisms would photosynthesize, gradually causing an increase in atmospheric O_2 concentration and a decrease in CO_2.

Data from the *Viking* missions to Mars suggest that limitations in the concentrations of available nitrogen and water appear to eliminate the possibility for ecosynthesis on Mars as a whole. Nevertheless, Kuhn, Rogers, and MacElroy (1979) developed an energy balance model to investigate how the Martian atmospheric environment could influence a community of photosynthetic microorganisms with properties similar to those of a blue-green algal mat and a lichen. They calculated that a colony of Martian photosynthetic organisms near the equator would have photosynthetic rates of 10^{-6}–10^{-7} mole $O_2 \cdot cm^{-2} \cdot y^{-1}$. However, the specific environmental parameters to include in the model were major uncertainties. Kuhn et al. (1979) suggested that data obtained by subjecting a colony of photosynthetic organisms (e.g., blue-green algal mat) to anaerobic conditions simultaneously with continued low-level UV radiation might improve their model.

Establishment of a human outpost on Mars in the early 2000s has been predicted (McKay 1985). While the *Viking* missions gave valuable insight into the biological potential on Mars, a more extensive search for extant Martian life forms must be conducted before humans land on the surface. The possibility that life (algae or otherwise) could exist in an undiscovered cryptic niche will continue to intrigue those who desire absolute answers from biology. Furthermore, the biological impact of a localized human presence on (as yet undiscovered) Martian organisms must be defined. It will be important to determine the required level of search using automated landing vehicles and samplers to insure that any undiscovered cryptic Martian community is so well hidden that it is unlikely to be affected by a human presence.

These considerations underscore the necessity of understanding the ecology of organisms that occupy cryptic niches in hostile environments on earth (McKay, 1986). Examples with relevance to Mars include the study of algae in endolithic habitats of the Antarctic dry valleys (Friedmann 1982; McKay and Friedmann 1985; Friedmann and Weed 1987) and on the surface of glaciers (Wharton et al. 1985). Do these habitats and the algae they contain represent analogues for life in rocks or in the polar ice caps on Mars? The cryptoendolithic microorganisms (including algae and fungi) in the Antarctic dry valleys survive in an inhospitable environment without actually adapting to its extremes (Friedmann 1982). Within a largely hostile macroclimate, these organisms find refuge inside rocks where conditions for life exist (Fig. 19-6). Depressions on the surface of glaciers, called cryoconite holes, also represent a potential habitat analogue for Martian life. Algae and other microorganisms are commonly found in cryoconite holes (Fig. 19-7). It is possible that life exists (or at one time existed) within the surface of Martian rocks and/or on the Martian polar ice caps.

Another area where algae may be important on Mars concerns the search for fossils of an ancient Martian biota. The Martian environment in the past was considerably less hostile biologically, and it might possibly have permitted the origin and transient establishment of a biota (Mazur et al. 1978). For example, the lower part of the Valles Marineris canyon system on Mars (Fig. 19-8), estimated to be 3.5 billion years old (Carr 1981), contains sedimentary deposits that exhibit rhythmic horizontal layering suggestive of deposition in standing water (Carr 1983; Lucchitta and Ferguson 1983; Nedell, Squyres, and Anderson, 1987). McKay et al. (1985) recently developed a model that predicts that paleolakes on Mars could have been ice-covered. The suggestion (Wallace and

Fig. 19-6. Cryptoendolithic lichen (dark band within sandstone matrix) from Antarctic dry valleys. December 1982. Magnification: ×4.5. (Courtesy of E. I. Friedmann.)

Sagan 1979; Carr 1983) that lakes on Mars would freeze solid is tempered by this model, which indicates that ice thickness could be limited by ablation. Ice-cover thicknesses could have ranged from 65 to 650 m using present Martian equatorial climate data and abalation rates of 1–10 $cm \cdot y^{-1}$. In addition to possible warmer temperatures in early Martian history, relatively higher ablation rates could result from downslope winds along canyon walls producing thinner ice covers (McKay et al. 1985). The possibility exists that sediments from these paleolakes contain fossils of life evolved between 3 and 4 billion years ago on Mars.

The only terrestrial analogue of these purported ancient Martian environments are the perennially ice-covered Antarctic lakes. They have a 3–5 m thick perennial ice cover that allows penetration of less than 3% of incident photosynthetically active radiation to the water below (Palmisano and Simmons, 1987). The presence of a perennial ice layer also restricts gas exchange between the water column and the atmosphere. Biologically important gases such as nitrogen and oxygen concentrate in the water column to supersaturated levels (Wharton et al. 1986, 1987). In addition to providing a relatively warm, liquid water environment, the process of concentrating atmospheric gases beneath the ice cover could have significantly affected the gas budget in a Mar-

Fig. 19-7. Cryoconite holes of the surface of Canada Glacier, Southern Victoria Land, Antarctica. Lake Hoare is in upper left of photo. Ice ax is 70 cm long. December 1979.

tian paleolake, possibly enhancing the levels of biologically important gases from the thin Martian atmosphere (Wharton et al. 1987).

The dominant life form in the Antarctic lakes is algae (Fig. 19-9) in the form of benthic mats (Wharton, Parker, and Simmons 1983). Do the sediments in Valles Marineris contain fossil algae? The answer may be forthcoming when sediment samples from the Valles Marineris are returned and analyzed from a proposed Mars sample-return mission in the 1990s (McKay 1985). In the meantime, research on algae in Antarctic lakes may provide valuable lessons for interpreting observations on Martian samples.

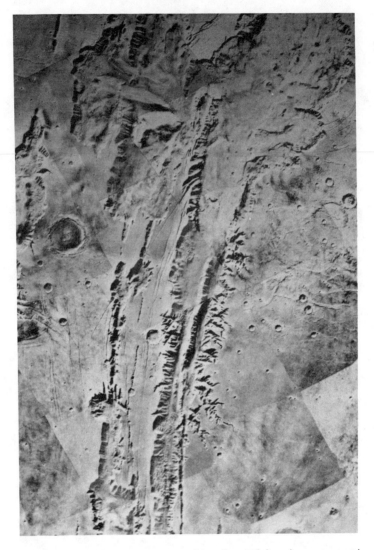

Fig. 19-8. Valles Marineris canyon system on Mars. Layered deposits are present in each of the canyons. Scale 1:2,000,000. (From USGS Map MC-19 NW. Composite image from *Viking 2* photographs, 1976.)

Algae also have been used to consider the biological potential of the Jovian moon Europa. Tidal and radiogenic heating of Europa leading to the formation and maintenance of a liquid water ocean overlaid by a relatively thin ice crust has been suggested (Squyres et al. 1983). If such conditions currently exist, the amount of calculated photosynthetically active radiation (0.05% of incident light between 400 and 700 nm at the

Fig. 19-9. Benthic algal mats in perennially ice-covered Lake Hoare, Southern Victoria Land, Antarctica. Columnar structures have a maximum vertical height of 0.5 m. December 1980.

ice surface) reaching the proposed liquid water ocean through the ice crustal fractures on Europa is sufficient to support algal mats growing in Antarctic lakes as well as sea-ice algal communities on earth (Reynolds et al. 1983). Using the characteristics of sea-ice diatoms, Reynolds et al. (1983) estimated the maximum biomass potentially supported by oxygenic photosynthesis in an ocean on Europa to be $1.1–3.4 \times 10^{-6}$ g· m^{-2}, or a total of $4.6 \times 10^{7}–1.4 \times 10^{8}$ g dry wt. Although these calculations cannot establish whether life forms exist or could exist on Europa, they do suggest that regions on Europa may provide physical conditions within the range of adaptation of life on earth.

IV. The future of algae in space

In years to come, the role of algae in space most likely will continue to focus on the use of algae in bioregenerative life support systems. As previously mentioned, algae as well as higher plants are envisioned to provide air revitalization and food. Some major obstacles, requiring both basic and applied research, must be overcome. These include increasing the efficiency of photosynthesis, developing an algal reactor system suitable to the space environment, and developing extraction and processing procedures that result in a nutritious and edible food. Eventually, research should be directed at integrating the "algal system" into the overall bioregenerative life support system, that is, determining how the algal system will link up with the waste-processing system, the higher plant system, and the crew's habitat.

Many of the developmental processes associated with utilizing algae in the space environment will require experimentation in space. A rigorous program of space flight experimentation aboard the space shuttle and space station will be necessary to obtain a "space rating" for any bioregenerative life support system to be used for flights with humans.

In the next 20 to 30 years, algal systems (combined with the rest of a life support system) may be engineered to support a small crew of less than 10 people for a period of up to five years. Such a system will be characterized by minimal reliance on resupply from earth, a high degree of control, and a minimal amount of adaptability. However, by the middle of the next century, the exploration and possible settlement of other planets (e.g., Mars) will require a greater diversity and adaptability for the algal systems utilized. For example, a settlement on Mars with 1,000 humans will probably require large-scale algal reactors, greenhouses, and animal farms. The role of algae conceivably could be greatly expanded. Algae might be required to fix atmospheric nitrogen and possibly serve as a nutrient-rich fertilizer, as animal feed, and as feedstock for the manufacture of industrial products. In this regard, genetic engineering of algae may play a key role in permitting successful settlement of a planet such as Mars. Will it be possible to engineer strains of blue-green algae genetically to fix N_2 under Martian conditions? Is it possible to utilize algae (and other microorganisms) to produce the building blocks (e.g., oils and waxes) of petrochemical products (e.g., fuels) from Martian natural resources? Meyer (1981) presented a conceptual design for the extraction and processing of consumables (e.g., O_2, H_2O, organic compounds) from Martian resources for a human research station. Algae may play a key role in this process.

In planetary biology studies, certain algal habitats, such as perennially ice-covered Antarctic lakes, rocks, and cryoconite holes on glaciers, will continue to serve as models for use in determining the site of future missions to the surface of Mars. Our knowledge of the diversity and adaptability of algae in unusual and extreme environments on earth is preliminary at best. Studies of these habitats also may be useful in the development of procedures for examining Martian samples for fossil evidence following a sample-return mission.

Another future role of algae in planetary biology might focus on adapting algae to space and extraterrestrial environments. For example, it may be possible to develop strains of algae capable of growing unprotected on Mars or Venus. These algae could then be used as a resource and periodically harvested as a "natural" resource. Over many thousands of years their metabolic activity could help to alter the extraterrestrial environment to a relatively more hospitable one. It is also conceivable, albeit somewhat controversial, that specially adapted algae would function as inocula for a human-directed panspermia. Imagine algal spores set free in the interstellar medium to drift where they may, or a space capsule full of blue-green algae heading toward a planet around a distant star. Upon impact the algae are spread over the planet's surface and begin fixing N_2, CO_2, and producing O_2. Over a period of years, the planet's environment becomes altered in such a way that it would be more suitable to colonization by humans and other forms of terrestrial life, possibly following later in another space vessel. The future of algae in space is limited only by our willingness to explore the universe and our ability to learn from it.

V. Acknowledgments

This work was supported in part by NASA Grants NCC-2-027 and NCA2-2, and a National Research Council/NASA Fellowship Award to Robert A. Wharton. We are grateful to C. P. McKay, R. Mancinelli and C. R. Fassi for timely review and discussion of the manuscript.

VI. References

Adams, C. C. 1963. Nutritional aspects of space flight. In *Medical and Biological Problems of Space Flight*, ed. G. H. Bourne, pp. 237–44. Academic Press, New York.

Averner, M., M. Karel, and R. Radmer. 1984. *Problems Associated with the Utilization of Algae in Bioregenerative Life Support Systems*. NASA Contractor Rep. 166615. 21 pp.

Averner, M. M., and R. D. MacElroy. 1976. *On the Habitability of Mars: An Approach to Planetary Ecosynthesis*. NASA Special Publication 414. 105 pp.

Averner, M. M., B. Moore, III, I. Bartholomew, and R. Wharton. 1984. Atmosphere behavior in gas-closed mouse-algal systems: an experimental and modelling study. *Adv. Space Res.* 4, 231–9.

Bovee, H. H., A. J. Pilgrim, L. S. Sun, J. E. Schubert, T. L. Eng, and B. J. Benishek. 1962. Large algal systems. In *Biologistics for Space Systems Symposium*, Report no. AMRL-TDR-62-116, Wright-Patterson Air Force Base, OH.

Bowman, N. J. 1953. The food and atmosphere control problem in space vessels. II. The use of algae for food and atmosphere control. *J. Brit. Interplanetary Soc.* 12, 159–67.

Note: NASA publications available from National
Technical Information Service, Springfield, VA
22161.

Bowman, N. J., and F. W. Thomae. 1960. An algae life support system. *Aerospace Eng.* 19, 26–30.

Bowman, N. J., and F. W. Thomae. 1961. Long term non-toxic support of animal life with algae. *Science* 134, 55–6.

Carr, M. H. 1981. *The Surface of Mars.* Yale University Press. New Haven, CT. 232 pp.

Carr, M. H. 1983. Stability of streams and lakes on Mars. *Icarus* 56, 476–95.

De Vincenzi, D. L.' 1984. NASA's Exobiology Program. *Origins of Life* 14, 793–9.

Doney, R. C., and J. Myers. 1958. Quantitative repetition of the Priestly experiment. *Plant Physiol.* 33 (Suppl.), xxv.

Eley, J. H., Jr., and J. Myers. 1964. Study of a photosynthetic gas exchanger, a quantitative repetition of the Priestly experiment. *Texas J. Sci.* 26, 296–333.

Friedmann, E. I. 1982. Endolithic microorganisms in the Antarctic cold desert. *Science* 215, 1045–53.

Friedmann, E. I., and R. Weed. 1987. Microbial trace fossil formation, biogenesis, and abiotic weathering in the Antarctic cold desert. *Science* 236, 703–5.

Gafford, R. D., and C. E. Craft. 1959. *A Photosynthetic Gas Exchanger Capable of Providing for the Respiratory Requirements of Small Animals.* USAF School Aviation Med. Report no. 58-124. Brooks Air Force Base, TX.

Garavelli, J. S. 1985. Airborne trace contaminants of possible interest to CELSS. In *Controlled Ecological Life Support System: CELSS 1985 Workshop*, ed. R. D. MacElroy, N. V. Martello, and D. T. Smernoff, pp. 253–62. NASA Technical Memorandum 88215.

Gitelson, I. 1977. *Problems of Creating Biotechnical Systems of Human Life Support.* NASA Technical Translation F-17533. 337 pp.

Gitelson, I., B. G. Kovrov, G. M. Lisovsky, Y. N. Okladnikov, M. S. Rerberg, F. Y. Sidko, and I. A. Terskov. 1975. *Problems in Space Biology.* Vol. 28, *Experimental Ecological Systems Including Man.* NASA Technical Translation F-16933. 347 pp.

Golueke, C. G., and W. J. Oswald. 1963. Closing of an ecological system consisting of a mammal, algae, and non-photosynthetic organisms. *Amer. Biol. Teacher* 25, 522–8.

Gustan, E., and T. Vinopal. 1982. *Controlled Ecological Life Support System: Transportation Analysis.* NASA Contractor Report 166420. 127 pp.

Kamarei, A. R., Z. Nakhost, and M. Karel. 1985. *Potential for Utilization of Algal Biomass for Components of the Diet in CELSS.* Society of Automotive Engineers, Tech. Paper Ser., no. 851388. Fifteenth Intersociety Conference on Environmental Systems, San Francisco, July 15–17. 10 pp.

Karel, M. 1980. Problems of food technology in space habitats. In *Human Factors of Outer Space Production*, ed. T. S. Cheston and D. L. Winter, pp. 147–56. AAAS Selected Symposium no. 50. Westview Press, Boulder, CO.

Karel, M., and A. R. Kamarei. 1984. *Feasibility of Producing a Range of Food Products from a Limited Range of Undifferentiated Major Food Components.* NASA Contractor Rep. 177329. 115 pp.

Kirensky, L. V., I. A. Terskov, I. I. Gitelson, G. M. Lisovsky, B. G. Kovrov, and Y. N. Okladnikov. 1968. Experimental biological life support system. II. Gas exchange between man and microalgae culture in a 30-day experiment. In *Life Sciences and Space Research,* ed. A. H. Brown and F. G. Favorite, pp. 37–40. North Holland, Amsterdam.

Kirensky, L. V., I. A. Terskov, I. I. Gitelson, G. M. Lisovsky, B. G. Kovrov, F. Y. Sidko, Y. N. Okladnikov, M. Antonyuk, V. N. Belyanin, and M. S. Rerberg. 1967. Gas exchange between man and a culture of microalgae in a 30-day experiment. *Kosmicheskaya Biol. Med.* 1, 23–8.

Krauss, R. W. 1968. The physiology and biochemistry of algae, with special reference to continuous-culture techniques for *Chlorella.* In *Bioregenerative Systems,* pp. 97–109, NASA Special Publication 165. 153 pp.

Krauss, R. W. 1978. Closed ecology in space from a bioengineering perspective. In *Life Sciences and Space Research,* ed. R. Holmquist, pp. 13–26. Pergamon Press, New York.

Kuhn, W. R., S. R. Rogers, and R. D. MacElroy. 1979. The response of selected terrestrial organisms to the martian environment: a modeling study. *Icarus* 37, 336–46.

Lucchitta, B. K., and H. M. Ferguson. 1983. Chryse Basin channels: low gradients and ponded flows. *J. Geophys. Res.* 88, 553–86.

MacElroy, R. D. 1983. Farming in space. In *Agriculture in the Twenty-First Century,* ed. J. W. Rosenblum, pp. 157–65. Wiley-Interscience, New York.

MacElroy, R. D., D. T. Smernoff, and H. Klein. (eds.). 1985. *Life Support Systems in Space Travel.* NASA Conference Publication 2378. 70 pp.

McKay, C. P. (ed.). 1985. *The Case for Mars.* American Astronautical Society vol. 62. Science and Technology Series, Univelt, San Diego, CA. 716 pp.

McKay, C. P. (ed.). 1986. Exobiology and future Mars missions: the search for Mars' earliest biosphere. *Adv. Space Res.* 6:269–85.

McKay, C. P., G. A. Clow, R. A. Wharton, Jr., and S. W. Squyres. 1985. Thickness of ice on perennially frozen lakes. *Nature* 313, 561–2.

McKay, C. P., and E. I. Friedmann. 1985. The cryptoendolithic microbial environment in the Antarctic cold desert: temperature variations in nature. *Polar Biol.* 4, 19–25.

Mazur, P., E. S. Barghoorn, H. O. Halvorson, T. H. Jukes, I. R. Kaplan, and L. Margulis. 1978. Biological implications of the Viking mission to Mars. *Space Sci. Rev.* 22, 33–4.

Meshcheryakova, L. K., M. G. Petrovnin, I. S. Sakovich, and V. A. Shevchenko. 1967. Study of the development of *Chlorella* during space flight. *Envir. Space Sci.* 1, 169–71.

Meyer, T. R. 1981. Extraction of Martian resources for a manned research station. *J. Brit. Interplanetary Soc.,* 34, 285–8.

Miller, C. W., and D. B. Heppner. 1985. *Space Station Environmental Control/Life Support System Engineering.* Society of Automotive Engineers Tech. Paper Ser., no. 851375, Fifteenth Intersociety Conference on Environmental Systems, San Francisco, July 15–17. 11 pp.

Miller, R. L., and C. H. Ward. 1966. Algal bioregenerative sytems. In *Atmosphere in Space Cabins and Closed Environments,* ed. K. Kammermeyer, pp. 186–222. Appleton-Century-Crofts, New York.

Myers, J. 1954. Basic remarks on the use of plants as biological gas exchangers in a closed system. *J. Aviation Med.* 24, 407–11.

Nedell, S., S. W. Squyres, and D. W. Andersen. 1987. Origin and evolution of the layered deposits in the Valles Marineris, Mars. *Icarus* 70:409–41.

Oleson, M., F. Slavin, R. Liening, and R. Olson. 1986. *Controlled Ecological Life Support Systems (CELSS): Physiochemical Waste Management Systems Evaluation.* NASA Contractor Report 177422. 133 pp.

Oswald, W. J., C. G. Golueke, and D. O. Horning. 1965. Closed ecological systems. *J. San. Eng. Div., ASCE* 91, 23–46.

Palmisano, A. C., and G. M. Simmons, Jr. 1987. Spectral downwelling irradiance in an Antarctic lake. *Polar Biol.* 7:145–51.

Phillips, J. N. 1962. Experiments with photosynthetic micro-organisms. In *Radiobiologic Experiments in Discoverer Satellite XVII,* ed. G. W. Crawford, USAF Report 62-67. Brooks Air Force Base, TX.

Radmer, R., P. Behrens, E. Fernandez, and K. Arnett. 1985. An analysis of the productivity of a CELSS continuous algal culture system. In *Controlled Ecological Life Support System: CELSS 1985 Workshop,* ed. R. D. MacElroy, N. V. Martello, and D. T. Smernoff, pp. 315–28. NASA TM-88215.

Reynolds, R. T., S. W. Squyres, D. S. Colburn, and C. P. McKay. 1983. On the habitability of Europa. *Icarus* 56, 246–54.

Schopf, J. W. (ed.). 1983. *Earth's Earliest Biosphere.* Princeton University Press, Princeton, NJ.

Schwartz, M. R., and S. I. Oldmark. 1986. *Analysis and Composition of a Model Trace Gaseous Mixture for a Spacecraft.* Society of Automotive Engineers Tech. Paper Ser., no. 860917, Sixteenth Intersociety Conference on Environmental Systems, San Diego, July 14–16. 12 pp.

Semenenko, V. Ye., and M. G. Vladimirova. 1963. Effect of spaceflight conditions in the satellite on the preservation of the viability of *Chlorella* cultures. In *Problems of Space Biology,* vol 1, ed. N. M. Sesakyan. USSR Academy of Science Publishing House, Moscow, and NASA Tech. Translation F-174, Washington, DC.

Smernoff, D. T. 1985. *Atmosphere Stabilization and Element Recycling in an Experimental Mouse-algal System.* M.S. thesis. University of San Francisco. 93 pp.

Smernoff, D. T., R. A. Wharton, Jr., and M. M. Averner. 1987. Operation of an experimental gas exchanger for use in a CELSS. *Adv. Space Res.* 7:17–27.

Squyres, S. W., R. T. Reynolds, P. M. Cassen, and S. J. Peale. 1983. Liquid water and resurfacing on Europa. *Nature* 301, 225–6.

Taub, F. B. 1974. Closed ecological systems. *Ann. Rev. Ecol. Syst.* 5, 139–60.

Wallace, D., and C. Sagan. 1979. Evaporation of ice in planetary atmospheres: ice-covered rivers on Mars. *Icarus* 39, 385–400.

Ward, C. H., and C. N. Guerra. 1962. Growth of photosynthetic microorganisms following orbital space flight. In *Biological Systems of Discoverer Satellites XXIX and XXX,* ed. J. E. Prince. USAF Report 62-62. Brooks Air Force Base, TX.

Ward, C. H., and J. N. Phillips. 1968. Stability of *Chlorella* following high-altitude and orbital space flight. *Dev. Ind. Microbiol.* 9, 345–54.

Ward, C. H., S. S. Wilks, and H. L. Craft. 1970. Effects of prolonged near weightlessness on growth and gas exchange of photosynthetic plants. *Dev. Ind. Microbiol.* 11, 276–95.

Wharton, R. A., Jr., C. P. McKay, G. M. Simmons, Jr., and B. C. Parker, 1985. Cryoconite holes on glaciers. *BioScience* 35, 499–503.

Wharton, R. A., Jr., C. P. McKay, G. M. Simmons, Jr., and B. C. Parker. 1986. Oxygen budget of a perennially ice covered Antarctic lake. *Limnol. Oceanogr.* 31, 437–43.

Wharton, R. A., Jr., C. P. McKay, R. L. Mancinelli, and G. M. Simmons. 1987. Perennial N₂ supersaturation in an Antarctic lake. *Nature* 325, 343–5.

Wharton, R. A., Jr., B. C. Parker, and G. M. Simmons, Jr. 1983. Distribution, species composition, and morphology of algal mats in Antarctic dry valley lakes. *Phycologia* 22, 355–65.

Zuraw, E. A. 1962. Algae-primate gas exchange in a gas closed system. *Dev. Ind. Microbiol.* 3, 140–9.

20. The genetic improvement of algae: progress and prospects

JOHN P. VAN DER MEER

Atlantic Research Laboratory, National Research Council of Canada, Halifax, Nova Scotia B3H 3Z1, Canada

CONTENTS

Contribution No. 29008 of the National Research Council of Canada.

[511]

I. Introduction

Before considering what impact genetics and plant breeding may have on algae in the next few decades, it is useful to put the subject into context, with respect to both historical patterns of crop development and the present state of algal domestication. The development of algae is likely to parallel that of terrestrial systems, and an appreciation of where algae currently fit into the general pattern of development will help in estimating where they are likely to be early in the next century.

A recurrent theme in the domestication of wild species is ever-increasing human control over reproduction, including phenotypic selection of the parents, which leads to extensive changes in the original genotype. The process generally proceeds as follows. An organism becomes a candidate for domestication because it is discovered to have a useful property. After the initial discovery, similar organisms are examined for that property, and eventually a certain species becomes recognized as the best source. This species then becomes the most desirable, and if natural supplies are not sufficient, managed propagation is initiated, which signals the beginning of domestication.

Informal selection of superior individuals in cultivated populations soon follows. In addition to selecting the most desirable genotypes found in the natural population, new genotypes arising through mutation may also become important. Under human management, a species no longer has the same constraints it had in the wild. Thus, in the early stages of domestication, mutations that would have been deleterious or even lethal in nature may be selected by growers because they are advantageous from a human perspective (e.g., the nonshattering cob of *Zea mays*). Selection of radical mutations in crop lines obviously has important consequences for the subsequent development of the domesticated varieties. Historically, it is at approximately this stage of development that more systematic genetic improvement of a species takes place, with selection for characteristics such as yield, disease resistance, uniformity, hardiness, and time to maturity. Domestication of many terrestrial crops has proceeded for thousands of years to arrive at the present condition; however, for some recently domesticated organisms, the process has been shortened to just a few decades, thanks to rapid progress made possible by twentieth-century knowledge of genetics and breeding.

With the advent of gene isolation and cloning techniques, the development of plasmid vectors for transferring genes from one organism to another, and related developments that permit the manipulation of single cells or protoplasts of multicellular organisms, exciting new possibilities, unforeseen a few decades ago, have suddenly emerged. Although these new techniques remain largely experimental, they are powerful tools with tremendous potential for rapid genetic manipulation of a species. Their ultimate impact is difficult to predict, but they undoubtedly will find many imaginative applications in future years.

Where then, do the algae fit into the historical pattern of crop domestication and genetic improvement that has just been described? In brief, they have been ignored almost completely. With a few recent exceptions, algal species have not been domesticated, a situation not altogether surprising. Only a few of the earth's total number of plant and animal species are used by humans for food or other products, still fewer have actually been domesticated, and essentially all of the latter are terrestrial. It is probable there was little motivation to domesticate algae while terrestrial resources were sufficient to satisfy basic needs and enough wild algae could be found to supply that more limited demand. Certainly, algal species would have been much more difficult to learn to manage than terrestrial plants and animals, which are more accessible and have more obvious reproductive structures. The complexity of algal life histories and their reproduction by means of microscopic spores probably placed them beyond the farming capabilities of ancient peoples.

What will the future bring? With the full awareness that predictions can be no more than imperfect projections, based on our knowledge of what has already occurred and what is happening at present, and that unforeseeable, unprecedented events may change everything, I shall (1) present an overview of where we are now with genetic manipulation of algae and (2) attempt to make some reasonable projections, with an emphasis on commercial utilization.

II. Classical plant breeding

A. *Porphyra*

Domestication of *Porphyra* strains in Japan has progressed rapidly during the past few decades, and commercial cultivars of *P. yezoensis* and *P.*

tenera now differ noticeably from wild plants in average size and growth rate. The general life history of this genus became understood about 35 years ago when the connection between *Porphyra* blades and shell-boring *Conchocelis* filaments was recognized (Drew 1949). Up to that point growers attempted to get good crops by obtaining their "seed" in localities where populations of *Porphyra* were known from experience to yield superior plants; however, significant strain selection was precluded due to the lack of knowledge. (See Mumford and Miura, Chapter 4, for references and a more complete discussion of the history of *Porphyra* cultivation.) After Drew's discovery it became possible to control reproduction more completely, and "seeding" of cultivation nets from conchocelis grown in shells in culture soon became normal practice for the industry (see Mumford and Miura, Chapter 4). With this maricultural revolution, selection of genetically superior strains was greatly facilitated (Miura 1975, 1979) because large, robust fronds discovered by growers during cultivation could become the source for improved lines of conchocelis. In addition, because the cultivated fronds of *Porphyra* are monoecious as well as haploid, direct selection through homozygous conchocelis could be obtained readily. The entire next generation of plants "seeded" from such conchocelis inherits the superior genotype, and the cumulative effect after several rounds of selection has been impressive. Fronds of some domesticated strains can reach a meter in length, which is three to five times the average length of their wild relatives. The increased size of these fronds apparently results from the selection of genotypes with delayed fertility (Miura 1975). This allows a longer growth period before the fronds begin to erode as a consequence of gamete and spore release.

Because genetic improvement of *Porphyra* is still quite recent in historical terms, the current breeding objectives remain similar to those of the past; however, the cultivation success has been such that the nori supply has exceeded demand for several years, and as a result the focus has shifted from increased production to increased efficiency and product quality. One manifestation of this is the selection of strains that produce monospores that can compensate for an initial inadequate seeding of the nets. Diseases remain a major concern, and thus disease-resistant lines are another obvious goal (see Mumford and Miura, Chapter 4). Japanese scientists and breeders have been collecting *Porphyra* species from around the world, some of them presumably with the objective of finding strains with commercially interesting properties; however, it is difficult to get information about selection progress. Strain selection at present remains largely in the hands of individual growers. Organized breeding efforts are not yet designed to benefit the industry as a whole, although this may change with the restructuring of the industry that is underway in Japan (see Mumford and Miura, Chapter 4). There has been some experimentation with cross-fertilization to obtain interspecific hybrids, but this appears an unlikely route for success since the hybrids are often unstable (Suto 1963). Recent new information on the position of meiosis in the life cycle of *Porphyra* (Miura and Kunifuji 1980; Ma and Miura 1984; Ohme, Kunifuji, and Miura 1986) helps to explain some of this hybrid instability and should also put future selection programs on a better footing.

B. Laminaria japonica

The Chinese have made dramatic improvements in cultivated strains of *Laminaria japonica*. (See Druehl, Chapter 5, for information on the history of kelp utilization in China and Japan.) In early quantitative genetic studies, they demonstrated that variations in frond length, frond width, frond thickness, and stipe length (important properties of cultivated kelp) were highly heritable (Fang, Jiang, and Li 1962, 1965, 1966; Tseng 1981). Since that time Chinese scientists and breeders have successfully tapped this genetic variation for the improvement of their cultivated strains (Fang et al. 1966; Anonymous 1976). Through recurrent cycles of inbreeding and selection, they have obtained stable varieties with very long fronds and increased iodine concentration. They are now widely grown along the coast of China (Anonymous 1976; Tseng 1981). Because these lines are inbred many generations, they breed reasonably true, making it possible to maintain the strain by simple sexual propagation.

Haploid sporophytes with normal morphology (but smaller than diploids) have been obtained from the Chinese strains of *Laminaria japonica* by parthenogenesis (Fang et al. 1978; Fang 1984). In addition, completely homozygous, diploid sporophytes can be obtained by spontaneous chromosome doubling during the early stages of parthenosporophyte development. These are normally fertile but have reduced growth rate. Because *Laminaria* has a heteromorphic life history that includes filamentous gametophytes, it is ordinarily not possible to determine the sporophytic frond characteristics of the haploid genome except by the "breeding value" calculated from growth trials of families of sporophytic progeny. With the ability to form monoploid or homozygous diploid sporophytic fronds by parthenogenesis, the breeding potential of the female genome can be estimated directly from the growth characteristics of the parthenosporophyte. This new monoploid breeding approach is still experimental but recently has been used in conjunction with other selection techniques to obtain a new variety that is claimed to be excellent (Fang et al. 1983).

The most recent new cultivar of *L. japonica* reported by the Chinese was obtained by hybridization between a female gametophytic clone derived from Chinese material and a male clone obtained from Japan

(Fang et al. 1985). This hybrid is reported to be extremely vigorous, with luxuriant growth, strong holdfasts, and high iodine content. The superior growth of the hybrid has been attributed to hybrid vigor, but it remains to be demonstrated that this is heterosis in the strict sense: that it is unfixable by inbreeding and selection. If it is true heterosis, commercial exploitation will require asexual propagation of the two component gametophytes in large amounts to obtain the gametes necessary for sporophyte production. This could be considerably more expensive and challenging than the maintenance of existing inbred lines. Nevertheless, it would probably be easier than vegetative propagation of desirable sporophytes by tissue culture techniques.

C. Other seaweeds

For the tropical, carrageenan-producing red alga *Eucheuma,* simple genetic selection was incorporated into farming recommendations made to growers, namely, that the largest, most robust plants should be used as inoculum for new nets (Doty 1978). A specific improvement attributable to this type of selection scheme was the successful spread of an extremely vigorous and nonsporulating clone of *Eucheuma* known in the Philippines as "Tambalang" (Doty and Alvarez 1975). Other commercially cultivated algae such as species of *Gracilaria* and *Undaria* must be under at least informal selection for agronomic traits at the present time (such selection is essentially unavoidable for cultivated strains), and there may be conscious efforts to select superior plants for propagation; however, no plant-breeding data for these species have appeared in the open literature.

Robust clones have been selected from wild populations of a few additional species that might be considered for farming under appropriate economic conditions; however, these selections have generally been made in the context of academic research projects and have not led to successful commercial cultivation or further attempts at genetic improvement. Examples of this type of selection exist for *Chondrus crispus* (Neish and Fox 1971; Cheney, Mathieson, and Schubert 1981), *Gigartina exasperata* (Waaland 1979), *Gracilaria tikvahiae* (Ryther et al. 1979), and *Gracilaria sjoestedtii* (Hansen 1984).

Attempts have been made to improve the growth of *Gracilaria tikvahiae* by constructing autopolyploids, but these experiments did not yield any lines that grew faster than normal plants (Patwary and van der Meer 1984). In general, triploid sporophytes grew as well as diploids whereas most tetraploids grow more slowly. No significant differences were observed between polyploids from a single line and hybrid polyploids with genomes from different populations. In another experiment, normal diploid hybrids obtained by crossing moderately inbred lines of *G. tikvahiae* also showed no evidence of hybrid vigor (Patwary and van der

Meer 1983b) raising the possibility that heterosis is of minor significance in some algae.

Mutants have proved useful for basic phycological studies (van der Meer 1979, 1986) but have not yet made any major contribution toward the commercial development of algae. Only some color variants of *Porphyra* are being cultivated commercially, and these on a very small scale (Aruga and Miura 1984). A much-branched Mendelian mutant of *Gracilaria tikvahiae* has somewhat faster growth than related plants with normal morphology and has some other interesting properties (Patwary and van der Meer 1982, 1983a, c; Craigie, Wen, and van der Meer 1984), but the yield improvements can be matched by selected, unrelated, wild-type lines and are not sufficient to justify commercial development.

In recent years there have been several studies on wide crosses of seaweed species. It appears to be difficult to obtain interspecific hybrids of most red algae, and the few that have been obtained are sterile or yield abnormal progeny. Most red algal species appear to have strict compatibility requirements between male and female gametes for successful fertilization. Even within currently recognized species, subgroupings sometimes do not hybridize with each other (e.g., *Laurencia pacifica* [Howard, Nonomura, and Fenical 1980], *Gigartina stellata* [Guiry and West 1983]) or produce hybrids with greatly reduced fertility (e.g., *Palmaria palmata* [van der Meer, 1987]). In contrast to the red algae, many interspecific and even intergeneric hybrids of brown algae have been reported (e.g., Sanbonsuga and Neushul 1978, 1979; Cosson and Olivari 1982); however, these hybrids usually grow rather poorly and often abnormally. When tested, they have been sterile. At present no indication exists that wide crosses of this type will yield hybrids of commercial interest for red or brown algae.

D. Microalgae

Cultivation of unicellular algae for human food or other products such as fine chemicals has been attempted a number of times in the past few decades but, generally speaking, these efforts have limited financial success owing to the high costs of production and harvesting in relation to the value of the product. However, there is continuing research and some commercial aquaculture activity for microalgae such as *Chlorella* (green) and *Spirulina* (blue-green), whose products range from fine chemicals to health foods (see Jassby, Chapter 7). Other microalgae such as *Dunaliella* (green) have received considerable attention in recent years as a possible source of carotenoids and petroleum substitutes (Borowitzka, Borowitzka, and Moulton 1984; Spencer, Chapter 10). The use of microalgae as feed in fish and shellfish farming has also led to substantial cultivation of certain species (De Pauw, Morales, and

Persoone 1984). In spite of this activity, the actual domestication of microalgae is still at the point of trying to identify species and strains with desirable characteristics. Genetic improvement of the cultivated species beyond simple strain selection does not appear to have been undertaken and indeed would have been nearly impossible for species lacking sexuality. Extensive genetic studies have been conducted on the green microalgal genus *Chlamydomonas* (e.g., Hudock and Rosen 1976), but these have not had commercial application thus far. Nevertheless, as will be discussed in the following section, these fundamental studies may have significant impact for practical breeding in future years.

E. Future directions for plant breeding

For the near future, classical plant breeding with selection under cultivation conditions will continue to be the most effective method for obtaining, evaluating, and improving varieties of algae. Even strains modified by genetic engineering (see Section III) will have to be selected properly if they are to be grown successfully. As long as the present use of algae continues and new uses are discovered, it can be expected that the wild plant resources will not be enough and that currently cultivated species will continue to be improved. Additional species also will be domesticated and improved with time. If government agencies or large private corporations become involved to a greater extent in the development and genetic improvement of algal crops (as they have for terrestrial crops and also to some extent for *Porphyra* and *Laminaria*), it is likely that commercial characteristics such as yield, product quality, resistance to adverse environmental conditions, and resistance to pathogens will be improved more rapidly than otherwise. Certainly the active participation of a well-positioned company would be necessary to bring sophisticated propagation techniques (such as vegetative amplification by means of isolated cells (see Section III) to market. For any domesticated algal species, varieties with different properties will probably be selected for different growing regions or different applications; this already has been done for *Porphyra*, and other species surely will follow. Terrestrial crops were, and continue to be, improved in these ways, and there is every reason to think that parallel developments will occur for algae with sufficient economic attraction.

III. Molecular genetics

Progress in molecular genetics over the past 10 years has been spectacular. New techniques using reliable and highly specific DNA probes that permit the identification and isolation of individual genes have led to breakthroughs in biological problems that seemed intractable only a few years ago. They also have provided direct new methods that do not depend on the existence of a sexual cycle for manipulating the genome. Although research on the molecular genetics of algae is presently restricted almost entirely to a small number of unicellular forms, particularly in the genus *Chlamydomonas* (e.g., Rochaix 1981; Lemieux et al. 1985), this situation is beginning to change. Experiments on other algae have begun and, although at present it is not clear what directions these will take, it can be expected that in the future molecular genetic techniques will be applied to commercially valuable species with increasing frequency.

A. Characterization of organelle DNA

The hypothesis that plastids and mitochondria evolved from free-living endosymbiotic prokaryotes (see Harlin and Darley, Chapter 1) has provided a focus for a substantial proportion of the current research. Restriction enzymes and radioactive DNA probes are being used by a number of groups to map specific genes and restriction sites on the organelle genomes of algae. Available data on the organization of organelle DNA were reviewed fairly recently (Loiseaux and Dalmon 1983; Palmer 1985a, b).

To date most of the molecular studies on chloroplasts used terrestrial plants or green algae, and tremendous progress has been made in understanding the organization of these extranuclear genomes (Palmer 1985). Very recently it was announced that a Japanese group determined the complete nucleotide sequence for chloroplast DNA of *Marchantia polymorpha* (Ohyama et al. 1986), with reference to the fact that another Japanese group had accomplished the same feat for tobacco chloroplast DNA! Breaking with tradition, some investigators have recently begun to examine the very different plastid systems found in nongreen groups of algae. This transfer of attention has been rewarded with some surprising and exciting new discoveries.

One of the most significant discoveries is that the two genes encoding the subunits of ribulose-1,5-bisphosphate carboxylase (RuBPCase), which fixes carbon dioxide in photosynthesis, are not always encoded as they are in green plants. The RuBPCase enzyme consists of small and large subunits, and in all higher plants and green algae that have been examined the gene encoding for the small subunit is located in the nucleus whereas the gene for the large subunit is found in the plastid DNA (Palmer 1985). Both genes are, of course, in the genome of cyanobacteria, the living relatives of ancient forms thought to be the ultimate origins of all photosynthetic organelles. In contrast to the situation for green plants, the RuBPCase small subunit gene is encoded in plastid DNA in such diverse systems as the plastids of the red algae *Porphyridium aerugineum* and *Cyanidium caldarium* (Steinmüller, Kaling, and Zetsche 1983), the cyanelle of the protist *Cyanophora paradoxica* (Heinhorst and Shively 1983), and the plastid of the chrysophyte *Olisthodiscus luteus* (Reith and Cattolico 1986). Thus, the transfer of the small subunit gene from plastid to nucleus, which occurred in the evolution of green plants, apparently did not occur in all lines of photosynthetic eukaryotes.

The genes for the subunits of the phycobiliproteins are now also known to be located in the plastid genomes of a number of the organisms mentioned above (Egelhoff and Grossman 1983; Steinmüller et al. 1983; Lemaux and Grossman 1984). DNA sequence data reveal that the genes for the two subunits of phycocyanin are located in adjacent coding regions within the genome of the blue-green alga *Agmenellum quadruplicatum* (de Lorimier et al. 1984; Pilot and Fox 1984). The same general arrangement of these genes was reported in the cyanelle DNA of *Cyanophora paradoxica* (Lambert et al. 1985; Lemaux and Grossman 1985), along with the locations of more than 20 different genes on the cyanelle plastid DNA (Lambert et al. 1985). Although there are some parallels with the chloroplast genomes of green plants for overall genome structure, the placement of specific genes is quite dissimilar. From the above and other recent data (e.g., Fain and Druehl 1984; Bergmann et al. 1985; Schantz 1985), it is becoming clear that a great deal of variability exists among plastid genomes of different groups.

Algal genomes will continue to have an impact as objects of research in the future, and it is in this area that algae may make their most meaningful, even if noncommercial, contribution. The evolution and interrelationship of living organisms make a fascinating subject area and remain an important focus for modern biological research. The characterization of nuclear and organelle genomes will play an increasing role in these studies because DNA is an enormous and still largely untapped reservoir of evolutionary information. For questions relating to the origins and evolution of organelles, the algae have the most diversified systems for exploration and, if we estimate from the speed with which data are becoming available, it is likely that a fairly complete picture of the origins and evolution of various plastids will be available in less than a decade. Information on the nuclear component of cells will take longer to accumulate but, by analyzing carefully selected genes, considerable progress can be expected here as well. It is to be anticipated that this new information will help resolve many of the taxonomic puzzles and disputes that exist today.

Although it is clear from the preceding that parts of some algal genomes are rapidly being characterized in detail, molecular genetic manipulation of algae has not yet arrived and particularly not for the larger commercially important seaweeds. In principle, there is no reason why a seaweed should not be suitable for the application of such techniques, but from a practical point of view too little is known about these systems at the present time. To achieve success at the molecular genetic level in multicellular algae, the ability to manipulate tissues, cells, and protoplasts will likely be of critical importance.

B. Tissues, cells, and protoplasts

The ability to manipulate cells and protoplasts of many higher plant species has improved greatly during the past two or three decades (Davey 1983). Although research on higher plants provides many insights to help guide similar research on algae, there will be many differences, particularly for the marine species. At present the main objective of research on marine algae is still to develop and improve techniques required for *in vitro* manipulations (e.g., Fries 1980; Saga and Sakai 1983; Cheney 1984; Polne-Fuller and Gibor 1984; Yan 1984; Fujita and Migita 1985; Chen 1986; Cheney et al. 1986). The objectives to which the new techniques will be applied remain to be defined, which is probably not surprising if we consider the early stage of the work.

One possibility is that populations of algal cells or protoplasts could be screened for mutations (see Maliga 1984) that confer resistance to such things as heavy metals, herbicides, or pathogens. In at least some higher plant species, spontaneous genetic variability called *somaclonal variation* arises by a number of different mechanisms during somatic growth (Evans, Sharp, and Medina Filho 1984). This somaclonal variation can be used to select improved varieties from cell cultures of established commercial lines without major changes to previously selected genotypes. If somaclonal variation is found in algae, such selection might also prove beneficial for improving established cultivars of seaweeds. In some instances the asexual propagation of an individual thallus by regeneration of plants from its constituent cells might be useful in itself, though it seems doubtful that this technology could be applied economically by growers for routine propagation of commercial cultivars.

Additional opportunities exist for species where plants can be regenerated from protoplasts. The *in vitro* fusion of protoplasts could permit more rapid hybridization in algal species that have long sexual cycles. It could also be used to perform crosses between individuals of asexual species, mutli- or unicellular, although the outcome of such fusions could not be predicted. Doubling the chromosomes in the absence of meiosis might create gene-dosage problems for the alga that would disrupt its normal regulatory and metabolic functions. Wide crosses between two sexually reproducing species that are genetically isolated from each other could be attempted, but here too a variety of problems could ensue if the species are too unrelated for harmonious functioning of the hybrid nucleus. With a well-established protoplast system, organelle substitution or direct gene transfer also becomes a reasonable possibility.

C. Future possibilities for molecular genetic modification of algae

A discussion of the business climate necessary for commercial development of algae and speculations on some of the possibilities "phycotechnology" will bring are presented by Nonomura in Chapter 21. Some of my own speculations on targets for molecular genetics are presented here.

Molecular genetic techniques could facilitate the transfer of genes between two different algal species. As already mentioned, wide crosses involving whole nuclei, whether through sexual crosses or protoplast fusion, often yield inviable or weak hybrids. The transfer of small pieces of DNA is likely to cause fewer difficulties and may be a better method for the introduction of new variability into a species. With modern DNA technology it is at least theoretically possible to make very wide crosses, that is, to transfer any gene of any organism to any other organism. In a recent example of this type of research a mutant gene conferring resistance to the herbicide glyphosate in the bacterium *Salmonella typhimurium* was transferred to tobacco plants using a T-DNA-based vector and was found to confer similar resistance in the new location (Comai et al. 1985). It is possible to imagine that algae could be made herbicide-resistant and then kept free of competing algae in the cultivation system. Similarly, it might be desirable to breed heavy-metal resistance into algae that are to be grown in polluted water. This type of resistance is available on certain bacterial plasmids, and perhaps it will be possible to transfer appropriately modified versions of these to algae. At present no "shuttle" vectors are sufficiently developed for use with eukaryotic algae, but a beginning has been made. Modified bacterial plasmids containing algal DNA sequences required for self-replication in *Chlamydomonas* (Rochaix, Van Dillewijn, and Rahire 1984) or in yeast (Vallet, Rahire, and Rochaix 1984; Loppes, Denis, and Bairiot-Delcoux 1985) have been constructed and partially characterized.

Unicellular algae (often asexual) used to feed shellfish or even algae used for human food could be genetically modified to improve their nutritional value. By amplifying the most limiting dietary component in the algal cell through the action of specific genetic modifications, the alga would become a richer food source. Microalgae cultivated for petroleum substitutes could be modified to yield a higher lipid content. In more general terms, a molecular approach could be used to improve the quality of an alga wherever the action of individual genes affecting the target characteristic can be identified. Much remains to be learned, however, before it becomes possible to insert a new gene into an algal genome in such a way that it functions properly and does not impair the activity of the host genome.

For some applications it would be beneficial to remove certain enzymatic activities from an alga. For example, when a commercially valuable product undergoes unwanted biochemical modifications or when too much of the cell's resources is diverted to storage products of no commercial interest, elimination of the unwanted enzymatic activity would be beneficial. Techniques now exist for the modification of isolated genes in vitro (Botstein and Shortle 1985). It would be possible to prevent formation of an unwanted enzyme by isolating the normal gene for that enzyme and replacing it with an inactive, mutated version.

Algae could become sources as well as recipients of genes in the biotechnology of tomorrow. They might be screened for useful genes that could improve the functioning of nonalgal organisms such as terrestrial crops. Genes coding for stable enzymes that could be used for extracellular modification of specific substrates might also be sought. It is unlikely that the alga itself would be cultivated in such instances since the gene coding for the desired enzyme could be isolated from a cloned DNA "library," maintained and amplified by standard molecular genetic techniques. Although it will be difficult for most phycologists to perceive the traditional beauty and complexity of their organisms in such a development, the "alga" of molecular genetics will likely be a population of DNA fragments packaged in bacteriophage coats!

IV. Summary

It is likely that many different genetic approaches, alone or in combination, will be used for future improvements of algae. The precise methods adopted will of course depend on the nature of the improvement being sought, on the sophistication of techniques available for a particular organism, and in some cases on the properties of the organism itself. An increasing contribution from molecular genetic techniques is to be expected but, for the near future, classical plant breeding will continue to make the largest contribution. The level of future genetic research activity on algae will depend to a large extent on economic opportunities and political factors that cannot be foreseen. It seems reasonable to predict that over the next few decades there will be a tremendous increase in the store of basic knowledge about algae of all sorts and that there will be continued but modest improvement, along with some further diversification, of the cultivated strains used in the algal mariculture industry.

V. References

Anonymous. 1976. The breeding of new varieties of Haidai (*Laminaria japonica* Aresch.) with high production and high iodine content. *Sci. Sin.* 19, 243–52.

Aruga, Y., and A. Miura. 1984. *In vivo* absorption spectra and pigment contents of the two types of color mutants of *Porphyra*. *Jap. J. Phycol.* 32, 243–50.

Bergmann, P., M. Schneider, G. Burkard, J. H. Weil, and J. D. Rochaix. 1985. Transfer RNA gene mapping studies on chloroplast DNA from *Chlamydomonas reinhardii*. *Plant Sci.* 39, 133–40.

Borowitzka, L. J., M. A. Borowitzka, and T. P. Moulton. 1984. The mass culture of *Dunaliella salina* for fine chemicals: from laboratory to pilot plant. *Proc. Int. Seaweed Symp.* 11, 115–21.

Botstein, D., and D. Shortle. 1985. Strategies and applications of *in vitro* mutagenesis. *Science* 229, 1193–201.

Chen, L. C. M. 1986. Cell development of *Porphyra miniata* (Rhodophyceae) under axenic culture. *Bot. Mar.* 29, 435–9.

Cheney, D. P. 1984. Genetic modification in seaweeds: applications to commercial utilization and cultivation. In *Biotechnology in the Marine Sciences*, ed. R. R. Colwell, E. R. Pariser, and A. J. Sinskey, pp. 161–75. John Wiley, New York.

Cheney, D. P., E. Mar, N. Saga, and J. van der Meer. 1986. Protoplast isolation and cell division in the agar-producing seaweed *Gracilaria* (Rhodophyta). *J. Phycol.* 22, 238–43.

Cheney, D., A. Mathieson, and D. Schubert. 1981. The application of genetic improvement techniques to seaweed cultivation. I. Strain selection in the carrageenophyte *Chondrus crispus*. *Proc. Int. Seaweed Symp.* 10, 559–67.

Comai, L., D. Facciotti, W. R. Hiatt, G. Thompson, R. E. Rose, and D. M. Stalker. 1985. Expression in plants of a mutant *aroA* gene from *Salmonella typhimurium* confers tolerance to glyphosate. *Nature* 317, 741–4.

Cosson, J., and R. Olivari. 1982. First results concerning the interspecific and intergeneric hybridization possibilities in Laminariales from the Channel coast. *C. R. Acad. Sci. Ser. III* 295, 381–4. (In French with English abstract.)

Craigie, J. S., Z. C. Wen, and J. P. van der Meer. 1984. Interspecific, intraspecific, and nutritionally-determined variations in the composition of agars from *Gracilaria* spp. *Bot. Mar.* 27, 55–61.

Davey, M. R. 1983. Recent developments in the culture and regeneration of plant protoplasts. In *Protoplasts 1983*, ed. I. Potrykus, C. T. Harms, A. Hinnen, R. Hutter, P. J. King, and R. D. Shillito, pp. 19–29. Birkhäuser Verlag, Stuttgart.

de Lorimier, R., D. A. Bryant, R. D. Porter, W.-Y. Liu, E. Jay, and S. E. Stevens, Jr. 1984. Genes for the α and β subunits of phycocyanin. *Proc. Natl. Acad. Sci. USA* 81, 7946–50.

De Pauw, N., J. Morales, and G. Persoone. 1984. Mass culture of microalgae in aquaculture systems: progress and constraints. *Proc. Int. Seaweed Symp.* 11, 121–34.

Doty, M. S. 1979. Status of marine agronomy, with special reference to the tropics. *Proc. Int. Seaweed Symp.* 9, 35–58.

Doty, M. S., and V. B. Alvarez. 1975. Status, problems, advances, and economics of *Eucheuma* farms. *Mar. Technol. Soc. J.* 9, 30–5.

Drew, K. M. 1949. Conchocelis-phase in the life-history of *Porphyra umbilicalis* (L.) Kütz. *Nature* 164, 748–9.

Egelhoff, T., and A. Grossman. 1983. Cytoplasmic and chloroplast synthesis of phycobilisome polypeptides. *Proc. Natl. Acad. Sci. USA* 80, 3339–43.

Evans, D. A., W. R. Sharp, and H. P. Medina Filho. 1984. Somaclonal and gametoclonal variation. *Amer. J. Bot.* 71, 759–74.

Fain, S. R., and L. D. Druehl. 1984. Isolation and characterization of the chloroplast DNA of *Macrocystis integrifolia*. *Proc. Int. Seaweed Symp.* 11, 603–5.

Fang, T. C. 1984. Some genetic features revealed from culturing the haploid cells of kelps. *Proc. Int. Seaweed Symp.* 11, 317–18.

Fang, T., J. J. Cui, Y. L. Ou, J. X. Dai, M. L. Wang, Q. H. Liu, and Q. M. Yang. 1983. Breeding of the new variety "Danhai No. 1" of *Laminaria japonica* by using a female haploid clone of the kelp. *J. Shandong Coll. Oceanol.* 13, 63–70. (In Chinese with English abstract.)

Fang, T. C., B. Y. Jiang, and J. J. Li. 1962. On the inheritance of stipe length in Haidai (*Laminaria japonica* Aresch.) *Acta Bot. Sin.* 10, 327–35. (In Chinese with English abstract.)

Fang, T. C., B. Y. Jiang, and J. J. Li. 1965. Further studies of the genetics of *Laminaria* frond length. *Oceanol. Limnol. Sin.* 7, 59–66. (In Chinese with English abstract.)

Fang, T. C., B. Y. Jiang, and J. J. Li, 1966. The breeding of a long-frond variety of *Laminaria japonica* Aresch. *Oceanol. Limnol. Sin.* 8, 43–50. (In Chinese with English abstract.)

Fang, T. C., C. H. Tai, Y. L. Ou, C. C. Tsuei, and T. C. Chen, 1978. Some genetic observations on the monoploid breeding of *Laminaria japonica*. *Sci. Sin.* 21, 401–10.

Fang, Z. X. [T. C. Fang], Y. L. Ou, and J. J. Cui. 1985. An experiment on the hybrid vigour in *Laminaria japonica*. *Collected Oceanic Works* 8, 42–5.

Fries, L. 1980. Axenic tissue cultures from the sporophytes of *Laminaria digitata* and *Laminaria hyperborea* (Phaeophyta). *J. Phycol.* 16, 475–7.

Fujita, Y., and S. Migita. 1985. Isolation and culture of protoplasts from some seaweeds. *Bull. Fac. Fish. Nagasaki Univ.* 57, 39–45. (In Japanese with English abstract.)

Guiry, M. D., and J. A. West. 1983. Life history and hybridization studies on *Gigartina stellata* and *Petrocelis cruenta* (Rhodophyta) in the North Atlantic. *J. Phycol.* 19, 474–94.

Hansen, J. E. 1984. Strain selection and physiology in the development of *Gracilaria* mariculture. *Proc. Int. Seaweed Symp.* 11, 89–94.

Heinhorst, S., and J. M. Shively. 1983. Encoding of both subunits of ribulose-1,5-bisphosphate carboxylase by organelle genome of *Cyanophora paradoxa*. *Nature* 304, 373–4.

Howard, B. M., A. M. Nonomura, and W. Fenical. 1980. Chemotaxonomy in marine algae: secondary metabolite synthesis by *Laurencia* in unialgal culture. *Biochem. Syst. Ecol.* 8, 329–36.

Hudock, G. A., and H. Rosen, 1976. Formal genetics of *Chlamydomonas rein-*

hardtii. In *The Genetics of Algae,* Botanical Monographs, vol. 12, ed. R. A. Lewin, pp. 29–48. Blackwell Scientific Publications, Oxford.

Lambert, D. H., D. A. Bryant, V. L. Stirewalt, J. M. Dubbs, S. E. Stevens, Jr., and R. D. Porter. 1985. Gene map for the *Cyanophora paradoxa* cyanelle genome. *J. Bact.* 164, 659–64.

Lemaux, P. G., and A. R. Grossman. 1984. Isolation of genes encoding phycobiliprotein subunits. *Plant Physiol.* 75 (Suppl. 1), 64.

Lemaux, P. G., and A. R. Grossman. 1985. Major light harvesting polypeptides encoded in polycistronic transcripts in a eukaryotic alga. *EMBO* 4, 1911–19.

Lemieux, C., M. Turmel, V. L. Seligy, and R. W. Lee. 1985. The large subunit of ribulose-1,5-bisphosphate carboxylase-oxygenase is encoded in the inverted repeat sequence of the *Chlamydomonas eugametos* chloroplast genome. *Curr. Genetics* 9, 139–45.

Loiseaux, S., and J. Dalmon. 1983. Some characteristics of plastid DNA from different algae. *Physiol. Veg.* 21, 775–84.

Loppes, R., C. Denis, and M. Bairiot-Delcoux. 1985. Isolation of *Cyanidium caldarium* and *Porphyridium purpureum* DNA fragments capable of autonomous replication in yeast. *J. Gen. Microbiol.* 131, 1745–51.

Ma, J. H., and A. Miura. 1984. Observations on the nuclear division in the conchospores and their germlings in *Porphyra yezoensis* Ueda. *Jap. J. Phycol.* 32, 373–8. (In Japanese with English summary.)

Maliga, P. 1984. Isolation and characterization of mutants in plant cell culture. *Ann. Rev. Plant Physiol.* 35, 519–42.

Miura, A. 1975. Studies on the breeding of cultivated *Porphyra* (Rhodophyceae). In *Third International Ocean Development Conference,* pp. 81–93, Aug. 5–8, Tokyo. Reprint vol. 3, Marine Resources.

Miura, A. 1979. Studies on genetic improvement of cultivated *Porphyra* (Laver). In *Proceedings of the Seventh Japan–Soviet Joint Symposium on Aquaculture,* pp. 161–8, Sept. 1978, Tokyo.

Miura, A., and Y. Kunifuji. 1980. Genetic analysis of the pigmentation types in the seaweed susabi-nori *(Porphyra yezoensis). Iden* 34, 14–20. (In Japanese.)

Neish, A. C., and C. H. Fox. 1971. Greenhouse experiments on the vegetative propagation of *Chondrus crispus* (Irish moss). *Tech. Rep. Atl. Reg. Lab. Nat. Res. Counc. Canada* [Halifax] 12, 1–35.

Ohme, M., Y. Kunifuji, and A. Miura. 1986. Cross experiments of the color mutants in *Porphyra yezoensis* Ueda. *Jap. J. Phycol.* 34, 101–6.

Ohyama, K., H. Fukuzawa, T. Kohchi, H. Shirai, T. Sano, S. Sano, K. Umesono, Y. Shiki, M. Takeuchi, Z. Chang, S. Aota, H. Inokuchi, and H. Ozeki. 1986. Chloroplast gene organization deduced from complete sequence of liverwort *Marchantia polymorpha* chloroplast DNA. *Nature* 322, 572–4.

Palmer, J. D. 1985a. Comparative organization of chloroplast genomes. *Ann. Rev. Genetics* 19, 325–54.

Palmer, J. D. 1985b. Evolution of chloroplast and mitochondrial DNA in plants and algae. In *Monographs in Evolutionary Biology: Molecular Evolutionary Genetics,* ed. R. J. MacIntyre, pp. 131–240. Plenum Press, New York.

Patwary, M. U., and J. P. van der Meer. 1982. Genetics of *Gracilaria tikvahiae* (Rhodophyceae). VIII. Phenotypic and genetic characterization of some selected morphological mutants. *Can. J. Bot.* 60, 2556–64.

Patwary, M. U., and J. P. van der Meer. 1983a. Improvement of *Gracilaria tikvahiae* (Rhodophyceae) by genetic modification of thallus morphology. *Aquaculture* 33, 207–14.

Patwary, M. U., and J. P. van der Meer. 1983b. An apparent absence of heterosis in hybrids of *Gracilaria tikvahiae* (Rhodophyceae). *Proc. Nova Scotia Inst. Sci.* 33, 95–9.

Patwary, M. U., and J. P. van der Meer. 1983c. Growth experiments on morphological mutants of *Gracilaria tikvahiae* (Rhodophyceae). *Can. J. Bot.* 61, 1654–9.

Patwary, M. U., and J. P. van der Meer. 1984. Growth experiments on auto-polyploids of *Gracilaria tikvahiae* (Rhodophyceae). *Phycologia* 23, 21–7.

Pilot, T. J., and J. L. Fox. 1984. Cloning and sequencing of the genes encoding the α and β subunits of C-phycocyanin from the cyanobacterium *Agmenellum quadruplicatum*. *Proc. Natl. Acad. Sci. USA* 81, 6983–7.

Polne-Fuller, M., and A. Gibor. 1984. Developmental studies in *Porphyra*. I. Blade differentiation in *Porphyra perforata* as expressed by morphology, enzymatic digestion, and protoplast regeneration. *J. Phycol.* 20, 609–16.

Reith, M., and R. A. Cattolico. 1986. Inverted repeat of *Olisthodiscus luteus* chloroplast DNA contains genes for both subunits of ribulose-1,5-bisphosphate carboxylase and the 32,000-dalton QB protein: phylogenetic implications. *Proc. Natl. Acad. Sci. USA* 83, 8599–603.

Rochaix, J.-D. 1981. Organization, function, and expression of the chloroplast DNA of *Chlamydomonas reinhardtii*. *Experientia* 37, 323–32.

Rochaix, J.-D., J. Van Dillewijn, and M. Rahire. 1984. Construction and characterization of autonomously replicating plasmids in the green unicellular alga *Chlamydomonas reinhardii*. *Cell* 36, 925–32.

Ryther, J. H., L. D. Williams, M. D. Hanisak, R. W. Stenberg, and T. A. De Busk. 1979. Biomass production by marine and freshwater plants. In *Third Annual Biomass Energy Systems Conference Proceedings*, pp. 13–24. Solar Energy Research Institute, Golden, CO.

Saga, N., and Y. Sakai. 1983. Axenic tissue culture and callus formation of the marine brown alga *Laminaria angustifolia*. *Bull. Jap. Soc. Sci. Fish.* 49, 1561–3.

Sanbonsuga, Y., and M. Neushul. 1978. Hybridization of *Macrocystis* (Phaeophyta) with other float-bearing kelps. *J. Phycol.* 14, 214–24.

Sanbonsuga, Y., and M. Neushul. 1979. Cultivation and hybridization of giant kelps (Phaeophyceae). *Proc. Int. Seaweed Symp.* 9, 91–6.

Schantz, R. 1985. Mapping of the chloroplast genes coding for the chlorophyll A-binding proteins in *Euglena gracilis*. *Plant Sci.* 40, 43–9.

Steinmüller, K., M. Kaling, and K. Zetsche. 1983. In-vitro synthesis of phycobiliproteids and ribulose-1,5-bisphosphate carboxylase by non-poly-adenylated-RNA of *Cyanidium caldarium* and *Porphyridium aerugineum*. *Planta* [Berlin] 159, 308–13.

Suto, S. 1963. Intergeneric and interspecific crossings of the lavers *(Porphyra)*. *Bull. Jap. Soc. Sci. Fish.* 29, 739–48.

Tseng, C. K. 1981. Marine phycoculture in China. *Proc. Int. Seaweed Symp.* 10, 123–52.

Vallet, J.-M., M. Rahire, and J.-D. Rochaix. 1984. Localization and sequence analysis of chloroplast DNA sequences of *Chlamydomonas reinhardii* that promote autonomous replication in yeast. *EMBO* 3, 415–22.

van der Meer, J. P. 1979. Genetics of *Gracilaria tikvahiae* (Rhodophyceae). VI. Complementation and linkage analysis of pigmentation mutants. *Can. J. Bot.* 57, 64–8.

van der Meer, J. P. 1986. Genetic contributions to research on seaweeds. *Prog. Phycol. Res.* 4, 1–38.

van der Meer, J. P. 1987. Experimental hybridizations of *Palmaria palmata* (Rhodophyta) from the northeast and northwest Atlantic Ocean. *Can. J. Bot.* 65:1451–8.

Waaland, J. R. 1979. Growth and strain selection in *Gigartina exasperata* (Florideophyceae). *Proc. Int. Seaweed Symp.* 9, 241–7.

Yan, Z. M. 1984. Studies on tissue culture of *Laminaria japonica* and *Undaria pinnatifida*. *Proc. Int. Seaweed Symp.* 11, 314–16.

21. A future of phycotechnology

ARTHUR MICHIO NONOMURA

Marine Biology Research Division, Scripps Institution of Oceanography, University of California at San Diego, La Jolla, CA 92093

CONTENTS

I. Introduction

In recent years, developments in all realms of biotechnology have swiftly transformed laboratory discoveries to commercially useful products. The same rapid progress is occurring in diverse areas of phycology as well. Algae offer promise as direct and indirect renewable biological resources such as chemicals, food, and transport fuel. A major challenge is to accelerate the translation of fundamental scientific advances into useful algal products or processes. Many discoveries in basic research, be they in systematics, ecology, physiology, or biochemistry of algae, can be applied to practical problems and result in the development of commercial products and enterprises. A recent example is the use of phycofluors from blue-green and red algae in cytology and medicine as highly efficient and specific fluorescent tags. Another example is the development of improved strains of economically important red algae for mariculture. Such advances are based on research of the biochemical, physiological, and ecological properties of algae with each application arising from different fundamental sources of information. With recent, rapid advances in biotechnology, our perspectives of the field are continually changing as algae are examined for their economic impacts. The goal of this chapter is to provide some guidelines for the future based on phycology and new technology. The impact of biotechnology on research goals, economics, and subjects selected for study is considered.

II. Biotechnology/phycotechnology

Biotechnology has become a familiar term only within the past few years. This new technology requires careful definition and separation from processes such as fermentation that have been known for centuries. Biotechnology differs from previous application of biology in that it involves exploitation of sophisticated biological processes that were largely unknown only a decade ago.

Both mass and technical media use the term *biotechnology* with broad meanings implied by context, but rarely has the term been defined pre-

[530]

cisely. Broad definitions have been offered by governmental agencies and various other organizations. The most widely accepted definition is from the Office of Technology Assessment (Anonymous 1984), and it emphasizes the commercial nature of the field: "Biotechnology: Commercial techniques that use living organisms, or substances from those organisms, to make or modify a product, including techniques used for the improvement of the characteristics of economically important plants and animals and for the development of micro-organisms to act on the environment."

Still another definition from the OTA document emphasizes the role of organisms in biotechnology: "Biotechnology, broadly defined, includes any technique that uses living organisms (or parts of organisms) to make or modify products, to improve plants or animals, or to develop micro-organisms for specific uses."

Both definitions are generally accepted, but the first one, emphasizing the commercial aspect of biology, is most suitable to our current perception of the field because biotechnology is having considerable economic impact.

Phycotechnology is defined as a subset of the more general term *biotechnology*, but with algae as the focus of interest. Other chapters in this book describe current algal technologies as forms of "applied phycology." In perhaps the first use of the term, Lewin (1983) described some areas of microbial genetics that might be aided through phycotechnology. We might consider that first use of the term as a convenient milestone for separating "applied phycology" from "phycotechnology," in which the science of phycology joins with the business of technology. We can then define the integration of phycology with the latest developments in cellular and molecular biology, chemical engineering, aquaculture, and other related disciplines for specific commercial uses as *phycotechnology*.

III. Benefits and features of algae

Algae present a diversity of form, chemistry, genetic complexity, and range of habitats that are exploitable and adaptable to commercial processes. Distinct benefits are derived from joining phycology with technology. Indeed, phycotechnology is already the first plant biotechnology to show commercial success in large-scale bioprocesses. Intensive mass culture systems for algae range from open sunlit cultures to enclosed heterotrophic fermenters. Photosynthetic culture systems are generally flat bioreactors spread out to maximize the surface exposed to light and often draw the analogy to acre-sized photosynthetic "leaves."

For industrial biosynthesis of specific compounds, algae feature high rates of growth and natural products chemistry distinct from other organisms. To date, biotechnologists have rarely utilized algae as production organisms largely because their experience is limited to standard fermentation methods; however, the biologic and economic advantages of algae should be considered as they have many favorable properties. Many natural products, including vitamins, organic acids, polypeptides and polysaccharides are secreted by algae. Furthermore, algal cells need not be free-floating but may be immobilized on surfaces where they can serve as biocatalysts. Algae generally are not pyrogenic, rarely producing effective quantities of toxic components, and many species of algae are edible (Watt and Merrill 1975; Nonomura, Woessner, and West 1980). The life cycles of many algal species can be controlled. Algal strains are relatively stable genetically and are likely to perform more accurately than bacteria in many processes. For production of proteins, algae provide subunit assembly, correct folding, and clipping to form the finished product. Pigmentation can be controlled in some species of algae. Harvest methods for large algal forms are often much simpler than those for bacteria or yeasts. In general, the freshwater to highly saline culture media and phototrophic or heterotrophic bioprocessing methods used for algae are far less complex than those required for mammalian or vascular plant tissue culture. Microalgae cultured in fermentation vessels show doubling times as rapid as four hours and with dry weight densities up to 40 to 50 grams per liter.

Many exploitable algal features can be derived from specific ecological traits (see Raven 1984 for listings of algal diversity). For example, if a hydrocarbon-producing photoheterotrophic alga is sought, then the colonial green alga *Botryococcus* (Wolf 1983; Chirac et al. 1985) might be useful. This genus leaks hydrocarbons and floats, providing a simple means of harvest. If, instead, production of an orange-red indole pigment from an edible benthic marine alga is required, then *Caulerpa* might be used. Some *Caulerpa* species grow as giant cells up to two meters long from which cytoplasmic contents can be squeezed for making cytoplasts (Lobban, Harrison, and Duncan 1985). In many instances, algae with features closely matching bioprocessing design needs can be selected, but first the right questions must be posed and the needs must be understood.

IV. The phycotechnologist

The transition from basic scientist to biotechnologist requires some understanding of a dynamic field that has its holdfasts in pure science and only now is branching into the business world. The problems of phycotechnology are different from those faced by marine agronomists several years ago (Doty 1978) in that since its start biotechnology has been in the high-finance arena, a place unfamiliar to most pure scientists. Thus, to integrate science with business requires a change of perspective.

The rapid increase of industrial demand for biologists that accompanied the past decade of development of biotechnology found many scientists and businesses unprepared to deal with each other. Employment of scientists and engineers in biotechnology grew to 8,000 people in 1986 in the United States, representing a 12% increase of job opportunities for the year (Anonymous 1987). Of necessity, industry has attempted to solve the short-term need for professional biological expertise by employing academically oriented scientists, practically the only source of such expertise. As employment in industry becomes more commonplace for biologists in the next decade and beyond, a new generation of business-oriented phycotechnologists may emerge.

The perspectives of the scientist and the technologist are different. Business and anthropological analyses have described differences in language, working methods, and cultures that cause communication problems and clashes between scientists and management (Dubinskas 1985). In order to advance science, academic or government laboratories operate independently in pursuit of relatively long-range goals. Research management is relatively open-ended, and communication with colleagues is open. In contrast, a private-sector employee is usually part of a team focused on a narrowly defined, short-term project with an eye to protection of the intellectual property of the company. The composition of the research team, the management of the project, and the goal of the project are all influenced by the financial position of the company. For example, since most business investments are reviewed in the short term, business managers often require quantification, which may be difficult for most biologists to produce within a monthly or quarterly review period. To better meet the requirements of a well-managed business, it is to the biologist's advantage to plan research with frequently placed milestones such that nonscientists who may control the financing of a research project can understand advancements.

In a phycotechnologically oriented enterprise, algae are the common bond joining what is usually a multidisciplinary team of scientists, engineers, and business experts. Prospective phycotechnologists must realize that the cooperative environment in which they will operate as company employees may be quite different from the relatively independent modus operandi they experienced in the academic environment that nurtured their scientific training. When biologists are consulted by business executives, they must be able to explain their projects and results in such a way that the executives, who may be unfamiliar with the usual scientific terminology, understand the substance, value, and essence of the work. Communication is important in all directions since ideas can come to fruition only when they are understood.

The industrialization of high technology in Japan is regarded as the most rapid and efficient business development of this type in the world. The transition from laboratory finding to commercial product is eased by close interactions maintained among the major universities, industries, and government (Dibner 1985). For example, in a feasibility study for the production of fuel from algae conducted under the auspices of the Japanese Ministry of International Trade and Industry (MITI), the study group included university, industry, and government representatives (Iwamoto 1985). These interactions, in combination with a long history of fermentation technologies and algae-based industries, mark Japan as a leader in the development of phycotechnology. Biotechnology has been identified by Japan as the last major technological revolution of this century, and phycotechnology is already an advanced participant. The advanced status of phycotechnology in Japan is evident in research on nori where, for instance, such a fundamental feature as the nucleotide sequence of 5S rRNA was determined for *Porphyra yezoensis* some years ago (Takaiwa et al. 1982).

V. Business purpose and intellectual property

Insistence by phycologists on using their organisms in biotechnology is healthy because it brings algae to the forefront of public attention (e.g., Schlender 1986). Research is further accelerated by new professionals entering the field, and the discipline gains breadth when more research can be funded from self-generated profits rather than from external government support.

The main goals of business enterprises are to profit from their products or services and to generate funds for new projects. To retain profit potential, new concepts may require legal protection. The commercial potential of biotechnology has stimulated reexamination of intellectual property rights. Two landmark decisions have been particularly important to biotechnology. The first was that of Ananda M. Chakrabarty, a bacteriologist for General Electric Company, who, while working with recombinant DNA techniques, developed a strain of the bacterium *Pseudomonas* sp. that consumed crude oil (*Diamond* v. *Chakrabarty* 1980). The company filed a patent application claiming the bacterium per se, the process of producing the strain, and the method of application of the bacterium in a straw carrier. The U.S. Supreme Court qualified the oil-consuming strain of *Pseudomonas* as a patentable subject matter since it was a distinctive product of human ingenuity. After the *Chakrabarty* decision was made, guidelines were issued by the Patent and Trademark

Office that set forth several tests for assessing the patentability of genetically engineered organisms (Anonymous 1983).

The second landmark case was based on patent claims for mutants of corn that produce high yields of tryptophan (Hibberd, Anderson, and Barker 1985). In this case, the question of preeminence of plant patents was raised, and the maize plant, its seed, its tissue cultures, and the genes for tryptophan production were patented under the basic patent statute Section 101 rather than under Section 161 Patent for Plants.

The matter of patents for algae deserves special attention because algae can be protected in the United States under Section 161 Patent for Plants (Williams 1984). Algae can be claimed as new varieties when protected under the plant patent. Since the Patent for Plants was intended to protect vascular plant horticultural innovations, the plant patent has rarely been used for algae; and, in fact, only two plant patents have been claimed in the United States for algae (Avron and Ben-Amotz 1980; Nonomura 1988 in Table 21-1). Plant patents lack international protection in most cases. These problems were resolved in the aforementioned test cases, which ruled that utility patents for plants are valid. The algae can, thus, be patented under Section 101 for utility or Section 161 for plant variety protection. In the U.S. Patent and Trademark Office, algae are currently placed in Class 435 Chemistry: Molecular Biology and Microbiology; Subclass 257 Algae. This categorization allows more freedom and greater protection for algal innovations than previously (see Neagley, Jeffrey, and Diepenbrock 1983 for a legal perspective).

Normally, for plant patents, new varieties are claimed that are named following the International Code of Botanical Nomenclature. To insure the validity of the innovation, one should pay careful attention to the taxonomy of a transformed alga that expresses the characteristics of an inserted foreign gene, especially to avoid synonymy. Biotechnological methods such as recombinant DNA transformations, monoclonal antibodies, culture isolation, and chemistry will be useful for elucidation of algal relationships (e.g., Howard, Nonomura, and Fenical 1980 established sibling species relationships in *Laurencia* with culture isolates).

Patents are viewed as indicators of corporate strength and health in biotechnology companies. Large biotechnology companies are beginning to consider patent issues early in their research plans (Klausner 1985). Although numerous algal process and product patents have been awarded over the past 30 years for many of the applications discussed in this volume, several new applications have been assigned to phycotechnology companies for inventions more recently. Patent applications cannot be released until they are approved by the Patent Office, but some proposed titles and filing dates are listed in Table 21-1 as examples of pending developments.

Table 21-1. *Recent patent applications*

AUSICH, RODNEY LEE. 1983. Method for introducing foreign genes into green algae utilizing T-DNA of agrobacterium. European Patent Application no. 83306603.8. Publication no. 0108580. Applied for by Standard Oil Company, 200 East Randolph Drive, Chicago, IL 60601.

CODER, DAVID M., and A. M. NONOMURA. 1988. Method for selecting high-carotene-productivity strains of microalgae by means of fluorescent activated cell sorting.

CYSEWSKI, GERALD. 1985. Pulsed wave flow algal culture. U.S. Patent and Trademark Office (PTO) ser. no. 06/559,162. Assigned to Cyanotech Corporation, 18748 142nd St. NE, Woodinville, WA 98072.

HUNTLEY, MARK E. 1988. Process and apparatus for enclosed mass-culturing of microalgae for manufacture of fish feed. Assigned to Aquasearch, Inc., 9535 Poole St., La Jolla, CA 92037.

METTING, BLAINE. 1987. A method for preserving microalgae and Cyanobacteria in a desiccated state and their recovery therefrom. Assigned to R & A Plant-Soil, Inc., 24 Pasco-Kahlotus Road, Pasco, WA 99301.

NONOMURA, ARTHUR M. 1987. Naturally derived carotene/oil composition. U.S. PTO ser. no. 4,713,398. Assigned to Microbio Resources, Inc., 6150 Lusk Blvd., Suite B-105, San Diego, CA 92121.

NONOMURA, ARTHUR M. 1987. Process for producing a naturally derived carotene/oil composition by extraction from algae. U.S. PTO ser. no. 4,680,314. Assigned to Microbio Resources, Inc., 6150 Lusk Blvd., Suite B-105, San Diego, CA 92121.

NONOMURA, ARTHUR M. 1988. Method for eliminating predators from algae cultures. U.S. PTO ser. no. 06/770,982. Assigned to Microbio Resources, Inc., 6150 Lusk Blvd., Suite B-105, San Diego, CA 92121.

NONOMURA, ARTHUR M. 1988. *Botryococcus braunii* var. *showa:* a new variety of alga with high hydrocarbon yields. U.S. PTO Application ser. no. 852,048. Patent for Plants assigned to University of California, Berkeley, CA 94720.

NONOMURA, ARTHUR M. 1988. A naturally derived high carotene concentrate/ oil composition from algae. Assigned to Microbio Resources, Inc., 6150 Lusk Blvd., Suite B-105, San Diego, CA 92121.

SKATRUD, THOMAS J. 1985. Product chemicals from algal cells. Assigned to Biotechnical Resources, Inc., Seventh and Marshall Streets, Manitowoc, WI 54220.

SPENCER, K. G. 1988. Process for industrial biosynthesis of oxycarotenoids from algae. Assigned to Microbio Resources, Inc., 6150 Lusk Blvd., Suite B-105, San Diego, CA 92121.

STRYER, LUBERT, AND ALEXANDER N. GLAZER. 1985. Novel phycobiliprotein fluorescent conjugates. U.S. PTO Application ser. no. 483,006. Assigned to University of California, Berkeley, CA 94720.

SZALAY, ALADAR A., AND JOHN WILLIAMS. 1985. Genetic engineering in prokaryotic organisms. U.S. PTO ser. no. 396, 595. Assigned to the Boyce Thompson Institute for Plant Research, Cornell University, Ithaca, NY.

YOKOYAMA, HENRY, and JAMES KEITHLY. 1987. A method for coloring algae. Assigned to the United States of America by the Secretary of Agriculture, Washington, DC.

VI. Regulation and release of algae

During the development of any new technology, regulations addressing the public concern for technical and scientific hazard assessment are tempered by the regulatory environment. Introduction of genetically altered organisms into the environment is of widespread public concern (e.g., Colwell, Brayton, et al. 1985) because of the potential for worldwide impacts by self-replicating organisms with novel biochemistries. Biotechnology activities to date have been safe and stringently monitored. Tight regulatory controls of environmental applications are being worked out for field release of rDNA transformed organisms. Industries in the United States have been self-regulating (Gore 1984), but recently a framework for regulation of biotechnology has been proposed by the U.S. Office of Science and Technology (Anonymous 1986a). The guidelines are the composite statements of the Food and Drug Administration, Environmental Protection Agency, U.S. Department of Agriculture, and National Institutes of Health. One advisory group to the EPA and USDA, the National Biological Impact Assessment Program, will evaluate the effects of genetically altered organisms on the environment. The Biotechnology Science Coordinating Committee, an advisory group to the president, is in charge of resolving questions of jurisdiction among the five agencies, defining terms, and identifying international issues.

Cooperative efforts in research and regulation of biotechnology are encouraged by international organizations, many of which are governed by the United Nations system. Some UN programs of particular interest include: MIRCEN (UNEP-UNESCO), a world network of microbiological resource centers with a marine and aquatic microbiology section under R. Colwell, Department of Microbiology, University of Maryland, College Park, MD 20742; Food and Agriculture Organization (UN); and International Center for Genetic Engineering and Biotechnology (UNIDO). Safety codes aimed primarily at developing countries are being assembled by the combined efforts of UNIDO, World Health Organization, and UN Environment Program. International nongovernmental organizations include: International Organization for Biotechnology and Bioengineering, Committee on Genetic Experimentation, World Federation of Culture Collections, International Union of Microbiological Sciences, International Federation of Science, International Cell Research Organization, International Foundation for Science, and the Federation of Institutes of Advanced Study (Dasilva and Heden 1985).

International policies developed within the next decade are likely to be based on ecosystem impact assessments. The development of these large-scale ecological evaluation methods will require advances equal in rapidity to those of the biotechnological revolution. Large aquariums can be useful research instruments for modeling ecosystems (Adey 1983). Massive enclosed marine ecosystems similar to the 1 million-liter kelp forest tank at the Monterey Bay Aquarium might be monitored during tests of organisms that are the result of genetic manipulations. As the scope of the potential ecological effects of some algae is modeled in detail, the use of supercomputers (Dagani 1985) or parallel processors (Denning 1987) and supercomputer networks (Denning 1985) will be necessary to integrate information on genetically manipulated species for global perspectives. Detection of these altered organisms will require novel and specific methods to monitor their release if conventional methods for quality assurance are inappropriate (Colwell, Norse, et al. 1985). Phycology and its technology must therefore move forward in congress to increase the quality of life and maintain the health of the biosphere (Odum 1985).

VII. The academic foundation for phycotechnology

As a field such as phycotechnology develops, problems must be solved while new opportunities arise. Perhaps the greatest technical problem that the field must face is the rapid advance and commercial popularity of biotechnology itself. As industries and nations seek biologists to fill their positions for biotechnology, specific courses, departments, and institutions must be created to fill the needs. Many organismal biology departments are being transformed or replaced by biotechnology sections rather than by adding separate biotechnology divisions to existing institutional structures. In the university, this results in many students developing a great appreciation for the gene but with little understanding of the new organism created by genetic engineering or of its role in nature. The great universities will maintain their organismal biology sections in balance with biotechnology divisions similar to the situation in chemistry in which pure and industrial divisions coexist. Funding trends (Chirichiello and Crowley 1984) indicate increases for the life sciences largely with emphasis on biotechnology, but it is important for universities to insist upon maintaining strong organismal biology programs because the technology, once developed, must return to nature's library.

Several countries, including Japan, Germany, France, and the United States, are developing separate biotechnological institutes in accordance with their national goals. For example, the Center of Marine Biotechnology (COMB) will be established as a national research institute with the University of Maryland. COMB will devote a portion of its research to marine algae (Mikulsky 1985). On the West Coast, the nine campuses of the University of California established a major biotechnology program, of which marine biotechnology research is identified as a key element in this systemwide approach (Rote 1985). Similarly, nine university campuses have joined forces under the North Carolina Biotechnology Center to insure that the state benefits economically from the commercialization of projects in several divisions, including marine biotechnology (Patterson 1986). At the Solar Energy Research Institute in Colorado, algal biotechnologies are funded to establish the feasibility of using algae for fuel (Barclay and McIntosh 1986). The British Technology Group assists in technology transfer for the commercialization of innovative academic research in the United Kingdom. In Japan, Tokyo University has initiated input to an international data base for all applications of algae that is accessed through Professor S. Miyachi of the Department of Applied Microbiology. At the University of California, Berkeley, citations of all algae are being undertaken in the *Index Nominum Algarum* (Silva, pers. commun.).

VIII. Application of some old and new plant technologies

Techniques specific to plant biotechnology (Mantell and Smith 1983) are applicable to algae (van der Meer, Chapter 20). Much of vascular plant biotechnology is concerned with improving food crops and in creating transformations of the whole plant. In contrast, because there are comparatively few species of algae used as major food crops, most phycotechnological research and applications will be involved in developing new products or algae that will synthesize commercial products competitively. A wider range of methods is available for algal systems as compared with vascular plant systems because of the diversity of biochemistry, life histories, and morphologies of the algae.

Production of protoplasts in algae is relatively simple in some forms and difficult in others (Berliner 1981, 1983). Formation of protoplasts can be mechanically induced in giant algal cells (O'Neil and La Claire 1984). Protoplasts of *Chlamydomonas* are useful tools for investigation of vegetative plant cell fusions. The fast regeneration times and well-documented genetics from protoplasts of this organism are exploitable to confirm and forecast vascular plant phenomena (Mantagne and Mathieu 1983).

Flow cytometry will be utilized to isolate desirable cells for regeneration to whole thalli. Monoclonal antibodies to specific algal biochemicals (Vreeland and Laetsch 1985) may be used to enhance the sensitivity of cell sorting, especially where the detection of small quantities is necessary. As an additional benefit of the development of the cell selection instrumentation, flow cytometry will be useful for isolating unialgal cultures, selecting for high-quality wild strains, sorting organelles, and selecting for nuclear characteristics.

Propagation of commercially important mariculture crops will increase in efficiency through use of clonal regenerants of tissue culture (Polne-Fuller, Biniaminov, and Gibor 1984; Polne-Fuller and Gibor 1986) and by selection of somaclonal variants (Evans and Sharp 1986). The isolates that are necessary for establishment of protoplast or callus cultures provide an excellent foundation for maintenance of pure stocks and gene banks. In addition to the commercial importance of the gene banks, the mass propagation potential of tissue cultures may be important for ecological reconstruction.

IX. New bioprocesses

The potential for use of phycotechnology is massive in both current commercial applications and basic and applied research. Industrial biotechnology conducted US $1.10 billion worth of research and development in the United States and grew 20% in 1986 (Anonymous 1987). Large, closed-system cultivation and processing units designed for chemical and food production are expected to increase in scale from acre to square-mile area coverage, as in the expansion of open-ocean culture of nori in Japan. Decreases in mass-culture energy inputs will increase commercial participation in phycotechnology as diversity and profits grow. Mechanical efficiencies will be improved by means of detailed engineering analyses.

Innovative uses of new materials have quickly been found for algal culture. Cultures of *Chlorella* covered with polyvinylchloride films showed increased growth rates because the films stop ultraviolet light (Manabe et al. 1986). Fiber optics have been used for illumination of enclosed bioreactors and may be used to increase algal productivity by transmitting selected wavelengths of sunlight to the deep ocean floor (Hane 1986). As the cost of glass fibers decreases and the transparency of the glass increases, mariculture will realize vast potentials for expansion. Materials science must provide low-friction surfaces and environments that are resistant to corrosive forces such as high salinities, pH, and temperature for maintenance of cultures that are free of contaminants (see also Evans, Chapter 17). As the scale of processes increases, factors that are negligible in the laboratory and pilot scale, such as the friction made by water movement, become major engineering considerations that are likely to have financial and biological impacts. Even the color of the bioreactor liner is important, as the dye may be toxic to the alga or, as with solar units, reflective colors may be advantageous.

As molecular mechanisms in algae are elucidated, parameters that control interactions will become tightly controlled. Site mutations will be used to tune fine scale processes to reach goals of increased overall production. Bioprocessing calculations will be made according to altered nucleotide sequences in the gene regulatory regions with gene copy numbers optimized for biosyntheses. Using the concepts of artificial intelligence, the structure of reaction networks will be found by screening which, of all possible metabolic pathways, will function if certain enzymes are added or deleted. These search algorithms for metabolic pathways will require exacting knowledge of algal genetics of the sort now available for *Chlamydomonas* (e.g., Mantagne and Mathieu 1983; Aldrich et al. 1985).

Large corporations such as chemical manufacturers have established sophisticated research facilities for recombinant DNA investigations. Some have programs for algal investigations (e.g., Sohio [U.S.], Kirin Brewery [Japan], DuPont [U.S.], Thapar Corporate R and D Center [India], and Dainippon Ink [Japan]). Most research in large corporations using algae is directed toward elucidation of fundamental biological properties (e.g., Aldrich et al. 1985). Fuel producers such as Amoco have supported research on hydrocarbon production from algae (Wolf 1983). Government agencies have used algal hydrocarbons as geochemical markers to predict the occurrence of extensive crude oil source beds (McKirdy et al. 1986; George, Chapter 13). Whether or not transport fuel production by photosynthesis in algae proves economical, developments in this area will drive improvements in bioprocessing toward high efficiencies and increases of biological yields. Deepwater algae show high photosynthetic efficiencies (Littler et al. 1985; Littler and Littler, Chapter 2), and use of chloroplasts or genetic information from these algae may increase bioprocessing efficiencies. Pigments from these deepwater algae may be useful where low light intensity detection is needed. Physiological and metabolic investigations will result in highyielding strains of algae that produce hydrocarbons (Wolf, Nonomura, and Bassham 1985; Nonomura 1986 in Table 21-1). Classical economic discussions may require reevaluation as new algal isolates with advantageous characteristics, such as reusable biomass with rapid hydrocarbon productivity, are discovered (e.g., Nonomura 1988 in Table 21-1). Genetic manipulations, such as importation of polypeptides into chloroplasts, may confer herbicide resistance or change other phenotypic characteristics (Cashmore et al. 1985) of algae that will increase commercial process efficiencies. Importation of genes into hydrocarbon producers may result in production of enzymes or accessory pigments of sufficient value to pay for the cost of fuel production many times over. *Botryococcus braunii* has been evaluated at Ciba-Corning Diagnostics Company for vegetable oil content as the price of this product would justify its production (Weetall 1985); and if the biosynthetic pathway is manipulated for additional production of fine chemicals (Spencer, Chapter 10), such as pharmaceutically valuable diterpenoids, excretion of transport fuel hydrocarbons as a by-product would provide economic motivation for fuel production by this species. The use of algae as a renewable source of hydrocarbons is likely to advance phycotechnology into the multibillion-dollar finance scene.

Microalgae are cultured on a large scale to supplement fish and shellfish diets (McVey 1983; see also Spencer, Chapter 10). Algae are used to improve food color and taste in most cases. Carotenoids in microalgae, for example, contribute to the pink coloration in rainbow trout. Three optical isomers of astaxanthin (the carotenoid form in the green alga *Haematococcus*) were found to be more efficacious than canthaxanthins in pigmenting the flesh of *Salmo gairdneri* (Foss et al. 1984). Commercial mariculture facilities for expensive shellfish such as abalone use films of adhering algae for settling larvae (Leighton 1985), and kelp, costing US $5 per 100 kg, is provided to the abalone as an essential food source (McDaniel and Garrett 1987). Large-volume continuous-flow algal pond systems have been developed by commercial aquaculture businesses to feed bivalves (Walsh et al. 1985).

In an exciting demonstration of how understanding ecological interactions can work to our benefit, novel uses of algal mariculture methods that exploit natural life histories of both fish and alga have been transferred from the People's Republic of China for use in San Francisco Bay. *Laminaria* thalli were cultured to provide a substrate for herring roe deposition (Hansen 1983). The kelp plants are started *in vitro* and then moved into herring spawning areas during the mating season. The roe-covered thalli are harvested by drawing in the culture ropes rather than by expensive diving and handpicking underwater.

X. New products

Exploitation of algae, especially for biotechnological products, occurs in established and new businesses. Knowledge of algal polysaccharides continues to grow, and new uses continue to be developed in microbial and cell culture, chromatography, and diagnostics (Renn 1984). A recent novel use for alginate fiber paper was in high-fidelity audio speakers developed by the Government Industrial Institute of Shikoku, Japan (Anonymous 1986b). Also in Japan, alginate fibers were polymerized into a fluffy, absorbent wound dressing.

Several new businesses process phycobiliprotein pigments from algae for use as sensitive fluorescent dyes, immunochemical reagents, and clinical diagnostics (Ong, Glazer, and Waterbury 1984; Glazer and Stryer 1984; Kronick 1986). As many as 40 diagnostic applications using phycobiliproteins have been developed to date (e.g., Perdomo 1986). Phycobiliproteins offer several advantages as immunofluorescent tags. They are extremely efficient fluorochromes, they function at wavelengths compatible with existing light sources, and they can be used in multiple color analyses (Lewis et al. 1985; Stryer, Oi, and Glazer 1985). Phycobiliproteins with different properties may be obtained by surveying the Rhodophyta, Cryptophyta, or Cyanophyta. These pigments derived from algae also have found use as food dyes in products ranging from chewing gum to yogurt (Ogawa et al. 1979; Tanabe and Iwamoto 1979; Langston and Maing 1983). Phycobiliproteins, coincidentally, enhance the fermentation process in yogurt.

The pharmaceutical industry should continue to benefit from phycotechnology. Use of phycocolloids as carriers, emulsifiers, suspending agents, and as clinical test substrata has long been standard practice (see reviews in Renn 1984; Lewis, Stanley, and Guist, Chapter 9). At least 30 different species of microalgae show antibacterial activity with many more species containing bioactive substances (Metting and Pyne 1986). Retinoids are important in treatment of acne, and those from algae may prove useful in other areas of clinical therapy. Identification of rhodopsin in *Chlamydomonas* (Foster et al. 1984) forecasts its use in vision models and research (Nakanishi 1985). Carotenoids are nontoxic sources of retinol and are valuable natural food colorants that occur in algae at the highest concentrations known in nature (Spencer, Chapter 10). Eicosapentaenoic acid (EPA), found in *Isochrysis* and other phytoplankton, was epidemiologically indicated as an antiarteriosclerotic food supplement (Bang, Dyerberg, and Nielsen 1971; Dyerberg 1986; Lands 1986), and a commercial method for enzymatic production of EPA was described (Kayama et al. 1986). An unsaturated wax from *Euglena* was developed as a new lubricant that can be used in cosmetics (Tani, Okumura, and Ii 1985). Specific antiviral activity was claimed for *Cryptosiphonia woodii* for treatment of herpes simplex virus (Nonomura and Pappo 1985), and an antitumor agent was described from *Chlorella* (Shimpo 1986). Multimillion-dollar investments in private sector research this year indicate concentrated efforts to exploit the algae for pharmaceutical uses. As examples, SeaPharm of Fort Pierce, Florida, is conducting deep-sea surveys costing several million dollars per year for marine pharmaceuticals; Microbio Resources, Inc., invested over US $15 million to develop facilities to produce and process carotenoids from *Dunaliella;* and a multimillion-dollar screening program for bioactive compounds has been initiated at Kirin Brewery, Tokyo. When indications of activity are discovered, computer-assisted drug design (Hopfinger 1985) will enhance the potential of algae in the pharmaceutical field.

Algal toxins have been used to elucidate neuronal and pharmacological mechanisms (e.g., Kao and Walker 1982), and the ecological impact of cyanobacterial toxins is reviewed in this volume (Gorham and Carmichael, Chapter 16). As a spinoff of the research on toxins from algae, pesticides have been developed. For example, Cheminova, Ltd. (Denmark), synthesized the insecticidal derivatives of dithiolane and trithiane that were originally described from *Chara globularis*. The insecticidal properties of these derivatives are similar to nereistoxin (Jacobsen and Pedersen 1983). Some other green algae produce compounds that inhibit mosquito larvae (Rich 1985), indicating still other pesticidal applications.

Of more than 2,000 different enzymes described in the literature, only 16 are available in industrial amounts and none is from algae. Biotechnological applications of enzymes from algae should be promoted, especially for unconventional catalytic properties (Klibanov 1983) such as digestion under organic solvents. For genetic recombinations, enzymes from algae have only recently been exploited, and the use of restriction enzymes for nucleic manipulations suggests a rich resource from blue-green algae (Kawamura et al. 1986).

New algal products are expected to result from business and academic research using the basic chemical and genetic techniques outlined in previous chapters. New firms are likely to be created in order to develop products that result from discoveries made at nearby universities. A recent example is Biosponge, Inc., whose cooperators from the Duke University Marine Laboratory seek essential oils and essences from microalgae for pharmaceuticals (J. Bonaventura, personal communication). Recombinant DNA manipulation in eukaryotes may benefit from basic phycological discoveries. For example, the first plaque assay for a plant virus was demonstrated with the chlorellaphage PBCV1, which showed a 1% genetic recombination yield (Tessman 1985). Vectors for recombinant DNA transformations may be found in algal viruses. A survey of cross-infections from vascular plants to algae might also identify useful vectors. The *Agrobacterium* Ti plasmid was used for transformation of *Chlamydomonas* (see Ausich 1983 in Table 21-1), but the stability of the transformant requires improvement. Transformation of the green alga *Chlamydomonas* with yeast DNA (Rochaix and van Dillewijn 1982) and the establishment of several shuttle vectors between cyanobacteria and the bacterium *Escherichia coli* have been reported (Kuhlemeier et al. 1984). Improvement of the gene expression system of blue-greens provides opportunity for gene transfer of valuable enzymes such as nitrogenase and nitrate reductase. Parasitic red algae show unusual penetration (Nonomura and West 1980) and nuclear transfer into host cells (Goff and Coleman 1985), which may ultimately help us to understand genetic compatibility mechanisms (Nonomura and West 1981). Biosynthetic pathways will be further elucidated by specific gene regulatory compounds. Onium compounds (Gausman et al. 1985), which are finding application in mass cultures of algae, cause specific metabolite responses in algae similar to those of vascular plants (see Yokoyama and Keithly 1987 in Table 21-1).

XI. New pioneers

As we view the future, we envision algae and their products as important market commodities. Exploiting biosynthetic pathways of algae in industrial processes has long-term economic and public health benefits. Use of algae in agriculture to close the nutrient cycle for livestock (e.g., Goh 1986; Lincoln et al. 1986) is an elegant demonstration of the application of ecosystems concepts. On a grand scale, the Ministry of Light Industry, People's Republic of China, is developing plans to use thousands of square kilometers of salt evaporation ponds to integrate cultivation of halophilic algae with shrimp mariculture and simultaneously, to improve the quality of salt products (Nonomura 1986). Such large-scale systems may benefit from the use of remote sensing techniques to monitor algal movement and growth (e.g., Horne and Nonomura 1976).

Algae have long been used as sensitive pollution indicators and detectors of the ecological impact of industries and processes that cause physical changes in the local environment or ecosystem alterations (e.g., Lebednik 1985). Algae are useful in actual water treatment as well as being good indicator organisms. Methods of water treatment in which algae are used to eliminate or inactivate heavy metals and toxic substances are currently in use in some civic water treatment systems and are being considered for others (Oswald, Chapter 11). The tendency for algae to bind transition metals is being exploited in immobilized systems for water treatment. Bioclean, Inc. (New Mexico), developed ion exchange columns that use microalgal cells polymerized to silica gel for removal of copper from electroplating wastewater. There is also potential for exploitation of similar rechargeable systems in mining for fine ores. Accumulation of elemental gold (Hosea et al. 1986), silver, mercury (Darnall et al. 1986), and uranium (Greene et al. 1986) on lyophilized preparations of *Chlorella vulgaris* was demonstrated, with binding capacity exceeding that of other microorganisms.

As more methods for genetic recombination of algae become available, the potential products that can be tapped will be as many as there are identified genes (van der Meer, Chapter 20). The bioprocessing methods for transformed algae are vast and may result in such systems as *Spirulina* or *Chlamydomonas* reactors producing enzymes or hormones or other chemicals of value.

The history of algal applications follows a long line of pioneering successes in which determined individuals saw their visions through to improve the quality of life. Their creative applications of algae have positive impacts on the economies of nations. This is particularly evident in certain Pacific island nations whose entries into the twentieth-century economy were based on phycocolloid production. The use of phycocolloids in agriculture as soil-conditioning agents provides still other opportunities (Metting, Rayburn, and Reynaud₃ Chapter 14). Phyco-technology is currently earthbound, but as we reach for the stars with one hand, the other hand may well hold a bottle of algae to inoculate and condition planets for human habitation (Wharton, Smernoff, and Averner, Chapter 19).

The celluloses of green algae are the same as those found in paper, and some algae have been used to produce handsheets (Kiran et al. 1980). Algae grow rapidly, and for countries with long coastlines and limited forest resources fiber-rich species might provide an inexpensive, alternative source of pulp. Initiation of a program to use marine algae as an alternative source of newsprint in areas with deforestation problems such as India (Rao 1986) may assist in reversing or slowing otherwise problematic ecological trends.

Phycotechnology is applied humanism as much as it is algal utilization. It is a source of careers and jobs with extraordinary potential. It is also the basis for personal satisfaction, creative scientific contribution, social interaction, and great fun. The common focus of phycotechnology is the alga, and the binding endeavor of phycotechnologists is to improve algae and thereby human affairs.

XII. Acknowledgments

The author thanks Takayasu Hirosawa of the Plant Bioengineering Laboratory, Kirin Brewery Co., Japan, for his helpful discussions and comments.

XIII. References

Adey, W. H. 1983. The microcosm: a new tool for reef research. *Coral Reefs* 1, 193–201.

Aldrich, K. J., B. Cherney, E. Merlin, C. Williams, and L. Mets. 1985. Recombination within the inverted repeat sequences of the *Chlamydomonas reinhardtii* chloroplast genome produces two orientation isomers. *Curr. Gen.* 9, 233–8.

Anonymous, 1983. *Patent Profiles: Biotechnology.* Office of Technology Assessment and Forecast. Government Printing Office, Washington, DC. 297 pp.

Anonymous. 1984. *Commercial Biotechnology: An International Analysis.* U.S. Congress, Office of Technology Assessment, OTA-BA-218. Government Printing Office, Washington, DC. 612 pp.

Anonymous. 1986a. Coordinated framework for regulation of biotechnology: announcement of policy and notice for public comment. *Fed. Reg.* 51, 23302–93.

Anonymous. 1986b. Speaker cone from alga. *Nikkan Kogyo Shimbun.* 1 Sept.

Anonymous. 1987. *Science resources studies highlights.* NSF 87-304. National Science Foundation, Washington, DC. 4 pp.

Avron, M., and A. Ben-Amotz. 1980. Alga strain. U.S. Plant Patent 4,511. Washington, DC.

Bang, H. O., J. Dyerberg, and A. B. Nielsen, 1971. Plasma lipid and liproprotein pattern in Greenlandic west-coast Eskimos. *Lancet* 7701, 1143–6.

Barclay, W. R., and R. P. McIntosh (eds.). 1986. Algal biomass technologies. *Nova Hedw., Beih.* 83, 1–273.

Berliner, M. D. 1981. Protoplasts of eukaryotic algae. *Int. Rev. Cytol.* 73, 1–19.

Berliner, M. D. 1983. Protoplasts of nonvascular plants. *Int. Rev. Cytol.* 16 (Suppl.), 21–31.

Cashmore, A., L. Szabo, M. Timko, A. Kausch, G. Van den Broeck, P. Schreier, H. Bohnert, L. Herrera-Estrella, M. Van Montagu, and J. Schell, 1985. Import of polypeptides into chloroplasts. *Bio/technology* 3, 803–8.

Chirac, C., E. Casadevall, C. Largeau, and P. Metzger. 1985. Bacterial influence upon growth and hydrocarbon production of the green alga *Botryococcus braunii. J. Phycol.* 21, 380–7.

Chirichiello, J., and M. Crowley. 1984. *National Patterns of Science and Technology Resources.* NSF 84-311. Government Printing Office, Washington, DC. 89 pp.

Colwell, R. R., P. R. Brayton, D. J. Grimes, D. B. Roszak, S. A. Huq, and L. M. Palmer. 1985. Viable but non-culturable *Vibrio cholerae* and related pathogens in the environment: implications for release of genetically engineered microorganisms. *Bio/technology* 3, 817–20.

Colwell, R. K., E. A. Norse, D. Pimentel, F. E. Sharples, and D. Simberloff. 1985. Genetic engineering in agriculture. *Science* 229, 111–12.

Dagani, R. 1985. Supercomputers helping scientists crack massive problems faster. *Chem. Eng. News* 63, 7–13.

Darnall, D. W., B. Greene, M. T. Henzi, J. M. Hosea, R. A. McPherson, J. Sneddon, and M. D. Alexander. 1986. Selective recovery of gold and other metal ions from an algal biomass. *Envir. Sci. Technol.* 20, 206–8.

Dasilva, E. J., and C. G. Heden, 1985. The role of international organizations in biotechnology: cooperative efforts. In *Comprehensive Biotechnology*, vol. 4, ed. M. Moo-Young, pp. 717–49, Pergamon Press, Elmsford, NY.

Denning, P. J. 1985. The science of computing supernetworks. *Amer. Sci.* 73, 225–7.

Denning, P. J. 1987. The science of computing multigrids and hypercubes. *Amer. Sci.* 75, 234–8.

Diamond v. *Chakrabarty.* 1980. 100 S. Ct. 2204, 206 U.S.P.Q. 193.

Dibner, M. D. 1985. Biotechnology in pharmaceuticals: the Japanese challenge. *Science* 229, 1230–5.

Doty, M. S. 1978. Status of marine agronomy, with special reference to the tropics. *Proc. Int. Seaweed Symp.* 9, 35–58.

Dubinskas, F. A. 1985. The culture chasm: scientists and managers in genetic engineering firms. *Tech. Rev.* 88, 24–30.

Dyerberg, J. 1986. Linolenate-derived polyunsaturated fatty acids and prevention of atherosclerosis. *Nutr. Rev.* 44, 125–34.

Evans, D. A., and W. R. Sharp. 1986. Applications of somaclonal variation. *Bio/technology* 4, 528–32.

Foss, P., T. Storebakken, K. Schiedt, S. Liaaen-Jensen, E. Austreng and K. Strieff. 1984. Carotenoids in diets for salmonids. I. Pigmentation of rainbow trout with the individual optical isomers of astaxanthin in comparison with canthaxanthin. *Aquaculture* 41, 213–26.

Foster, K. W., J. Saranak, N. Patel, G. Zarilli, M. Okabe, T. Kline, and K. Nakanishi, 1984. A rhodopsin is the functional photoreceptor for phototaxis in the unicellular eukaryote *Chlamydomonas*. *Nature* 311, 756–9.

Gausman, H. W., J. D. Burd, J. Quisenberry, H. Yokoyama, R. Dilbeck, and C. R. Benedict. 1985. Effect of DCPTA on cotton plant growth and phenology. *Bio/technology* 3, 255–7.

Glazer, A. J., and L. Stryer. 1984. Phycofluor probes. *Trends Biochem. Sci.* 9, 423–7.

Goff, L. J., and A. W. Coleman. 1985. The role of secondary pit connections in red algal parasitism. *J. Phycol.* 21, 483–508.

Goh, A. 1986. Production of microalgae using pig waste as a substrate. *Nova Hedw.* 83, 235–44.

Gore, A., Jr. 1984. *The Environmental Implications of Genetic Engineering.* Committee on Science and Technology, U.S. House of Representatives, 98th Congress, 2nd sess., ser. V. Government Printing Office, Washington, DC. 51 pp.

Greene, B., M. T. Henzi, J. M. Hosea, and D. W. Darnall. 1986. Elimination of bicarbonate interference in the binding of U(VI) in mill-waters to freeze-dried *Chlorella vulgaris*. *Biotech. and Bioeng.* 28, 764–7.

Hane, Y. 1986. Ocean Fisheries Farm System. Japanese Patent Application no. 61-124333. Assigned to Shimizu Kensetu Co., Ltd.

Hansen, J. E. 1983. Mariculture of marine plants: critical pathway approach. In *Proceedings of Oceans,* pp. 890–4.

Hibberd, K. A., P. C. Anderson, and M. Barker. 1985. Tryptophan overproducer mutants of cereal crops. U.S. Patent and Trademark Office Application ser. no. 647,008. Washington, DC.

Hopfinger, A. J. 1985. Computer assisted drug design. *J. Med. Chem.* 28, 1133–9.

Horne, A. J., and A. M. Nonomura. 1976. *Drifting Macroalgae in Estuarine Water: Interactions with Salt Marsh and Human Communities.* SERL Report no. 76-3. Sanitary Engineering Research Laboratory, University of California, Berkeley. 76 pp.

Hosea, M., B. Greene, R. McPherson, M. Henzel, M. D. Alexander, and D. W. Darnall, 1986. Accumulation of elemental gold on the alga *Chlorella vulgaris*. *Inorg. Chim. Acta* 123, 161–5.

Howard, B. M., A. M. Nonomura, and W. Fenical. 1980. Chemotaxonomy in marine algae. II. Secondary metabolite synthesis by *Laurencia nipponica* (Rhodophyta, Ceramiales) in unialgal culture. *J. Exp. Biochem. Ecol.* 8, 329–36.

Iwamoto, H. 1985. *A Feasibility Study for the Production and Utilization of Fuel by Biotechnology.* Japanese Association of Fermentation Industries, 2-5-5 Atago, Minato-ku, Tokyo. 101 pp.

Jacobsen, N., and E. K. Pedersen. 1983. Synthesis and insecticidal properties of derivatives of propane-1,3 dithiol (analogues of the insecticidal derivatives of dithiolane and trithiane from the alga *Chara globularis* Thuillier). *Pestic. Sci.* 14, 90–7.

Kao, C. Y., and S. E. Walker. 1982. Active groups of saxitoxin and tetrodotoxin as deduced from actions of saxitoxin analogues on frog muscle and squid axon. *J. Physiol.* 323, 619–37.

Kawamura, M., M. Sakakibara, T. Watanabe, H. Obayashi, S. Hiraoka, and K. Kita. 1986. A new restriction enzyme and methods for its production. Japanese Patent Application no. 61-40789. Assigned to Dai Nippon Ink Co., Tokyo.

Kayama, H., M. Makura, Y. Yawata, and K. Kuwata. 1986. Method for enzymatic production of eicosapentaenoic acid. Japanese Patent Application no. 61-31092. Assigned to Ikeda Tohka Kogyo Co., Ltd., Tokyo.

Kiran, E., I. Teksoy, K. C. Guven, E. Guler, and H. Guner. 1980. Studies on seaweeds for paper production. *Bot. Mar.* 23, 205–8.

Klausner, A. 1985. Biotech's first steps into the business world. *Bio/technology* 3, 869–72.

Klibanov, A. M. 1983. Unconventional catalytic properties of conventional enzymes: applications in organic chemistry. In *Basic Biology of New Developments in Biotechnology,* ed. A Hollaender, A. I. Laskin, P. Rogers, S. Dagley, R. Hanson, L. McKay, and J. Messing, pp. 497–518. Plenum Press, New York.

Kronick, M. 1986. The use of phycobiliproteins as fluorescent labels in immunoassay. *J. Immunol. Methods* 92, 1–13.

Kuhlemeier, C. J., V. J. P. Teeuwsen, M. J. T. Janssen, and G. A. van Arkel. 1984. Cloning of a 3rd nitrate reductase gene from the cyanobacterium *Anacystis nidulans* R-2 using a shuttle cosmid library. *Gene* [Amsterdam] 31, 109–16.

Lands, W. E. M. 1986. Renewed questions about polyunsaturated fatty acids. *Nutr. Rev.* 44, 189–95.

Langston, M., and I. Maing. 1983. Acid soluble *Spirulina*-blue colorant preparation by treating *Spirulina*-blue with protease and alkaline medium and acidifying. U.S. Patent and Trademark Office no. 4,400,400. Washington, DC.

Lebednik, P. A. 1985. *Thermal Effects Monitoring Program, 1984 Annual Report, Diablo Canyon Power Plant.* PG&E Report no. E-85-04. Pacific Gas and Electric Company, 77 Beale Street, San Francisco. 23 pp.

Leighton, D. L. 1985. Early growth of green abalone in hatchery and field. In *Proceedings of the Joint International Scientific Diving Symposium.* American Academy of Underwater Sciences, La Jolla, CA. 12 pp.

Lewin, Ralph, A. 1983. Phycotechnology: how microbial geneticists might help. *BioScience* 33, 177–9.

Lewis, D. E., J. M. Puck, G. F. Babcock, and R. R. Rich, 1985. Disproportionate expansion of a minor T cell subset in patients with lymphadenopathy syndrome and acquired immunodeficiency syndrome. *J. Infect. Dis.* 151, 555–9.

Lincoln, E. P., B. Koopman, L. O. Bagnell, and R. A. Nordstedt. 1986. Aquatic system for fuel and feed production from livestock wastes. *J. Aquat. Eng. Res.* 33, 159–69.

Littler, M. M., D. S. Littler, S. M. Blair, and J. N. Norris. 1985. Deepest known plant life discovered on an uncharted seamount. *Science* 227, 57–9.

Lobban, C. S., P. J. Harrison, and M. J. Duncan. 1985. *The Physiological Ecology of Seaweeds.* Cambridge University Press, Cambridge. 242 pp.

McDaniel, J. B., and E. M. Garrett. 1987. Four approaches to aquaculture. *Venture* 9, 59–62.

McKirdy, D. M., R. E. Cox, J. K. Volkman, and V. J. Howell. 1986. Botryococcane in a new class of Australian non-marine crude oils. *Nature* 320, 57–9.

McVey, J. P. 1983. *Crustacean Aquaculture*. Vol. 1. CRC Press, Boca Raton, FL. 456 pp.

Manabe, E., T. Hirosawa, M. Tsuzuki, and S. Miyachi, 1986. Effect of solar near ultraviolet radiation on growth of *Chlorella* cells. *Physiol. Plant.* 67, 598–603.

Mantagne, R. F., and D. Mathieu. 1983. Transmission of chloroplast genes in triploid and tetraploid zygospores of *Chlamydomonas reinhardtii:* roles of mating-type gene dosage and gametic chloroplast DNA content. *Proc. Natl. Acad. Sci. USA* 80, 4780–3.

Mantell, S. H., and H. Smith (eds.). 1983. *Plant Biotechnology.* Cambridge University Press, Cambridge. 334 pp.

Metting, B., and J. W. Pyne. 1986. Biologically active compounds from microalgae. *Enzyme Microbial Technol.* 8, 386–94.

Mikulsky, B. 1985. *Marine Biotechnology: Hearings before the Subcommittee on Oceanography,* July 15, 1–73. Committee on Merchant Marine and Fisheries, U.S. House of Representatives, Baltimore, MD.

Nakanishi, K. 1985. Bioorganic studies with rhodopsin. *Pure and Appl. Chem.* 57, 769–76.

Neagley, C. H., D. D. Jeffrey, and A. B. Diepenbrock. 1983. Section 101 Plant Patents: panacea or pitfall? In *American Patent Law Association: Selected Legal Papers* 1, A1–A34.

Nonomura, A. M. 1986. *Brine Biotechnology in Salt Ponds.* CPR/85/066/A/01/99. United Nations Development Programme, 2 Sanlitun Dongqijie, Beijing, People's Republic of China. 33 pp.

Nonomura, A. M., and R. Pappo. 1985. Composition of matter from *Cryptosiphonia woodii* useful for the treatment of herpes simplex virus. U.S. Patent no. 4,522,814. Washington, DC.

Nonomura, A. M., and J. A. West. 1980. Ultrastructure of the parasite *Janczewskia morimotoi* and its host *Laurencia nipponica* (Ceramiales, Rhodophyta). *J. Ultrastruct. Res.* 73, 183–98.

Nonomura, A. M., and J. A. West. 1981. Host specificity of *Janczewskia* (Ceramiales, Rhodophyta). *Phycologia* 20, 251–8.

Nonomura, A. M., J. Woessner, and J. A. West. 1980. *Making Seaweeds Worth Eating.* Carolina Biological Supply Co., 2700 York Rd., Burlington, NC 27215. 10 pp., 76 figs.

Odum, Eugene P. 1985. Biotechnology and the biosphere,. *Science* 229, 1338.

Ogawa, K., S. Tezuka, Y. Tsucha, Y. Tanabe, and H. Iwamoto. 1979. Blue chewing gum colored by phycocyanin obtained by extraction of *Spirulina.* Japanese Patent no. 79138156, Tokyo.

O'Neil, R. M., and J. W. La Claire II. 1984. Mechanical wounding induces the formation of extensive coated membranes in giant algal cells. *Science* 225, 331–3.

Ong, L. J., A. J. Glazer, and J. B. Waterbury. 1984. An unusual phycoerythrin pigment from a marine cyanobacterium. *Science* 224, 80–83.

Patterson, R. J. 1986. *News from the North Carolina Biotechnology Center. Techne*, NCBC, PO Box 13547, Research Triangle Park, NC 27709. 10 pp.

Perdomo, J. 1986. Phycobiliproteins as fluorescence tracers. *Biomeda News* 1, 1–12. Biomeda Corporation. PO Box 8045, Foster City, CA 94404.

Polne-Fuller, M., M. Biniaminov, and A. Gibor. 1984. Vegetative propagation of *Porphrya perforata. Hydrobiologia* 116/117, 308–13.

Polne-Fuller, M., and A. Gibor, 1986. Calluses, cells, and protoplasts in studies towards genetic improvement of seaweeds. *Aquaculture.* 57, 117–23.

Rao, R. 1986. Rising above forest decline. *Science* 323, 284–5.

Raven, J. A. 1984. *Energetics and Transport in Aquatic Plants.* MBL Lectures in Biology, vol. 4. Alan R. Liss, New York. 587 pp.

Renn, D. W. 1984. Marine algae and their role in biotechnology. In: *Biotechnology in the Marine Sciences*, ed. R. R. Colwell, E. R. Pariser, and A. J. Sinskey, pp. 191–5. John Wiley, New York.

Rich, R. 1985. Chlorococcales, Tetrasporales inhibit mosquito larvae. *Phycol. Soc. Amer. Newsletter* 21, 10.

Rochaix, J.-D., and J. van Dillewijn. 1982. Transformation of the green alga *Chlamydomonas reinhardtii* with yeast DNA. *Nature* 296, 70–2.

Rote, J. 1985. *Review of Federal and State Regulations Affecting the California Biotechnology Industry.* 068-A. Joint Publications Office, Box 90, State Capitol, Sacramento, CA 95814. 81 pp.

Schlender, B. R. 1986. New uses for algae improve image of a lowly plant group. *Wall Street Journal* 115, 23.

Shimpo, K. 1986. Antitumor agent. Japanese Patent Application no. 61-71358. Assigned to Chlorella Kogyo Co., Ltd., Tokyo.

Stryer, L., V. Oi, and A. N. Glazer. 1985. Fluorescent conjugates for analysis of molecules and cells. U.S. Patent no. 4,520,110. Washington, DC.

Takaiwa, F., M. Kusuda, N. Saga, and M. Suguira. 1982. The nucleotide sequence of 5S rRNA from a red alga *Porphyra yezoensis. Nucleic Acids Res.* 10, 6037–40.

Tanabe, Y., and H. Iwamoto. 1979. Phycocyanin for fermented milk production. Japanese Patent no. 79095770. Tokyo.

Tani, Y., M. Okumura, and S. Ii. 1985. Production of unsaturated wax ester by *Euglena. Jap. Soc. Agric. Chem.* (Abstract) 60, 477.

Tessman, I. 1985. Genetic recombination of the DNA plant virus PBCV1 in a *Chlorella*-like alga. *Virology* 145, 319–22.

Vreeland, V., and W. M. Laetsch. 1985. Monoclonal antibodies to seaweed carbohydrates. In *Biotechnology of Marine Polysaccharides*, ed. R. R. Colwell, E. R. Pariser, and A. J. Sinsky, pp. 399–428. Hemisphere Publishers, New York.

Walsh, D. T., C. A. Withstandley, R. A. Kraus, and E. J. Petrovitz. 1985. Mass culture of selected marine microalgae for the nursery pond production of bivalve seed. *Aquaculture Digest* 10, 8–12.

Watt, B. K., and A. L. Merrill. 1975. *Composition of Foods.* Agriculture Handbook no. 8. Government Printing Office, Washington, DC. 190 pp.

Weetall, H. H. 1985. Studies on the nutritional requirements of the oil-producing alga *Botryococcus braunii. Appl. Biochem. Biotech.* 11, 377–91.

Williams, S. B., Jr. 1984. Protection of plant varieities and parts as intellectual property. *Science* 225, 18–23.

Wolf, F. R. 1983. *Botryococcus braunii:* an unusual hydrocarbon-producing alga. *Appl. Biochem. Biotech.* 8, 249–60.

Wolf, F. R., A. M. Nonomura, and J. A. Bassham. 1985. Growth and branched hydrocarbon production in a strain of *Botryococcus braunii* (Chlorophyta). *J. Phycol.* 21, 388–96.

INDEXES

Author index

[555]

Taxonomic index

Subject index

acrolein, 471

activated sludge, 258

adhesives of marine-fouling algae, 446–7

adulteration, *see* contaminants; extraneous materials in algal products; metals as adulterants

aesthetic nuisances from *Cladophora*, 70, 458

agar, 14, 95, 138, 141–2; chemistry of, 229–30; extraction and processing of, 230; future for, 232; history of, 229; market for, 231–2; production of, 230; products from, 231; properties of, 230–1

agarophytes, 144–5

agarose, 142

agrichemicals, cyanobacterial growth in soils and, 340–1

agriculture: algal uses in, 4, 337; *Azolla* and, 345–50; ecological costs of, 22; free-living cyanobacteria and, 337–45; microalgal plant growth regulators and, 355–7; microalgal soil conditioners and, 350–3; rangeland ecology and algal crusts and, 353–5; seaweed use in, 357–60

air revitalization, 490–4

algae: advantages of, as production organisms, 532; domesticated, 513; characteristics of major groups, 13t; doubling time (in integrated algal-bacterial waste treatment systems), 266; fossil, 321–2, 325–6, 327f, 328; molecular genetic manipulation of, 520–1; new uses for, 540–6; origins of, 5–10; the term, 4

algae in space: future of, 503–5; life support systems and, 485–8, 489f, 490–7

algal coals, 325

algal growth reactor for use in space, 497

algal innovations, 540–4; patent protection for, 535

algalization trials (rice fields), 344t

algal kerogen in petroleum and coal, 323–5

algal mats, 459, 468, 503f

algal oils, cost of, 242; *see also* lipid *entries;* triacylgylerols

algal plant breeding: future of, 518; *Laminaria japonica,* 515–16; micro-algae, 517–18; *Porphyra,* 513–15; seaweeds, 516–17

algal polysaccharides, 353

algal pond systems, 542

algal productivity in wastewater treatment systems, 266–7, 271–2; *see also* aquaculture; biomass; mariculture

algatrons, 489f

algicides: tolerance to, 473; in weedy algae management, 471–2; *see also* biocides for prevention of marine biofouling

alginate products, 215, 216t, 217

alginates: commercially available, 214; extracting and processing of, 213–14; in foods, 215, 216t, 217; harvesting methods for, 212–13; history of, 211; industrial applications of, 217, 218t; pharmaceutical applications of, 217, 218t; and precipitation with polyvalent salts, 217; production of, from seaweeds, 211–12; properties of, 214, 216f

alginic acid, 13

allergic responses to microalgae, 190

All-India Coordinated Project on Algae, 342

amanori, 89, 139; *see also* nori

amino acids, 125t; *see also* protein

Amoco (company), 541

anatoxin-*a*, 408

animal food: blue-green algae as, 422; seaweed as, 357

animal manures in integrated feedlots, 277–8

Antarctic dry valleys, microorganisms in, 499, 500f

Antarctic lakes, dominant life form in, 501, 503f

antibiotic activity of cyanobacteria and eukaryotic microalgae, 341

antx-*a*, 406–8